In Situ Bioremediation of Petroleum Hydrocarbon and Other Organic Compounds

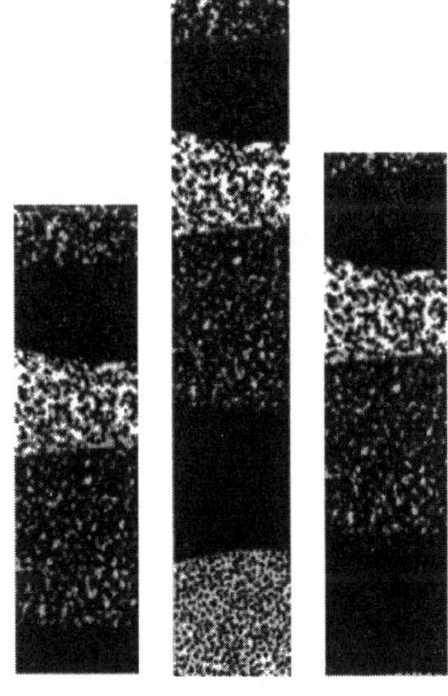

Editors

Bruce C. Alleman
and Andrea Leeson
Battelle

The Fifth International In Situ and On-Site Bioremediation Symposium

San Diego, California, April 19–22, 1999

BATTELLE PRESS
Columbus • Richland

Library of Congress Cataloging-in-Publication Data

In situ bioremediation of petroleum hydrocarbon and other organic compounds / editors, Bruce C. Alleman and Andrea Leeson
p. cm.
Proceedings from the Fifth International In Situ and On-Site Bioremediation Symposium, held April 19–22, 1999, in San Diego, California.
Includes bibliographical references and index.
ISBN 1-57477-076-4 (hardcover : alk. paper)
1. Petroleum chemicals--Biodegradation Congresses. 2. Hydrocarbons--Biodegradation Congresses. 3. In situ bioremediation Congresses.
I. Alleman, Bruce C., 1957– . II. Leeson, Andrea, 1962– .
III. International Symposium on In Situ and On-Site Bioremediation (5th : 1999 : San Diego, Calif.)
TD196.P4I5 1999
628.5'2--dc21 99-23331
CIP

Printed in the United States of America

Battelle Press
505 King Avenue
Columbus, Ohio 43201, USA
614-424-6393 or 1-800-451-3543
Fax: 1-614-424-3819
Internet: press@battelle.org
Website: www.battelle.org/bookstore

For information on future environmental conferences, write to:

Battelle
Environmental Restoration Department, Room 10-123B
505 King Avenue
Columbus, Ohio 43201-2693
Phone: 614-424-7604
Fax: 614-424-3667
Website: www.battelle.org/conferences

CONTENTS

FOREWORD

The Fifth International In Situ and On-Site Bioremediation Symposium was held in San Diego, California, April 19–22, 1999. The program included approximately 600 platform and poster presentations, encompassing laboratory, bench-scale, and full-scale field studies being conducted worldwide on a variety of bioremediation and supporting technologies used for a wide range of contaminants.

The author of each presentation accepted for the program was invited to prepare a six-page paper, formatted according to specifications provided by the Symposium Organizing Committee. Approximately 400 such technical notes were received. The editors conducted a review of all papers. Ultimately, 389 papers were accepted for publication and assembled into the following eight volumes:

Natural Attenuation of Chlorinated Solvents, Petroleum Hydrocarbons, and Other Organic Compounds – Volume 5(1)
Engineered Approaches for In Situ Bioremediation of Chlorinated Solvent Contamination – Volume 5(2)
In Situ Bioremediation of Petroleum Hydrocarbon and Other Organic Compounds – Volume 5(3)
Bioremediation of Metals and Inorganic Compounds – Volume 5(4)
Bioreactor and Ex Situ Biological Treatment Technologies – Volume 5(5)
Phytoremediation and Innovative Strategies for Specialized Remedial Applications – Volume 5(6)
Bioremediation of Nitroaromatic and Haloaromatic Compounds – Volume 5(7)
Bioremediation Technologies for Polycyclic Aromatic Hydrocarbon Compounds – Volume 5(8)

Each volume contains comprehensive keyword and author indices to the entire set.

This volume deals with in situ bioremediation of petroleum hydrocarbon and other organic compounds. From supertanker oil spills to the leaking underground storage tank at the corner gas station, contamination from such compounds affects nearly every small hamlet and major metropolis throughout the world. Moreover, the world's rivers, estuaries, and oceans are threatened by contamination from petroleum leaks and spills. Fortunately, most petroleum hydrocarbons are amenable to biodegradation, and a considerable body of experience has been built up over the past two decades in applying in situ bioremediation to a variety of contaminants in all media. Good progress is being made in terms of developing innovative, cost-effective in situ approaches to bioremediation. The articles in this volume provide a comprehensive guide to the latest technological breakthroughs in both the laboratory and the field, covering such topics as air sparging, cometabolic biodegradation, treatment of MTBE, real-time control systems, nutrient addition, rapid biosensor analysis, multiphase extraction, and accelerated bioremediation.

We would like to thank the Battelle staff who assembled the eight volumes and prepared them for printing. Carol Young, Lori Helsel, Loretta Bahn, Gina Melaragno, Timothy Lundgren, Tom Wilk, and Lynn Copley-Graves spent many hours on production tasks—developing the detailed format specifications sent to each author; examining each technical note to ensure that it met basic page layout requirements and making adjustments when necessary; assembling the volumes; applying headers and page numbers; compiling the tables of contents and author and keyword indices, and performing a final check of the pages before submitting them to the publisher. Joseph Sheldrick, manager of Battelle Press, provided valuable production-planning advice and coordinated with the printer; he and Gar Dingess designed the covers.

The Bioremediation Symposium is sponsored and organized by Battelle Memorial Institute, with the assistance of a number of environmental remediation organizations. In 1999, the following co-sponsors made financial contributions toward the Symposium:

Celtic Technologies
Gas Research Institute (GRI)
IT Group, Inc.
Parsons Engineering Science, Inc.
U.S. Microbics, Inc.
U.S. Naval Facilities Engineering Command
Waste Management, Inc.

Additional participating organizations assisted with distribution of information about the Symposium:

Ajou University, College of Engineering
American Petroleum Institute
Asian Institute of Technology
Conor Pacific Environmental Technologies, Inc.
Mitsubishi Corporation
National Center for Integrated Bioremediation Research & Development (University of Michigan)
U.S. Air Force Center for Environmental Excellence
U.S. Air Force Research Laboratory Air Base and Environmental Technology Division
U.S. Environmental Protection Agency
Western Region Hazardous Substance Research Center (Stanford University and Oregon State University)

The materials in these volumes represent the authors' results and interpretations. The support of the Symposium provided by Battelle, the co-sponsors, and the participating organizations should not be construed as their endorsement of the content of these volumes.

Bruce Alleman and Andrea Leeson, Battelle
1999 Bioremediation Symposium Co-Chairs

BIODEGRADATION OF MTBE UTILIZING A MAGNESIUM PEROXIDE COMPOUND: A CASE STUDY

Stephen T. Defibaugh (Block Environmental, Laguna Hills, California)
Daniel S. Fischman (TOSCO Marketing Company, Costa Mesa, California)

ABSTRACT: A patented magnesium peroxide compound was recently employed at an underground storage tank facility to increase the available oxygen in the groundwater for the biodegradation of fuel hydrocarbons. The objective of this remedial effort was to substantially reduce the dissolved-phase concentrations of BTEX and MTBE and to arrest further migration of the dissolved fuel hydrocarbon plume. Analysis of chemical groundwater parameters following injection of the magnesium peroxide compound, indicated that over the short term, dissolved MTBE as well as the dissolved BTEX constituents were significantly degraded.

INTRODUCTION

In recent years, commercially available chemical compounds designed to slowly release oxygen have demonstrated some success in aiding the in situ biodegradation of certain fuel hydrocarbon constituents such as benzene, toluene, ethylbenzene, and xylenes (BTEX). These commercial compounds are designed to increase the dissolved oxygen content in groundwater, increasing aerobic biodegradation of hydrocarbons by naturally occurring microorganisms that would otherwise be oxygen limited. While methyl tertiary butyl ether (MTBE) had been thought to be recalcitrant to bioremediation, several recent studies have noted MTBE attenuation rates higher than those anticipated by dispersion alone (Barker et al, 1998). Other studies suggest that in mixed contaminant systems MTBE is consumed, but only after the BTEX constituents have been depleted.

Reports by the manufacturer of a patented magnesium peroxide compound called Oxygen Release Compound (ORC®) that their product has been successful in biodegrading fuel release sites with significant MTBE concentrations dissolved in the groundwater (Regenesis, 1998a) led to its use at the subject site. The following is an examination of the effectiveness of ORC® in accelerating the biodegradation of fuel hydrocarbons in the groundwater in general, and specifically its effectiveness in affecting the biodegradation of MTBE. Although not implemented as a research project, sufficient data was collected to warrant discussion.

Site Description. The site is an active service station dispensing gasoline stored in two underground tanks (Figure 1). The service station is located along the coast of southern California at the mouth of the San Juan Creek Valley, approximately one-third mile (0.5 km) to the north of the Pacific Ocean.

Site Background. Following replacement of the underground storage tanks in the early 1990's, hydrocarbon-affected soil was encountered beneath the tank cavity and one of the dispenser islands. Subsequent site assessment activities performed following overexcavation of the tank cavity area indicated that total petroleum hydrocarbon (TPHg) concentrations of up to 9,000 mg/kg remained in the soil and extended to groundwater, present at a depth of approximately 18 feet (5.5 m) below grade.

Dissolved-phase hydrocarbon concentrations of up to 25,000 µg/L benzene and 13,000 µg/L MTBE have been present in the groundwater beneath the site. The direction of groundwater flow is toward the west and southwest at an average gradient of 0.0037 ft/ft. Both the dissolved-phase BTEX and MTBE plumes appear to be stable.

Soil vapor extraction was employed at the site for approximately 13 months during the mid 1990's but did not have a lasting impact on dissolved-phase hydrocarbon concentrations in the saturated zone. Due to the fine grained nature of the materials within the vadose zone, a significant mass of fuel hydrocarbons are suspected to remain adsorbed to the soil.

Geology and Hydrogeology. Subsurface materials encountered in the vadose zone and capillary fringe beneath the service station are composed of silts and clays, from the surface to a depth of approximately 20 feet (6.1 m), and silty, fine-grained sands to medium-grained sands in the saturated zone.

Though used intermittently by a local water district, the groundwater quality for this area of the basin is listed as poor due to high concentrations of dissolved solids (Woodside, 1986). Groundwater beneath the service station contains a chloride concentration of 950 mg/L and a total dissolved solids concentration of 4,700 mg/L. The darcy velocity of the groundwater was calculated to range from 0.067 to 0.24 ft/day (0.020 to 0.073 m/day).

TECHNICAL APPROACH TO USING MAGNESIUM PEROXIDE

While the oxygen release rate of ORC® is dependent upon the level of contaminant flux, this reaction is normally expected to last about six months. However, due to the brackish nature of the groundwater, it was anticipated that the reaction time could be decreased by as much as 50% (Regenesis, 1998b). The subsequent increase in available dissolved oxygen is intended to metabolize the dissolved fuel hydrocarbons by stimulating aerobic biodegradation.

Other factors that were considered prior to implementation of this approach were the groundwater velocity, biological oxygen demand (BOD), soil texture, porosity, pH, temperature, and the concentrations of alternate electron acceptors. Although the salinity of the groundwater caused some concern, most other factors appeared to be within an acceptable range for the use of this product.

Oxygen Demand Calculations. A conservative estimate of the oxygen to carbon stoichiometry of 3:1 has been employed by the manufacturer of ORC®. Based on the dimensions and average concentration of the dissolved fuel hydrocarbon plume (BTEX + MTBE) in the groundwater and the estimated porosity of the saturated soils, a mass of 3.1 pounds (1.4 kg) of fuel hydrocarbons were estimated to be dissolved in the groundwater. An additional oxygen demand factor of 8.075 was also used to account for oxygen that would be used by competing, non-targeted, organic material such as organic carbon in the matrix other than the fuel hydrocarbons and by fuel hydrocarbons adsorbed on the soil particles within the saturated zone and capillary fringe. Based on these factors the amount of oxygen required to metabolize the dissolved hydrocarbon was calculated using equation 2.

(3 lbs Oxygen/1 lb Carbon)(3.1 lbs HC)8.075 DF = 75 lbs (34 kg) Oxygen

Where HC = fuel hydrocarbons
DF = additional oxygen demand factor

Injection Point Field Design. Injection into direct-push bore holes was chosen as the preferred application for this site so that the entire amount of ORC® could be applied evenly across the source area of the dissolved-phase fuel hydrocarbon plume. As an added measure to prevent further migration of the dissolved phase hydrocarbon plume, an oxygen barrier was placed at the downgradient property boundary using an additional application of direct injected ORC® in that area.

A total of 750 pounds (340 kg) of ORC® was injected through 18 direct-push bore holes located within the source area by pushing hollow rods with removable tips to a depth of 28 feet (8.5 m), then pumping the slurry through the rods as they were raised from a depth of 28 feet (8.5 m) to 18 feet (5.5 m).

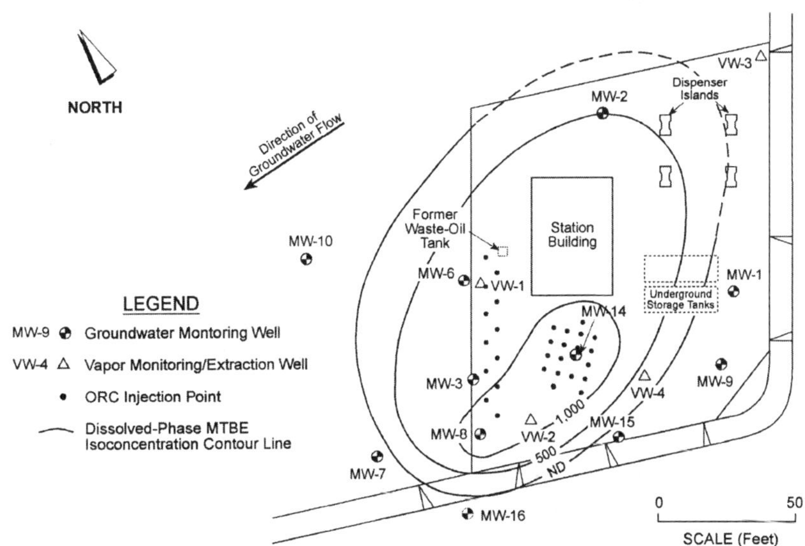

FIGURE 1. Site Plan Showing ORC® Injection Points and Baseline MTBE Plume

An additional 150 pounds (46 kg) of ORC® was divided between 12 direct-push bore holes located at the property line and injected as in the source area (Figure 1).

VERIFICATION SAMPLING

To monitor the effectiveness of the magnesium peroxide compound in biodegrading the dissolved fuel hydrocarbons, baseline groundwater samples were collected from representative groundwater monitoring wells just prior to injection, and then monthly for a period of five months after the injection. The groundwater samples were analyzed for TPHg, BTEX, and MTBE. Prior to collection of the groundwater samples, field measurements of depth to water, dissolved oxygen, temperature, pH, oxidation/reduction potential, and conductivity were collected.

Dissolved Hydrocarbons. Dissolved-phase MTBE and BTEX concentrations in Monitoring Well MW-14 (the well located within the source area) were plotted over time both prior to and following injection of the magnesium peroxide compound (Figure 2). Along with an 82% reduction of dissolved BTEX baseline concentration of 16,600 μg/L within the first month after the injection of ORC®, dissolved MTBE concentrations were reduced from the baseline concentration of 2,200 μg/L by 77%. Dissolved phase BTEX and MTBE concentrations dropped an additional 1% and 20%, respectively, during the second month after injection. By the third month after injection, however, dissolved phase BTEX concentrations had rebounded to 151% of the baseline results and dissolved MTBE had rebounded to 82% of the baseline. Subsequent analysis of groundwater samples collected during the next two months indicated that dissolved BTEX and MTBE concentrations had returned to pretreatment levels.

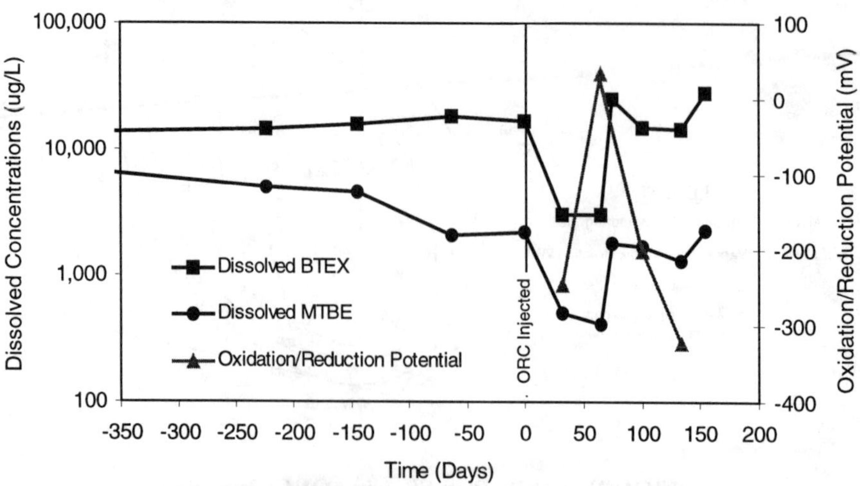

Figure 2. Dissolved MTBE and BTEX Concentrations and Oxidation/Reduction Potential vs. Time in MW-14.

Dissolved Oxygen and Oxidation/Reduction Potential. Dissolved oxygen readings taken at the beginning and following injection of ORC® never varied more than 1 mg/L (between 0 and 1 mg/L) for the selected wells tested.

Oxidation/reduction potential readings were not collected prior to injection but were collected in the months following. As expected the lowest potential was observed in Monitoring Well MW-14, located in the center of the dissolved fuel hydrocarbon plume. The oxidation/reduction potential peaked for MW-14 during the second month following ORC® injection and then steadily declined (Table 1 and Figure 2). The oxidation/reduction potential for the two wells located within the oxygen barrier peaked during the third month after injection and then declined.

Table 1. Chemical and Physical Groundwater Properties of MW-14

Time (Days)	TPHg (µg/L)	B (µg/L)	T (µg/L)	E (µg/L)	X (µg/L)	MTBE (µg/L)	Ground Water Elevation (Feet)	DO (mg/L)	O/R Potential (mV)
-369	ND	9,000	1,000	1,600	2,100	6,700	9.22	--	--
-223	35,000	9,900	2,000	1,400	1,100	5,000	9.20	--	--
-145	34,800	9,640	2,530	2,020	1,390	4,600	9.37	--	--
-63	38,000	10,000	4,800	1,600	1,700	2,100	9.99	--	--
0	**33,000**	**11,000**	**1,200**	**1,900**	**2,500**	**2,200**	**10.02**	**0.0**	**--**
31	5,200	2,400	12	260	350	510	10.78	0.0	-246
65	3,800	2,500	6.4	200	280	410	8.95	0.0	33
74	3,600	13,000	6,300	2,600	3,200	1,800	9.78	--	--
100	27,000	9,400	3,000	920	1,600	1,700	9.48	0.0	-201
134	28,000	9,600	1,800	1,200	1,700	1,300	9.18	0.0	-322
155	47,000	15,000	6,400	2,400	4,100	2,300	9.20	--	--

Notes: TPHg = total petroleum hydrocarbons as gasoline
B = benzene, T = toluene, E = ethylbenzene, X = total xylenes
MTBE = methyl tertiary butyl ether
DO = dissolved oxygen, O/R Potential = oxidation/reduction potential

DISCUSSION AND CONCLUSIONS

Reduction of MTBE and BTEX Concentrations. Both dissolved BTEX and MTBE concentrations decreased significantly during the first two months after injection of the magnesium peroxide compound. These decreases appear to have coincided with the time of greatest activity in the reaction between the ORC® and the groundwater, as evidenced by the increase in the oxidation/reduction potential during that time period. According to the manufacturer of ORC®, it is not unusual to see no substantial increase in the dissolved oxygen content of the groundwater following injection (the usual range is between 0 and 5 mg/L).

One or all of several factors may have caused the rebound of dissolved MTBE and BTEX concentrations seen along with the decrease in the oxidation/reduction potential: 1) actual dissolved fuel hydrocarbon concentrations may be greater than indicated by the monitoring wells currently at the site (in which case the oxygen was exhausted before all of the hydrocarbons were metabolized), 2) the salinity of the groundwater may have caused the oxygen to be released at a rate greater than could be used by the in situ microbes (although if this were the case, a more significant increase in the dissolved oxygen content would have been expected), and 3) fuel hydrocarbons which remain adsorbed to the soil particles in the capillary fringe may have provided a source to reintroduce hydrocarbons in the dissolved-phase.

Although no empirical evidence was collected to confirm an increase in microbial activity, the coincidence of greater oxygen availability and the reduction of dissolved hydrocarbon concentrations lends support to the theory of biodegradation at the service station site.

Decreases in dissolved MTBE concentrations were observed to coincide with decreases in BTEX concentrations in both time and magnitude, demonstrating that the same processes that worked to reduce the dissolved BTEX concentrations acted upon the dissolved MTBE.

ACKNOWLEDGMENTS

The authors would like to thank Dr. Stephen Koenigsberg and Kim Sakata (Regenesis) for their material and technical support during the planning and implementation of the field work and help in the preparation of this paper.

REFERENCES

Barker, J. F., M. Schirmer, B. J. Butler, and C. D. Church. 1998. "Fate and Transport of MTBE in Groundwater – Results of a Controlled Field Experiment in Light of Other Experience." *The Southwest Focused Ground Water Conference: Discussing the Issue of MTBE and Perchlorate in Ground Water*: 10-14.

Edgington, W. J. 1974. *Geology of the Dana Point Quadrangle, Orange County, California.* California Division of Mines and Geology. Special Report No. 109.

Regenesis. 1998a. *Potential for the Bioremediation of Methyl Tertiary Butyl Ether (MTBE).* ORC Technical Bulletin # 2231.

Regenesis. 1998b. *Performance in Regions of Higher Salinity.* ORC Technical Bulletin # 2432.

Woodside, R. D. and G. D. Woodside. 1986. "San Juan Ground Water Basin in Orange County, California." *Hydrogeology of Southern California, Part One – Hydrogeologic Basins and Their Management.* Geological Society of America: 35-41.

PHYSIOLOGICAL AND ENZYMATIC FEATURES OF MTBE-DEGRADING BACTERIA

Michael Hyman (North Carolina State University, Raleigh, North Carolina)
Kirk O'Reilly (Chevron Research and Technology Co., Richmond, California)

ABSTRACT: We have examined the cometabolic MTBE-degrading activity of two groups of bacteria. In one study we determined the effects of hydrocarbon growth substrate on subsequent MTBE degradation by various alkane-utilizing strains. Only a subset of these bacteria were capable of MTBE degradation and these organisms were distinguished by their ability to grow on both gaseous *n*-alkanes (*e.g.* propane) and simple branched hydrocarbons (*e.g.* *iso*-butane). In a second study we examined the MTBE-degrading ability of other cometabolically-active microorganisms grown on ammonia, propylene, methane and toluene. None of these microorganisms were capable of detectable rates of MTBE oxidation despite the fact that many of these bacteria are known to oxidize propane/and or propylene and can catalyze O-dealkylation reactions similar to those thought to be involved in the initial steps of MTBE oxidation. Together these results suggest that MTBE-degrading activity can be most consistently predicted on the basis of the ability of microorganisms to grow on simple branched alkanes.

INTRODUCTION

Methyl *tert*-butyl ether (MTBE) is included in many gasoline formulations as both an octane enhancer and as an oxygenate. The extensive use of MTBE in gasoline has inevitably resulted in a widespread distribution of this compound in the environment. The uncertain health effects and low taste and odor thresholds for MTBE have led to considerable concern over the presence of MTBE in aquifer-based supplies of human drinking water. Several studies have demonstrated that MTBE is largely resistant to biodegradation in groundwater under a variety of redox conditions when it is present as a sole source of carbon and energy for microbial growth. Although several strains of aerobic bacteria have been isolated that are capable of growth on MTBE, these processes are typically slow. In contrast to growth-based approaches, we have focused on cometabolic processes in which MTBE is fortuitously degraded by microorganisms previously grown on other substrates. We have previously reported that MTBE can be cometabolically degraded by both fungi (Hardison *et al.*, 1997) and bacteria (Hyman *et al.*, 1998) after growth on simple gaseous *n*-alkanes such as propane or *n*-butane. We have also demonstrated that the major gasoline component *iso*-pentane can serve as effective inducing substrate for MTBE cometabolism by undefined microorganisms in soil (Hyman *et al.*, 1998). These studies suggest that cometabolic MTBE degradation in the presence of gasoline hydrocarbons is a possible mechanism for the natural attenuation of this compound. Some of our recent studies that are described here have continued to explore the physiological and enzymatic features that confer MTBE-degrading ability on alkane-utilizing microorganisms.

MATERIALS AND METHODS

All of the microorganisms discussed in Table 1 were grown in batch cultures in sealed glass serum vials (160 ml) containing mineral salts medium (60 ml) (Ensign *et al.*, 1992). Gaseous hydrocarbon growth substrates (propane, *n*-butane and *iso*-butane) were added to the vials as an overpressure (30 ml added to 100 ml gas phase). All other hydrocarbon growth substrates were liquids and were added directly to the reaction medium to an initial concentration of 0.1% (v/v). The inoculated vials were incubated on an orbital shaker (200 rpm) in an unlit constant temperature room (30° C). Organisms capable of growth on individual hydrocarbons were cultivated for up to 5 d and the cells were harvested when A_{600} measurements were between 0.3 and 0.8. The inability of a microorganism to grow on a particular substrate was recorded if A_{600} measurements made after incubation for 12-14 d were less than 0.1. The microorganisms described in Table 2 were grown on the following media: *Nitrosomonas europaea*, phosphate-buffered (pH 8.0) mineral salts medium (50mM P_i) containing $(NH_4)_2SO_4$ (25 mM) (Hyman and Arp, 1992): *Xanthobacter* Py2, mineral salts medium (Ensign *et al.*, 1992) using propylene (50% [v/v] gas phase) as sole source of carbon and energy: *Methylosinus trichosporium* OB3b, on mineral salts medium (Whittenbury *et al.*, 1970) +/- 20 µM $CuSO_4$) with methane (50% [v/v] gas phase) as sole source of carbon and energy: *Burkholderia cepacia* G4, *Pseudomonas mendocina* and *Pseudomonas putida* FI were grown on mineral slats medium containing toluene (Yeager *et al.* 1999). In all cases the cells were harvested by centrifugation and were resuspended in sodium phosphate buffer (50 mM, pH 7.0 [or pH 7.8 for *N. europaea*]). The cells were sedimented again by centrifugation and were resuspended to a final protein concentration of between 4 and 15 mg/ml.

The degradation of MTBE and diethyl ether (DEE) by each cell type was monitored under standard conditions. The reactions were conducted in glass serum vials sealed with butyl rubber stoppers and aluminum crimp seals. The vials contained sodium phosphate buffer and 1.1 µmoles of MTBE or 1 µmole DEE. Both substrates were added from aqueous saturated solutions. The reactions were initiated by the addition of cells (*ca.* 100 µl) to give a final reaction volume of 1 ml and a protein concentration of between 0.4 to 1.5 mg/ml. The reaction vials were incubated in a shaking (200 rpm) heated (30° C) water bath. Samples (2 µl) were removed from reaction vials at 10 min intervals over a 1 h period. The samples were directly injected into a gas chromatograph equipped with and FID detector and a 6 ft stainless steel column containing Porapak Q (Hardison *et al.*, 1997). The GC analysis allowed us to follow MTBE consumption and the accumulation of *tert*-butyl formate (TBF) and *tert*-butyl alcohol (TBA) as reaction products. Rates of MTBE oxidation were determined on the basis of the total mass of MTBE consumed over the 1 h reaction period. The GC analysis also allowed us to follow the consumption of DEE and the production of ethanol and acetaldehyde as oxidation products. The oxidation of propylene by *B. cepacia* G4 was conducted, as described by Yeager *et al.* (1999)

RESULTS AND DISCUSSION:

In the first series of experiments a variety of alkane-utilizing bacteria were initially examined for their growth substrate range. If growth on a particular hydrocarbon substrate was established the microorganisms were then examined for their ability to cometabolically degrade MTBE in the absence of any other exogenous compounds. The organisms chosen for this study included a previously described MTBE-degrading species (*Mycobacterium vaccae* JOB5) (Hyman *et al.*, 1998) as well as several other previously identified hydrocarbon-oxidizing bacteria (*Alcaligenes eutrophus*, *Pseudomonas mendocina* and *Rhodococcus rhodocrous*). These results are summarized in Table 1.

Table 1. The relationship between *n*- and *iso*-alkane growth substrate range and the rate of MTBE-degradation by selected alkane-utilizing bacteria.

Hydrocarbon Growth Substrates

Microorganism	*n*-C_3	*n*-C_4	*n*-C_5	*n*-C_6	*n*-C_8	*iso*-C_4	*iso*-C_5	*iso*-C_6
M. vaccae JOB5	++ 9.9	++ 16.3	++ 11.1	++ 21.0	++ 5.5	++ 13.6	++ 12.6	++ 17.1
A. eutrophus	++ 12.3	++ 15.0	++ 9.0	++ 7.3	++ 5.9	++ 8.9	++ 14.9	++ 8.7
P. mendocina	-	+	++ <1.0	++ <1.0	++ <1.0	-	-	+ 4.0
R. rhodocrous	-	-	-	++ <1.0	++ <1.0	-	-	-

Symbols: (++) abundant growth; (+) limited growth; (-) no growth after 14 d
Rates are expressed in nmoles MTBE consumed /min/mg protein

The microorganisms studied here fall into two reasonably distinct physiologically groups. The first group, consisting of *M. vaccae* and *A. eutrophus,* were capable of growth on all of the *n*- and *iso*-alkanes examined. These microorganisms were also capable of oxidizing MTBE after growth on all of these hydrocarbons and in all cases TBF and TBA were detected as MTBE oxidation products. The second group of microorganisms, (*P. mendocina* and *R. rhodocrous*) exhibited a much narrower hydrocarbon growth substrate range and were largely consistent in their inability to grow on gaseous *n*-alkanes (propane or *n*-butane) or *iso*-alkanes. In most instances these microorganisms were unable to oxidize MTBE at rates that were detectable above background abiotic losses. However, low concentrations of TBA were detected in MTBE-containing incubations conducted using cells of *P. mendocina* grown on *n*-octane. Low, but detectable rates of both MTBE consumption and concurrent TBA production were also observed for cells of *P. mendocina* after limited growth of this organisms on 2-methylpentane.

The results presented in Table 1 suggest that the cometabolic degradation of MTBE can be largely predicted by two physiological characteristics; the ability of a microorganism to grow on gaseous *n*-alkanes and the ability of a microorganism to grow on branched alkanes. As part of this present study we have also examined whether these themes also extend to the filamentous fungus in which propane and *n*-butane-dependent cometabolism of MTBE was first described (Hardison *et al.*, 1997). We have established that this unusual microorganism is capable of growth on *iso*-butane and expresses an oxygenase capable of the oxidation of propane, *n*-butane and MTBE after growth on this compound, albeit it slower rates than those we have described for mycelia grown on *n*-alkanes (Hardison *et al.*, 1997).

An important distinction between growth-related metabolism of pollutants and cometabolism is that the former process involves the concerted activities of numerous enzymes while the latter can usually be attributed to the fortuitous activity of a single enzyme within a pathway. For example, in the case of the bacteria described in Table 1, MTBE-oxidizing activity, at least to the level of TBF and TBA, appears to be the result of the activity of non-specific oxygenase enzymes expressed to initiate the oxidation of selected groups of structurally related alkane growth substrates. This "single enzyme" feature of cometabolism suggests that predictors of MTBE-oxidizing activity can be largely reduced to considerations of oxygenase substrate specificty and enzyme induction, two features that essentially merged during growth on a particular substrate. If this argument is accepted, as it is for other recalcitrant compounds, we reasoned that it might be possible to oxidize MTBE through the activity of other oxygenase enzymes that are used by bacteria to oxidize substrates other than *n*-alkanes. In the second element of this study we therefore examined the MTBE-degrading activity of a series bacteria that have been previously reported to express highly non-specific oxygenase enzymes in response to well-defined growth substrates. In many cases the lack of oxygenase specificity has been utilized or identified as part of studies of the cometabolic degradation of another important ground water pollutant, trichloroethylene. Various physiological and catalytic features of these microorganisms that have been established as part of this study, or are described in the existing literature, are summarized in Table 2.

None of the bacteria described in Table 2 exhibited detectable rates of MTBE oxidation under the conditions tested and in no case did we observe the accumulation of TBA and TBF, the characteristic MTBE oxidation products observed with the organisms described in Table 1. The only exception to this generalization was the methane-oxidizing bacterium *M. trichosporium* OB3b. When this bacterium was grown under copper limited conditions we observed a transient accumulation of low concentrations of both TBF and TBA over the first 20 min of the reaction time course. These characteristic products suggest limited MTBE oxidation was catalyzed by this microorganisms although subsequent experiments using sodium formate (4 mM) as a supplemental electron donor did not increase or sustain the rate of product accumulation.

Table 2. Physiological and cometabolic activities of selected oxygenase-expressing bacteria.

Microorganism	Growth Substrate	MTBE Oxidation	Diethyl ether Oxidation	Propane Oxidation	Propylene Oxidation
N. europaea	NH_3	--	++	+	+
***Xanthobacter* Py2**	Propylene	--	++	-	+
***B. cepacia* G4**	Toluene	--	+	?	++
***P. putida* F1**	Toluene	--	--	?	?
P. mendocina	Toluene	--	++	?	?
***M. trichosporium* (-Cu)**	CH_4	--	+	+	+
***M. trichosporium* (+Cu)**	CH_4	--	?	+	+

Symbols: (-) no oxidation described in literature; (+) oxidation described in literature (?) unknown
(- -) no oxidation established in this study; (++) oxidation established in this study

Although the microorganisms examined in Table 2 were largely unreactive towards MTBE, we did establish that several of these organisms *(N. europaea, Xanthobacter* Py2 and *P. mendocina*) were capable of rapidly oxidizing diethyl ether to mixtures of ethanol and acetaldehyde. These oxidation products are characteristic of an O-dealkylation reaction, a reaction that has been proposed to initiate the oxidation of MTBE in a variety of systems. In addition to this, Table 2 summarizes several previous reports that have demonstrated *N. europaea* and *M. trichosporium* OB3b are capable of cometabolic propane oxidation, an activity that also extends to numerous other *n*-alkanes. These two bacteria, in addition to *Xanthobacter* Py2 and *B. cepacia* G4, are also capable of oxidizing propylene, the alkene analog of propane.

Taken together the observations summarized in Table 2 do little to support the argument that the ability on an oxygenase enzyme to oxidize MTBE is closely correlated with the ability of the same enzyme to oxidize gaseous *n*-alkanes like propane or propane analogs such as propylene. For instance if this enzymatic theme were more substantial we might have expected *N. europaea* to readily oxidize MTBE. In fact the inability of *N. europaea* to oxidize MTBE is noteworthy in itself. The ammonia monooxygenase enzyme responsible for the cometabolic activity of *N. europaea* has a substrate range that is almost as extensive as the highly non-specific soluble methane monooxygenase, another enzyme which we have shown here to be only marginally reactive towards MTBE. If, as we suggest, the connection between MTBE degradation and propane oxidation is not absolute, we then have to look elsewhere for more consistent predictors of MTBE-degrading activity. Although our present study has been limited to only four alkane-utilizing bacteria, our results (Table 1) suggest that the ability of a microorganism to grow on simple branched hydrocarbons may be the most unequivocal predictor of MTBE-degrading activity available at the moment. With this point in mind it is worth noting that *M. vaccae* JOB5, the most extensively characterized MTBE-degrading bacterium, is often regarded as a typical propane-oxidizer. However, this organism was originally isolated on *iso*-butane as a sole source of carbon and energy (Ooyama and Foster, 1965). It is also important to note that our conclusion concerning the role of branched hydrocarbons as a predictor of MTBE degrading activity compliments our previous observation that cometabolic MTBE-degrading activity can be rapidly established in pristine soils using *iso*-pentane as a stimulant (Hyman *et al.*, 1998). Finally, this conclusion further suggests that the numerous branched hydrocarbons that are important components of gasoline formulations may represent an important resource for promoting the enhanced or natural attenuation of MTBE in the field. Our current research is aimed at examining these possibilities.

ACKNOWLEDGMENTS: This research was supported by funding from Chevron Research and Technology Company and the EPA-sponsored Western Region Hazardous Substances Research Center.

REFERENCES:

Ensign, S. A., M. R. Hyman, and D. J. Arp. 1992. Cometabolic Degradation of Chlorinated Alkenes by Alkene Monooxygenase in a Propylene-Grown *Xanthobacter* Strain. *Appl. Environ. Microbiol.* 58 (9): 3038-3046.

Hardison. L. K., S. S. Curry, L. M. Ciuffetti, and M. R. Hyman. 1997. "Metabolism of Diethyl Ether and Cometabolism of Methyl *tert*-Butyl Ether by a Filamentous Fungus, a *Graphium* sp". *Appl. Environ. Microbiol.* 63 (8): 3059-3067.

Hyman, M. R., P. Kwon, K. Williamson, and K. O"Reilly. 1998. "Cometabolism of MTBE by Alkane-Utilizing Microorganisms. In G. B. Wickramanayake and R. E. Hinchee (Eds.) *Natural Attenuation: Chlorinated and Recalcitrant Compounds*, pp 321-326, Battelle Press, Columbus, OH.

Hyman, M. R., and D. J. Arp. 1992 "$^{14}C_2H_2$- and $^{14}CO_2$-labeling Studies of the *de Novo* Synthesis of Polypeptides by *Nitrosomonas europaea* during Recovery from Acetylene and Light Inactivation of Ammonia Monooxygenase". *J. Biol. Chem* 267 (3) 1534-1545.

Ooyama, J. and J. W. Foster. 1965. "Bacterial Oxidation of Cycloparaffinic Hydrocarbons" *Antonie van Leeuwenhoek* 31: 45-65.

Whittenbury, R., K. C. Phillips, and J. F. Wilkinson. 1970. "Enrichment, Isolation and Some Properties of Methane-utilizing Bacteria" *J. Gen. Microbiol.* 61: 205-218

Yeager, C. M., P. J. Bottomley, D. J. Arp, and M. R. Hyman. 1999 "Inactivation of Toluene-2-Monooxygenase in *Burkholderia cepacia* G4 by Alkynes" *Appl. Environ. Microbiol.* 65 (2) xxx-xxx (In Press)

PEROXYGEN MEDIATED BIOREMEDIATION OF MTBE

Stephen Koenigsberg (Regenesis, San Juan Capistrano, California)
Craig Sandefur (Regenesis, San Juan Capistrano, California)
William Mahaffey (Pelorus EnBiotech, Evergreen, Colorado)
Marc Deshusses (University of California, Riverside)
Nathalie Fortin (University of California, Riverside)

ABSTRACT: Evidence is mounting, from laboratory and field studies, that oxygen plays a role in the bioremediation of methyl tertiary butyl ether (MTBE). Our experience with a number of commercial applications of peroxygen, in the form of Oxygen Release Compound (ORC®), correlates the provision of enhanced aerobic conditions with enhanced MTBE biodegradation. Subsequent laboratory and field studies have documented this phenomenon while showing it 1) follows Michaelis-Menten kinetics, 2) is subject to interference by background hydrocarbons as represented by benzene and xylene and that, 3) this interference dynamic may be specifically described in terms of classical competitive inhibition. Most recently, evidence for benzene mediated co-oxidation of MTBE has been demonstrated. Given that requisite enzymatic inductions will have already taken place in a co-contaminated aquifer, enhanced aerobic bioremediation with ORC, represents a cost-effective approach to facilitate both the direct bioremediation of MTBE and the inhibitory hydrocarbons that need to be eliminated as well.

INTRODUCTION

MTBE complicates remediation and closure of properties contaminated with BTEX and other fuel hydrocarbons. A number of parties are becoming increasingly concerned about the environmental impact of MTBE. Several factors are responsible for the heightened level of concern as folows: 1) MTBE tends to degrade very slowly, 2) MTBE is highly soluble in water and does not easily sorb onto the aquifer matrix; retardation of MTBE is therefore minimal and plume dimensions are enhanced, 3) Due to its Henry's Law constant, MTBE is slow to volatilize out of groundwater, 4) Taste and odor thresholds for MTBE are very low – approximately 35 ppb, 5) MTBE toxicity and carcinogenicity profiles are largely undetermined.

Some of these characteristics compromise active remediation methods such as air sparging and pump and treat systems. In the latter case, stripping inefficiencies encountered with extracted water have caused many consultants to evaluate other treatment options. One of these options is in-situ aerobic bioremediation and we have been demonstrating for several years that the bioremediation of MTBE is enhanced by ORC.

As early as 1994, Regenesis noticed that MTBE concentrations decreased at an unusually high rate, relative to the literature (Howard, et al., 1991), in monitoring wells containing ORC filter socks. With a series of laboratory experiments eliminating abiotic chemical and physical mechanisms as the cause, emphasis was placed on bioremediation as the operant mechanism. Eventually, more compelling field evidence became available using monitoring wells downgradient of ORC injection fields. Furthermore, there was as suggestion that background hydrocarbon contaminants repressed MTBE degradation; on a majority of the sites investigated MTBE degradation only occurred after decreases in BTEX levels (Koenigsberg, 1997). Currently there are number of examples of ORC-mediated MTBE degradation in Regenesis files; a representative case is presented below. Similar results have been published that involve the sparging of air or oxygen (Javanmardian and Glasser ,1997; Carter et al., 1997).

RESULTS AND DISCUSSION

MTBE Bioremediation Field Results. A service station in Lake Geneva, Wisconsin was contaminated with high levels of MTBE and BTEX due to a leaking underground storage tank (UST). Measurements indicated MTBE and BTEX concentrations reached levels up to 800 ppb and 14,000 ppb, respectively. Though project engineers removed the UST and excavated the contaminated soil, MTBE and BTEX still persisted in the groundwater.

Project geologists injected 1,700 pounds of ORC as a slurry (via direct-push technology) to enhance aerobic degradation in the saturated zone. MTBE degradation results are presented in Figure 1. Over nine months following ORC injection, results from two downgradient monitoring wells indicate that MTBE concentrations degraded to less than two ppb. The site has since been submitted for closure to the state of Wisconsin.

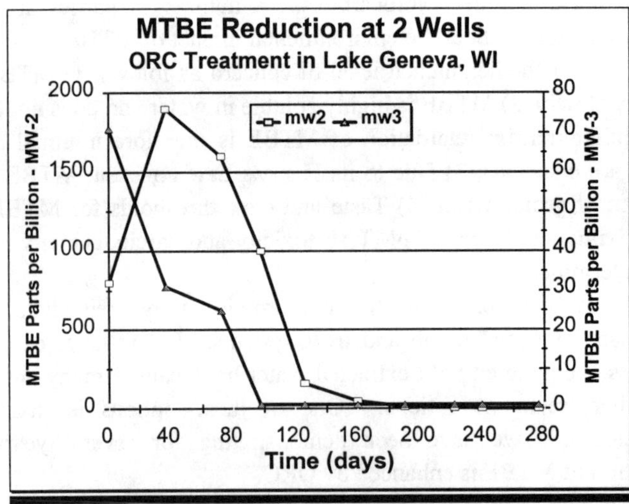

FIGURE 1. MTBE bioremediation field rcsults.

MTBE Bioremediation as a Function of Dissolved Oxygen. An independent laboratory study by Fortin and Deshusses at the University of California, Riverside (supported by Regenesis) investigated the biodegradation of MTBE by respirometry using a mixed culture. In the experiment, oxygen uptake rates at various dissolved oxygen concentrations were used to quantify the influence of dissolved oxygen concentration on the rate of MTBE biodegradation. Results of the experiment, are presented in Figure 2 and demonstrate 1) the rate of MTBE biodegradation was proportional to the concentration of dissolved oxygen in water and 2) MTBE uptake followed a Michaelis-Menten kinetics with respect to dissolved oxygen.

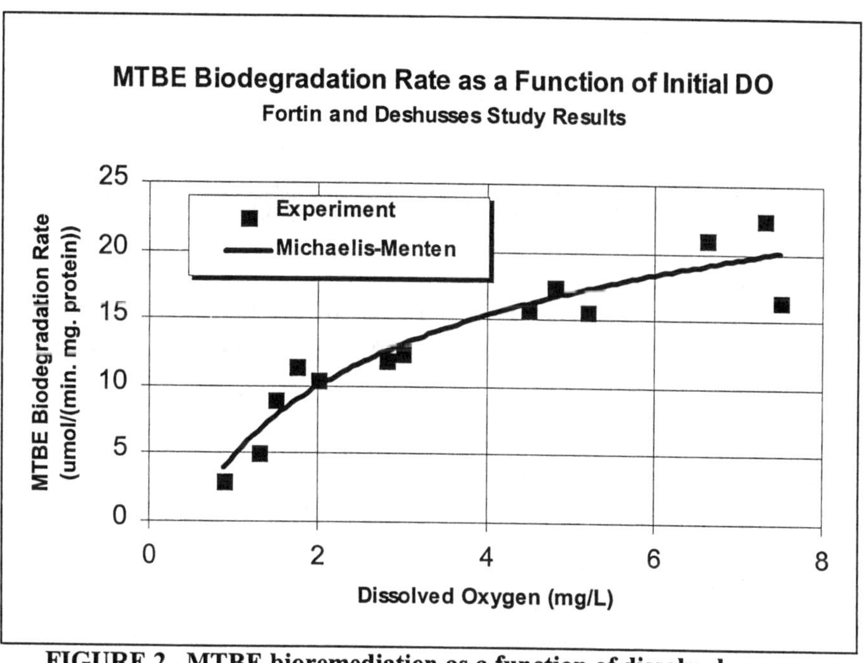

FIGURE 2. MTBE bioremediation as a function of dissolved oxygen.

Does Competitive Inhibition Play a Role in MTBE Bioremediation?

Competitive inhibition is a term used to describe enzymatic activity in which two or more different substrates compete for the same enzyme. One indication that competitive inhibition may be occurring is when the degradation of one substrate is repressed in the presence of another substrate.

Field observations suggest that background hydrocarbons may repress MTBE degradation and vice versa. As presented in Figure 3, data from a site in Michigan show that in the presence of ORC, MTBE degradation occurs *after* BTEX concentrations subside. This effect has been documented at the majority of MTBE-impacted sites using ORC.

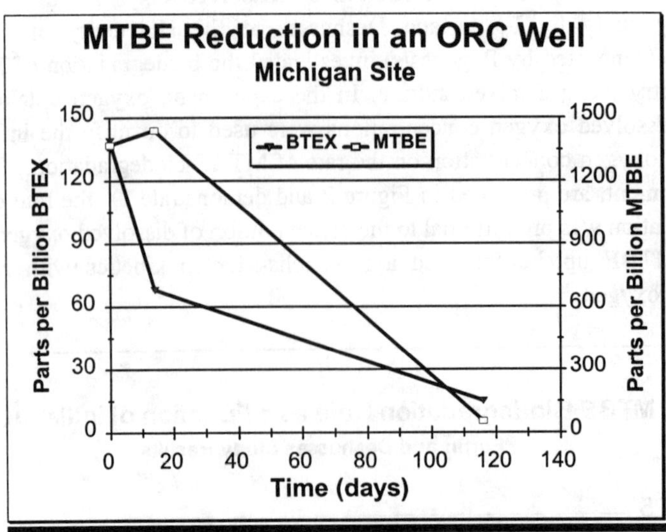

FIGURE 3. Competitive inhibition field results.

Prompted by such field results, a series of laboratory experiments were conducted to test whether background hydrocarbons interfere with MTBE degradation. In an in-vitro experiment with aerobic bacteria (known to be capable of degrading MTBE and BTEX), results suggest that MTBE metabolism is inhibited by background hydrocarbons. MTBE degradation was measured in the presence of 1) MTBE only and 2) MTBE and xylene during a seven day period. Results indicated a 52% reduction of MTBE in the absence of xylene versus a 9% reduction of MTBE with xylene present.

Independent experiments were then performed for Regenesis, by Pelorus EnBiotech Corporation, that explored the hypothesis that MTBE biodegradation is 1) an aerobic co-oxidative process and 2) that competitive inhibition could exist between a primary substrate and MTBE. The most likely primary substrates involved in co-oxidation and competitive inhibition are compounds found at the aerobic fringe of a petroleum hydrocarbon plume. Initial studies, using resting cell transformation tests, demonstrated that substantial removal of MTBE was achieved with cultures that were acclimated to benzene, camphor, o-xylene and cyclohexanone. In those tests a specific benzene acclimated mixed culture, designated PEL-B201, was most efficient in degrading MTBE (58% removal). This established the possibility that a single organism could metabolize both MTBE and alternate substrates and therefore be under the influence of competitive inhibition dynamics. The competitive inhibition hypothesis was bolstered by demonstrating both MTBE inhibition of benzene metabolism and the inhibition of MTBE metabolism with increasing benzene concentrations.

The benzene utilizing culture (PEL-B201) used in the experiments was grown in basal salts media on benzene vapors. Growth and activity experiments were performed to determine optimum conditions for biomass production. MTBE biotransformation experiments were performed in 160 ml Wheaton serum bottles containing oxygen saturated phosphate buffer supplemented with MTBE (3.35 mg/L). The bottles were sealed with Teflon lined serum septa. To evaluate the effects of benzene on MTBE degradation a stock solution of benzene in DMF was added to achieve final concentrations of 1.9 mmM and 3.8 mmM respectively. PEL-B201 acclimated cell suspensions were added to each test reactor to a cell density of approximately 2.0 x 10E8 cells/ml. Controls were inoculated to the same level with un-acclimated PEL-B201 cells grown on succinate. Over a 48 hour test period, samples were removed from each reactor and placed in 2.0 ml GC vials. Headspace samples were analyzed for MTBE by gas chromatography (GC/PID).

Optimum growth conditions established for strain PEL-B201 were developed through growth curve and oxygen uptake studies on benzene. Optimum degradative activity and cell yield were achieved when optical densities reached a nominal value of approximately 1.10 (OD 600). Results of oxygen uptake (OU) tests are shown in Table 1. These tests clearly indicate that MTBE inhibits oxygen uptake associated with benzene metabolism.

Results of the biotransformation experiments with PEL-B201are presented in Figure 4. Benzene induced cell suspensions degrade >99% of the added MTBE. Increasing levels of benzene (1.9 uM and 3.8 uM) result in a significant reduction in the rates of MTBE degradation. No degradation of MTBE was observed with cells grown on the non-inducing substrate succinate. The lack of MTBE degradation on succinate grown cells demonstrates that the MTBE metabolism occurs with an enzyme system associated with benzene metabolism and reaffirms the hypothesis that MTBE is metabolized by co-oxidation.

REFERENCES

Howard, P.H., Boethling, R.S., Jarvis, W.F., Meylan, W.M. and E.M. Michaelenko. 1991. *Handbook of Environmental degradation Rates.* Lewis Publishers, Boca Raton, FL.

Koenigsberg, 1997. "MTBE Wildcard in Ground Water Cleanup". *Environmental Protection.* 8(11): 26-28

Javanmardian, M and H.A. Glasser. 1997. "In Situ Biodegradation of MTBE Using Biosparging". In: Proceedings of the American Chemical Society Division of Environmental Chemistry, April 13-17, San Francisco, CA. Pp 424.

Carter, S.R., Bullock, J.M. and W.R. Morse. 1997. "Enhanced Biodegradation of MTBE and BTEX using Pure Oxygen Injection". In: B.C. Alleman and A. Leeson (Eds.), In Situ and On Site Bioremediation 4(4):147. Battelle Press,Columbus,OH

TABLE 1. Oxygen Uptake Rates (OUR) with resting cell suspensions of the benzene degrading bacterial culture PEL-B201.

μM Benzene	μM MTBE	nMoles-O_2/min	Percent Inhibition	Comments
50		23.80	-	Primary substrate activity
100		27.50	-	"
250		30.80	-	:
50	100	12.50	47.5 %	MTBE Inhibition of primary substrate activity
100	100	13.90	49.5 %	"
250	50	17.00	44.8 %	"

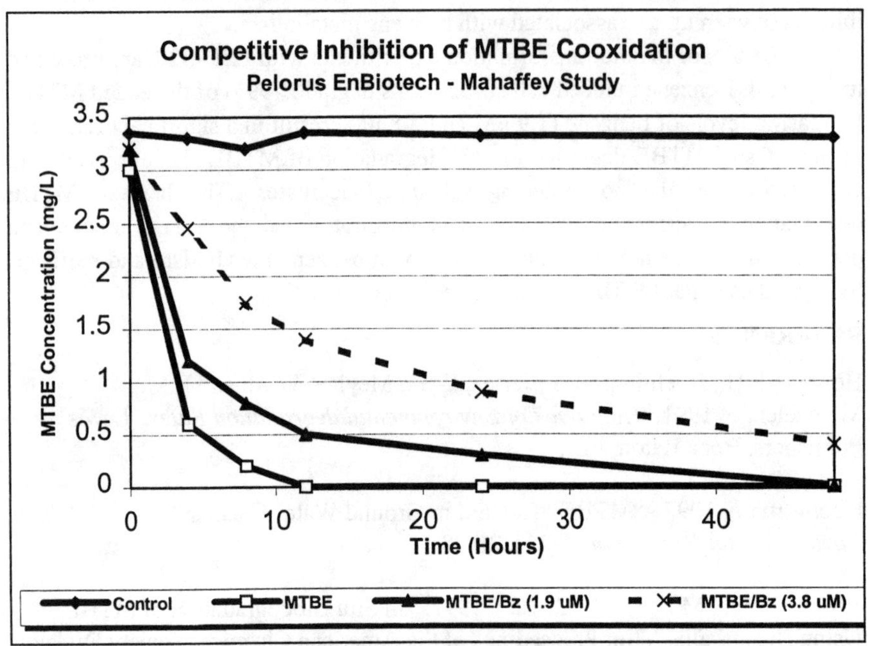

FIGURE 4. Competitive inhibition of MTBE cooxidation Pelorus EnBiotech-Mahaffey Study

AN EMPIRICAL STUDY OF MTBE, BENZENE, AND XYLENE GROUNDWATER REMEDIATION RATES

Tom R. Peargin (Chevron Research and Technology Co., Richmond, CA)

ABSTRACT: This study evaluates whether groundwater remediation rates for MTBE, benzene, and xylenes were consistent with Raoult's Law equilibrium partitioning from non-aqueous phase liquid (NAPL). Time series analytical data were used to compare remediation rates from eight retail service stations where UST releases had occurred in the late 1980's and early 1990's. Remediation systems combined vapor extraction with groundwater extraction to dewater smear zone soils and volatilize NAPL. A first order decay function was fit to 71 time series data sets collected from wells completed and screened in the smear zone to calculate the initial concentration (C_0) and decay constant (k) for each well. The mean and standard deviation of the set containing all regression k's for MTBE, benzene, and xylene were compared and were found to be statistically similar. Remediation rates were modeled for each site assuming equilibrium NAPL partitioning, and using actual remediation performance data. Due to it's higher relative solubility and volatility, the model predicts MTBE should have been remediated significantly faster than other analytes; approximately 5 times faster than benzene and 63 times faster than xylene. However, on average, MTBE was actually remediated 0.7 times slower than benzene and 0.9 times slower than xylene.

INTRODUCTION

Raoult's Law predicts that under conditions of local chemical equilibrium MTBE should partition from NAPL into air or groundwater at concentrations significantly higher than BTEX, and as a result, should be more rapidly depleted from NAPL. Rixie (1998) performed fixed bed, 1-D dissolution studies of a multi-component residually trapped NAPL to verify that both MTBE and BTX partition into water at concentrations predicted by Raoult's Law. Rates of depletion from NAPL were also consistent with assumptions of local chemical equilibrium as observed over a concentration range of at least 4 orders of magnitude. This paper examines eight remediation sites where UST releases occurred in the late 1980's or early 1990's and both MTBE and BTEX were included as groundwater analytes. Groundwater remediation systems targeting NAPL removal from the smear zone were installed and operated at these sites for several years. The resulting groundwater analytical database provides a unique opportunity to evaluate remediation rates for MTBE, benzene, and xylene in the context of equilibrium NAPL partitioning kinetics.

REMEDIATION SITES

Table 1 summarizes pertinent hydrogeologic information for the eight remediation sites included in the study. Groundwater concentrations had been monitored for a period of up to 3 years prior to remediation and were reasonably stable, with no indication of preferential depletion of MTBE from smear zone NAPL through natural processes. Remediation targeted NAPL removal from smear zone soils through dewatering via groundwater extraction, and volatilization via vapor extraction. Volatilization accounted for 87% of all mass removal during remediation. Table 2 lists pertinent remediation system design and performance information.

Average groundwater concentration reduction factors ($C_{initial}/C_{end}$) were 27 for MTBE, 69 for benzene, and 23 for xylene. Average post-remediation concentration rebound ($C_{post\ remediation}/[C_{initial} - C_{end}] \times 100$) was nonexistent for MTBE and benzene (-3% and -9%, respectively) and modest for xylene (23%).

TABLE 1. Site characteristics

Site	Pre-remediation Depth to Water Table (ft.)	Pre-remediation Water Table Fluctuation (ft.)	K (cm/s)*	Release Date
A	31-37	2.5	?	1989
B	2-7	5	$1.9\ 10^{-3}$	Pre-2/93
C	?	?	$1.32\ 10^{-4}$	Pre-4/88
D	21-23	?	$2.67\ 10^{-3}$	Pre-2/91
E	8-12	13	$3.5\ 10^{-4}$	Pre-9/89
F	25-28	2	$7.05\ 10^{-5}$	5/90
G	21-25	4	$7.4\ 10^{-4}$	Pre-12/92
H	28-40	?	$8.96\ 10^{-4}$	Pre-1/93

* Pump testing at sites C, D, E, and G; slug testing at sites B, F, and H

TABLE 2. Remediation system information

Sites	VE* Wells	Mean System Vacuum ("H2O)	Ave Air Flow/Well (SCFM)	GE* Wells	Mean Water Yield/well (GPM)	Dissolved Mass Removed (kg)	Vapor Mass Removed (kg)
A	10	36	9.8	6	0.24	120	6,901
B	8	39	4.6	8	0.33	12	1,367
C	5	?	?	5	?	?	1,936
D	4	37	18.5	3	0.67	91	3,524
E	7	?	15.2	2	?	?	2,399
F	6	40	13.2	6	0.11	91	3,714
G	17	41	6.9	8	0.38	524	95,648
H	5	22	17.4	5	0.49	3,256	13,723

* VE = vapor extraction; GE = groundwater extraction

REGRESSION ANALYSIS

Analytical data was collected from 113 wells for this study, including active remediation wells and groundwater monitoring wells. Sampling was performed quarterly during remediation, with analysis performed by EPA Method 8020.

Regression analysis was performed on smear zone wells only; dissolved plume wells were excluded. Smear zone well selection criteria included historical presence of liquid hydrocarbon, and/or high headspace concentrations from soil samples collected at or below the water table during well installation. The final regression database consisted of 71 wells.

A first order decay function (equation 1) was fit to each time series data set using least squares regression to characterize the rate of groundwater concentration change and starting concentration.

$$C_t = C_o\, e^{-kt} \tag{1}$$

Where C_t = dissolved phase concentration at time t
C_o = calculated concentration at time t=0
k = decay constant (expressed as %/day concentration reduction).
In order to use regression analysis as a meaningful tool to characterize and compare remediation rates, a screening process was necessary to exclude wells with data sets containing excessive natural or manmade "noise" superimposed on remediation-driven concentration reduction trends. Standard Error of the Estimate (*SeY*) was calculated for each data set, and an arbitrary variability standard of 0.6 (*SeY* <0.6) was used. Wells with analytical data sets meeting the *SeY* selection criteria had low regression residuals such that decay constants (*k's*) could be reliably compared to evaluate relative remediation rates. A total of 39 MTBE, 35 benzene, and 30 xylene wells met the *SeY* screening standard.

The sets of all regression *k's* for a given analyte were compared to evaluate the degree of variability in remediation rate for all wells passing the *SeY* selection criteria. The *k* data sets are statistically similar, with means of 0.30 for MTBE, 0.35 for benzene, and 0.22 for xylene, and standard deviations of 0.20 for MTBE, 0.30 for benzene, and 0.21 for xylene.

Remediation rate prediction. A simple model was used to predict the maximum remediation rate which could have been observed for each site, assuming equilibrium NAPL partitioning is the sole rate-limiting factor. NAPL-liquid equilibrium partitioning concentration was calculated by equation 2.

$$C_{id} = X_i C_{is} \tag{2}$$

Where C_{id} = equilibrium concentration in water of *i*
X_i = mole fraction of *i* in NAPL
C_{is} = pure phase solubility of *i*.
NAPL-vapor partitioning was calculated using equation 3.

$$C_{iv} = X_i P_i^v M_{w,i} (RT)^{-1} \tag{3}$$

Where C_{iv} = equilibrium concentration in vapor of *i*
X_i = mole fraction of *i* in NAPL
P_i^v = pure-phase vapor pressure
$M_{w,i}$ = molecular weight of *i*.

Since equilibrium concentration is driven by mole fraction per equations 2 & 3, it was necessary to estimate a site-specific NAPL composition at the outset of remediation. NAPL mole fractions for MTBE and BTEX were derived using the highest pre-remediation groundwater concentration, assuming these values approximate local NAPL-liquid equilibrium. The remaining NAPL composition was based on a 58 component weathered gasoline published by Johnson (1990) that was combined into 8 normal alkane fractions for input to the multi-component partitioning model. Table 3 lists calculated mole fractions for MTBE and BTEX from five sites with pre-remediation data sets.

TABLE 3. Estimated mole fractions based on pre-remediation maximum groundwater concentration

Site	A	B	D	F	G	MEAN
MTBE	0.15%	0.04%	0.03%	0.35%	0.02%	0.12%
Benzene	1.04%	0.26%	0.43%	0.23%	0.92%	0.58%
Toluene	4.56%	1.57%	2.48%	2.82%	4.35%	3.16%
Ethylbenzene	3.01%	1.78%	0.98%	1.79%	0.94%	1.70%
Xylene	3.78%	5.89%	6.13%	10.18%	5.79%	6.35%

The model solved equation 4 (Johnson et. al., 1990).

$$M_i = \frac{Z_i P \phi V}{RT} + X_i M^{HC} + y_i M^{H2O} + k_i y_i \frac{M_{soil}}{M_{w,H2O}} \quad (4)$$

Where Mi = total moles of contaminant i in all phases
Zi = mole fraction of i in the vapor phase
P = vapor pressure
ϕ = porosity
V = volume of contaminated soil
RT = universal gas constant and absolute temperature, respectively
Xi = mole fraction of i in NAPL
M^{HC} = total moles NAPL
y_i = mole fraction of i in water
M^{H2O} = total moles in dissolved phase.

The sorbed phase term (4th term) was ignored, as was any contribution to mass removal through biodegradation. P was calculated as pure-phase vapor pressure at $T = 20^0$C. V was estimated for each site based on smear zone dimensions which, along with an estimated ϕ of 0.35 and residual oil saturation (S_{or}) of 0.1, established Mi at the outset of remediation. A fixed proportionality factor (0.5) between air and water saturations was used to simulate the average fraction of dewatered smear zone soil accessible to the vapor extraction system. Each pore volume of water and air removed from the smear zone was assumed to contact all NAPL equally, and each pore volume of water and air entering the smear zone was assumed to contain no contaminants. Pore volume and NAPL mole fractions were recalculated for each time step to account for NAPL mass loss through both volatilization and dissolution.

Remediation rate comparison. Six sites were modeled using cumulative groundwater and vapor extraction recovery volumes, and smear zone volume, to predict MTBE, benzene, and xylene k's.

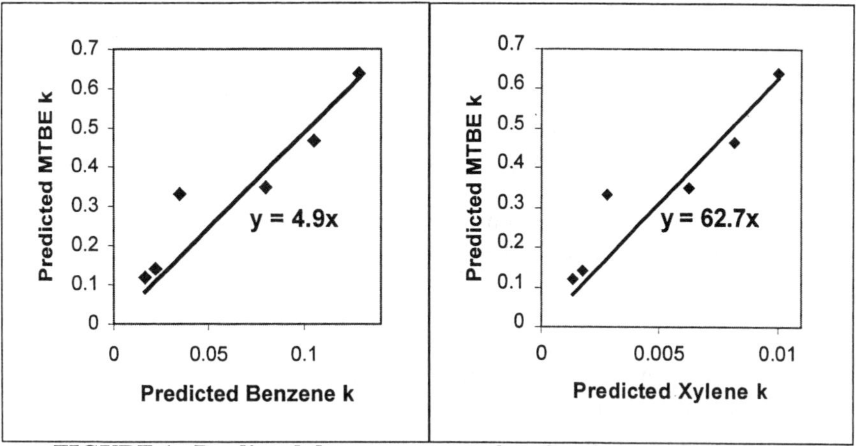

FIGURE 1. Predicted decay constant plots for six remediation sites.

Plots of model predicted MTBE:benzene and MTBE:xylene *k* ratios (Figure 1) demonstrate a linear trend due to similar initial NAPL compositions and high air:water cumulative recovery ratios. The slopes of the linear fits indicate the average ratio between MTBE remediation rate versus benzene or xylene. The model predicted that, on average, MTBE should have been remediated approximately 5 times faster than benzene and 63 times faster than xylene.

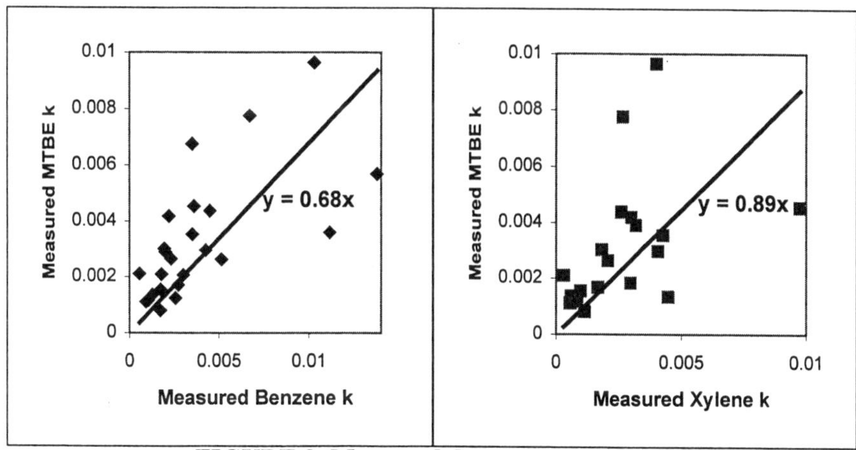

FIGURE 2. Measured decay constant plots.

Figure 2 plots measured MTBE:benzene and MTBE:xylene *k* ratios for wells where at least two analytical data sets satisfied the *SeY* screening standard. The slope of the linear fits through the measured *k* ratio plots indicate the average rate of reduction in MTBE concentration was actually 0.7 times slower than benzene, and 0.9 times slower than xylene.

CONCLUSIONS

Relatively uniform groundwater remediation rates were observed for MTBE, benzene, and xylene at sites included in this study. Equilibrium partitioning modeling predicts very different relative remediation rates for these compounds than were observed in the field, especially as regards a MTBE:xylene comparison. The effect of hydrogeologic heterogeneity limiting contact between mobile remediation fluid (pumped water or air) and stationary NAPL is an attractive explanation for the observed discrepancy between predicted and observed remediation rates. However, minimal post-remediation rebound observed at these sites suggest mass transfer efficiencies were high enough overall to deplete remaining NAPL of MTBE, benzene, and xylene. This work suggests that for heterogeneous hydrogeologic settings, mass transfer limitations on the scale of the subsurface smear zone may prove to be the dominant controlling factor determining remediation effectiveness.

REFERENCES

Johnson, P. C., M. W. Kemblowski, and J. D. Colthart. 1990. "Quantitative Analysis for the Cleanup of Hydrocarbon-Contaminated Soil by In-Situ Soil Venting." *Ground Water*. 28(3): 413-429.

W. G. Rixey. 1998. Presentation at the 1998 National Ground Water Association Petroleum Hydrocarbons and Organic Chemicals in Groundwater Symposium. November 11-13. Houston, TX.

MINERALIZATION OF ETHYL *t*-BUTYL ETHER BY DEFINED MIXED BACTERIAL CULTURES.

Françoise FAYOLLE
Françoise LE ROUX, Guillermina HERNANDEZ and Jean-Paul VANDECASTEELE
(Institut Français du Pétrole, Rueil-Malmaison, France)

ABSTRACT: Aerobic biodegradation of ethers was investigated. An activated sludge from a waste water treatment plant was able to mineralize ethyl *tert*-butyl ether (ETBE) when provided as sole carbon and energy source but not methyl *tert*-butyl ether (MTBE) or *tert*-amyl Methyl ether (TAME). From the sludge two strains were isolated : *Gordona terrae* IFP 2001 and *Rhodococcus equi* IFP 2005 able to utilize ETBE as sole carbon and energy source and leading to *tert*-butanol (TBA) accumulation. A third strain, *Pseudomonas* sp. IFP 2003 able to grow on TBA as sole carbon and energy source was also isolated from the same source. The specific activity of ETBE degradation by *G. terrae* IFP 2001 was 174 mg of ETBE degraded.h^{-1}.g^{-1}biomass and the specific activity of TBA degradation by *Pseudomonas* sp. was 18 mg of TBA degraded.h^{-1}.g^{-1}biomass. A mixed culture of the two strains was found able to mineralize ETBE at a rate of about 4 mg.h^{-1}. L^{-1} at an initial ETBE concentration of 250 mg.L^{-1}without providing any other substrate.

INTRODUCTION

Due to new regulations, addition of oxygenated compounds such as methyl *tert*-butyl ether (MTBE), *tert*-amyl Methyl ether (TAME) or ethyl *tert*-butyl ether (ETBE) in gasoline has been decided in order to maintain their octane index. MTBE is now the second chemical manufactured in the United States (Davidson and Parsons, 1996). The fate of such compounds in soils and aquifers is then of great concern because of their high water solubility (4.3% for MTBE, 3.5% for ETBE) and their relative recalcitrance to biodegradation. During the last years, several studies dealt with the aerobic biodegradation of MTBE. An aerobic consortium able to degrade MTBE was first described (Salanitro et al., 1994). More recently, different studies (Hardison et al., 1997, Steffan et al., 1997) showed that under certain conditions MTBE was biodegraded by cometabolism.

We recently described the first microorganisms *Gordona terrae* IFP 2001 and *Rhodococcus equi* IFP 2005 able to grow on ETBE as sole carbon and energy source isolated from activated sludge (Fayolle et al., 1998) and producing *tert*-butanol (TBA).

Here, we report the isolation from the same activated sludge of a *Pseudomonas* sp. strain IFP 2003 able to grow on TBA and its utilization in a reconstituted mixed culture with *G. terrae* IFP 2001 for ETBE mineralization.

MATERIALS AND METHODS

Microorganisms. *Gordona terrae* IFP 2001 and *Rhodococcus equi* IFP 2005 were isolated from an activated sludge of an urban waste water treatment plant near Paris, France. They were identified and deposited at the CNCM, Pasteur Institute, Paris, France.

Growth Conditions. *G. terrae* IFP 2001 and *R. equi* IFP 2005 were cultivated on the mineral medium I containing in g per liter of deionized water : KH_2PO_4, 6.8; K_2HPO_4, 8.7; Na_2HPO_4. $2H_2O$, 0.334; $MgSO_4$. $7H_2O$, 0.04; NH_4Cl, 1.5; $CaCl_2$. $2H_2O$, 0.04; $FeCl_3$. $6H_2O$, 0.0012 and 1 mL of vitamins solution containing in mg per liter of deionized water : biotin, 200; riboflavin, 50; nicotinamic acid, 50; pantothenate, 50; p-aminobenzoic acid, 50; folic acid, 20; thiamin, 15: cyanocobalamin, 1.5. The pH was 6.95.

Pseudomonas sp. IFP 2003 and the reconstituted mixed culture were cultivated on the mineral medium II containing in g per liter of deionized water : KH_2PO_4, 1.4; K_2HPO_4, 1.7; $MgSO_4$. $7H_2O$, 0.5; $NaNO_3$, 1.5; $CaCl_2$. $2H_2O$, 0.04; $FeCl_3$. $6H_2O$, 0.0012 and 1 mL of the vitamins solution cited above. The pH was 7.0.

The solid medium used for control purity of cultures and isolation of *Pseudomonas* sp. IFP 2003 was the Luria-Bertani medium supplemented with 20 $g.L^{-1}$ of pure agar.

Evaluation of TBA Biodegradability by Respirometry. Assays were performed in the liquid medium described in the ISO 9408 international standard (AFNOR, 1991). Time courses of O_2 demand were obtained by electrolytic respirometry with a D-12 Sapromat apparatus (Voith, Heidenheim, Germany).

ETBE or TBA Degradation Test. Degradation tests were carried out at 30°C in shaked cultures using closed flasks containing the mineral liquid media mentioned above with ETBE or TBA as sole carbon and energy source. After inoculation, culture samples were filtered (0.22 µm) and concentrations of residual ETBE or TBA were determined by GC analysis.

Analytical Procedure. ETBE and TBA were analysed using a VARIAN 3500 gas chromatograph equipped with a flame ionization detector on a 0.32mm x 30m DB 624 column. The temperature of the column was initially 100°C during 0.5 min, then increased to 150°C at $10°C.min^{-1}$ in a first step and up to 250°C at $50°C.min^{-1}$ in a second step. The carrier gas was helium ($1.6 mL.min^{-1}$). Filtered samples of cultures were directly injected on the column.

RESULTS AND DISCUSSION

ETBE Degradation by *G. terrae* IFP 2001 and *R. equi* IFP 2005. The two strains were isolated from an activated sludge (Fayolle et al.,1998). The time course of ETBE degradation by *G. terrae* IFP 2001 and *R. equi* IFP 2005 was measured on mineral medium I containing ETBE as sole carbon and energy source. The production of biomass was measured and the specific activity of ETBE degradation was calculated for both strains. TBA was stoechiometrically produced from degraded ETBE (Table 1).

From these results, it appeared that *G. terrae* IFP 2001 and *R. equi* IFP 2005 were able to grow at the expense of the ethyl group of the molecule. As the presence of oxygen was found necessary to observe ETBE degradation and TBA production with resting cells (results not shown), we assumed that an oxygenase was implicated in the reaction.

TBA Mineralization by Activated Sludge. Activated sludge was then tested using TBA as sole carbon and energy source. It was tested in two ways : directly on TBA or after a previous cultivation on ETBE as a growth substrate which led to the complete mineralization of ETBE. The percentage of TBA mineralization and the lag time observed before adaptation are presented in Table 2.

TABLE 1. ETBE degradation by *G. terrae* IFP 2001 and *R. equi* IFP 2005

Strain	Specific activity for ETBE degradation (mg ETBE degraded. $h^{-1}.g^{-1}$dry weight)	Molar conversion yield (moles TBA formed/ moles ETBE degraded)
Gordona terrae IFP 2001	181	1.08
Rhodococcus equi IFP 2005	42	0.98

TABLE 2. TBA mineralization by activated sludge

Mode of culture of activated sludge *	Percentage of TBA mineralization **	Lag time (days)
Direct cultivation on TBA	90	15
Cultivation on TBA after previous cultivation on ETBE	95	No

*The inoculum was 380 $mg.L^{-1}$ activated sludge.
**The extent of TBA mineralization was calculated from O_2 consumption values using a value of 2.59 mg $O_2.mg^{-1}$ TBA for theoretical oxygen demand. Initial TBA concentration was 100 $mg.L^{-1}$.

TBA was largely mineralized in both cases. When the activated sludge was previously tested using ETBE as sole carbon and energy, we observed no TBA accumulation. However, as shown above (Table 1), when the strains isolated from the sludge were growing on ETBE, we observed TBA accumulation. So, the previous cultivation of the activated sludge on ETBE indirectly led to the selection of microorganisms growing on TBA and we observed no lag time when transferring the sludge on TBA as sole carbon and energy source. In the case of the direct cultivation on TBA, a period of selection or adaptation of the microorganisms growing on TBA was necessary.

Isolation of the Microorganism Growing on TBA. The activated sludge was diluted and spread on Luria-Bertani solid medium. Each colony formed was tested in liquid culture on the mineral medium II containing TBA as sole carbon and energy source. Residual TBA concentrations of each culture were determined. Among all the microorganisms tested, only one was able to use TBA as carbon and energy source. After purification, it was identified as *Pseudomonas* sp. IFP 2003.

The time course of TBA degradation by *Pseudomonas* sp. IFP 2003 is shown in Figure 1. The biomass was measured and TBA degradation rate of 18 mg of TBA degraded.$h^{-1}.g^{-1}$ dry weight was found.

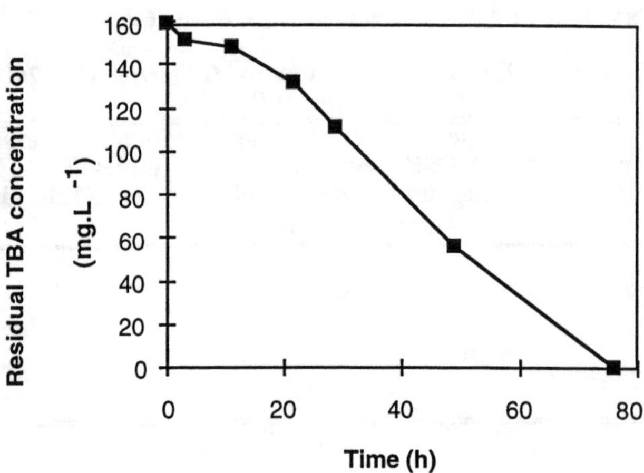

FIGURE 1. Time course of TBA degradation by *Pseudomonas* sp. IFP 2003

Complete Degradation of ETBE by a Reconstituted Mixed Culture. After caracterisation of the strain growing on TBA, mixed cultures of the two strains, *G. terrae* IFP 2001 and *Pseudomonas* sp. IFP 2003 were prepared. With an inoculum ratio *Pseudomonas* sp./*G. terrae* of about 2, two initial ETBE concentrations were tested. Results are shown on Figures 2 and 3.

In both cases, a lag time was observed. The reconstituted mixed culture was able to degrade nearly completely ETBE with a transient accumulation of TBA even when it was provided at a high concentration. When ETBE was provided at a 250 mg.L^{-1} initial concentration, it was degraded at more than 97% in 143 hours at a rate of 4 mg.h^{-1}.L^{-1}and at more than 91% in 235 hours when it was provided at a higher concentration.

CONCLUSION

ETBE was clearly found more biodegradable than MTBE or TAME as shown in biodegradability tests using activated sludge (Fayolle et al., 1998).

Comparing to the reports on MTBE degradation (Hardison et al., 1997; Steffan et al. 1997), the main point in favour of ETBE is that its degradation can clearly sustain bacterial growth instead of relying on cometabolism.

It was possible to isolate some of the different strains contained in the activated sludge responsible for ETBE mineralization. Two strains, *G. terrae* IFP 2001 and *R. equi* IFP 2005, were able to degrade ETBE to TBA by using the ethyl group of the molecule as growth substrate. The third one, *Pseudomonas* sp. IFP 2003, was able to use TBA as carbon and energy source.

It was thus possible to reconstitute a bacterial consortium with *G. terrae* IFP 2001 and *Pseudomonas* sp. IFP 2003 that was able to efficiently mineralize ETBE even at a high concentration.

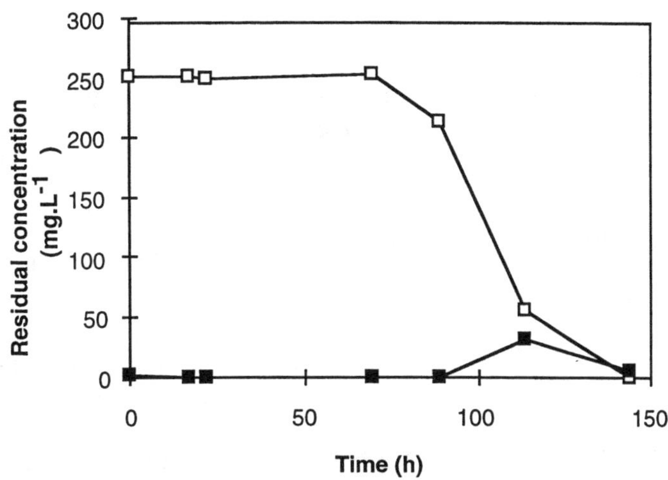

FIGURE 2. Time course of ETBE degradation by a reconstituted *G.terrae/Pseudomonas* sp. mixed culture at low ETBE concentration (250 mg.L^{-1}).

Legend : □, ETBE concentration; ■, TBA concentration.

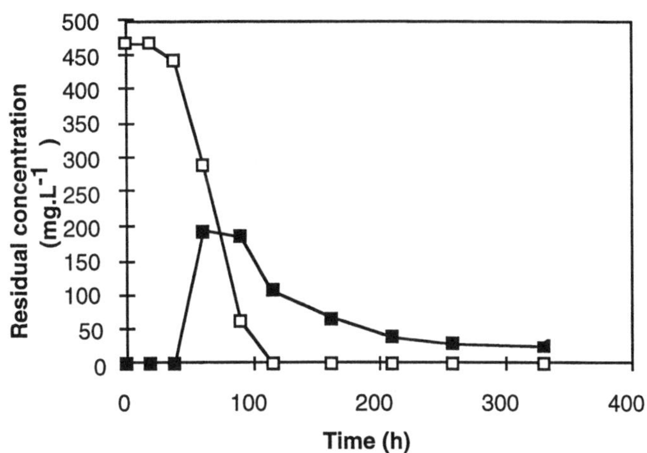

FIGURE 3. Time course of ETBE degradation by a reconstituted *G.terrae/Pseudomonas* sp. mixed culture at high ETBE concentration (465 mg.L^{-1}).

Legend : □, ETBE concentration; ■, TBA concentration.

REFERENCES

AFNOR. 1991. "NF ISO 9408 Standard. "Évaluation en Milieu Aqueux de la Biodégradabilité Aérobie Ultime des Composés Organiques". AFNOR, Paris, France.

Davidson, J.M. and R. Parsons. 1996. "Remediating MTBE with Current and Emerging Technologies". In : *Proceedings of the petroleum hydrocarbons and organic chemicals in ground water*, pp. 15-29, Ground Water Publishing, Westerville,OH.

Fayolle, F., G. Hernandez, F. Le Roux and J.-P. Vandecasteele. 1998. "Isolation of Two Aerobic Bacterial Strains that Degrade Efficiently Ethyl *t*-Butyl Ether (ETBE)". *Biotechnol. Lett.*, *20*, 283-286.

Hardison, L.K., S.S. Curry, L.M. Ciuffetti and M.L. Hyman. 1997. "Metabolism of Diethyl Ether and Cometabolism of MTBE by a Filamentous Fungus, a *Graphium* sp.". *Appl. Environ. Microbiol.*, *63*, 3059-3067.

Salanitro, J.P., L.A. Diaz, M.P. Williams and H.L. Wisniewski. 1994. "Isolation of a Bacterial Culture that Degrades Methyl *t*-Butyl Ether". *Appl. Environ. Microbiol.*, *60*, 2593-2596.

Steffan, R.J., K. Mac Clay, S. Vainberg, C.W. Condee and D. Zhang. 1997. "Biodegradation of the Gasoline Oxygenates MTBE, ETBE and TAME by Propane-Oxidizing Bacteria. *Appl. Environ. Microbiol.*, *63*, 4216-4222.

COMETABOLIC BIODEGRADATION OF METHYL T-BUTYL ETHER BY A SOIL CONSORTIUM

Patrice Garnier, Richard Auria (IRD, France)
Miguel Magaña and **Sergio Revah** (UAM-Iztapalapa, Mexico City, Mexico)

ABSTRACT: A soil consortium was adapted to degrade reformulated gasoline containing methyl-t-butyl ether (MTBE). The gasoline was rapidly degraded to completion. However, MTBE, when tested alone, was not degraded. A screening to identify compounds that participate in cometabolism with MTBE showed that the linear alkanes found in gasoline (pentane, hexane, heptane) did enable elimination of MTBE. Pentane was the most efficient (0.200 mg/day). Upon depletion of pentane, the consortium stopped degrading MTBE. A bacterial strain identified as *Pseudomonas aeruginosa* was isolated from the consortium *P. aeruginosa* could eliminate MTBE in presence of pentane as sole carbon and energy source. Pentane had an inhibitory effect on growth of *P. aeruginosa*, at a concentration as low as 0.085 mg/l. MTBE presence in the media reduced pentane uptake suggesting competitive inhibition.

INTRODUCTION

Fuel oxygenates are organic additives designed to increase the oxygen and octane content of gasoline. The most common are methyl-t-butyl ether (MTBE) and ethanol. Others include ethyl-t-butyl ether (ETBE), t-amyl methyl ether (TAME), isopropyl ether, and t-butanol (TBA). The massive production of MTBE, combined with its mobility, persistence, and toxicity, makes it the second most common ground water pollutant in the USA (Squillace et al. 1996). Oxygenates are used in the gasoline sold in Mexico in a proportion of around 10%. MTBE has also attracted attention because of its presence in urban air.

Considerable research has been undertaken on the biodegradability of MTBE and other fuel oxygenates under both aerobic (Cowan and Park 1996; Mo et al. 1997; Salanitro et al. 1994) and anaerobic (Mormile et al. 1994; Suflita and Mormile 1993) conditions. MTBE is chemically stable and its tertiary carbon structure and ether linkage are two characteristics that theoretically hinders biological attack. It has been reported (Salanitro et al. 1994) that a mixed bacterial culture is able to mineralize MTBE. Degradation of MTBE was shown to proceed through the formation of TBA which was then also degraded by the culture. Mo et al. (1996) reported the biodegradation of MTBE, ETBE and TAME by mixed cultures as well as by pure isolates. Studies on enrichment of mixed cultures degrading MTBE, ETBE, TAME, TBA and t-amyl alcohol (TAA) as sole carbon and energy source have also been reported (Cowan and Park 1996). Kinetic parameters and stoichiometric characteristics of the degradative ability were measured by Cowan and Park (1996). Recent studies by Fortin and Deshusses (1998) have demonstrated that an adapted consortium was capable of high degradation rates in a biotrickling filter, the structure of the consortium was difficult to elucidate.

More than 80% of the compounds present in reformulated gasoline are aromatic, alkanes and MTBE. Among the former compounds are those with a tertiary carbon structure (MTBE; 2,2,4-tri methyl pentane; 2,2-di methyl butane; etc....). They represent a large fraction (20-40%, [v/v]) of those found in reformulated gasoline. Cometabolic biodegradability of MTBE has recently been demonstrated. Pure microorganisms are able to degrade MTBE in the presence of propane (Steffan et al. 1997), butane and ether (Hardison et al. 1997). As MTBE pollution in water or air is often associated with gasoline spills or evaporation it is of considerable interest to study the effects of different gasoline compounds on microbial MTBE degradation.

MATERIALS AND METHODS

Microorganisms. Soil was obtained from contaminated gasoline sites in Mexico and further enriched with frequent addition of gasoline (5 µl) in 125 ml microcosms stoppered with Teflon Mininert valves. A mineral media (Whittenbury et al. 1970) was used.

Chemicals. Chemicals were reagent grade: Benzene, Cyclohexane, Di ethyl ether (DEE) 2,2-Dimethylbutane (22 DMB), Heptane, Hexane, Isopropanol MTBE, t-Butyl Alcohol (TBA), Toluene, 2,2,4-Trimethylpentane (224 TMP), and Xylenes. Lead free reformulated gasoline (Magna Sin) was from PEMEX.

Microcosms. For cometabolic studies in microcosms a mineral salt solution (20 ml) was added to each 125 ml bottle and autoclaved. After inoculation the bottles were stoppered and each growth substrate and/ or MTBE were injected with a gas tight syringe and incubated at 30°C on a rotatory shaker at 250 rpm. The degradation of substrates, MTBE, O_2, and the production of CO_2 were monitored by gas chromatography. The kinetic analyses of the data is described by Garnier et al. (1999).

Analyses. Organic volatile compounds (from headspace of each microcosm) were measured by taking a 200 µl gas sample and injecting in a FID-GC (Hewlett Packard (HP), model 5890-II) with a HPI column (Methyl Silicon Gum) 5 m x 0.53 mm. Oxygen and CO_2 concentrations in respirometric studies were measured using a TCD-GC (GOW MAC Series 550 with a concentric column CTR-1, Alltech, USA) with helium as carrier gas.

RESULTS

A consortium enriched from mixed polluted soils was able to mineralize reformulated gasoline, which included MTBE, when used as sole source of carbon and energy and producing CO_2. After 6 days of incubation, 97% of total reformulated gasoline was degraded. When only MTBE was added to this consortium, no significant degradation occurred within one week.

In order to identify some of the potential compounds, present in gasoline that foster MTBE cometabolism, a screening was carried out as shown in Table 1. Complete substrate utilization was observed for benzene, toluene and xylenes but cyclohexane was recalcitrant for the consortium. Xylenes were removed very rapidly (about 80% in 24 hours), followed by toluene (100% in two days) and benzene (100% in 3 days). No MTBE utilization was observed.

Complete substrate utilization was observed for the three alkanes tested. These compounds are present in gasoline at concentrations (v/v) of 4.7, 2.7 and 1.1%, respectively. Pentane favored the highest MTBE utilization rate (0.200 mg/day) as compared to hexane (0.042 mg/day) and heptane (0.087 mg/day). CO_2 production was coupled to substrate utilization. Compounds with similar structures to MTBE were tested for cometabolic activity, no utilization was observed for 224 TMP, 22 DMB and TBA as shown in table 1. Production of CO_2 was the lowest when compared to the other compounds tested. Di-ethyl ether was tested for cometabolism with MTBE. DEE is an industrial solvent which is not present in gasoline. Over a period of 17 days, 85% of DEE was degraded (around 0.175 mg/day) which was much slower than the alkanes which were fully degraded in three days. MTBE was cometabolized with an average degradation rate of 0.083 mg/day. Results for cometabolic studies with MTBE have been reported by Hardison and coworkers (1997) with butane and by Steffan et al. (1997) in the presence of propane.

TABLE 1: Microcosm screening studies with MTBE (2 µl/microcosm): Effect of different cometabolic substrates (5µl/microcosm).

Growth substrates	Degradation in four days (%)	MTBE degradation (mg/bottle)	Final CO_2 (%)
Aromatic and Cyclic			
Benzene	100%	0	6.5 +/- 0.4
Toluene	100%	0	5.9 +/- 0.1
Xylenes	100%	0	5.7 +/- 0.0
Cyclohexane	0%	0	0.9 +/- 0.1
Alkanes			
Pentane	100%	0.225 +/- 0.005	3.8 +/- 0.1
Hexane	100%	0.054 +/- 0.003	3.4 +/- 0.1
Heptane	100%	0.104 +/- 0.019	3.4 +/- 0.4
Structurally similar			
TBA	ND	0	0.4 +/- 0.0
22 DMB	0%	0	0.9 +/- 0.0
224 TMP	0%	0	0.3 +/- 0.0
Other			
Isopropanol	100%	0	3.5 +/- 0.1
DEE	85 % (17 days)	1.230 +/- 0.230	5.3 +/- 0.2

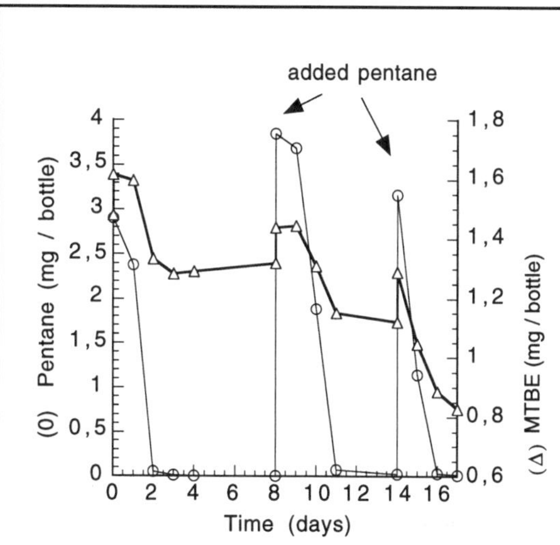

Figure 1: Cometabolic degradation of MTBE in presence of pentane.

Figure 1 shows the degradation of MTBE and pentane by the consortium. A reduction in the concentration of MTBE (initial concentration of 2 µl per bottle) was observed along with the degradation of pentane during the first four days of incubation. As soon as the pentane was consumed, MTBE degradation was halted and the concentration remained constant thereafter. The bottle was respiked with pentane and the cometabolic degradation of MTBE resumed. Once again, reduction of MTBE was observed as pentane was respiked.

A cometabolic ratio of 0.1 mg MTBE/mg pentane was observed. A control experiment showed no further MTBE degradation after the first addition and for 3 weeks. These experiments suggest that MTBE was degraded by an alkane-induced enzyme probably produced in the early steps of the pentane degradation pathway.

From this consortium a pentane degrading microorganism was isolated. It was identified as *Pseudomonas aeruginosa* from cellular fatty acid analysis. As with the consortium, the isolated strain was able to degrade MTBE when cometabolized in presence of pentane as can be seen in Figure 2. A higher MTBE degradation rate of 0.530 mg/day was found as compared to the consortium suggesting that this was the most relevant organism. No MTBE degradation was observed in the absence of pentane. For this strain a maximum growth rate μ_{max} of 0.19 h^{-1} on pentane was calculated which is equivalent to a biomass doubling time, t_d, of 3.65 h. From independent experiments a growth yield of 0.9 g/g was obtained.

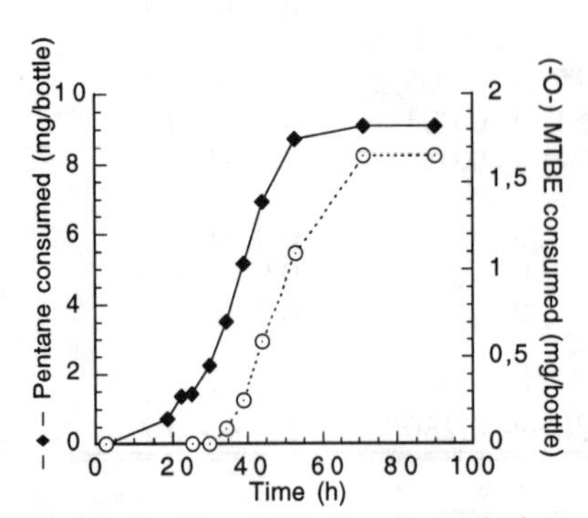

Figure 2: Cometabolic degradation of MTBE by *P. aeruginosa* grown on pentane.

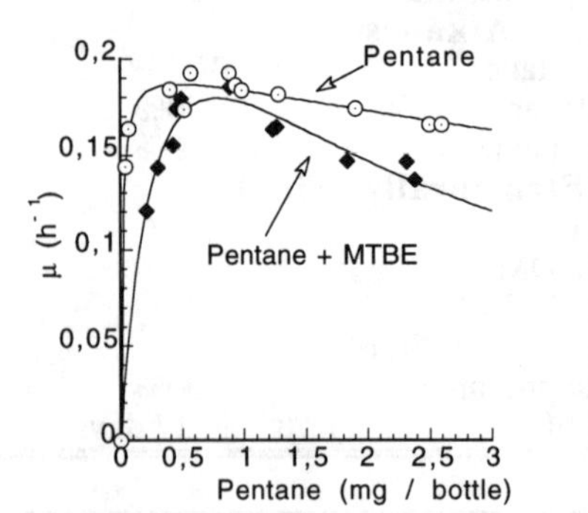

Figure 3: µ values for *P. aeruginosa* with pentane with and without MTBE (10 µl by bottle) addition.

To study the relation between pentane and MTBE uptake an inhibition model described by Pirt (1975) was used. A K_s (0.016 mg/bottle) and the K_i (13.69 mg/bottle) of *P. aeruginosa* in the presence of pentane were calculated with a regression coefficient of 0.995 (figure 3) (Garnier et al. 1999). Considering the gas liquid equilibrium for pentane as expressed by Henry's constant (128 x 10^3 Pa m^3 mol^{-1} at 25°C), saturation and inhibitory constants were calculated in the liquid as K_s equal to 0.0029 mg/l and the K_i equal to 3.5 mg/l. The growth rate increased up to a critical value S_{crit} equal to 0.47 mg/bottle or 0.085 mg/l as shown in figure 3. The same experiment with fixed initial MTBE concentration (10 µl/bottle) showed that at pentane concentrations lower than 0.5 mg/bottle, the calculated µ is lower with MTBE. After S_{crit}, a constant maximum growth rate (μ_{max} around 0.19 h^{-1}) between 0.5 and 1 mg of pentane per bottle was detected. This result indicates a competitive inhibition between pentane and MTBE. At higher concentration of pentane, the growth rate is lower with MTBE indicating an inhibition by MTBE of pentane degradation. For MTBE the Henry constant at 25°C is 55 Pa m^3 mol^{-1}, about 2000 times lower than pentane. The K_s of MTBE as determined by kinetic modeling was about 4.5 mg/bottle or 185 mg/l which is 65,000 higher than the value for pentane (Garnier et al. 1999).

With this strain a specific degradation rate of 3.9 nmol min^{-1} mg^{-1} of cell protein was determined. Analyses from produced CO_2 suggested a MTBE mineralization of only 20% (1/5 of carbon to CO_2) presumably resulting only from the complete oxidation of the methoxy methyl group of MTBE. As seen in Table 1 TBA was not utilized. Complete mineralization would require the presence of other strains.

CONCLUSIONS

With ever increasing concern about air and water contamination, a search for bioremediation alternatives is compulsory. The possibility of fostering the cometabolic activity to accelerate MTBE degradation remains an interesting alternative to complement current efforts in MTBE control in the environment.

REFERENCES

Cowan RM and K. Park 1996. "Biodegradation of the gasoline oxygenates MTBE, ETBE, TAME, TBA and TAA by aerobic mixed culture". *Proceedings of the 28th Mid-Atlantic Industrial and Hazardous Waste Conference*, Buffalo, NY. pp 523-530.

Fortin N. and M. Deshusses 1998. "Gas phase biotreatment of MTBE. "*Proceedings of the 1998 USC- TRG Conf. on Biofiltration.* USC- TRG, 123.

Garnier P., R. Auria, C. Augur and S. Revah. 1999. "Cometabolic biodegradation of Methyl t-Butyl Ether by *Pseudomonas aeruginosa* grown on pentane". *Appl. Microbiol. Biotechnol.* (accepted).

Hardison L. K., S. S. Curry, L. M. Ciuffetti and M. R. Hyman 1997. "Metabolism of diethyl ether and cometabolism of methyl *ter*-butyl ether by a filamentous fungus, a *Graphium* sp." *Appl. Environ. Microbiol.* 63: 3059-3067.

Mo K, C. O. Lora, A. E. Wanken, M. Javanmardian, X. Yang and C. F. Kulpa 1997. "Biodegradation of methyl t-butyl ether by pure bacterial cultures". *Appl. Microbiol. Biotechnol.* 47: 69-72.

Mormile M. R., S. Liu, and J. M. Suflita. 1994. "Anaerobic biodegradation of gasoline oxygenates: extrapolation of information to multiple sites and redox conditions". *Environ. Sci. Technol.* 28: 1727-1732.

Pirt S. J. 1975. *Principles of microbe and cell cultivation,*. Halsted Press Book, Wiley. UK.

Salanitro J. P., L. A. Diaz, M. P, Williams and H. L. Wisniewski. 1994. "Isolation of a bacterial culture that degrades methyl t-butyl ether". *Appl. Environ. Microbiol.* 60: 2593-2596.

Squillace P. J., J. S. Zogorski, W. G. Wilber and C. V. Price. 1996. "Preliminary assessment of the occurrence and possible sources of MTBE in ground water in the United States 1993-1994." *Environ. Sci. Technol.* 30: 1721-1730.

Steffan R. J., K. McClay, S. Vainberg, C. W. Condee and D. Zhang. 1997. "Biodegradation of the Gasoline Oxygenates Methyl *ter*-Butyl Ether, Ethyl *tert*-Butyl Ether, and *tert*-Amyl Ether by Propane-Oxidizing Bacteria". *Appl. Environ. Microbiol.*, 63: 4216-4222.

Suflita J. M. and M. R. Mormile. 1993. "Anaerobic biodegradation of known and potential gasoline oxygenates in the terrestrial subsurface." *Environ. Sci. Technol.* 27: 976-978.

Whittenbury R. W., K. C. Phillips and J. F. Wilkinson. 1970. "Enrichment, isolation and some properties of methane-utilizing bacteria." *J. Gen. Microbiol.* 61: 205-218.

DEMONSTRATION OF THE ENHANCED MTBE BIOREMEDIATION (EMB) IN SITU PROCESS

J. P. Salanitro, G. E. Spinnler, C. C. Neaville, P. M. Maner, S. M. Stearns
(Equilon Enterprises LLC [Shell/Texaco], Houston, Texas)
P. C. Johnson, C. Bruce (Arizona State Univ., Tempe, Arizona)

ABSTRACT: MTBE (methyl tert-butyl ether), a common gasoline additive, is a ground water chemical of concern at many fuel release sites. The MTBE plume at the USN Port Hueneme, California NEX service station is currently over 4000 ft (1200 m) long.

A field pilot test at Port Hueneme was conducted to assess the efficacy of an enhanced MTBE bioremediation (EMB) using the BC-4 culture (a microbial consortium degrading MTBE to CO_2). Essential features of the EMB process include an in situ biobarrier of MTBE-degraders, a network of O_2 injection wells near the seeded transect, and an array of monitoring wells upstream and downstream of the treatment zone. Three test plots located in an MTBE-only portion of the plume included a control plot (no treatment), an oxygenation-only plot (intermittent O_2-injection), and an oxygenated + BC-4 seeded plot.

Monitoring wells distributed throughout each plot showed that initial MTBE concentrations prior to treatment varied from 2000-8000 µg/L. After five months of operation, MTBE declined to 0-200 µg/L in wells just downstream of the BC-4 biobarrier. In the O_2-only plot, MTBE levels remained relatively stable for the first four months, and then decreased to 400-1100 µg/L in the fifth month, suggesting stimulation of natural MTBE-degraders in the aquifer. MTBE concentrations in the control plot varied from 1300-3900 µg/L, with some points being affected by the O_2 injection system as evidenced by increased dissolved oxygen (DO) levels. The intermittent O_2 injection system increased dissolved oxygen concentrations in the target treatment zones from <1 mg/L to 10 to ≥ 20 mg/L. These pilot-scale field results show the EMB process can be used effectively as a biobarrier to advancing MTBE ground water plumes.

INTRODUCTION

Reformulated gasoline (RFG) is marketed in the U.S. in many non-attainment cities which do not meet the ambient air quality standards for CO and ozone-forming compounds. Under the 1990 Clean Air Act Amendment (National Air Pollutant Trends, 1993), motor fuels are required to contain an oxygenate chemical (up to 2 - 2.7% oxygen w/w) to reduce engine emissions of such pollutants. The most widely used oxygenate in RFG is MTBE (methyl tert-butyl ether) although fuel in some cities contains ethanol as the required oxycompound additive. It is now widely recognized that the presence of MTBE in ground water monitoring wells is caused by the accidental release of RFG from the underground storage/delivery systems at fuel station facilities. MTBE is very mobile in aquifer systems because of its low Kow and soil sorption, relatively high water solubility (43,000 mg/L) and poor biodegradability (Squillace et al., 1996). Therefore, MTBE plumes may migrate farther than BTEX plumes, and its distribution will be influenced by the local hydrogeology (soil heterogeneities, gradient, water table fluctuations, etc.) and advective dispersion and dilution along the ground water flow path (Rice et al., 1995; Mace & Choi, 1998; Buscheck et al., 1997; Squillace et al., 1996).

The aerobic biodegradation of MTBE has been reported in mixed cultures derived from activated sludge (Cowan and Park, 1996; Salanitro et al., 1994), pure cultures of alkane-oxidizing bacteria (Hyman et al., 1998; Steffan et al., 1997), a fungus (Hardison et al., 1997), and a Sphingomonas species (Hanson et al., 1998). The BC-1 mixed bacterial culture described by Salanitro et al. (1994) is a unique consortium which can metabolize MTBE completely to CO_2.

Only a few studies have shown any significant MTBE bioattenuation in aquifers (Borden et al., 1997; Schirmer and Barker, 1998). We have also observed that indigenous MTBE-degraders may be present in subsoils and ground water at fuel spill sites and that their growth and metabolic activity are significantly reduced by low DO in aquifers (Salanitro et al., 1998). The natural attenuation of aromatic compounds like BTEX has been a very effective tool for controlling plume migration at many sites because of the widespread distribution of naturally-occurring aromatic hydrocarbon-degrading microbes. In contrast, the presence of difficult-to-degrade compounds in aquifers like MTBE may require specific bioaugmentation regimes with specialized microbes to reduce a contaminant below regulated concentrations.

In this study we describe a method, the enhanced MTBE bioremediation (EMB) process for stimulating the biodegradation of fuel ethers in situ. The process is being demonstrated at the USN Port Hueneme, California MTBE plume site. Essential features of the EMB process include: a) a region of the aquifer seeded with MTBE-degraders (the biobarrier), b) a network of oxygen (O_2) injection wells for maintaining optimal conditions for aerobic biodegradation in the target treatment zone, and c) a network of monitoring wells upstream and downstream of the treatment zone. This combination of microbial inoculation, O_2 delivery, and appropriate monitoring may be used as a cost-effective in-situ treatment process for controlling the leading edge of a MTBE plume or degrade source areas with residual hydrocarbons and oxygenate.

MATERIALS AND METHODS

The MTBE Plume at Port Hueneme. During 1984-1985 several thousand gallons of leaded gasoline containing MTBE were released from the storage tanks at the NEX Gasoline Station of the Port Hueneme, California Naval Base. Analyses of samples from ground water monitoring wells in 1997-1998 indicated that the MTBE plume had traveled over 4000 ft (1200 m) of the aquifer from the source. The plume is at least 400 ft (120 m) wide, and about 75% of the soluble plume is only MTBE. The BTEX components of the spilled gasoline have been attenuated within the remaining residual phase nearer the source. The water table is roughly 10 ft (3 m) bgs, and the thickness of this upper aquifer is 10 ft (3 m). Cone penetrometer strain gauge analysis of the Port Hueneme upper aquifer sediment indicated upper and lower portions consisting of silty loam and fine-medium sandy soils, respectively. The apparent ground water velocity in the upper and lower segments of this upper aquifer varied from about 0.1 – 0.3 ft [0.03 - .09 m] (upper) and 0.3 – 0.5 ft [0.09 - .15 m] (lower)/day.

Description of Field Test Plots. The test site was located approximately midway down the advancing MTBE plume. In this region MTBE is the only soluble fuel constituent, and it is present at concentrations varying from 2000-8000 µg/L. The DO levels in this area were < 1 mg/L prior to the test. The test plan consisted of 3 plots: 1) O_2 only, 2) O_2 + BC-4 seeded biobarrier, and 3) a control, no treatment zone. Each test plot was aligned parallel to the initial estimated ground water flow direction and had dimensions of 20 ft (6 m) wide X 40 ft (12 m) long as diagrammed in Figure 1. The spacing between test plots was 10 ft (3 m). The O_2 + BC-4 seeded plot contained a treatment zone composed of a BC-4 inoculated area and an oxygen delivery system. The BC-4 microbial consortium was grown with MTBE as the sole carbon source, and it had a specific MTBE removal rate of 20-30 mg/g cells/hr. It was injected at points across and throughout the depth of the target treatment zone. The oxygen delivery system consisted of oxygen injection wells that were designed, placed, and operated so as to optimize oxygen delivery efficiency while ensuring the necessary conditions for aerobic biodegradation of MTBE by BC-4 (DO > 2 mg/L). The O_2-only plot consisted of only O_2-injection and monitoring wells that were constructed, placed, and operated in a manner similar to those used in the O_2 + BC-4 seeded plot. The control plot was maintained as a no treatment cell and contained only monitoring wells. O_2 was generated on site with a Matrix Technologies pure oxygen-generating system using an Air Sep AS 80 pressure swing adsorption.

Monitoring wells (1 inch [2.54 cm] PVC pipe) were installed with a direct-push soil coring system. They were screened over 5 ft (1.5 m) intervals (10-15 ft [3 – 4.5 m] and 15-20 ft [4.5 – 6 m] bgs) and were designated as either "shallow" or "deep" sampling points. Ground water samples were taken after purging one well volume from the monitoring points and the samples were then analyzed for MTBE, tert-butyl alcohol and numbers of MTBE-degraders.

Figure 1 presents a schematic of the test site layout.

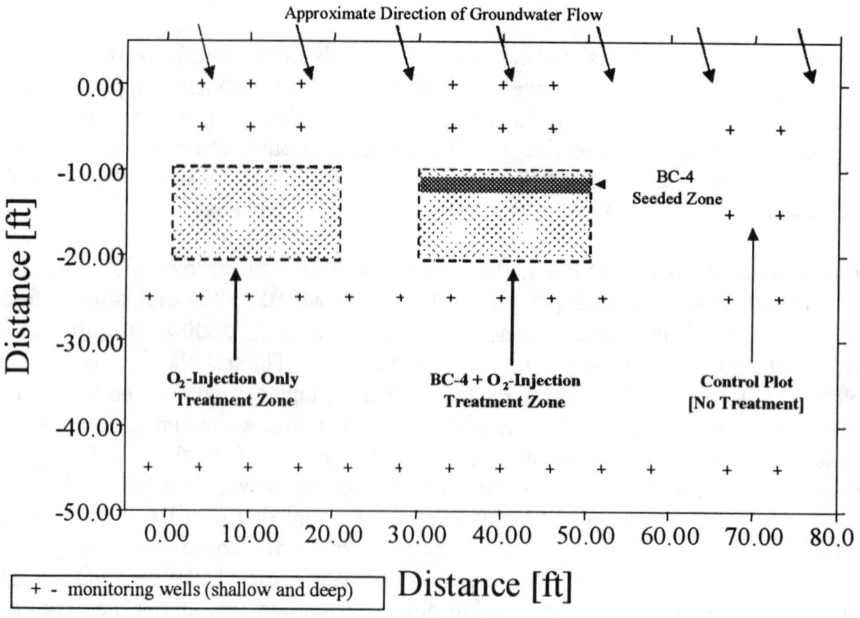

FIGURE 1. Layout of EMB Field Test Site. Initial MTBE concentrations 2000 – 8000 ug/L and initial dissolved oxygen concentrations <1 mg/L.

RESULTS AND DISCUSSION

Performance of the Oxygen Delivery System. The oxygen delivery system was started approximately 6 weeks prior to BC-4 seeding to increase dissolved oxygen (DO) in the target treatment zone to levels better-suited for aerobic biodegradation. Apparently the system was effective at increasing DO levels above the <1 mg/L initial condition; throughout the target treatment zones the dissolved oxygen level was generally >10 mg/L and in many cases was >20 mg/L. Oxygen levels also increased in areas up-gradient, down-gradient, and cross-gradient to the target treatment zones. Thus, some areas of the Control Plot were affected by oxygen injection. The intermittent operating mode appeared to be an efficient alternative to continuous O_2 injection, with sufficient oxygen remaining in the aquifer between injection cycles. Figures 3 and 4 present contour plots of the DO distribution for the shallow and deep monitoring points, respectively.

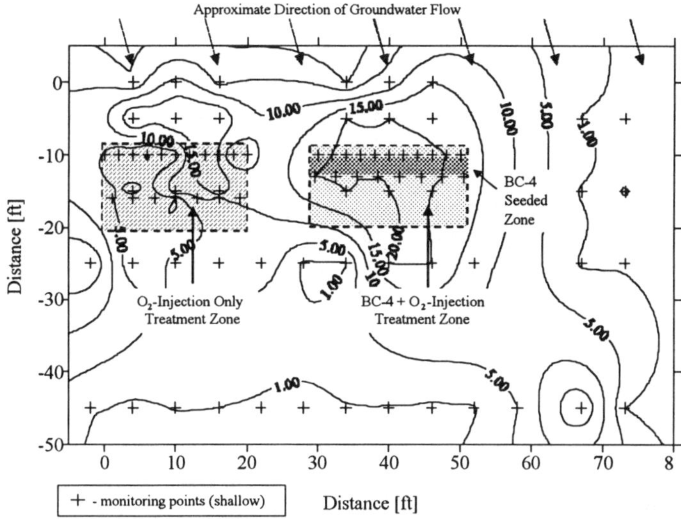

FIGURE 2. Dissolved oxygen concentrations in shallow (14 – 15 ft BGS) monitoring points after 110 days of O_2 injection. Contours labeled "20 mg/L" define regions where DO>20 mg/L.

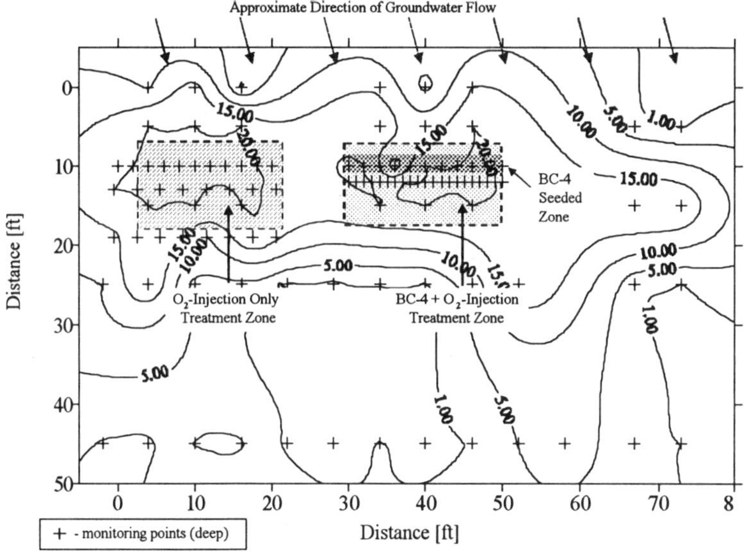

FIGURE 3. Dissolved oxygen concentrations in deep (18 – 20 ft BGS) monitoring points after 110 days of O_2 injection. Contours labeled "20 mg/L" define area where DO>20 mg/L.

Performance of the BC-4 BioBarrier. Figures 4a, 4b, 4c and 5a, 5b, 5c display MTBE concentrations vs. distance from the top of each treatment plot for various times prior to, and after BC-4 injection. For reference, the times shown are measured relative to the date of BC-4 seeding; thus, t = -44 d corresponds to the start of the O_2 delivery system and t=0 corresponds to BC-4 seeding. Each figure depicts a transect along the direction of groundwater flow for one of the study plots (i.e., O_2 injection-only, BC-4 + O_2 injection, and the control plot). For each distance along a given transect, the geometric mean of the measured concentrations is displayed in Figures 4 and 5.

For all the plots and all distances along the transects, MTBE concentrations are relatively stable for all times prior to the BC-4 seeding (t=0 in Figures 4 and 5). The operation of the O_2 injection system increased DO to concentrations similar to those shown in Figures 2 and 3; however, O_2 injection had little effect on MTBE concentrations in any plot.

Following BC-4 seeding and continued operation of the O_2 injection system for the next 67 days, decreases in MTBE concentrations are observed in the BC-4 + O_2 injection plot. Immediately down-gradient of the BC-4 seeded zone MTBE concentrations decrease >90%, and other less dramatic declines in MTBE were observed up-gradient of the seeded region. In comparison, no significant changes are observed in MTBE concentrations in the O_2 injection-only or control plots.

Between 67 d and 129 d after BC-4 seeding (173 d following the start of O_2 injection), other interesting changes appear in the data. First, MTBE concentrations continue to decline down-gradient of the BC-4 seeded zone. By this time MTBE was not detected in many of the samples at the closest down-gradient monitoring points, and several samples had MTBE concentrations in the 0.01 – 0.05 mg/L range. In addition, in the BC-4 + O_2 plot, MTBE concentrations continue to decline in the up-gradient monitoring points as well. Even more interesting is the sudden appearance in declines in MTBE concentration in the O_2-injection-only and control plots. Upon review of Figures 2 and 3 it can be seen that these declines are occurring in regions where the O_2 delivery system has caused increases in DO concentrations. This observation, along with other data not presented here, has led to the hypothesis that the O_2 injection system has stimulated the growth of naturally-present MTBE-degraders.

In summary, the time series data clearly show the impact of the EMB process, and effectiveness of it as a biobarrier to MTBE migration. To our knowledge, this is the first successful field-scale demonstration of any engineered <u>in-situ</u> MTBE biodegradation process.

ACKNOWLEDGEMENTS

The authors would like to thank NFESC personnel at the USN Port Hueneme National Environmental Test Site for their support; in particular Ernie Lory, Karen Miller, Gale Pringle, James H. Osgood, and Dorothy Cannon.

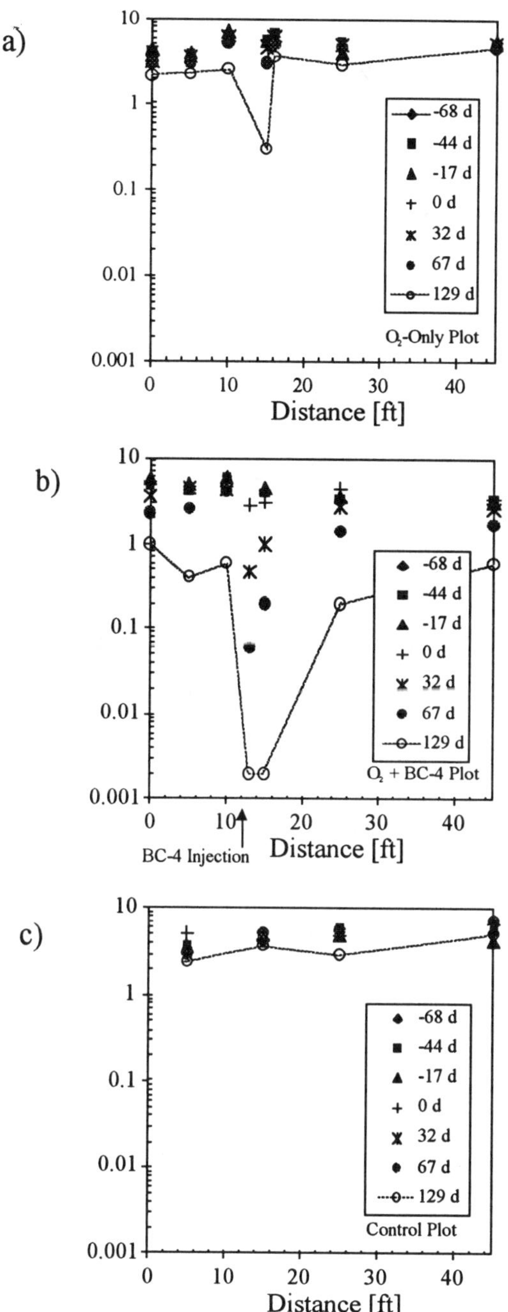

FIGURE 4. Mean MTBE concentrations in shallow monitoring points: a) O_2-injection-only plot, b) O_2 injection + BC-4 seeded plot, and c) the control plot.

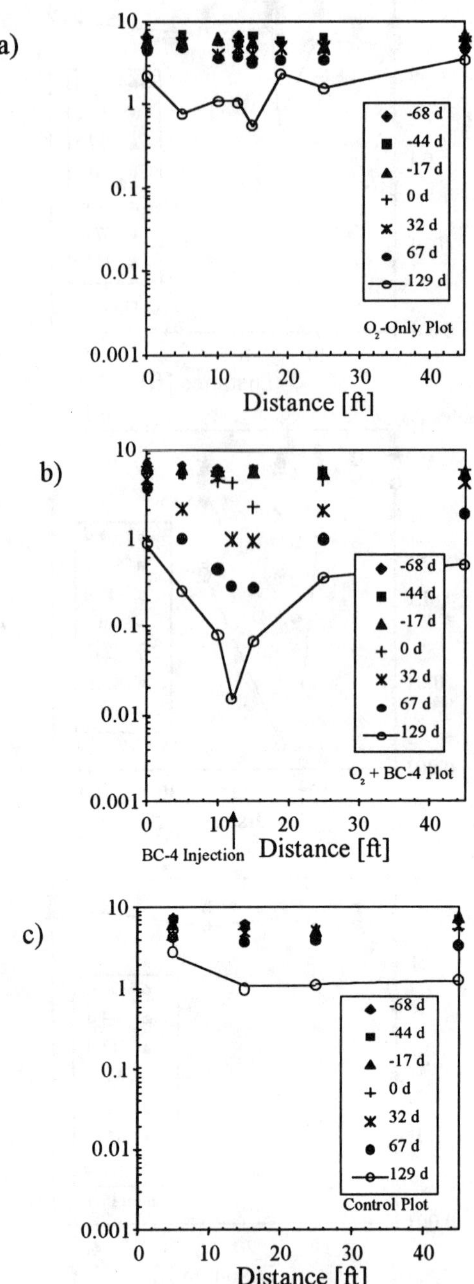

FIGURE 5. Mean MTBE concentrations in deep monitoring points: a) O_2-injection-only plot, b) O_2 injection + BC-4 seeded plot, and c) the control plot.

REFERENCES

Borden, R. C., R. A. Daniel, L. E. LeBrun IV, and C. W. Davis. 1997. "Intrinsic biodegradation of MTBE and BTEX in a gasoline-contaminated aquifer." *Water Resour. Res.* 33:1105-1115.

Buscheck, T. E., D. J. Gallagher, D. L. Kuehne, and C. R. Zuspan. 1997. "Occurrence and behavior of MTBE in ground water", in Proceedings, Petroleum Hydrocarbons and Organic Chemicals in Ground Water, NGWA/API, Houston, TX, Nov. 12-14.

Cowan, R. M. and K. Park. 1996. "Biodegradation of the gasoline oxygenates MTBE, ETBE, TAME, TBA, and TAA by aerobic mixed cultures," in Proceedings 28th Mid-Atlantic Indust. and Haz. Waste Conf., (ed. A.S. Weber), July, pp 523-530.

Hanson, J., K. M. Scow, M. Bruns, and T. Brethour. 1998. "Characterization of MTBE-degrading bacterial isolates and associated consortia." MTBE Workshop Abstracts, Univ. Calif., Davis, June 16.

Hardison, L. K., S. S. Curry, L. M. Ciuffetti, and M. R. Hyman. 1997. "Metabolism of diethyl ether and cometabolism of methyl tert-butyl ether by a filamentous fungus, a Graphium sp." *Appl. Environ. Microbiol.* 63:3059-3067.

Hyman, M. 1998. API proposal, "Cometabolism of gasoline oxygenates by alkane-utilizing bacteria."

Mace, R. E. and W. J. Choi. 1998. "The size and behavior of MTBE plumes in Texas," in Proceedings, Petroleum Hydrocarbons and Organic Chemicals in Ground Water, NGWA/API, Houston, TX, Nov. 11-13.

National Air Pollutant Trends 1900-1992. 1993. Office of Air Quality Planning and Standards, USEPA, Research Triangle Park, N.C.

Rice, D. W., R. D. Grose, J. C. Michaelson, B. P. Dooher, D. H. MacQueen, S. J. Cullen, W. E. Kastenberg, L. E. Everett, and M. A. Marino. 1995. *California Leaking Underground Fuel Tank (LUFT) Historical Case Analyses*, Lawrence Livermore National Laboratory, Livermore, California. (UCRL-AR-122207).

Salanitro, J. P. 1993. "The role of bioattenuation in the management of aromatic hydrocarbon plumes in aquifers." Ground Water Monit. Remed. 13:150-161.

Salanitro, J. P., L. A. Diaz, M. P. Williams, and H. L. Wisniewski. 1994. "Isolation of a bacterial culture that degrades methyl-t-butyl ether." *Appl. Environ. Microbiol.* 60:2593-2596.

Salanitro, J. P., C. C. Chou, H. L. Wisniewski, and T. E. Vipond. 1998. "Perspectives on MTBE biodegradation and the potential for in situ aquifer bioremediation." Southwestern Regional Conf. of the Natl. Ground Water Assn., Anaheim, California, June 3-4.

Schirmer, M. and J.F. Barker. 1998. "A study of long-term MTBE attenuation in the Borden aquifer, Ontario, Canada." Ground Water Monit. Remed. 18:113-22.

Squillace, P.J., J.S. Zogorski, W. G. Weber, and C. V. Price. 1996. "Preliminary assessment of the occurrence and possible sources of MTBE in ground water in the United States, 1993-1994." *Environ. Sci. Technol.* 30:1721-30.

Steffan, R. J., K. McClay, S. Vainberg, C. W. Condee, and D. Zhang. 1997. "Biodegradation of the gasoline oxygenates methyl tert-butyl ether, ethyl tert-butyl ether and tert-amyl methyl ether by propane-oxidizing bacteria." *Appl. Environ. Microbiol.* 63:4216-4222.

GROUNDWATER FLOW SENSOR MONITORING OF AIR SPARGING

David A. Wardwell (Mission Research Corporation)
Albuquerque, New Mexico, USA

ABSTRACT: A field experiment was conducted to determine the efficacy of a single air sparging point at different injection rates. Seven HydroTechnics Inc. In Situ Groundwater Velocity Sensors were installed around a single air injection well to study the groundwater flow regime when air was injected into the saturated zone at 5, 10, and 20 standard cubic feet per minute (scfm). Horizontal groundwater flow velocities were significantly reduced 15 ft up gradient at each injection rate of 5, 10 & 20 scfm. In fact, at 20 scfm, the horizontal flow velocity was reduced to 0.02 feet/day (ft/day), and the groundwater flow direction was deflected away from the sparge area. This indicates the possibility that once air is injected into the subsurface at rates greater than 10 scfm, contaminated groundwater in shallow areas of the aquifer is diverted from the most effective regions of an air sparging system. Groundwater flow velocity changes during mounding or groundwater displacement episodes lasted less than 24 hours, which had little impact on thc distribution or spreading of hydrocarbon contamination (2-3 feet dispersion).

INTRODUCTION

Air sparging is the process of injecting clean air directly into an aquifer for remediation of contaminated groundwater. In situ air sparging remediates groundwater through a combination of volatilization and enhanced biodegradation. The induced air transport through the groundwater removes the more volatile and less-soluble contaminants by physical stripping. Increased biological activity is stimulated by increased oxygen availability.

The United States Air Force (USAF) in conjunction with the Environmental Protection Agency (EPA) initiated an air sparging experiment at the National Test Site (NTS) located at the Naval Base in Port Hueneme, CA. The Air Force has initiated these actions with the goal of creating the Department of Defense (DOD) In Situ Air Sparging (IAS) Manual to be used in future remediation activities. The Department of Energy (DOE) had previously created a similar manual called the Parsons Document. The document includes 16 heterogeneous hydrogeological scenarios catalogued for preliminary air sparging site evaluation, design, and utilization. The reasoning behind the Parsons Document is the belief that air sparging success, and more importantly air distribution in the subsurface, is controlled by stratigraphy. The purpose of the DOD IAS manual is to combine the information provided in the Parsons Document and further correlate generic hydrogeological settings with air sparging design and degree of success. The USAF plans to accomplish this with a pair of experiments utilizing intense and advanced monitoring systems.

A team of environmental professionals was gathered to conduct in depth in situ air sparging research. This team, organized by the Air Force Research Laboratory (AFRL), included Mission Research Corporation (MRC), Battelle Inc., Arizona State University (ASU), Oregon Graduate Institute (OGI), and Parsons Engineering.

The primary objective of this experiment is to verify air sparging performance in a series of experiments under different geologic conditions. The results of these studies will be used to further refine existing air sparging models and to support the development of the Air Force Air Sparging Manual.

As part of this experiment, MRC was tasked to identify and install state-of-the-art, in situ sensors to monitor the physical parameters of the air sparging experiment. In situ groundwater velocity sensors were deployed that are capable of detecting very low three-dimensional flow velocities from a single location. This technology helped to determine water flow characteristics (3-D velocity) surrounding the sparge area. While there were many chemical and physical tests performed during the air sparging experiment at Port Hueneme, this analysis focused on the results of changing groundwater flow characteristics, surrounding the air sparging zone, while air was injected into the subsurface at regular intervals of 5, 10, and 20 scfm.

TECHNOLOGY DESCRIPTION

HydroTechnics Inc. In Situ Groundwater Flow Sensors were used to monitor and define the three-dimensional groundwater flow regime surrounding the air sparging process. The flow sensors are new, state-of-the-art instruments, which use a thermal perturbation technique to directly measure the three-dimensional groundwater flow velocity vector in unconsolidated, saturated, porous media. Sensors are installed in direct contact with the subsurface formation and are capable of continuous remote monitoring and data retrieval.

The instrument consists of a cylindrical heater 30 inches long by $2^{3/8}$ inches in diameter with an array of 30 carefully calibrated temperature sensors on its surface. When the probe is installed directly in contact with unconsolidated saturated sediments, at the point where the measurement is to be made, and the heater activated with approximately 70 watts of continuous power, the sediments and groundwater surrounding the probe are warmed by 20 to 30 °C.

The temperature distribution on the probe surface is independent of azimuth and symmetric about the vertical midpoint of the probe in the absence of any flow. When there is flow past the tool, the surface temperature distribution is perturbed as the heat emanating from the probe is advected around the probe by the moving fluid.

Relatively cool temperatures are observed on the upstream side of the tool while relatively warm temperatures are observed on the downstream side. PC-based software is available which converts the measured probe surface temperature distribution into flow velocity (3D magnitude and direction). Darcy velocities in the range of 0.01 to 2.0 ft/day (3×10^{-6} to 3×10^{-4} cm/s) can be accurately measured with this technology. Measurement resolution is (0.001 ft/day or 3 x 10^{-7} cm/s).

The flow sensor probe is a $2^{3/8}$" diameter by 30" long cylinder (FIGURE 1). The probe is engineered to accommodate the expected deployment conduit of a 2.5" ID borehole casing, provided by any standard hollow-stem auger. The probe is connected to the surface by cables housed in 2" SCH40 PVC, Tri-Lock monitoring well casing. TABLE 1 contains the flow sensor specifications.

FIGURE 1. HydroTechnics Groundwater Velocity Sensor

The heater on the probe is a 40-ohm electrical resistance heater, which is normally operated at 70 watts of power output. This requires 57 volts DC power supply at 1.4 amps. The surface of the flow sensor is covered with an array of 30 thermistors that have a nominal resistance at 1 megaohm at 25°C. The resistance of the thermistors varies from about 2.5 megaohm at 10°C to about 125 killiohm at 70°C.

Outer sensor material is PVC and high-density plastic. The sensors are safe for environmental use. Sensors have been successfully installed to depths of 400 ft below ground surface, and are very durable.

TABLE 1. Specifications

Description	Specification
Operating Range	0.01 to 2.0 ft/day
Accuracy	0.01 ft/day
Resolution	0.001 ft/day
Probe Length	30 inches
Probe Diameter	$2^{3/8}$ inches
Power Required	57 Volts at 1.4 amps
Data Collection	CR10 data logger
Data Processing	HTFLOW© software
Depth	>400 ft bgs

SITE DESCRIPTION/SENSOR ARRAY

A total of seven in situ groundwater velocity sensors were installed at the test air sparging site at Port Hueneme in July 1996. FIGURE 2 illustrates all subsurface installations at the site, not including vapor extraction wells that are only in the vadose zone. In situ flow sensors are designated with the code MRC-F-0#. Cross Section SW-NE depicts the flow sensor cross section illustrated in FIGURE 3.

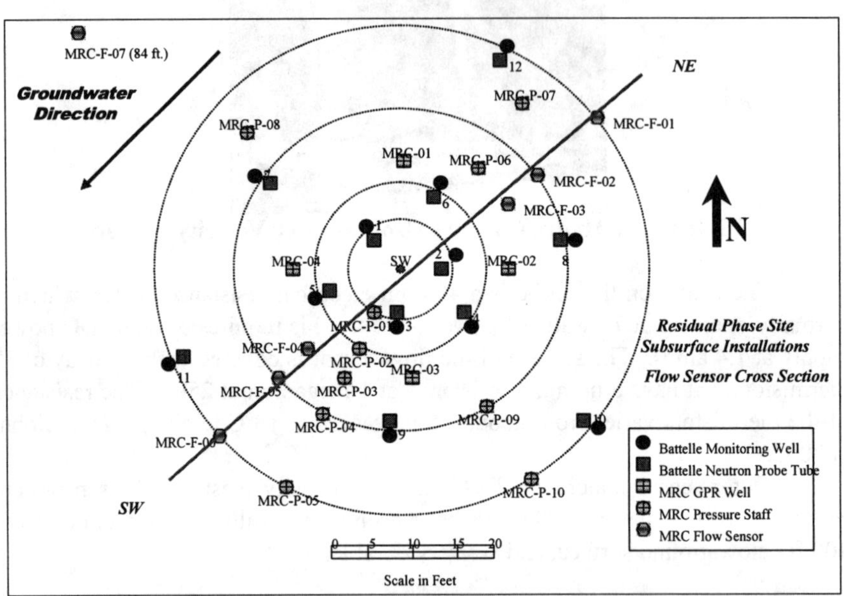

FIGURE 2. Port Hueneme Residual Phase Site Subsurface Installations

FIGURE 3 illustrates the depths and locations of the flow sensors in relation to the sparge well, and the geological formations the flow sensors are installed within.

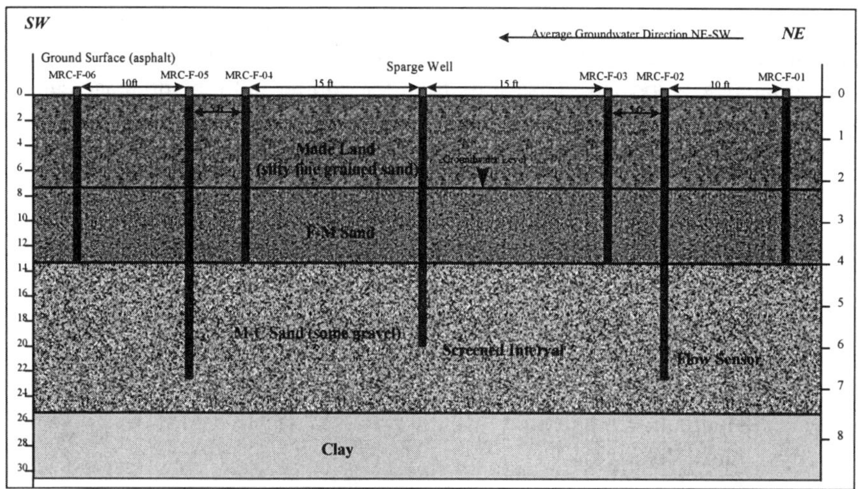

FIGURE 3. Port Hueneme Residual Phase Site SW-NE Flow Sensor Cross Section

The test site is covered with 4-6 inches of asphalt, which prevents precipitation from entering the site vertically through the vadose zone. From the asphalt to a depth of 7 feet below ground surface (bgs) there is a silty fine-grained sand called made-land. This subsurface unit is a fine-grained fill dirt that is less permeable than the lower sand formations. The made-land also occupies most of the vadose zone, with the groundwater level near the contact between the made land and the underlying sand formations. Below the made land, begins the upper aquifer consisting of a fining upward sand and gravel sequence that extends to a clay layer located at 25-26 feet bgs. The clay layer runs the entire length of the site and acts as an aquitard. The entire sand and gravel formation is saturated. The flow sensors are installed within the sand formations at either 22 or 13-ft bgs depths.

The uppermost, fine-grained silty sand unit has been interpreted to be fill material dredged from the Channel Island Harbor and Port Hueneme Harbor, but that interpretation has not been confirmed[1]. Some investigators have interpreted the uppermost silty units as a confining layer to the semi-perched aquifer, which may be true for some portions of the site. However, in the area containing the plume, the aquifer appears to be unconfined. The groundwater levels are well below the contact between the uppermost silty unit and the underlying sandy unit.

The surrounding site contains a total of 34 groundwater monitoring wells. All boreholes and completed monitoring wells have been drilled to evaluate the semi-perched aquifer to depths about 25 ft bgs. The depth to water and lithology relationship has created a semi-perched upper conterminous aquifer above the silty clay aquitard at 25 ft bgs. This clay aquitard is laterally extensive across the entire naval base. The groundwater flow direction of the uppermost aquifer is to the southwest. The semi-perched aquifer is brackish to saline with an elevated nitrite concentration due to irrigation practices in the rural portions of the Oxnard Plain. Groundwater movement within the semi-perched aquifer is influenced by tidal fluctuations over some portions of the base, but not in the northeast quadrant, where the NEX is located. Discharge from the aquifer is to the drainage canals, harbors, and beaches on and around the base. TABLE 2 presents additional hydrogeological investigation results from the previous site literature.

TABLE 2. Port Hueneme Hydrogeological Values

Hydrogeological Parameter	Value
Depth to Water	8-9 ft
Groundwater Direction	Southwest
Transmissivity	10,000 - 44,000 gpd/ft
Storativity	0.001 - 0.92, ~0.05
Hydraulic Conductivity	1,267 - 3,000 gpd/ft
Flow Velocity	694 - 1,643 ft/year
Porosity	30%
Free Product Migration Velocity	0.90 ft/day
Total Dissolved Solids (TDS)	1,212 mg/L

Flow sensors in the cross section were installed at either 13 or 22 feet bgs. Flow sensors at 13 feet bgs were installed to monitor groundwater flow characteristics within the upper realms of the aquifer, where most of the dynamic changes of groundwater flow were expected to occur during air sparging. The flow sensors were installed at least 13 feet bgs to ensure that with seasonal water table fluctuation the sensors would be submerged in the saturated zone at all times and that data collection would occur in the more porous sand formation below the made land silty sand.

Flow sensors installed at 22 feet bgs, were placed to monitor flow characteristics in a more porous media at a greater depth. In addition, the placement of flow sensors at this lower depth was to observe, if possible, any groundwater convection cell currents that may occur once sparging was initiated. Sensors were also installed at 22 feet bgs to monitor groundwater flow characteristics that could be correlated with Naval Facilities Engineering Service Center (NFESC) groundwater level and flow measurements.

DATA COLLECTION AND ANALYSIS

All flow sensors were installed in late July 1996. The data acquisition system was in place and flow data collection began on July 27, 1996. The groundwater flow sensors were programmed to collect a velocity reading every 15 minutes. The data represented in the following analysis subsamples data into one hour average data points. Data was collected for one full year, with the last data transfer from the site including data collected on 11 August 1997. Due to funding restrictions, data analysis has been limited to June 1997.

Raw data collected from the HydroTechnics flow sensors is mV (millivolt) data converted to temperature data with the HydroTechnics HTFLOW© software. Once the mV data is quickly converted to temperature data, the data is then mathematically processed within the HTFLOW© software to resolve three-dimensional groundwater velocities. The entire data reduction process can be done relatively quickly, and can be processed automatically once raw data has been downloaded from the flow sensors. Processed sensor data was loaded into spreadsheet software for analysis and presentation.

Groundwater velocity sensors generate three data sets consisting of horizontal speed, vertical speed, and azimuth. All three data sets were evaluated in order to generate a three-dimensional model of changes in groundwater flow patterns. FIGURE 4 illustrates the recorded changes in horizontal groundwater flow velocities as the air sparging injection rate was stepped from 0 to 5 to 10 and to 20 scfm.

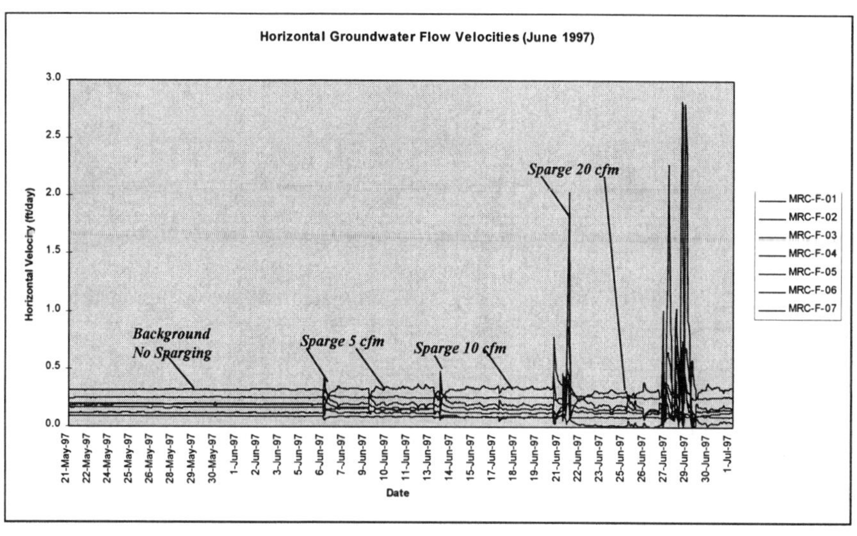

FIGURE 4. Horizontal Groundwater Flow Velocity History

RESULTS

Background Flow Velocities. Background groundwater flow velocities remained constant for the time period in May 1997 prior to the initiation of air sparging. Flow sensors installed at 13 ft bgs recorded horizontal velocities ranging from 0.19 to 0.32 ft/day. These velocities compared closely to velocities calculated from the length of a benzene plume, which exists at the same site. Based on the length of the benzene plume, and the estimated date of release, horizontal groundwater flow velocities were estimated from 0.27 to 0.35 ft/day[2].

Horizontal Groundwater Velocities. FIGURE 5 illustrates the mean horizontal groundwater velocities for each flow sensor at each sparge injection rate. The means have been calculated without the inclusion of outlier data points recorded during periods of transition between air sparging rates, where large velocities are recorded for a few hours when air injection begins, ends, or is increased.

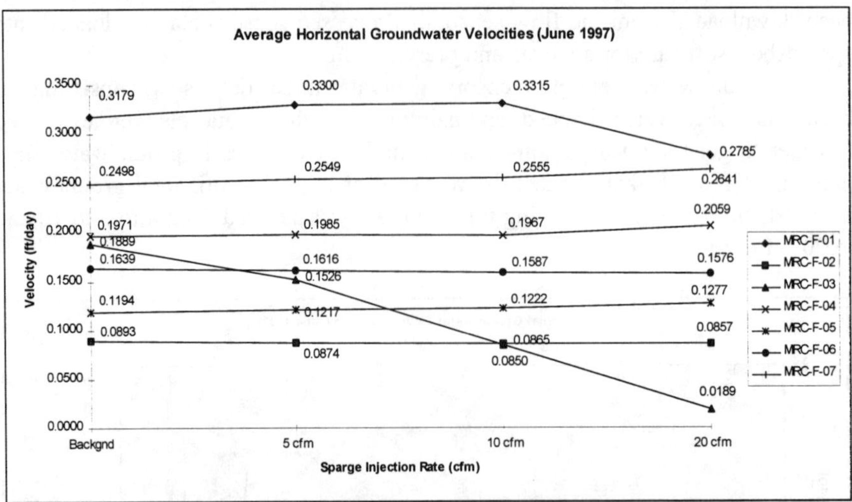

FIGURE 5. Mean Horizontal Groundwater Velocities

Horizontal groundwater velocities remain relatively constant at most flow sensor locations except for flow sensor 1 and 3. Flow sensor 3 is located 15 ft up-gradient from the sparge point and 13 ft bgs. This sensor recorded a constant decrease in horizontal velocity at each injection rate of 5, 10, and 20 scfm. At 20 scfm, horizontal groundwater velocity at flow sensor 3 has been reduced by an order of magnitude from background velocity. Flow sensor 1, located 30-ft up-gradient from the sparge point remains relatively constant until air is injected into the formation at 20 scfm.

Vertical Groundwater Velocities. FIGURE 6 illustrates the mean horizontal groundwater velocities for each flow sensor at each sparge injection rate. The means have been calculated without the inclusion of outlier data points recorded during periods of transition between air sparging rates, where large velocities are recorded for a few hours when air injection begins, ends, or is increased.

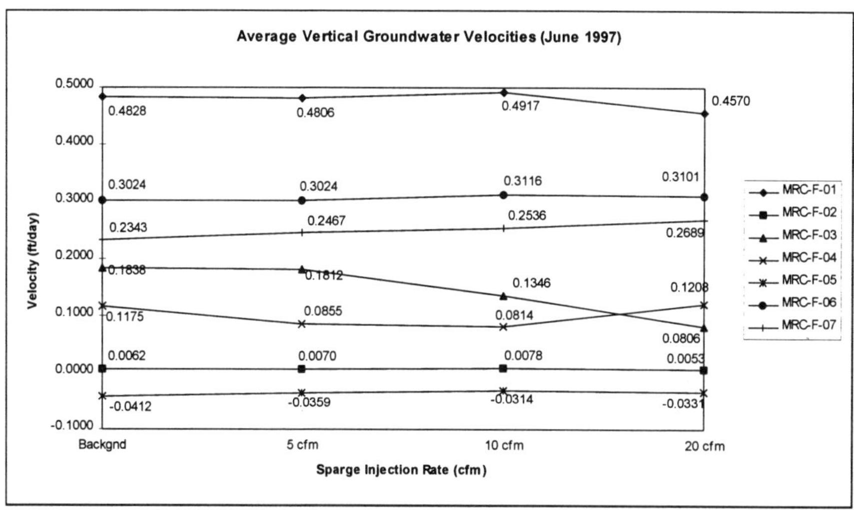

FIGURE 6. Mean Vertical Groundwater Velocities

Vertical groundwater velocities remain relatively constant at any air injection rate. There is a noticeable change in vertical velocities at flow sensors 3 and 4 when air injection is 20 scfm. While the vertical velocities at flow sensors 3 and 4 remain positive, there is evidence that air sparging at 20 scfm has an influence on the vertical component of groundwater flow. Flow sensor 3, which is located 15 ft upgradient from the sparge point and 13 ft bgs is effected by a negative component of vertical flow at 10 and 20 scfm injection rates. This is a result of the up-gradient flow or pressure generated by air injection and water displacement near the surface of the aquifer. Down-gradient at flow sensor 4, located 15 ft southwest and 13 ft bgs, a small positive vertical component is recorded. This is a result of the down-gradient movement of displaced water with the background flow velocity rising through the aquifer under flow sensor 4.

Flow sensors were initially installed in the array pictured in FIGURE 3 in order to monitor for groundwater circulation cells created by the air sparging process. No groundwater circulation cells were recorded.

Azimuth. FIGURE 6 illustrates the mean groundwater directions for each flow sensor at each sparge injection rate. The means have been calculated without the inclusion of outlier data points recorded during periods of transition between air sparging rates, where large changes in direction are recorded for a few hours when air injection begins, ends, or is increased.

Groundwater directions remain constant for all groundwater sensors except flow sensor number 3, located 15 ft up-gradient from the sparge point and 13 ft bgs. Flow sensor 3 records a gradual change in flow direction at 5 and 10 scfm injection rates, and a large change or deflection in azimuth at 20 scfm injection rate. Groundwater flow at sensor number 3 deflects to the west away from the sparge point and the original groundwater flow direction by more than

65 degrees. This azimuth indicates that groundwater flow 15-ft upgradient from the sparge point is deflected around the zone of aeration created by the sparging process.

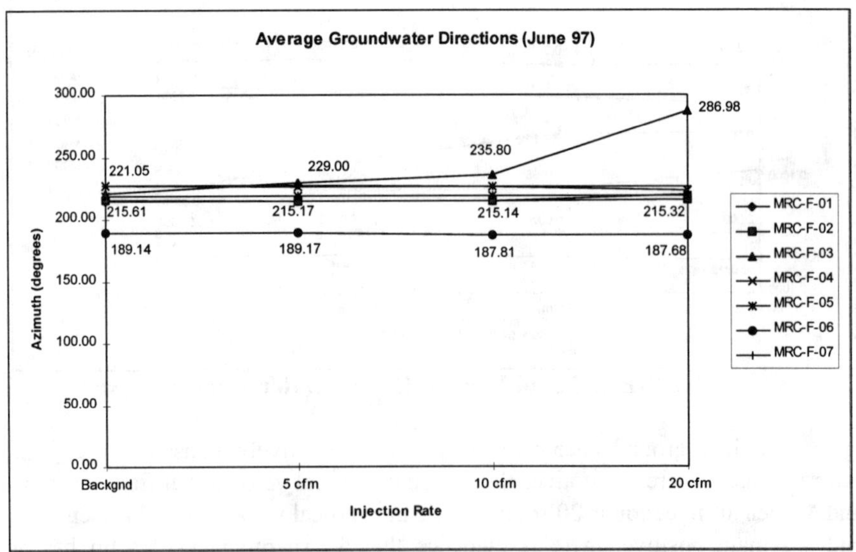

FIGURE 7. Mean Groundwater Directions

CONCLUSIONS

The original intent of this research was to monitor the physical parameters and groundwater flow regime associated with different air sparging operating parameters. Other areas of interest included the existence and effect of mounding on the possible spread of contamination, and the presence of groundwater circulation cells caused by the air sparging process.

Groundwater velocity sensor data has indicated that mounding is created by the air sparging process, but it is a short event that lasts only 12-24 hours. Mounding is created by the initial displacement of water by air injected into the saturated formation. Once displacement is complete, the groundwater flow regime returns to equilibrium close to background levels. The mounding event does occur, and will spread contamination away from the sparge area, but the length of the mounding event is so short, groundwater movement is less than one foot. This indicates that mounding is not a serious contributor to the spread of groundwater contamination.

Groundwater velocity sensors were positioned in an array that could have recorded groundwater circulation cells created by the air sparging process. There was no evidence that any groundwater circulation cells were created at the Port Hueneme test site.

FIGURE 8 illustrates the changes in the groundwater flow regime up-gradient from the sparge point at different air injection rates. Flow sensor 3, located 15-ft up-gradient from the sparge well, was the sensor most effected by air injection at all three injection rates. Horizontal groundwater flow velocities were significantly reduced 15 ft up gradient at each injection rate of 5, 10 & 20 scfm. In fact, at 20 scfm, the horizontal flow velocity was reduced to 0.02 ft/day, and the groundwater flow direction was deflected away from the sparge area. This indicates the possibility that once air is injected into the subsurface at rates greater than 10 scfm, contaminated groundwater in shallow areas of the aquifer is diverted from the most effective regions of an air sparging system.

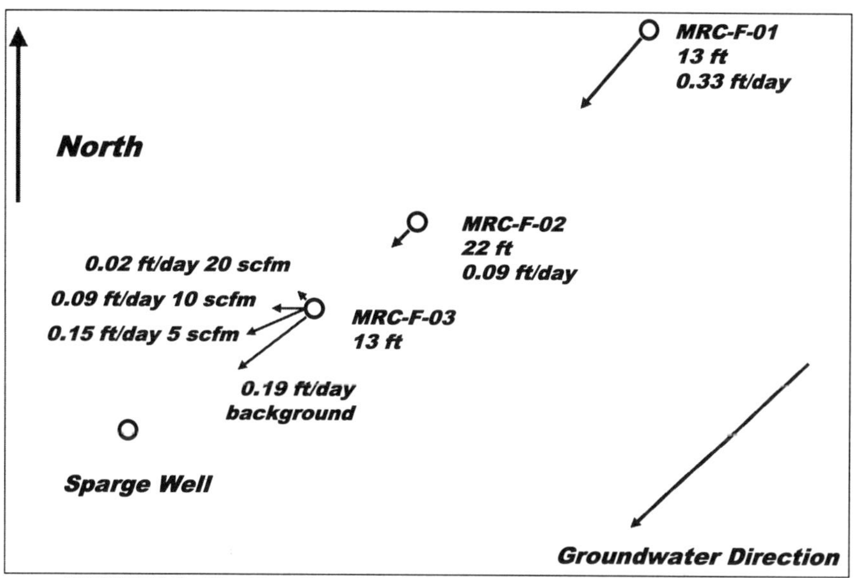

FIGURE 8. Up-Gradient Horizontal Groundwater Flow Regime

REFERENCES

ERC Environmental and Energy Services Company (ERC)., "Final Report, Predesign Studies at Naval Exchange Gas Station, Naval Construction Battalion Center, Port Hueneme, California," Report prepared for Naval Energy and Environmental Support Activity, Port Hueneme, California (1989).

Naval Facilities Engineering Service Center (NFESC)., "Naval Exchange Gasoline Station Fourth Quarter 1995 Ground Water Monitoring Report," Monitoring and Reporting Program Number 7485 (1995).

FIELD PILOT STUDY OF TRENCH AIR SPARGING FOR REMEDIATION OF PETROLEUM HYDROCARBONS IN GROUNDWATER

Jinshan Tang, Gary R. Walter, and Mark W. Kuhn (Hydro Geo Chem, Tucson, AZ)

ABSTRACT: Trench air sparging is performed by allowing contaminated groundwater to flow through a gravel-filled trench in which air is injected to enhance the volatilization and aerobic biodegradation of dissolved and separate-phase organic compounds (Pankow et al., 1993). A simple mathematical model of the trench sparging process was developed to analysis of pilot test results. The model treats the sparging trench as a well-mixed reactor where air-stripping is the only process removing volatile organic compounds (VOCs) from groundwater. A field pilot study of trench air sparging was conducted at a site where groundwater was contaminated with dissolved petroleum hydrocarbons. The gross sparging efficiency, defined as the reduction in hydrocarbon concentrations, was greater than 99% for benzene, toluene, xylenes, and total petroleum hydrocarbons (TPH). The dynamic partitioning factor (λ), which measures the deviation from equilibrium of air-water partitioning described by Henry's Law, was estimated to vary between 0.5 at low sparging rates and 0.05 at high sparging rates.

INTRODUCTION

Trench air sparging (Pankow et. al., 1993) is being considered for remediation of groundwater contaminated by fuel hydrocarbon releases at an active service station. The prototype trench sparging system is shown in Figure 1. Contaminated groundwater flows into the sparging trench under the influence of the local hydraulic gradient. Air is injected into the saturated gravel-fill and flows upward volatilizing VOCs and potentially enhancing biodegradation. The effectiveness and final design of the trench sparging system depends primarily on the rate of hydrocarbon removal as a function of the air injection rate. The pilot test was performed to determine the removal rate at various air injection rates to optimize the final sparging trench design.

THEORY

A simple mathematical model of the trench sparging process was developed. This model extends the steady-state model of Pankow et. al. (1993) by considering transient sparging effects during short-term pilot tests and cyclical injection modes. The model treated the sparging trench as a well-mixed reactor where air-stripping is the only process removing VOCs from groundwater. During sparging, groundwater is assumed to flow into the trench at a rate determined by Darcy's law. Air is injected into the horizontal, perforated pipe at the base of the gravel-fill and is assumed to be uniformly distributed along the length of the injection pipe and throughout the gravel-fill. Neither adsorption of VOCs on the gravel-fill nor biodegradation within the trench were considered because of the very low organic content of the fill and because of the short residence of the groundwater in the trench. Some biodegradation may have occurred during the pilot test, however.

The mathematical model was developed based on the mass balance of dissolved VOCs in the saturated portion of the trench. The rate of change of VOC mass in the trench (***dM/dt***) is described by:

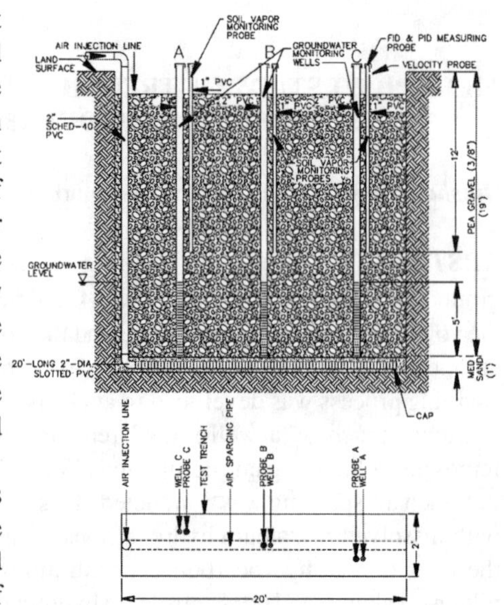

FIGURE 1: Prototype trench sparging system.

$$\frac{dM}{dt} = \varphi XYH_b \frac{dC}{dt} = YH_b qC_i - Q_g C_g - YH_b qC \tag{1}$$

where φ is the saturated porosity of the trench fill [dimensionless]; X is the width of the trench [L]; Y is the length of the trench [L]; H_b is the saturated thickness in the trench [L]; C is the average VOC concentration in the trench represented by water samples collected from the centerline of the trench [M/L^3] and equal to VOC concentrations in water leaving the trench; t is the elapsed sparging time [t]; q is the specific flux [L/t]; C_i is the upgradient concentration [M/L^3]; Q_g is the volumetric air flow rate [L^3/t]; and C_g is the concentration in the off-gas [M/L^3], which can be related to the average water VOC concentration using the dimensionless Henry's Law coefficient (H_D) and a dynamic partitioning factor ($C_g = \lambda\ H_D\ C$). The dynamic partitioning factor (λ) accounts for non-equilibrium behavior of VOCs partitioning between the air and water.

Integrating (1) gives the following solution for the average dissolved VOC concentrations (C) in the trench when C^O is the starting trench VOC concentration [M/L^3]:

$$C = \frac{YH_b qC_i}{Q_g \lambda H_D + YH_b q} - \left[\frac{YH_b qC_i - (Q_g \lambda H_D + YH_b q)\ C^0}{Q_g \lambda H_D + YH_b q}\right] \bullet \exp\left[-\frac{(Q_a \lambda H_D + YH_b q)}{\varphi XYH_b}t\right] \tag{2}$$

Equation 2 is applicable to sparging tests even when the air injection rate varies. At steady-state the dissolved VOC concentration (C) is given by:

$$C = \frac{Y\,H_b\,q\,C_i}{Q_g\,\lambda\,H_D + Y\,H_b\,q} \tag{3}$$

The gross sparging efficiency (E) for a given VOC concentration can be defined in terms of the physical and chemical sparging conditions, such as:

$$E = \frac{Q_g\lambda H_D}{Q_g\lambda H_D + YH_b q} \tag{4}$$

The build-up of dissolved VOC concentration after sparging stops is given by:

$$C = C_i - (C_i - C_s) \cdot \exp[-\frac{q\,t'}{\varphi\,X}] \tag{5}$$

where C_s is the average concentration in the trench when sparging was stopped [M/L^3] and t' is the time since sparging stopped [t]. The groundwater flux into the trench can be estimated from (5)

FIELD PILOT TEST

Site Conditions. The pilot test was performed within a plume of dissolved TPH downgradient from an active service station (Figure 2). The site is underlain by unconsolidated sand, gravel, and cobbles interbedded with silty clay. The clastic sediments are underlain by granitic bedrock at a depth of 20 to 40 feet below land surface (bls). Groundwater occurs under unconfined at an average depth of 15 feet bls. Figure 2 shows the groundwater flow and the TPH plume with concentrations determined by modified Method 8015. Dissolved constituents in the plume consist primarily of benzene, toluene, ethylbenzene, and xylenes (BTEX).

Pilot Test Procedures and Results. The trench was centered within the TPH plume perpendicular to the direction of groundwater flow (Figure 2). The pilot test was conducted for 23 hours. In order to determine BTEX removal efficiency, sparging was performed at three air injection rates, 7, 25, and 53 scfm for 165, 117, and 1,080 minutes, respectively. Recovery of groundwater VOC concentrations in the trench was monitored for 8 days after sparging had stopped. Groundwater samples were collected from the three trench monitoring wells A, B, and C, and analyzed for BTEX and TPH during sparging and recovery, the average concentrations of which are listed in Table 1.

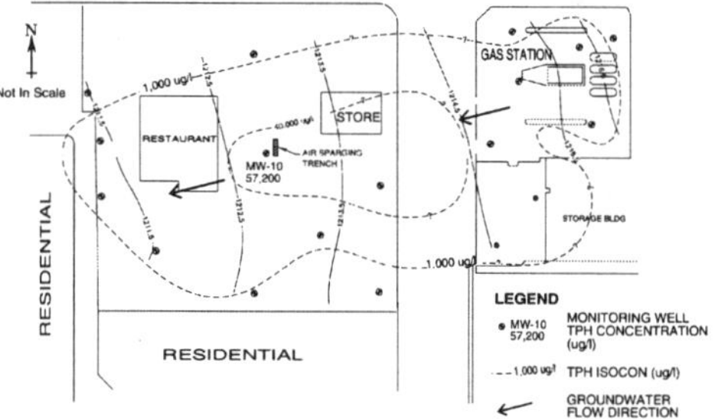

FIGURE 2: Site Map

TABLE 1. Average Groundwater VOC Concentrations

	Time (minutes)	Air Flow (scfm)	Benzene (µg/l)	Ethylbenzene (µg/l)	Toluene (µg/l)	Xylenes (µg/l)	TPH (µg/l)
Sparging	0	0	1,663	116	3,967	3,359	12,500
	22	7	1,733	<1	3,630	2,978	11,970
	51	7	817	<1	1,591	1,780	7,520
	89	7	445	<1	830	669	3,750
	101	7	248	<1	47	400	2,397
	196	25	114	<1	224	145	1,420
	254	25	71	<1	142	130	1,097
	339	53	41	<1	77	67	57
	461	53	20	<1	32	24	265
	703	53	10	<1	8	5	143
	1,195	53	13	<1	1	7	120
Recovery	1,473	0	1	--	6	6	225
	1,883	0	43	7	85	85	950
	2,643	0	110	4	235	235	4,434
	3,272	0	218	21	367	367	2,803
	7,509	0	1,307	216	1,610	1,610	11,307
	13,185	0	1,667	367	4,633	4,633	33,033

ANALYSIS OF RESULTS

The primary purpose of the pilot test was to determine the efficiency of the sparging process in removing VOCs from groundwater and to evaluate the operational performance of the trench design. The physical and chemical constants used in the following analyses were: 1) Trench Width (X), 2 feet; 2) Trench Length (Y), 20 feet; 3) Saturated Thickness (H_b), 5 feet; 4) Henry's Coefficient (H) for benzene, toluene, and xylenes, 5.6×10^{-3}, 6.4×10^{-3}, and 7.0×10^{-3} atm-m^3/mol, respectively; 5) Gas Constant (R), 8.2 x 10-5 atm-m3/mol-k; 6) Gravel-fill Porosity (φ), 0.3; and 7) temperature, 298 °K. Analyses were performed for benzene, toluene, and xylenes. The sparging efficiency was evaluated in terms of the air-water partitioning factor (2) at various flow rates and in terms of the overall sparging efficiency factor (E) at steady-state conditions.

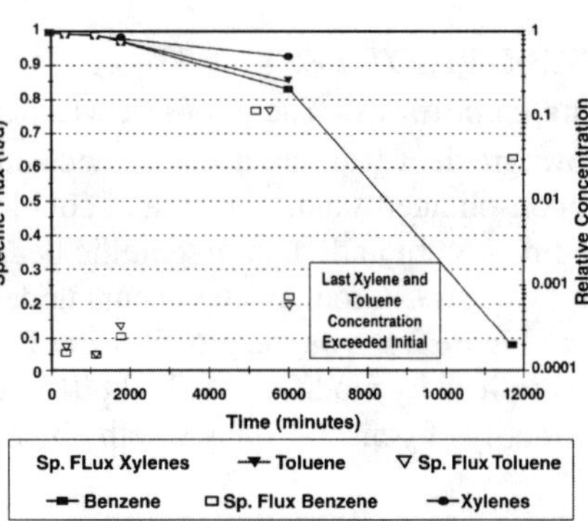

FIGURE 3: Specific flux estimated from recovery data for benzene and xylenes.

Specific Flux (q). The specific flux was estimated in two ways: 1) the observed hydraulic gradient, and the hydraulic conductivity from an aquifer test; and 2) using build-up concentrations in (5). The specific based on Darcy's law was 0.5 ft/d.

The specific flux from the buildup concentrations of benzene and xylenes was estimated from a regression analysis based on (5). The resulting curves and calculated velocities for benzene and xylenes are shown in Figure 3. The apparent

q increases with recovery time for both benzene and xylenes (Figure 3). The benzene, toluene, and xylene recovery data indicate a q of approximately 0.05 ft/d during the first 1,500 minutes of recovery. The q's calculated from the data at a later time are regarded as unreliable because the calculated q is strongly influenced by small errors in the reported concentrations. A q of 0.05 ft/d based on the early portion of the recovery test rather than that calculated from Darcy's Law was used in the following calculations because it resulted in a conservatively low estimate of the efficiency of the sparging process.

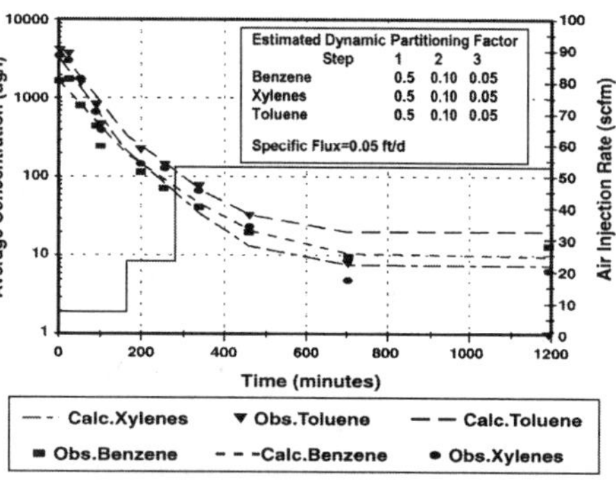

FIGURE 4: Estimate for transient aqueous concentrations of benzene, xylenes, and toluene.

Dynamic Partitioning Factor (λ). The dynamic partitioning factor (λ) is a measure of the extent to which the sparging air is in equilibrium with the aqueous-phase VOC concentrations in the sparging trench. It allows the sparging removal efficiency (E) to be estimated for sparging trench designs other than that actually used for the pilot test. Physical reasoning suggests that λ depends on the size of the air-filled channels in the gravel-fill and on the residence time of the air in the trench. Likewise, λ would be expected to approach 1 at very low sparging rates and to decrease as the sparging rate increases. The dynamic partitioning factor can be estimated from either the transient trench aqueous concentrations during the early portion of the sparging test or from the stabilized concentrations. Estimates of λ were made for benzene, toluene, and xylenes under both conditions.

The estimates for transient conditions were made using the concentration data for the first 600 minutes of the sparging test by adjusting the value of λ in (2) so that computed concentrations matched detected concentrations. The resulting estimates are shown in Figure 4 along with the correspondence between the detected and calculated concentrations. The relationship of the variation of the logarithmic λ to the specific air flux (air injection rate divided by rate of water flow through the trench) is shown in Figure 5. Although, only three estimates are available, the line through the first two data points suggests that λ is 1 at zero air flow, which is physically reasonable.

The steady-state estimates of λ were made the average of the last two concentrations measured during the sparging period. The resulting estimates are listed in Table 2. The steady-state estimate for benzene (0.04) is similar to that estimated from the transient data. The estimates of λ for toluene and xylenes are higher by factors of 3 and 2, respectively, than for benzene. The gross efficiency factor is also listed in Table 2.

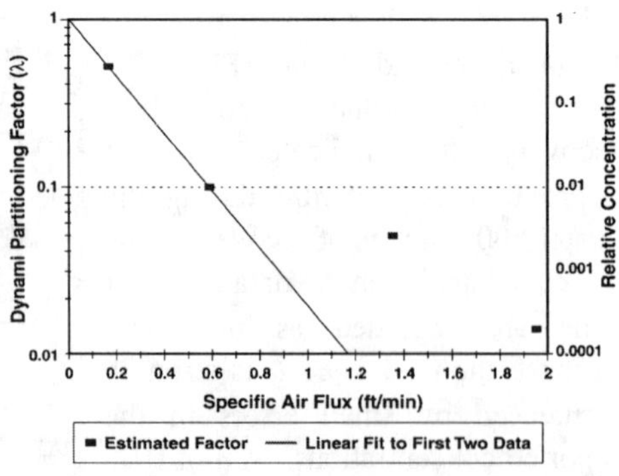

FIGURE 5: Variations of the logarithmic dynamic partitioning factor versus the specific flux.

TABLE 2. Steady-State Estimated of the Dynamic Partitioning Factor (λ) and Sparging Efficiency Factor (*E*) Calculated form Stabilized Concentrations

Constituent	H_D	C_i (μg/l)	C (μg/l)	λ	E (%)
Benzene	0.23	1,663	11.5	0.04	99.3
Toluene	0.26	3,967	8.5	0.16	99.8
Xylenes	0.29	3,359	7.0	0.11	99.8
TPH	NA	12,500	120	NA	99.0

SUMMARY

The pilot study results indicate that concentrations of BTEX and TPH can be reduced by greater than 99% in trench sparging systems. Despite the high removal efficiencies, the sparging process is less efficient than would be predicted by gas-water partitioning calculations using Henry's Law. The dynamic partitioning factor, varied 0.5 at the lowest air flow to 0.05 at the highest rate. Even at the lowest air flow rate, the ratio air flow to water flow was over 5,000 indicating that the sparging trench was relatively inefficient compared to a properly designed air stripper.

REFERENCES

Pankow, J.F., R.L Johnson, and J.A. Cherry. 1993. Air Sparging in Gate Well in Cutoff Walls and Trenches for Control of Plumes of Volatile Organic Compounds (VOCs). Ground Water. v. 31, no. 4, pp. 654-663.

A CASE HISTORY OF ENHANCED BIOREMEDIATION UTILIZING PURE OXYGEN INJECTION

Lawrence D. Zamojski, P.E. (Acres International Corp., Amherst, NY)
James R. Stachowski (Acres International Corp., Amherst, NY)
Sean R. Carter (Matrix Environmental Technologies, Orchard Park, NY)

ABSTRACT: Investigations indicated petroleum hydrocarbons within the zone of saturation in glacial silt, sand, and gravel deposits to a depth of 5.5 m (18 ft). Groundwater was encountered approximately 2.7 m (9 ft) below grade with the groundwater gradient causing flow to the northwest toward the property boundary. Approximately 3,500 m^3 (4,600 yd^3) of soil and 694,000 liters (183,500 gal) of groundwater were impacted by the petroleum. Enhanced biodegradation using pure oxygen injection was selected for the site. The remediation included the installation of 12 oxygen injection points. Pure oxygen was delivered to these points via an 2,265 SLPH (80 SCFH) oxygen generating system (patented process). Baseline volatile organic and dissolved oxygen conditions within the groundwater are presented, and progressive reduction in contaminant levels and increased dissolved oxygen concentrations were observed. Oxygen injection appears to have enhanced the biodegradation of organics which were biodegradable under aerobic conditions.

INTRODUCTION

Leakage from an underground gasoline storage tank, discovered in 1993, resulted in contamination of surrounding soil and groundwater by volatile organic compounds. The affected site is an active facility located in a developed commercial district in New York State and the study area is principally covered by asphalt or building structures. Interim remedial action at the site involved removal and off-site disposal of the leaking tank and approximately 545 metric tons of contaminated soil. Not all contaminated soil and groundwater was removed due to constraints posed by a fuel dispensing area, public utility conveyances and property boundaries.

INVESTIGATION

Subsurface investigations identified groundwater between 2.7 to 3.4 m (9 to 11 ft) below ground surface under semi-confined conditions within a silt and sand layer. Petroleum contaminated soils were identified within the zone of saturation at levels exceeding New York State Department of Environmental Conservation (NYSDEC, 1992) guidance values for groundwater protection. This impacted zone is bounded by an underlying dense till layer and overlain by asphalt or buildings which individually act as horizontal confining layers.

Approximately 3,500 m^3 (4,600 yd^3) or 6,760 metric tons (7,450 U.S. tons) of contaminated soil and 694,000 L (183,500 gals) of contaminated groundwater were estimated to exist at the site based on the investigations.

OXYGEN INJECTION SYSTEM

A patented oxygen injection remediation system developed by Matrix Environmental Technologies, Inc. (Matrix) of Orchard Park, New York was selected for the site. The system consists of oxygen injection points and mechanical system components (listed below) contained within a cargo trailer. The system consists of an AirStep Module AS-80 pressure swing adsorption (PSA) oxygen generator with a 450 L (120 gal) surge tank and regulator, an Atlas Copco rotary screw air compressor with air dryer, vertical tank with autodrain, and low sound enclosure, a Manifold for twelve (12) injection points including an individual variable area flowmeter with needle control valve and pressure gage, and an adjustable timer with solenoid valve for each set of six (6) injection points to control oxygen flow for pulse injection. The Matrix system safely and cost effectively produces and stores high purity oxygen gas for metered injection into the subsurface.

Twelve injection points, shown on Figure 1, were individually constructed with a 0.3 m (1-ft) long section of 1.9 cm (¾ in) schedule 40 slotted PVC screen and installed within saturated sand and silt strata at depths ranging between 4.0 to 4.8 m (13.25 to 15.75 ft). Each point was connected to schedule 40 riser pipe and completed with #01 Morie sand and bentonite annular backfill. All injection points were covered using a flush mounted roadboxes. Each injection point was connected to a two-foot square header box via 1.9 cm (3/4 in) diameter polyethylene header piping with 690 kPa (100 psi) rating. The pipe was installed between 0.6 to 0.9 m (2 to 3 ft) below grade.

KEY ADVANTAGES OF THE SYSTEM

Advantages of the pure oxygen injection system include:

1. Increased transfer efficiency associated with injecting pure oxygen rather than air.
2. Higher solubility in groundwater results in more oxygen mass available for biodegradation. Conditions change from oxygen limited to hydrocarbon limited.
3. Oxygen permeation through the subsurface soils is approximately 40 times greater than groundwater based on viscosity.
4. Operation and maintenance costs are relatively low as there is no collection or treatment of groundwater or vapors necessary.
5. After installation of the injection points there is minimal disturbance of the area being treated.

Drawbacks of the system include:

1. Needs a power source and electricity for operation.
2. Requires confining soil layers to effectively control the dispersion of injected oxygen.
3. Contaminants must be aerobically biodegradable.

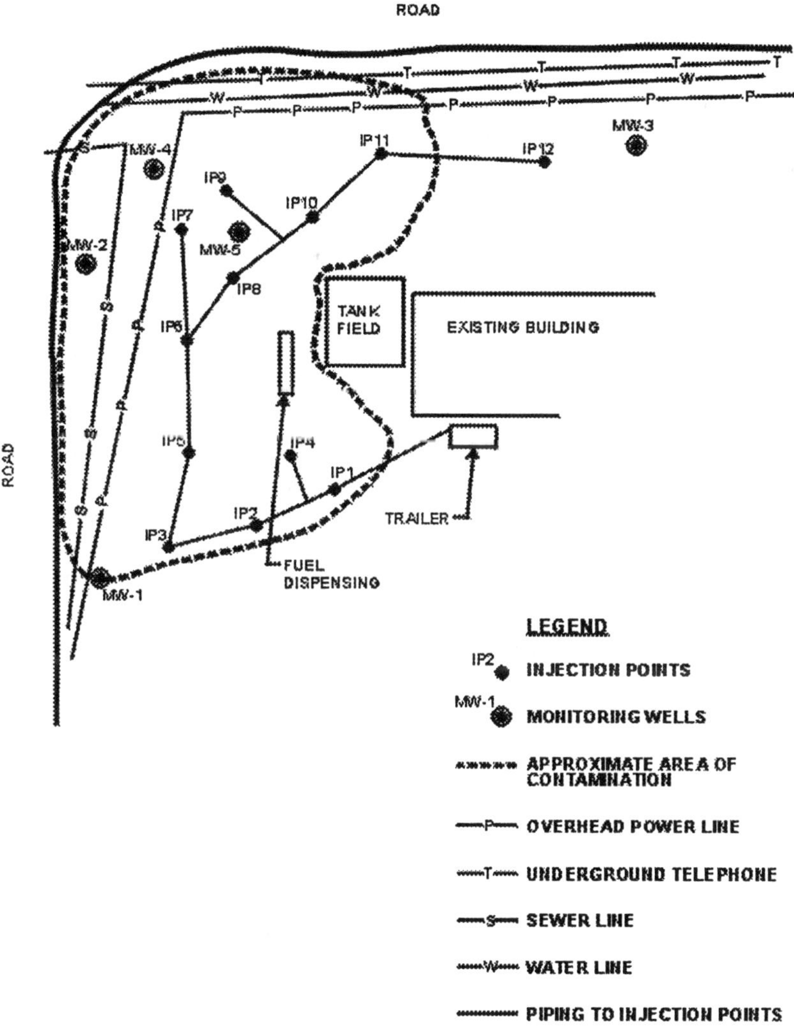

FIGURE 1. Investigation Locations and Remediation System Layout

SYSTEM OPERATION

Five monitoring wells are used at the site to record dissolved oxygen (DO), oxidation reduction potential (ORP) and petroleum organics in the groundwater. Baseline data for these parameters were obtained prior to any onsite construction activities.

The PSA oxygen generator produces oxygen (90 to 95 percent purity) separated from compressed air. The unit consists of two vessels filled with synthetic zeolite molecular sieves which act as adsorbers. Nitrogen is adsorbed as compressed air passes through the first sieve, allowing the remaining oxygen to exit. Before the sieve becomes saturated with nitrogen, inlet air flow switches to the second adsorber, allowing the first to be regenerated by depressurization and oxygen purging. The complete cycle is then repeated. Under normal operating conditions, the molecular sieve should be almost completely regenerative and last almost indefinitely.

Oxygen is delivered to the subsurface through injection points. An oxygen surge (storage) tank provides the required injection pressure, thus eliminating a need for mechanical pumping. The solubility of oxygen gas results in high groundwater dissolved oxygen levels at the point of injection.

The system was designed to deliver oxygen cyclically to two separate groups of 6 injection points each at intervals of 10 minutes on and 70 minutes off. An initial flow rate of 566 SLPH (20 SCFH) per point was used to remove silt accumulation in the recently installed injection points. The operating parameters were subsequently changed to a 10 minute on and 50 minute off cycle at a flow rate of 849 SLPH (30 SCFH) or 1.5 SLPH (5SCFH). Routine adjustments are made to raise transfer efficiency to the groundwater. At this flow, the injection pressure at each point will generally be in the 13.8 to 27.6 kPa (2 to 4 psi) range, sufficient to displace water at the injection point and force oxygen bubbles into the surrounding formation. The oxygen flow rate and cycle time were set to increase dissolved oxygen concentrations to the saturation level without causing volatization of hydrocarbons.

REMEDIATION PROCESS

The dissolved oxygen is transported by advection and dispersion resulting in increased downgradient dissolved oxygen concentrations. Due to increased levels of dissolved oxygen naturally occurring aerobic microorganisms biodegrade the contaminants in place. The increased oxygen levels also produce a geochemical change in the subsurface by increasing the oxygen-reduction potential and reducing the production of ferrous iron by bacteria.

RESULTS

The system has been in operation for approximately six months. Periodic DO and ORP measurements taken at the 12 injection points and 5 monitoring wells show a general increasing trend at each injection point.

Groundwater analytical data has been compiled for two sampling episodes (Table 1). Contaminated wells show a nearly 50 percent reduction in total volatile organic concentrations. Similar percentage decreases are observed for benzene and m&p xylene. Projecting the same degradation rates over time, it is anticipated that the system may be required to operate for 18 to 24 months to reach closure values.

SUMMARY

As noted, overall oxygen levels have increased and volatile concentrations have decreased in half. At this point the remediation appears headed for successful closure within the next two years.

ACKNOWLEDGMENTS

Thanks to Mr. B. Kieffer and his organization for allowing use of the site data in this paper. Special thanks to A. Klimek, S. Baumann, L. Iannone and S. Marchetti for their assistance with this paper.

REFERENCES

New York State Department of Environmental Conservation, Division of Construction Management, Bureau of Spill Prevention and Response. 1992. *Spill Technology and Remediation Series* (STARS) *Memo #1 Petroleum-Contaminated Soil Guidance Policy*, Albany, New York.

TABLE 1. Comparative Analytical Results

Parameters	Results (ug/L)									
	MW-1		MW-2		MW-3		MW-4		MW-5	
	Initial	Post	Initial	Post	Initial	Post	Initial	Post	Initial	Post
Dissolved oxygen (mg/L)	0.4	11	0.3	0.9	1.2	2.8	0.7	2.3	0.3	2.7
Benzene	1200	580	24	36	<0.5	<0.5	920	950	8300	3100
Ethylbenzene	910	<10	19	2	<0.5	<0.5	410	<25	590	17
Toluene	31	230	1.6	4	<0.5	<0.5	31	<25	2300	990
(m&P) Xylene	3400	1400	8.6	1	<0.5	<0.5	1300	220	2400	1100
o-Xylene	70	800	1.2	4	<0.5	<0.5	170	56	1000	570
Isopropylbenzene	56	<10	6.2	<1	<0.5	<0.5	110	<25	<50	<10
n-Propylbenzene	180	<10	13	1	<0.5	<0.5	520	<25	130	<10
p-Isopropyltoluene	<12	<10	<0.5	<1	<0.5	<0.5	14	<25	<50	<10
1,2,4-Trimethylbenzene	1400	110	22	<1	<0.5	<0.5	1700	290	1100	440
1,3,5-Trimethylbenzene	400	320	2.6	<1	<0.5	<0.5	680	97	360	160
n-Butylbenzene	59	52	5.9	3	0.6	<0.5	460	83	120	52
sec-Bitylbenzene	<12	<10	1.6	<1	<0.5	<0.5	31	<25	<50	<10
t-Butylbenzene	<12	<10	<0.5	<1	<0.5	<0.5	<12	<25	<50	<10
Naphthalene	140	<10	2.6	<1	<0.5	<0.5	100	<25	96	53
MTBE	<12	<200	2.8	<20	1.4	<10	350	<500	490	390

STUDIES OF *IN SITU* BIOREMEDIATION OF AN AQUIFER POLLUTED BY TOLUENE

A. Iacondini [1], F. Abbondanzi[1], F. Malaspina[1], S. Gagni[1], M. Carnevali[2], H. Hannula[3], and ***R. Serra***[1]
[1]Centro Ricerche Ambientali Montecatini (Montecatini Environmental Research Center), Marina di Ravenna (RA), ITALY; [2]Univ. Bologna ITALY, [3]Univ. Tampere FINLAND

Abstract: A study was conducted to determine the potential for in situ bioremediation of an aquifer contaminated by toluene. The results showed that the indigenous microbial population was able to degrade toluene at concentrations up to 180 mg/L. Kinetic studies were performed in two different types of reactors: sealable bottles, total volume 160 mL; and 1200-mL modified Drechsel-type bottles. Different oxygen sources were tested (none, air, molecular oxygen, and hydrogen peroxide). Biodegradation occurred only in aerobic conditions, with oxygen availability and nitrogen supply, independently from the delivery of oxygen. In these conditions toluene removal reached 100% in about 50 days. For these reasons, supplying oxygen (e.g., by H_2O_2 injection or air sparging) appears to be an appropriate technique for *in situ* bioremediation. Selective enrichment techniques were applied to the microbial consortium to enhance toluene degradation. A gas chromatographic identification technique was set up to identify toluene metabolites to obtain a microbial characterization on the basis of bacterial metabolic pathways. This method, based on Solid Phase Micro Extraction (SPME), showed the presence of *Pseudomonas* strains. Further analytical studies are required to improve the microbial characterization of the identification technique.

INTRODUCTION

Toluene is an industrial chemical produced in very large quantities. Because it is slightly soluble and very toxic, there is a major concern that this compound could reach aquifers (Bouwer, 1992; Karlson and Frankenberger, 1989). Toluene biodegradation, as that of other hydrocarbons, often occurs in aerobic conditions. Oxygen is the preferred electron acceptor because aerobic reactions are the most efficient from an energetic viewpoint. In the presence of molecular oxygen, the initial reaction of toluene is microbial-mediated oxidation. The methyl substituent may be oxidized by a monooxygenase or the aromatic nucleus may be directly attacked by monooxygenase or dioxygenase, depending on how many oxygen atoms are delivered. In both cases, the result is a substrate for the ring cleavage (Gibson and Subramanian, 1984; Mikesell et al., 1993).

A study of a possible intervention of *in situ* bioremediation in an aquifer contaminated by aromatic hydrocarbons was carried out with particular attention to toluene. Furthermore, to identify the toluene biodegradation pathway, the nonvolatile metabolites of the five main aerobic biodegradation routes were selected, and a gas chromatographic method to detect them was set up.

MATERIALS AND METHODS

Site Description. Groundwater samples were taken from an aquifer located under an abandoned industrial site contaminated mainly by toluene (90 to 190 mg L^{-1}) and, to a lesser extent, by other aromatic and chlorinated hydrocarbons. The groundwater was pressurized. Its temperature was about 16°C at 30- to 40-m depth and its oxygen concentration was close to detection limits. The water flowrate was poor (~3.5 m^3/h) and was heavily affected by rain, which was causing variations in the toluene concentrations.

Subsurface soil consisted of a top layer of heterogeneous clay (20 to 50 m), underlain by a layer of fissured conglomerate (20 to 35 m) containing the water bed, and then a layer of semi-impermeable clay.

Test Description. Kinetic studies were performed in two different kinds of reactors. Anoxic conditions and injection of molecular oxygen were tested in 160-mL total volume bottles, each sealed with an aluminum cap and a polytetrafluoroethylene (PTFE)-lined rubber septum (reactors M).

Further kinetic studies and metabolite detection were performed in reactors D, obtained from modified 1200-mL total volume Drechsel-type bottles, sealed with PTFE-lined screw caps. The inlets and outlets were closed with PTFE filters (0.22 μm) and sealed with PTFE tape. Sampling from reactors D was carried out in sterile conditions with a sterile syringe through the outlet of the reactor.

To investigate the role of oxygen in aquifer toluene biodegradation, different oxygen sources were tested:

1. Air oxygen in the headspace volume was used in both types of reactors.
2. Different quantities of O_2 gas were injected in reactor M at different time intervals.
3. Hydrogen peroxide as the oxygen source (Pardieck et al., 1992) was tested at different concentrations (50 to 1000 mg l^{-1}), in multiple or single additions.

Tests for the identification of the metabolites of toluene degradation were carried out in the reactors D with inoculum on Minimum Salt Medium, using toluene as the sole carbon source. Metabolites were extracted with SPME and detected with gas chromatography/mass spectronscopy (GC/MS).

Analytical Method for Toluene Determination. Toluene concentrations were determined following extraction from the water samples. Each extraction was performed with hexane and analyzed with a GC (HP 5890 II) equipped with an AT-Wax 0.53-mm inside diameter (ID) × 30-m-long column and a flame ionization detector (FID).

Analytical Method for Metabolic Intermediates Determination. Metabolite concentrations were determined first with standard chromatographic methods (U.S. EPA, 1986). Then, to detect microconcentrations, a new method using Solid Phase Micro Extraction (SPME) and GC/MS for analysis was devised. SPME uses a fused silica fiber coated with an immobilized phase fixed inside the needle of a syringe. The fiber is exposed to the liquid and the analytes are accumulated in the stationary phase until equilibrium is reached. Then the fiber is removed and the extracted organic substances are thermally desorbed in a GC injector. This new technique was chosen because it is well suited for detecting microconcentrations of environmental pollutants (Pawliszyn, 1997).

NaCl (0.5 g) and HCl (17 µL) were added to the aqueous sample (1.5 mL). An 85-µm polyacrylate SPME fiber was immersed directly into the stirred solution for 15 minutes. The samples were then desorbed for 10 minutes at 250°C into the GC (HP5890II) injector port and analyzed by MS. Separation was accomplished using a fused silica capillary column DB% MS (30 m x 0.25 mm × 0.25 µm thickness). The initial oven temperature was 70°C for 2 minutes, then increased to 230°C at a rate of 10°C/min for 10 minutes. The flowrate of the helium carrier gas was 0.8 mL/min. The samples were analyzed by selective ion monitoring (SIM).

RESULTS AND DISCUSSION

Figure 1 compares the headspace of the aerobic and anoxic conditions. Oxygen was supplied as air in the reactors headspace. Complete toluene degradation occurred only in reactors with oxygen (presence of air in headspace), and it started when air was supplied, whereas no degradation activity was observed in the anoxic conditions (without headspace).

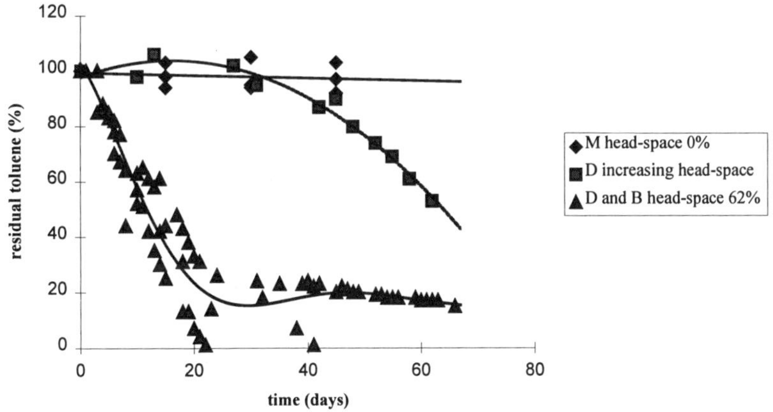

FIGURE 1. Comparison between different headspace tests.

A comparison of the oxygen sources (Figure 2) confirms that oxygen, supplied in different ways, plays a key role in toluene biodegradation.

Moreover, in Figure 2 it can be seen that, with adequate oxygen availability, toluene removal reached 100% in about 50 days.

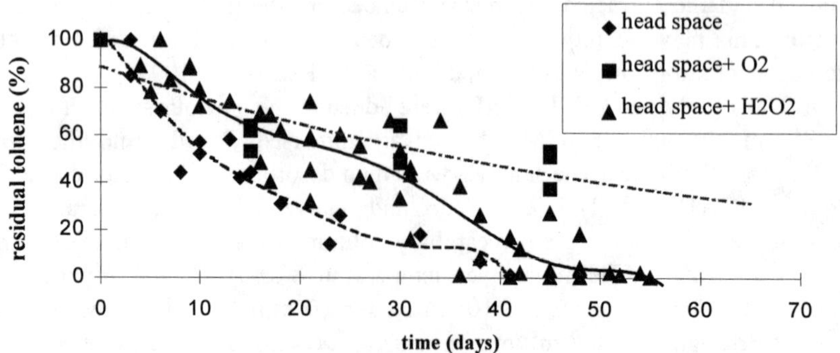

FIGURE 2. Comparison between different oxygen sources

Tests carried out for some time at low oxygen concentration showed an increase in toluene removal when more oxygen was added, as shown in Figure 3 (arrows indicate the addition time).

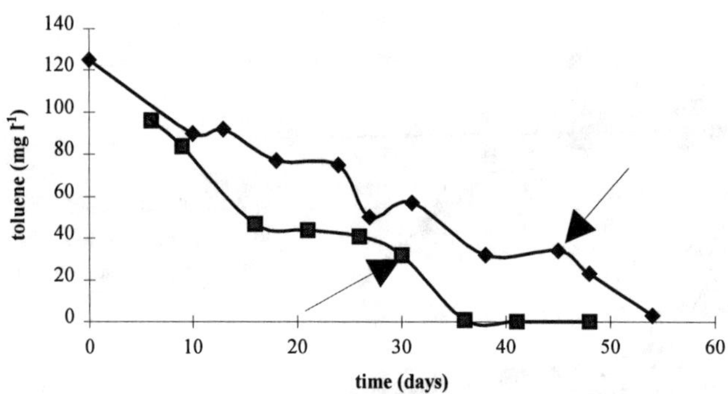

FIGURE 3. H_2O_2 addition in reactors D

Nevertheless, when oxygen availability was sufficient for toluene biodegradation, the presence of inorganic nutrients (nitrogen and phosphorous) might have been a limiting factor for the process (Nyer, 1993). In fact, the C:N:P ratio in the groundwater was not equilibrated for biomass growth, so that N and P should be supplied in a ratio near 100:5:1 (Vismara, 1988) to improve bacterial activity.

The primary metabolic intermediates checked and the analytical techniques used for their determination are reported in Table 1.

TABLE 1. Toluene degradation metabolites

Metabolite	GC/FID	GC/MS	HPLC*
Cathecol		no	yes
Protocatechuate			yes
Benzyl alcohol	yes	yes	yes
3-methylcathecol		yes	yes
Benzaldehyde	no	yes	yes
Benzoic acid	yes		yes
o-cresol	yes	yes	yes
m-cresol	no	yes	
p-cresol	no	yes	

*HPLC = high performance liquid chromatography

From the kinds of metabolites that were detected, it was possible to ascertain that aerobic toluene passes through at least the 2-monoxygenase pathway during biodegradation, thus leading to the supposition that *Pseudomonas spp* strains were involved in the process.

Figure 4 shows the trends of the two main intermediates involved (*o*-cresol and 3-methylcathecol). Other metabolites were detected and tracked, but the identification technique needs further analytical studies to improve the microbial characterization.

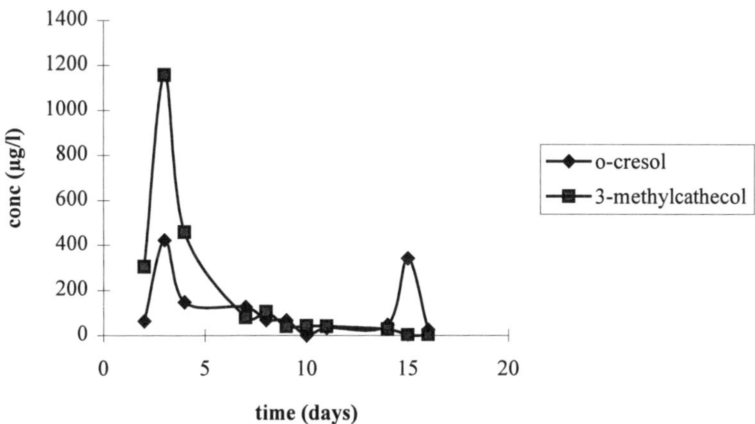

FIGURE 4. Toluene 2-monoxygenase pathway

CONCLUSIONS

The experimental studies carried out allow us to state that toluene biodegradation occurred only under aerobic conditions and that how oxygen was provided did not have a strong influence on toluene removal rates. Therefore, *in situ* aquifer bioremediation should be possible by oxygen addition (e.g., by H_2O_2 injection or air sparging). Supplying nutrients should further improve the process.

Bacteria belonging to the *Pseudomonas* genus appear to be involved in toluene biodegradation, but further studies are necessary to improve the microbial characterization.

REFERENCES

Bouwer, E.J. 1992. "Bioremediation of organic contaminants in the subsurface." In Mitchell, R. (Ed.), *Environmental Microbiology*. Wiley-Liss, Inc. New York, NY, 287-318.

Gibson, D.T. and Subramanian, V. 1984. "Degradation of Aromatic Hydrocarbons." In Gibson, D.T. (Ed.)., *Microbial Degradation of Organic Compounds*. Marcel Dekker, Inc. New York, NY. 181-294.

Karlson, U. and Frankenberger, W.T. 1989. "Microbial degradation of benzene and toluene in groundwater." *Bull. Environ. Contam. Toxicol.* 43: 505-510.

Mikesell, M.D., Kukor, J.J., and Olsen, R.H. 1993. "Metabolic diversity of aromatic hydrocarbon-degrading bacteria from a petroleum-contaminated aquifer." *Biodegradation*. 4: 249-259.

Nyer, E. K. 1993. *Practical Techniques for Groundwater and Soil Remediation*. Lewis Publishers. 186-188.

Pardieck, D.L., Bouwer, E.J., and Stone, A.T. 1992. "Hydrogen peroxide use to increase oxidant capacity for in situ bioremediation of contaminated soils and aquifers: a review." *Journal of Contaminant Hydrology*. 9: 221-242.

Pawliszyn, J. 1997. *Solid Phase Microextraction*. Wiley-VCH, Inc.

U.S. Environmental Protection Agency. 1986. *Test Methods for Evaluating Solid Waste. Vol. IB: Laboratory Manual Physical/Chemical Methods*. Met. 3650A; 8040A.

U.S. EPA, see U.S. Environmental Protection Agency.

Vismara, R. 1988. *Depurazione Biologica*. Hoepli Milano. 82.

CYCLED AIR SPARGING: FIELD RESULTS IN A HIGH-VELOCITY AQUIFER

James D. Hartley, PE, CH2M HILL, Sacramento, California
Lillian J. Furlow, PG, CH2M HILL, Atlanta, Georgia
William N. Zwiren, PE, CH2M HILL, Atlanta, Georgia

Abstract: A 21-well sparging system has operated since October 1997 to remediate dissolved hydrocarbons in groundwater upgradient of an oxbow marsh of the Black Warrior River in central Alabama. The groundwater velocity (greater than 3m/day) led to the preference of sparging over passive oxygenation. The width of the treatment zone (200m) would have required a system flow of 17 m^3/min (600 scfm) to sparge all wells continuously. The groundwater treatment system was based on six groups of sparging wells, with each group operated in synchronized cycles. The cycles consisted of 2 to 4 hours sparging, followed by up to12 hours quiescence. Synchronizing the cycles to avoid overlapping reduced the required compressor capacity to 3 m^3/min (100 scfm). By focusing the sparging on groundwater oxygen replenishment, rather than transferring volatiles to the vadose zone, the need for active air treatment was avoided. The groundwater responded with an increase in dissolved oxygen (DO) to over 9 mg/L, and a reduction of benzene, toluene, ethylbenzene and xylenes (BTEX) in compliance point wells to less than 0.002 mg/L. Synchronized cycling of the sparging wells was found to be as effective as continuous sparging at this site, and less expensive to build and operate.

INTRODUCTION

Dissolved hydrocarbons were found in the mid-1990's in groundwater downgradient of two petroleum pipeline terminals in northern Hale County, Alabama. The facilities are located on terrace deposits adjacent to an oxbow marsh of the Black Warrior River, to which contaminated groundwater discharges into seeps approximately 800m downgradient. The subsurface geologic profile upgradient of the seeps consists of 3-5 m of clayey sands, above 20m of a poorly graded fine to medium grained sand. The groundwater is unconfined in this sand unit, with the water table about 7m below ground surface. Groundwater velocities are estimated to be 0.3 to 3 m/day, which has led to a long, relatively narrow plume (800m long by up to 200m wide). Concentrations of BTEX compounds in the center of the plume exceeded 10 mg/L; discharge concentrations in seeps ranged as high as 5 mg/L of total BTEX. Figure 1 presents a site plan of the source and discharge area, with concentration contours as originally detected.

Reducing conditions in the discharging groundwater were inferred from the presence of rust (iron) staining on swamp mud and water at the seeps. The groundwater sampling program initiated the collection of dissolved oxygen and other natural attenuation parameters, which confirmed reducing conditions in the main portion of the groundwater containing dissolved hydrocarbons. These observations led to the selection of an oxygenation system to achieve the criteria for discharge concentrations at the seeps.

DESIGN BASIS

The remediation system design began with an evaluation of the relative efficiencies of passive and active oxygenation. This evaluation was presented at the 1997 Petroleum Hydrocarbons Conference in Houston, TX (Zwiren and Woodford, 1997).

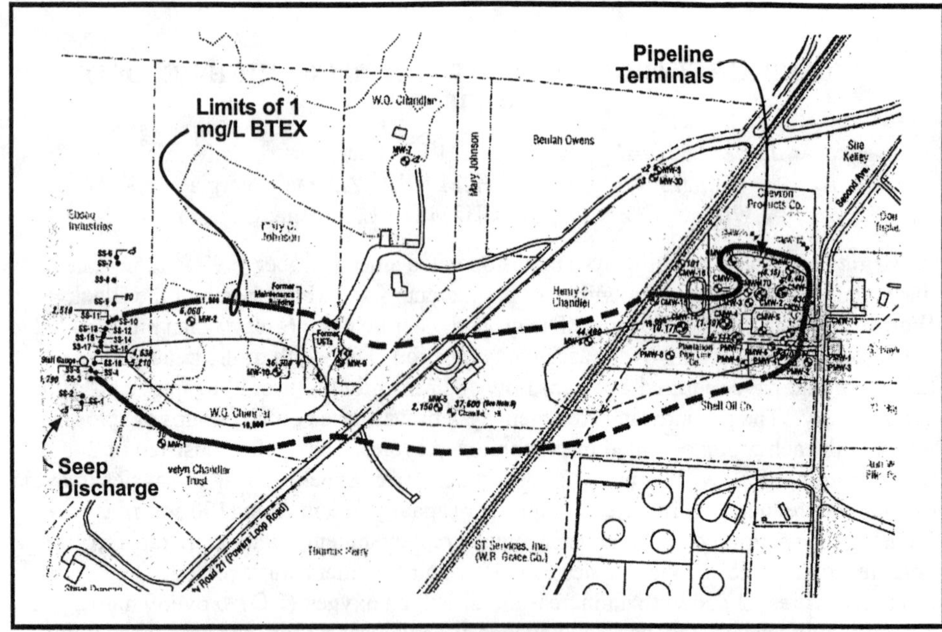

FIGURE 1. Site plan of groundwater contamination plume.

Largely because of the groundwater velocities, sparging proved to be preferred at this location. The initial design placed wells in two rows, spaced at 10m, operated continuously at 0.7 m^3/min (20 scfm). A sparging pilot test was conducted to establish the feasibility of this approach, and to collect data on design parameters. Among other results, the pilot test confirmed the need for collection and control of vadose zone vapors, for continuously operated sparging wells.

The cost estimate for this initial concept exceeded the budget established for this project, so the design team endeavored to seek improvements. Given the surficial clay unit, the pilot test data were revisited to explore the potential for reducing the sparge airflow, and consequently eliminating the need for active vapor collection. In addition, the concept of pulsed sparging was explored. The final design incorporated active sparging of a limited duration (2 to 4 hours, to be confirmed by field measurements of vapor piezometers) followed by a period of quiescence. Useful guidance on the expected shape and distribution of the sparge zone was found in an article in *Ground Water Monitoring and Remediation* by McKay and Acomb (1996); and in the same periodical, Rutherford and Johnson (1996) provided helpful insight on the expected transfer efficiencies of oxygen during pulsing.

The travel time of groundwater through the sparge zone and the time in which the transferred oxygen in the sparge zone might be consumed bound the period of allowable quiescence. For an assumed total petroleum concentration of 30 to 50 mg/L, and groundwater flow rates of 1.5 to 3 m/day, the theoretical range of maximum quiescence was found to be between 5 and 14 hours. This was found to be in the range of durations that could accommodate a totally synchronized pulsing pattern, with six groups of wells operating for 2 to 4 hours per cycle, powered by a single, smaller compressor. In the conceptual synchronization represented by Figure 2, Groups 3 and 4 are in the center of the plume (highest concentration areas).

Sparge Wells	Hourly Operating Sequence																							
	1	2	3	4	5	6	7	8	9	10	11	12	13	14	15	16	17	18	19	20	21	22	23	24
Group 1																								
Group 2																								
Group 3																								
Group 4																								
Group 5																								
Group 6																								

FIGURE 2. Conceptual sparging group rotation sequence.

The injection period of 2 to 4 hours was recognized to represent the period of maximum oxygen transfer efficiency, inasmuch as this represents the duration of the transient phase of the sparging cycle (greatest proportion of air volume, greatest surface area, greatest water table mounding). Beyond this duration continued airflow was not believed to contribute significantly to the transfer of oxygen to the groundwater. Counterproductively, continued airflow was believed to increase the proportion of hydrocarbon mass being transferred to the vadose zone, rather than biodegraded in situ. Continuous airflow is also believed to create a lower permeability zone that could divert groundwater around the treatment zone. Because the design modifications sought to maximize the proportion of remediation achieved through biodegradation, and minimize that achieved by phase transfer, the duration of injection was initially limited to 2 hours in the peripheral groups, and to no more than 4 hours in the center (more concentrated) areas. Vapor piezometers were installed and monitored during startup to confirm the maximum allowable injection period, as a function of detected pressure and vapor concentrations. Following initial operation, all wells were cycled to a 2 hour-on period.

INSTALLATION AND OPERATION

The subsurface system, as installed, consisted of 21 sparge wells, 11vapor piezometers, and 5 groundwater piezometers downgradient of the sparging wells. The layout of the wells and monitoring points is shown on Figure 3.

The sparge wells were installed with 1 m screens placed at the base of the saturated sand unit, approximately 16 m below the water table. Each well was connected to individual HDPE pipe, selected for its durability under pressurized conditions and for its ease of installation and local availability. Each well was plumbed separately to the compressor building. In the compressor building, the individual pipes were valved with solenoids, instrumented with rotameters, and connected to a single manifold. The solenoids were partitioned into six groups, each representing the groups of wells to be sparged together in sequence.

The compressors were selected for their ability to match the supplied airflow with the demand, depending on the number of wells actually in service. The compressors are of a low discharge pressure rotary-screw type, with an automatic throttling inlet to match well demand and improve efficiency. In comparison to a reciprocating compressor, necessitating a larger receiver and requiring a maximum duty factor of 50% to ensure

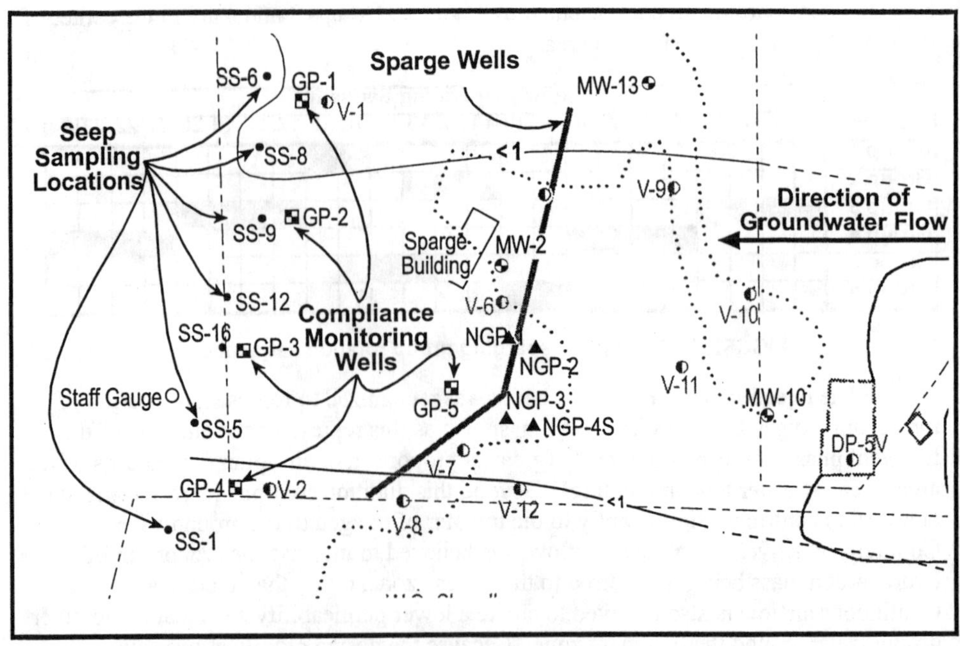

FIGURE 3. Detail map of sparging treatment zone.

normal lifespan, the rotary screw compressors are able to supply the needed flow at nearly half the power demand.

In a further effort to reduce cost, in this case with respect to operation, the six solenoid wire groups were connected to a six-station sprinkler controller. This unit was available from a local irrigation supply store. In particular, the programming requirements for this unit were well within the capabilities of the local system operator, which was not the case with the Programmable Logic Controller (PLC) originally considered.

FIELD PERFORMANCE

The system began operation in October 1997 and has operated continuously, except for periodic maintenance, since that date. In the compliance monitoring wells located approximately 30 m downgradient of the treatment zone, initial concentrations of BTEX ranged from 0.35 to 6.0 mg/l. After 15 months of system operation, these concentrations have been reduced to less than 0.002 mg/l in groundwater, and less than 0.015 mg/l in the seeps. The trends of BTEX and DO concentrations over time are shown in Table 1. In this table, the concentrations represent the average quantified values for groundwater upgradient of the treatment zone, in groundwater at the compliance point downgradient of the treatment zone, and in grab samples collected from the seep sampling stations. The upgradient values are presented to show the trend of concentrations entering the treatment zone. The values in the first three columns are prior to operation of the sparging system; the values in the right-hand five columns are during the sparge system operation.

TABLE 1. Average BTEX and DO concentrations in ground and surface water.

Monitoring Location (Parameter)	Average BTEX and DO Concentrations (mg/L)							
	3/96	3/97	9/97	12/97	3/98	6/98	9/98	12/98
Upgradient Plume	11	9.4	5.0	5.0	0.36	4.0	2.7	6.8
Treatment Zone			15	1.7	0.12	0.026	0.012	0.001
Discharge Seeps	1.4	0.2	0.6	0.5	0.5	0.042	0.018	0.017
Treatment Zone (DO)			1.8	7.1	8.4	8.2	8.3	9.1

CONCLUSIONS

These data provide strong confirmation that the synchronized cycling of the sparging system is providing adequate oxygenation of the groundwater, resulting in the desired concentration reduction of the dissolved contaminant plume. These results also suggest that synchronized cycling of sparging groups may be employed at other multiple well sites with little or no reduction in oxygen transfer effectiveness.

This result came directly from the focus, during design, to apply sparging primarily as an oxygen delivery system, and to minimize the effects of phase transfer into the vadose zone. Moreover, this project demonstrated that this design focus enabled the reduction of operational costs in terms of power demand, operational training, and air treatment operation and maintenance.

REFERENCES

Zwiren, W.N., and Woodford, J. 1997. "Evaluation of Oxygen Delivery Systems for a Downgradient Treatment Zone in a High Groundwater Velocity Aquifer." In Proceedings, 1997 Petroleum Hydrocarbons and Organic Chemicals in Groundwater, Prevention, Detection, and Remediation Conference, Nov. 12-14, 1997, Houston, TX: 656-667.

McKay, D.J. and Acomb, L. 1996. "Neutron Moisture Probe Measurements of Fluid Displacement During In Situ Air Sparging." Ground Water Monitoring and Remediation 16(4): 86-94.

Rutherford, K.W. and Johnson, P.C. 1996. "Effects of Process Control Changes on Aquifer Oxygenation Rates During In Situ Air Sparging in Homogeneous Aquifers." Ground Water Monitoring and Remediation 16(4): 132-141.

BIODEGRADATION AND VOLATILIZATION DURING IAS - A PUSH PULL TEST

Emile C.L. Marnette (Tauw bv, Deventer, The Netherlands)
Charles G.J.M. Pijls (Tauw bv, Deventer, The Netherlands) and
Derk van Ree (Delft Geotechnics, Delft, The Netherlands)

ABSTRACT: Push pull tests were carried out to investigate the contribution of biodegradation and volatilization to the removal of groundwater contamination during in situ air sparging (IAS). The push pull tests consisted of a controlled injection of a prepared test solution into two single wells at diferent distances from the sparge well, followed by periodical extraction of a mixture of the test solution and groundwater. Different tracers were added to the test solution: bromide to account for dilution effects, PCE as a volatile aerobically non biodegradable tracer and acetate as a non volatile aerobically biodegradable tracer.

Sparging has a substantial mixing effect on the groundwater within the radius of influence; at 2 m distance from the sparge well bromide was diluted to approx. 10% of the initial concentration due to sparging. Volatilization is most pronounced in the first hours after the start of sparging (PCE concentration down to approx. 20% of thc input concentration at 2 m distance from the sparge well). After this period the volatilization rate decreases.

The results indicate that a large fraction of the volatile contaminants is removed shortly after the start of sparging. Therefore this study provide important information on the design of an IAS system and on the dimensioning of an off gas treatment system.

INTRODUCTION

In situ air sparging (IAS) refers to a process in which air is injected into the saturated zone. IAS is used to strip volatile components from groundwater and to enhance aerobic biodegradable contaminants.

Quantative data on the contribution of volatilization and biodegradation during IAS are scarce (Veenis, 1995, Johnson, 1998). There is no general agreement on the relative significance of the individual processes that are occurring during IAS (Johnson, 1998). Johnson (1998) presented a theoretical analysis of the mechanisms contributing to the overall performance of IAS systems.

The extent at which volatilization and biodegradation occur is strongly dependent on the chemical and physical properties of the component and on the efficiency of the IAS system.

Objective. The objective of this study is to estimate the contribution of volatilization and biodegradation to the removal of contaminants during IAS. This goal is established as a part of the Dutch Research for Biotechnological In-

Situ Remediation (NOBIS) project Bioventing and Biosparging (Marnette, 1998).

Site description. The experimental site is a former petroleum hydrocarbon storage and distribution facility in Nijmegen, The Netherlands. The site was contaminated with BTEX and petroleum hydrocarbon and has been subjected to a full scale SVE/IAS remediation for 3 years. One injection well of the IAS system has been used for the experiments.

MATERIALS AND METHODS

Setup. To estimate the extent of volatilization and biodegradation, a series of push pull tests is performed. A push pull test consists of the controlled injection of a prepared test solution into a single well followed by periodical extraction of a mixture of the test solution and groundwater (Figure 1). Volatilization and biodegradation are individually assessed using two different tracers: a non volatile aerobically biodegradable tracer (sodium acetate) and a volatile aerobically non biodegradable tracer (PCE, Perchloroethylene). The advantage of using two tracers instead of one tracer which is both biodegradable and volatile, is that both processess can be followed separately. A disadvantage of using two tracers is that translation of the test results to volatilization and biodegradation of one single compound is difficult, since physical and chemical properties of the two separate tracers differ from that of the single compound.

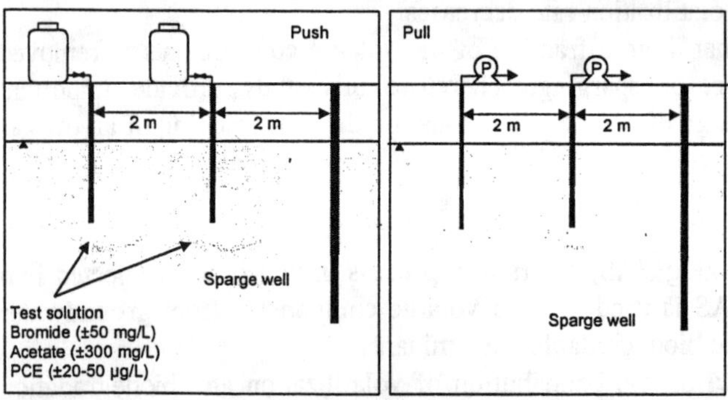

FIGURE 1. Setup of a push pull test

To account for dilution effects (mixing of the test solution with ambient groundwater) a third, conservative, tracer (bromide) is added.

Two series of push pull tests were carried out. Each series consisted of two push pull tests, one at 2 m and one at 4 m distance from a sparge well (Figure 1). One series was performed without sparging. These tests were considered as a reference. The second series was performed while the IAS system was in pulsed operation (1 hour on, 1 hour off). Only one sparge well was in service. During pulsed sparging, sampling was performed in the middle of the periods the IAS system was turned off.

Preparation of the test solution. A vessel was filled with 70 L of groundwater collected on site. An amount of acetate and sodiumbromide was added yielding concentrations of approx. 50 mg/L Br^- and 300 mg/L acetate. The solution was purged with N_2 for 2 days.. A concentrated PCE solution was added to the groundwater-tracer solution resulting in a concentration of approx. 20 - 50 µg PCE/L. The vessel was tightly closed prior to transport.

Introduction of the test solution (push). 50 L of the test solution was introduced into two discrete wells (l=20 cm, diameter = 30 mm, 3,5-3,7 m bg) each. About 20 L of the test solution was left in the vessel to minimize the chance of volatilization of PCE from the solution that is introduced into the soil to the headspace in the vessel. It took about 30 minutes before 50 L of the test solution was introduced into the monitoring well.

Sampling and analyses (pull). Shortly after the introduction of the test solution, the first "pull" was performed by sampling the discrete filters. Due to strict site regulations, samples could only be taken during the period from 7 AM to 5. During the day each 1,5 - 2 hours a groundwater samples were collected.

Each sample was analyzed on PCE, acetate (gaschromatography) and Br^- (ionchromatography). After both sessions, the introduced solution was recoved by extracting >80 L groundwater. After extraction, a sample was taken to determine possible residual concentrations of PCE.

Calculation. The amount of contamination removed by in situ processes is expressed as the decrease of the concentrations as a fraction of the input concentration.

To account for dilution effects during the push pull tests with and without sparging, PCE and acetate concentrations are corrected with a dilution factor f_{dil}:

$$f_{dil} = C^{Br}_{t} / C^{Br}_{o} \qquad (1)$$

where C^{Br}_{t} = concentration of Br^- in a groundwater sample at time = t
C^{Br}_{0} = concentration of Br^- in a groundwater sample at time = 0, i.e. the input concentration of Br^-.

The effect of sparging on the volatilization process (decrease of PCE concentration) can be calculated by substracting the decrease of PCE without sparging (push pull session 1, reference test without sparging) from the decrease of the PCE concentration during sparging (push pull session 2, with sparging). The decrease of the concentrations are normalized on the input concentrations at start of the push pull test.

When a zero concentration is reached during sparging (100% removal), the sparge effect will remain constant, since there is no acetate or PCE left to be susceptible for non-sparging processes as e.g. adsorption to the soil matrix.

RESULTS AND DISCUSSION

Mixing Effect of Sparging - Bromide. The bromide concentrations without sparging decreased gradually to 96% of the input concentration at 2 m distance from the sparge well (Figure 2.a) and to 82% at 4 m distance. Bromide concentrations during pulsed sparging decreased to 6% of the input concentration at 2 m distance from the sparge well and to 34% at 4 m distance. The results indicate that the mixing effect of sparging is considerable and that the effect is smaller at greater distance from the sparge well.

Biodegradation - Acetate. During the non-sparging session, acetate concentration decreased to 75% and 86% of the input concentration at 2 respectively 4 m distance from the sparge well (Figure 2.b). These data are corrected for the dilution effect. The decrease of acetate without the supply of oxygen may be due to aerobic as well anaerobic degradation. Before use, the test solution was purged with N_2 for 24h. The test solution however still contained about 5 mg/L dissolved oxygen before it was introduced into the soil. Also there may be some trapped air in the soil from which oxygen can be released to the test solution after its introduction into the soil. So aerobic degradation was still feasible.
During the sparging session, a lag phase occurred in which the acetate concentrations did not decrease significantly. After 24 hours, at 2 m distance from the sparge well, all the acetate was depleted, caused by biological degradation. As mentioned before, the data gap between 7,5 hours and 23,5 hours is due to the strict site regulations; between 7 AM and 5 PM entering the site is prohibited, so no samples could be collected. Therefore, there is no insight in the rate of acetate consumption between the lag phase and the point at which zero concentrations were reached. At 4 m distance from the sparge well, the acetate concentrations decreased more gradually to approx. 40% of the input concentration.
The effect of IAS on the biodegradation of acetate is significantly larger at smaller distance from the sparge well.

Volatilization -PCE. During the non-sparging session, the concentration of PCE decreased to approx. 80 and 90% of the input concentration at 2 respectively 4 m distance from the sparge well (Figure 2.c). Concentrations are corrected for dilution effects. The decrease may be attributed to adsorption of PCE to organic material in the soil.

During the sparging session, at 2 m distance from the sparge well, the PCE concentrations decreased within the first 7.5 hours to approx. 10% of the input concentration. During the sampling period between 24 and 32 hours after starting the session, the decrease of concentrations seems to stagnate between 15 and 25%. It must be noted that the duration of the test was rather short. It is likely that after one more day of sparging, PCE levels will slowly decrease by volatilization. At 4 m distance, sparging results in an almost instantaneous (1.5

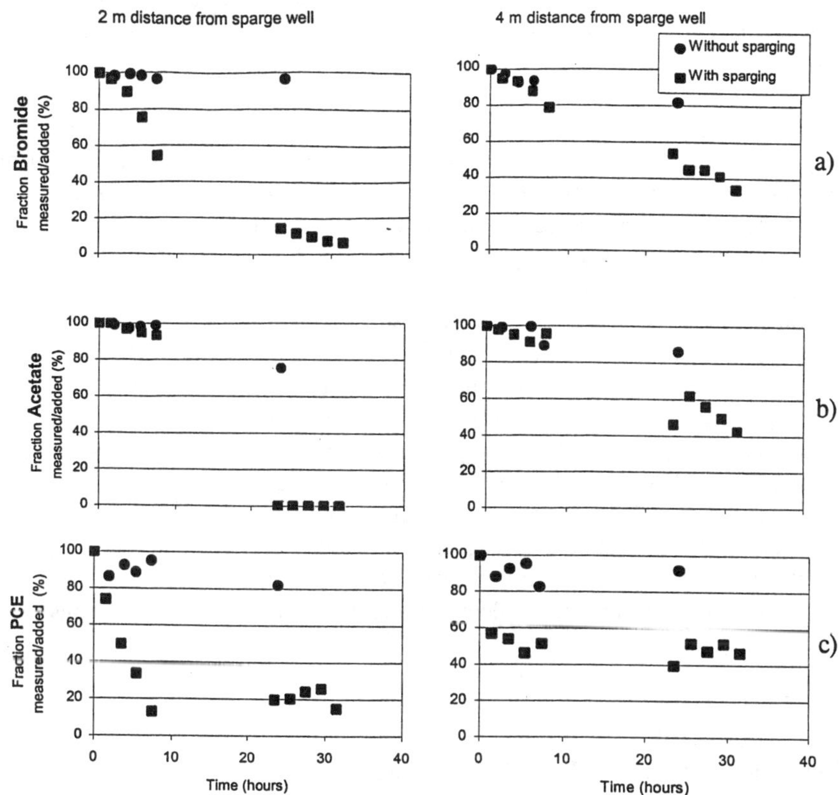

FIGURE 2. Fraction of measured/added tracer at 2 meter and 4 meter distance of the sparge well without (dots) and with (squares) sparging as function of time. a) Bromide. b) Acetate and c) PCE. The concentrations are corrected for dilution effects.

hour) decrease of the PCE concentration to approx. 60% of the input concentration. Further during the test, the concentration fluctuates around 40 - 50% of the input concentration.

The results of the sparge session indicate that after short term sparging (± 30 h) volatilization is not complete and concentrations seem to stagnate at a certain level. This residual level is higher at larger distance from the sparge well.

The results of the test agree with common knowledge that IAS is more effective in the removal of contaminants closer at the sparge well than at greater distance from the well. These observations should generally also apply for the rate of removal of contaminants.

Since the process of volatilization and biodegradation is assessed separately by using one tracer specifically for degradation and one specifically for volatilization, the simultaneous effect of biodegradation and volatilization on a compound that is both aerobically biodegradable as well volatile, is hard to assess. The results of this research indicate the importance of the initial period of the remediation, when a large fraction (40-60%) of the contaminant concentration is removed.

CONCLUSIONS

Sparging has a substantial mixing effect on the groundwater within the radius of influence; at 2 m distance from the sparge well bromide was diluted to 6% of the initial concentration due to sparging and at 4 m distance to 34%.

Volatilization is most pronounced in the first hours after the start of sparging. After this period the volatilization rate decreases. It is not clear whether this phenomenon is related to the concentration level of the volatile compound. In the test relatively low concentrations of PCE ($\pm$ 20-50 μg L^{-1}) were used;

Especially aerobic biodegradation and mixing of the groundwater is strongly dependent on the distance from the sparge well. At 2-m distance, aerobic biodegradation of acetate ($\pm$ 300 mg L^{-1}) was complete after 1 day, whereas at 4-m distance about 40% of the input concentration was depleted.

The results indicate that a large fraction of the volatile contaminants is removed at the very start of sparging. The results therefore provide important information on the design of an IAS system and an off gas treatment system.

The research in general demonstrates the wide applicability of push pull tests.

ACKNOWLEDGEMENT

This work was financially supported by NOBIS (Dutch Research Biotechnological In Situ Remediation. Karin van de Brink is acknowledged for technical assistance.

REFERENCES

Amerson, I.L. 1997. "Diagnostic tools for monitoring and optimization of in situ air sparging systems." M.S. Thesis, Arizona State University, Tempe, AR.

Johnson, P.C. 1998. "Assessment of the contributions of volatilization and biodegradation to in situ air sparging performance." *Environmental Science & Technology*. *32*(2): 276-281.

Marnette, E.C.L. 1998. *Technical report Push Pull test - discrimination between volatilization and biodegradation*. CUR/NOBIS, Gouda, The Netherlands.

Veenis.Y. (Ed). 1995. *In-Situ Bioventing Pilot Project*. Groundwater Technology, Inc. 985-001-6670.

DESIGN, MONITORING, AND CLOSURE STRATEGY FOR AN AIR SPARGING SITE

Keith A. Fields (Battelle, Columbus, Ohio)
Thomas C. Zwick, Andrea Leeson, and Godage B. Wickramanayake (Battelle, Columbus, Ohio); Herb Doughty (Southwest Division Naval Facilities Engineering Command, San Diego, California); Tracy Sahagun (Assistant Chief of Staff Environmental Security, Camp Pendleton, California)

ABSTRACT: This study presents the design, monitoring, and closeout strategy for an in situ air sparging/soil vapor extraction (IAS/SVE) system at a gas station site (Site 43286) at Marine Corps Base (MCB) Camp Pendleton. Battelle has conducted pilot-scale tests, designed, and installed a full-scale IAS/SVE system at Site 43286. The full-scale system design was based on the Draft Air Sparging Design Paradigm (Leeson et al., 1998) and results from pilot-scale testing. The system consists of 18 sparge wells, 5 vapor extraction wells, 12 soil and groundwater monitoring point locations, and 10 groundwater monitoring wells. Results of the pilot-scale testing activities indicate that sparge well spacing of 4.5 meters (15 feet) should be adequate for full-scale operations. The optimal air injection flowrate was determined to be 2.3 L/s (5 cfm) per well. This flowrate was achieved at an injection pressure of approximately 69 kPa (10 psi); therefore, a rotary positive blower will be used to deliver air to the subsurface rather than a compressor, which results in significant cost savings to the project. Also, pilot-scale testing activities indicated that an air extraction flowrate of 14 L/s (30 cfm) per well would be sufficient to prevent vapor migration into Building 43286.

The monitoring and closure strategy consists of daily and weekly collection of system operational data as well as quarterly groundwater monitoring events at the site. On a quarterly basis, the system will be evaluated to determine technical and economic efficiency based on contaminant removal from soil and groundwater. When system operation efficiency drops and contaminant removal rates level off over at least 2 consecutive quarters, the system will be shut down and monitoring for remediation by natural attenuation (RNA) will commence. After shutdown, the groundwater will continue to be monitored on a quarterly basis and if significant rebound occurs the IAS/SVE system will be reinitiated. RNA will be implemented to meet final cleanup goals at the site.

INTRODUCTION

MCB Camp Pendleton is located almost entirely in San Diego County, California between the cities of Los Angeles and San Diego. The northwestern border of MCB Camp Pendleton is located in Orange County. The Base covers approximately 50,600 ha and is bordered on the west by the Pacific Ocean, with roughly 27 km of coastline. Rolling hills and valleys extend inland 19 to 29 km from the coastline to the northeastern limits of the Base.

Site 43286 is located in the Las Flores Groundwater Basin near the center of MCB Camp Pendleton. The gas station at the site was constructed in 1958. The station did not operate from 1993 to 1997. Renovation of the gas station took place in late 1996, and the station is currently operational. The site is bounded by undeveloped land to the north and south, Las Flores Creek to the east, and Las Pulgas Road to the west (Figure 1).

Two 38-m^3 (10,000-gallon) gasoline underground storage tanks (USTs) and a 2-m^3 (550-gallon) waste oil tank were removed from the site in October 1993. An initial site assessment was conducted following tank removal and included the advancement of nine boreholes and the installation of three groundwater monitoring wells. This investigation indicated that fuel hydrocarbons, including benzene, toluene, ethylbenzene, and total xylenes (BTEX), had impacted both soil and groundwater at the site. Additional soil and groundwater investigations were conducted from 1995 through 1997, including a joint effort by the MCB Camp Pendleton and Lawrence Livermore National Laboratory (LLNL). These investigations, which included additional soil borings, monitoring wells, and Hydropunch™ sampling, further examined the extent of soil and groundwater contamination.

Based on these additional site investigations, there appears to be at least two related, but hydrogeologically distinct, groundwater zones, as revealed by the well pairs that were installed in the downgradient portion of the site. The shallow zone occurs at depths of 4 to 6 m (13 to 20 feet) bgs. The deep zone has an upper boundary of 6 to 7.5 m (20 to 25 feet) bgs and appears confined by a 1.2-m-thick clay layer. Benzene was detected in shallow-zone groundwater at levels significantly above maximum contaminant levels (MCLs). Groundwater in the deep zone does not appear to be significantly impacted, although minor concentrations of fuel hydrocarbons have been detected. Volatile organic compound (VOC) and methyl tert-butyl ether (MTBE) concentrations in the shallow groundwater zone appear to indicate that contaminants are either discharging into Las Flores Creek or are flowing beneath it.

A Final Corrective Action Plan (CAP) was submitted by OHM and M&E on May 1, 1998. This CAP summarized previous environmental activities conducted at the site between 1993 and April 1997, proposed remedial cleanup goals, presented an analysis of potential remedial alternatives, and recommended a remedial alternative. IAS/SVE was determined to be the preferred alternative for Site 43286 (OHM/M&E, 1998).

CLEANUP GOALS

Cleanup goals were established as part of the CAP (OHM/M&E, 1998). A dual cleanup standard for soil was accepted, and remediation shall be considered successful if confirmation soil samples from previously contaminated zones contain non-detectable concentrations of leachable BTEX and MTBE (using synthetic precipitation leaching procedure [SPLP] test), or if their total concentrations are less than the values proposed in Table 1. These thresholds define concentrations in soil that pose minimal risk to underlying groundwater resources.

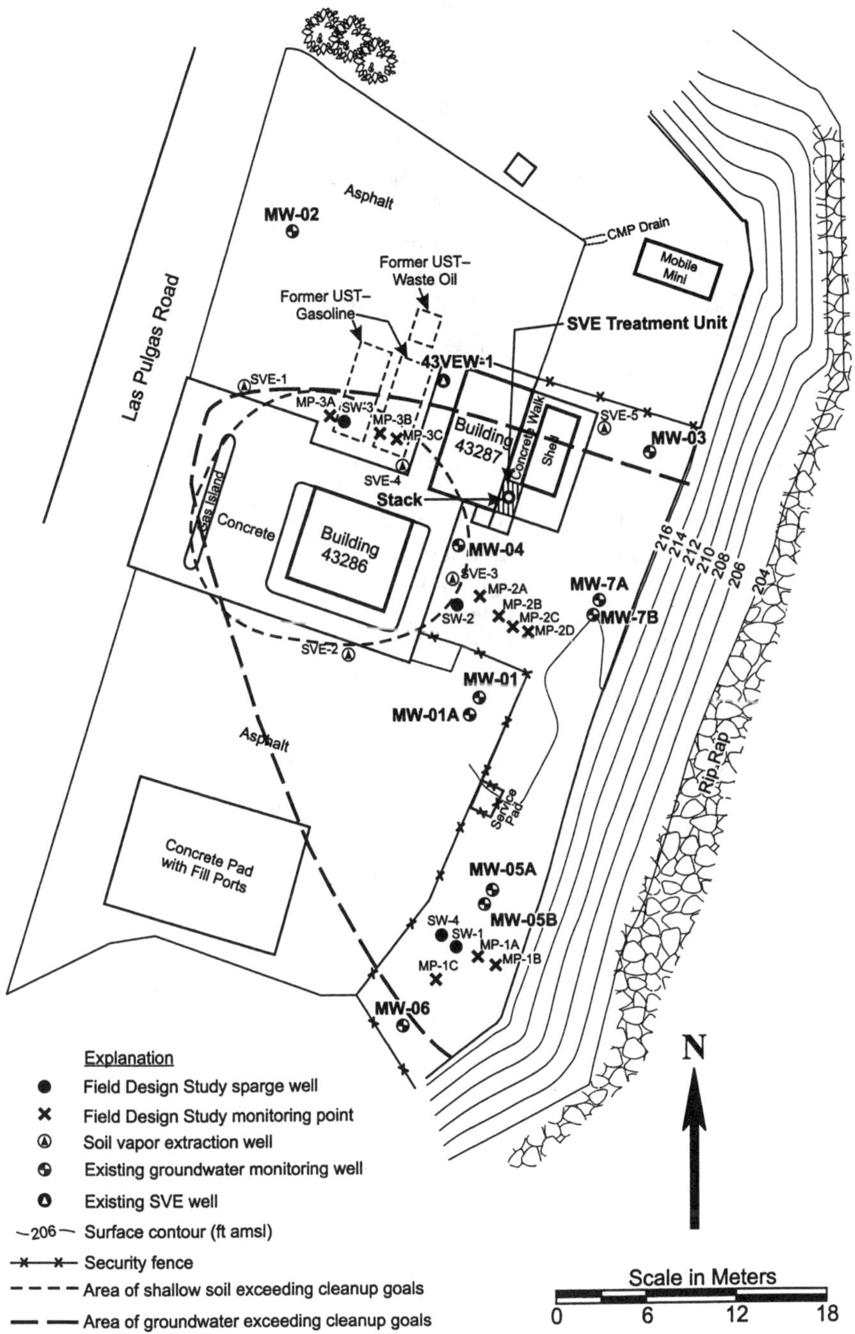

Figure 1. Site Map of Site 43286, MCB Camp Pendleton

TABLE 1. Final Cleanup Goals

Constituent	Soil	Groundwater[(a)]
Benzene	ND SPLP or 0.1 mg/kg[(b)] total	1 μg/L
Toluene	ND SPLP or 15 mg/kg total	150 μg/L
Ethylbenzene	ND SPLP or 68 mg/kg total	680 μg/L
Xylenes	ND SPLP or 175 mg/kg total	1,750 μg/L
MTBE	ND SPLP or 3.5 mg/kg total	35 μg/L[(c)]

(a) Groundwater cleanup goals correspond to drinking water MCLs.
(b) Assumes soil attenuation factor of 100.
(c) Based on draft EPA health advisory and San Diego RWQCB guidelines; subject to change.
ND = non-detect

Figure 1 shows areas of shallow unsaturated soils and groundwater that exceed the cleanup criteria and will be addressed by the IAS/SVE system.

DESIGN, MONITORING, AND CLOSEOUT STRATEGY

The proposed design, monitoring, and site closeout strategy for Site 43286 is presented in Figure 2. Design of the IAS/SVE system was based on the Standard Design Approach presented in the Draft Air Sparging Design Paradigm (Leeson et al, 1998). Also, pilot-scale testing was performed to determine injection and extraction flowrates and pressures and to obtain an approximate radius of influence (R_I) for the SVE wells and the sparge wells. Additional activities associated with the field design study included startup and shakedown of the SVE system, air sampling inside Building 43286, and baseline groundwater sampling from monitoring wells and monitoring points at the site.

The results of the pilot-scale testing activities are summarized in Table 2.

In Situ Air Sparging. It is recognized that there are physical limitations on contaminant removal using IAS in heterogeneous soil. For example, in the case study by Brown et al. (1998), only 12 out of 28 (43%) air sparging sites achieved an average permanent reduction of contaminant concentrations greater than 95%. Assuming remediation is relatively successful and 95% permanent reduction is achieved at Site 43286, benzene concentrations (observed concentrations have exceeded 2 mg/L) would still exceed the cleanup goals. Therefore, it is important that a reasonable close-out strategy for IAS be developed.

When the IAS/SVE system reaches a point at which the level of economic or technical efficiency is low, and thus is no longer substantially reducing contaminant levels and toxicological risk, the Navy will ask the Regional Water Quality Control Board (RWQCB) to consider natural attenuation to complete site cleanup. Natural attenuation, verified by periodic groundwater monitoring, will continue until final cleanup goals are achieved. The IAS/SVE system will be left in place and "on standby" during a transition period of approximately 1 year, so that the system can be easily reinitiated if rebound of contaminant concentrations occur.

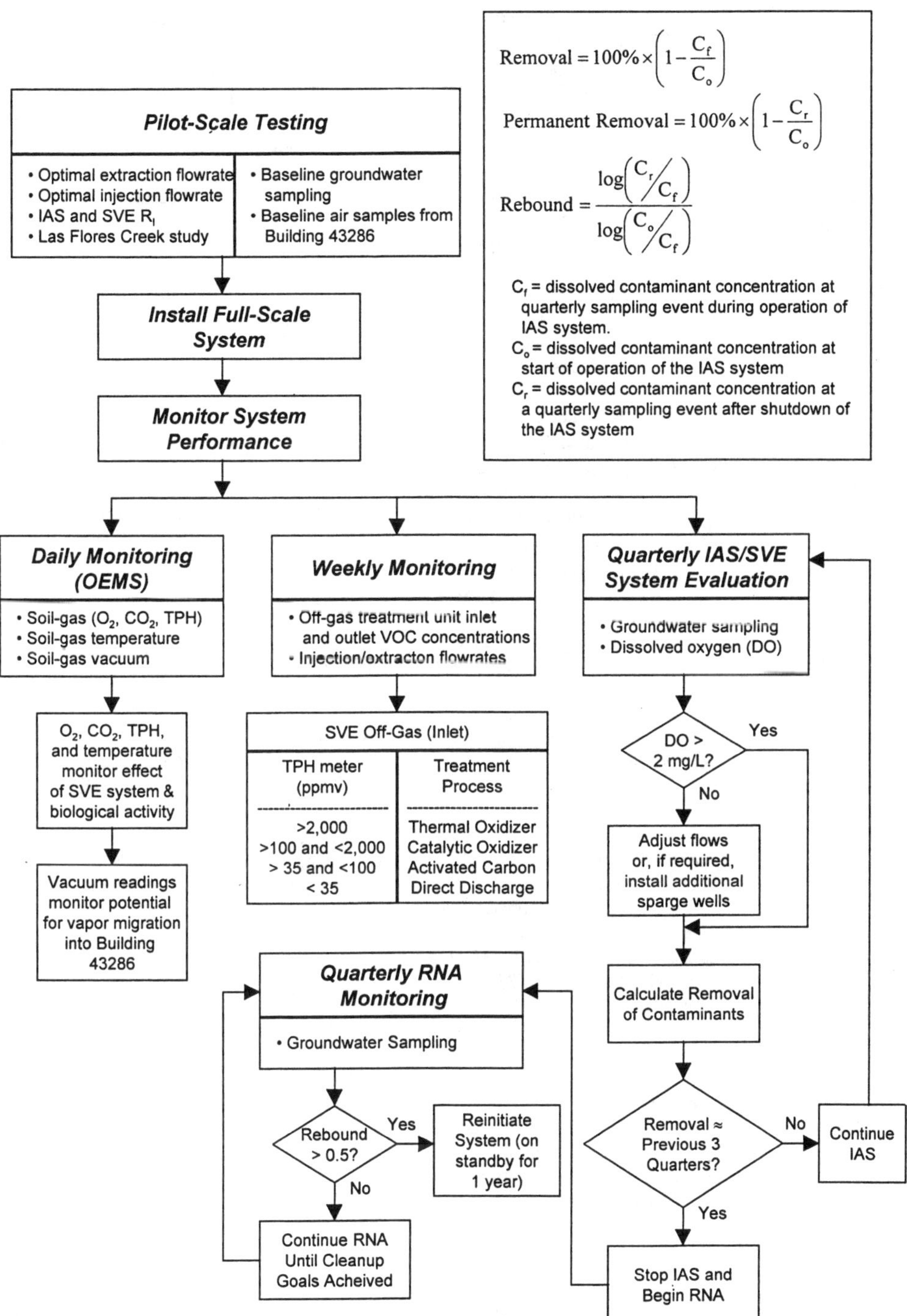

$$\text{Removal} = 100\% \times \left(1 - \frac{C_f}{C_o}\right)$$
$$\text{Permanent Removal} = 100\% \times \left(1 - \frac{C_r}{C_o}\right)$$
$$\text{Rebound} = \frac{\log\left(C_r / C_f\right)}{\log\left(C_o / C_f\right)}$$
C_f = dissolved contaminant concentration at quarterly sampling event during operation of IAS system.
C_o = dissolved contaminant concentration at start of operation of the IAS system
C_r = dissolved contaminant concentration at a quarterly sampling event after shutdown of the IAS system
Pilot-Scale Testing
• Optimal extraction flowrate
• Optimal injection flowrate
• IAS and SVE R_I
• Las Flores Creek study
• Baseline groundwater sampling
• Baseline air samples from Building 43286
Install Full-Scale System
Monitor System Performance
Daily Monitoring (OEMS)
• Soil-gas (O_2, CO_2, TPH)
• Soil-gas temperature
• Soil-gas vacuum
O_2, CO_2, TPH, and temperature monitor effect of SVE system & biological activity
Vacuum readings monitor potential for vapor migration into Building 43286
Weekly Monitoring
• Off-gas treatment unit inlet and outlet VOC concentrations
• Injection/extracton flowrates
SVE Off-Gas (Inlet)
TPH meter (ppmv)
Treatment Process
>2,000
>100 and <2,000
> 35 and <100
< 35
Thermal Oxidizer
Catalytic Oxidizer
Activated Carbon
Direct Discharge
Quarterly IAS/SVE System Evaluation
• Groundwater sampling
• Dissolved oxygen (DO)
DO > 2 mg/L?
Yes
No
Adjust flows or, if required, install additional sparge wells
Calculate Removal of Contaminants
Removal ≈ Previous 3 Quarters?
No
Continue IAS
Yes
Stop IAS and Begin RNA
Quarterly RNA Monitoring
• Groundwater Sampling
Rebound > 0.5?
Yes
Reinitiate System (on standby for 1 year)
No
Continue RNA Until Cleanup Goals Acheived

FIGURE 2. Design, Monitoring, and Closeout Strategy for Site 43286

TABLE 2. Pilot-Scale Testing Results and Subsequent Design Decision

Aspect of Study	Pilot-Scale Testing Results	Design Decision
Soil-Gas R_I Testing	R_I approximately 6 m (20 feet)	Vapor extraction rate of 14 L/s should be sufficient to control vapor migration into Building 43286.
Dissolved Oxygen Monitoring	Dissolved oxygen concentrations increased in monitoring points greater than 3 m from sparge well.	Well spacing = 4.5 m (Leeson et. al., 1998).
Water Level Monitoring	Groundwater elevations in monitoring wells increased at startup of the IAS and returned to pre-injection conditions within 2 hours.	Air is not trapped in aquifer (i.e., not forced to move laterally, which could potentially impact creek).
Soil-Gas O_2, CO_2, TPH, and vacuum monitoring	• Average O_2 concentrations increased from 2% to 19% • Average CO_2 concentrations decreased from 9% to 1% • Initial TPH concentrations were high (typically >2,000 ppmv) • Soil-gas vacuum readings were negative near Building 43286	Biological degradation in the vadose zone will be enhanced by vapor extraction and air injection. Also, the SVE system was able to maintain a negative pressure around the building (reducing possibility of vapor migration)
Extraction Flowrates and Pressures	14 L/s/well extracted at 20 kPa	Adequate extraction rates can be achieved.
Injection Flowrates and Pressures	3.8 L/s per well injected at approximately 69 kPa	Adequate flowrates can be achieved. Use of blower instead of compressor applicable (significant cost savings).
Thermal Oxidizer Operations	Inlet VOC concentrations at startup exceeded 5,000 ppmv. After 307 hours of operation the inlet VOC concentrations had dropped to less than 1,000 ppmv. Outlet VOC concentrations were below 10 ppmv. Approximately 104 kg (230 lb) of contaminants removed during pilot-scale testing activities.	Thermal oxidizer effective at destroying contaminants. Catalytic oxidation may be appropriate for full-scale treatment in a relatively short time period.
Air Monitoring Inside Building 43286	Samples collected before and during operation of IAS/SVE system. Minor concentrations were detected before and during operation; however, these concentrations were well below regulated levels.	SVE appears to be able to control vapor migration into Building 43286.

System performance will be evaluated on a quarterly basis following collection of groundwater samples. Operation of the system will be terminated when the system reaches a low level of economic or technical efficiency, as measured by the average contaminant removal based on groundwater samples collected during quarterly monitoring. Removal will be calculated using the following equation (Brown et al., 1998):

$$\text{Removal} = 100\% \times \left(1 - \frac{C_f}{C_o}\right)$$

Where C_f = dissolved contaminant concentration at a quarterly sampling event during operation of the IAS system

C_o = dissolved contaminant concentration at start of operation of the IAS system

A low level of economic and technical efficiency is reached when the average removal of contaminants levels off to a point where quarterly removal percentages become relatively constant. At this point, system operation will stop, and quarterly monitoring for natural attenuation will be initiated. Before the IAS/SVE system operation is terminated, the removal percentages should be relatively constant for at least 2 consecutive quarters.

After the IAS system is shut down, sample concentrations will continue to be evaluated on a quarterly basis to assess the permanence of the contaminant removal. Permanent removal will be calculated based on the following equation (Brown et al., 1998):

$$\text{Permanent Removal} = 100\% \times \left(1 - \frac{C_r}{C_o}\right)$$

Where C_r = dissolved contaminant concentration at a quarterly sampling event after shutdown of the IAS system

It is proposed that the overall system performance be evaluated based on permanent removal percentages. Similarly, rebound of contaminant concentrations will be used to evaluate system performance. Rebound will be calculated based on the following equation (Brown et al., 1998):

$$\text{Rebound} = \frac{\log\left(C_r / C_f\right)}{\log\left(C_o / C_f\right)}$$

A rebound value of less than 0.2 reflects permanent reduction, and a value greater than 0.5 reflects a substantial rebound (Brown et al., 1998). Therefore, if the calculated rebound value stays below 0.2, the system will remain on standby. If the rebound value is greater than 0.2 and less than 0.5, the Navy will evaluate the need to reinitiate system operation. Otherwise, if the rebound value is greater than 0.5, system operation will be reinitiated.

Soil Vapor Extraction. The air discharge permit requires the SVE system to be operated whenever the IAS system is operating. However, in order to accommodate reductions in the soil vapor concentrations with time, the air discharge permit was developed using a life-cycle design approach. The air discharge permit has been approved for 4 modes of off-gas treatment: thermal oxidation, catalytic oxidation, activated carbon, and direct discharge. The approved concentration range for each process is provided in Table 3.

TABLE 3. Off-Gas Treatment

Process	Inlet VOC Concentration Range (ppmv)	Allowable VOC Discharge Concentration (ppmv)
Thermal Oxidation	>2,000	350
Catalytic Oxidation	>100 and <2,000	100
GAC	>35 and <100	35
Direct Discharge	<35	35

CONCLUSIONS

IAS is a proven technology for remediation of VOC contaminated groundwater. It is important to develop strategies for design and monitoring to provide for the most cost-effective system. Equally important is the development of an appropriate site closure strategy. Based on a study of 28 IAS systems by Brown et al. (1998), 82% of the systems had average permanent removal of contaminants greater than 80%, while 43% had greater than 95%. Although most IAS systems achieve significant contaminant removal, it is reasonable to assume that relatively strict cleanup goals, such as MCLs, may not be met using an IAS system. For example, at Site 43286, observed benzene concentrations were as high as 2,000 µg/L. Assuming our system is relatively successful and 95% removal is achieved, residual concentrations of benzene are on the order of 100 µg/L. This value is 100 times greater than our cleanup goal, the MCL. Therefore, it is necessary to develop a site closure strategy that takes into account the limitations of an IAS system and does not require continued operation after the majority of the contamination has been removed. The strategy presented in this paper suggests the use of RNA to meet final cleanup goals after IAS system operation becomes technically and economically inefficient.

REFERENCES

Brown, R.A., D.H. Bass, and W. Clayton. 1998. "An Analysis of Air Sparging for Chlorinated Solvent Sites." Proceedings of First International Conference on Remediation of Chlorinated and Recalcitrant Compounds.Battelle Press, Columbus, OH.

OHM/M&E. 1998. *Final Corrective Action Plan, 43 Area MWR Gas Station, Marine Corps Base Camp Pendleton, CA.* Contract No. N47408-92-D-3042. May 1.

Leeson, A., P.C. Johnson, R.L. Johnson, R.E. Hinchee, and D.B. McWhorter. 1998. Air Sparging Design Paradigm. Draft version. August 27.

MODELING OF OXYGEN MASS TRANSFER DURING BIOSPARGING

Lisa E. Stright (Michigan Technological University, Houghton, Michigan)
John S. Gierke (Michigan Technological University, Houghton, Michigan)

ABSTRACT: Predictive models of biosparging may need to account for mass transfer limitations of oxygen dissolution and diffusion or else biodegradation and oxygen delivery rates could be overestimated. To evaluate the importance of coupling oxygen and contaminant mass transfer limitations for biosparging we developed a deterministic laboratory-scale model. The model was used to ascertain the relative amounts of volatilization and biodegradation as a function of contaminant properties and sparging conditions. Results show that as the channel width increases, the contaminant is increasingly limited by diffusion toward the channel and therefore is more susceptible to biodegradation. Consequently, oxygen diffuses into the interchannel region faster than the contaminant diffuses out toward the channel. The degradation rate in the interchannel is controlled by oxygen diffusion.

INTRODUCTION

In-situ air sparging (IAS) is an innovative remediation technique which has been employed in field settings despite the fact that quantitative performance data is lacking. The rather sparse data that exist suggest that mass transfer limitations, which are currently ignored in selection and design of IAS, significantly reduce the effectiveness of IAS to volatilize contamination. Nevertheless, proponents of IAS rationalize that even if volatilization rates are less than ideal, remediation will proceed because sparge air introduces oxygen that will encourage and increase biodegradation rates of many organic contaminants, commonly referred to as "biosparging". However, if volatilization rates of contaminants are mass transfer limited, then so, too, will be the dissolution and diffusion of oxygen.

Quantifying the transport mechanisms involved in a contaminated system is complicated by the non-equilibriums effects. Johnson (1998) developed an analytical model to understand air sparge mechanisms under idealized conditions. However, Johnson did not investigate the rate of oxygen diffusion into the contaminated zone. Rabideau and Blayden (1998) developed an analytical model that treats IAS as a completely mixed system with advection, volatilization and first-order decay. Rabideau and Blayden use a lumped rate coefficient, which incorporates volatilization, biotransformation, and advection as a "sparge constant". Rafai and Bedient (1990) developed the numerical model BIOPLUME III. BIOPLUME III incorporates advection, dispersion and reaction to simultaneously track contaminants and oxygen in a groundwater system. The system reaction is tracked with Monod kinetics. However, this model cannot be used for systems where non-equilibrium transport processes are important, as in IAS.

In this study a deterministic model was created to model volatile chemical transport in sparge air channels and contaminant and oxygen transport in the region between air channels (interchannel). This deterministic model was used to quantify contaminant removal rates and mass transfer limitations of oxygen during biosparging.

MODEL CONCEPTUAL PICTURE

O_2 Mass Transfer (O_2MT) models a single, representative air channel in a biosparging process where contamination exists in gaseous, dissolved and sorbed phases (Figure 1). This approach assumes that the heterogeneous effects of the sparge system can be represented by an "average" air channel. The contaminant transport within the air channel is described by advection (retarded by linear partitioning) and dispersion. The gas flow in the channel is assumed to be at steady state. The interchannel portion of the model accounts for diffusion of oxygen and contaminant and linear sorption of contaminant. The interchannel water is assumed to be immobile. A first-order mass transfer term links the air channel and the interchannel for both the oxygen and contaminant. Finally, the degradation of a contaminant and the depletion of oxygen are described using Monod kinetics. The concentration of actively degrading microorganisms is assumed to be constant.

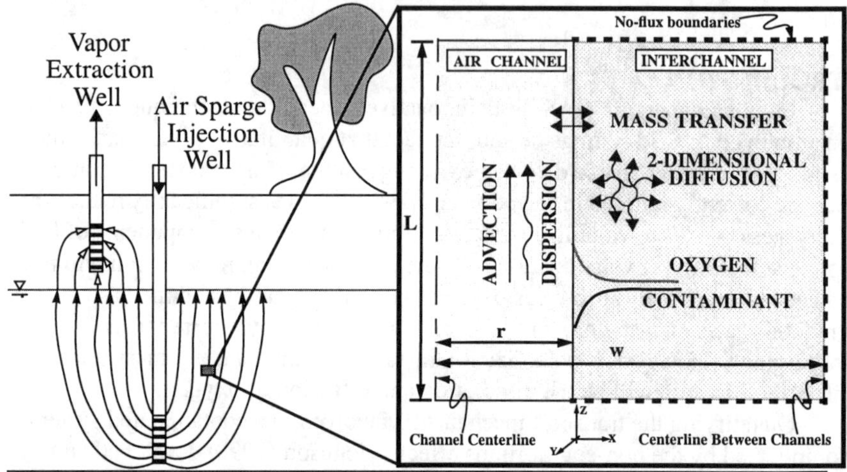

FIGURE 1: Conceptual picture of oxygen and contaminant mass transfer model (O_2MT) with a single, representative air channel and interchannel

The terms on the left side of mass balance equations 1-3 represent the transport mechanisms depicted in Figure 1 (gas dispersion, gas advection, and air-water mass transfer in equation 1 and aqueous diffusion in equations 2 and 3). The terms on the right side of the mass balance equations account for accumulation and biodegradation. Due to the relatively rapid movement of the sparge air, the oxygen concentration in the air channel is assumed to be at atmospheric levels instantaneously and continuously. Therefore, no mass balance equation is required for oxygen in the air channel. The initial condition for the contaminant is that it is uniformly distributed throughout, and the aqueous phase is assumed to be devoid of oxygen initially. No-flux boundary conditions are employed around the outside boundaries of the interchannel. Dankwerts-like boundary conditions are used for the air channel.

Contaminant in air channel. dispersion - advection - mass transfer = accumulation + rxn

$$D_L \frac{\partial^2 C_g(z,t)}{\partial z^2} - v_z \frac{\partial C_g(z,t)}{\partial z} - \frac{K_g a}{\theta_g}[C_g(z,t) - HC_w(x=r,z,t)] = R_{d,m} \frac{\partial C_g(z,t)}{\partial t} + \frac{(\mu_{max} M_t)}{\theta_g}\left(\frac{C_w(z,t)}{K_c + C_w(z,t)}\right)\left(\frac{P_o}{K_H K_o + P_o}\right) \quad \textbf{(1)}$$

Contaminant in interchannel. $\text{diffusion}_x + \text{diffusion}_y$ = accumulation + rxn

$$D_{x,c} \frac{\partial^2 C_w(x,z,t)}{\partial x^2} + D_{z,c} \frac{\partial^2 C_w(x,z,t)}{\partial z^2} = R_{d,im} \frac{\partial C_w(x,z,t)}{\partial t} + \frac{\mu_{max} M_t}{\theta_w}\left(\frac{C_w(x,z,t)}{K_c + C_w(x,z,t)}\right)\left(\frac{C_o(x,z,t)}{K_o + C_o(x,z,t)}\right) \quad \textbf{(2)}$$

Oxygen in internal channel. $\text{diffusion}_x + \text{diffusion}_y$ = accumulation + rxn

$$D_{x,o} \frac{\partial^2 C_o(x,z,t)}{\partial x^2} + D_{z,o} \frac{\partial^2 C_o(x,z,t)}{\partial x^2} = \frac{\partial C_o(x,z,t)}{\partial t} + \frac{\mu_{max} M_t F}{\theta_w}\left(\frac{C_w(x,z,t)}{K_c + C_w(x,z,t)}\right)\left(\frac{C_o(x,z,t)}{K_o + C_o(x,z,t)}\right) \quad \textbf{(3)}$$

Where C_g = contaminant vapor concentration in gas channel (mg/L)
C_o = oxygen aqueous concentration in interchannel (mg/L)
C_w = contaminant aqueous concentration in interchannel (mg/L)
$D_{i,c}$ = effective contaminant aqueous diffusion coefficient, i=x,z (cm^2/s)
$D_{i,o}$ = effective oxygen aqueous diffusion coefficient, i=x,z (cm^2/s)
D_L = gas dispersion coefficient (cm^2/s)
F = stoichiometric ratio of O_2 moles required to degrade a mole of contaminant [1 mole of C_XH_{2Y} + F(=X+Y/2)O_2 → XCO_2 + YH_2O]
H = contaminant Henry's partition coefficient (dimensionless)
K_c = contaminant half-saturation constant (mg/L)
$K_g a_j$ = overall contaminant gas-water mass transfer coefficient (1/s), where subscript j = c(contaminant) and o(oxygen)
K_H = Henry's constant for oxygen (mg/L-atm)
K_O = oxygen half-saturation constant (mg/L)
L = distance air travels from sparge point to the groundwater table (cm)
M_t = average biomass concentration (mg/L)
r = half-width of air channel (cm)
$R_{d,im}$ = contaminant retardation factor for interchannel
$R_{d,m}$ = contaminant retardation factor for air channel (includes partitioning into residual water and sorption within air channel)

v_z = gas velocity (cm^2/s)
w = half-width between channel centerlines (cm)
μ_{max} = maximum contaminant utilization rate per unit mass of microorganism (1/s)
θ_g = volumetric gas content in air channel
θ_w = volumetric water content in interchannel

NUMERICAL SOLUTION VERIFICATION

An analytical solution does not exist for the complex conceptual picture described herein. Therefore a numerical approximation method (Galerkin Finite Elements) was employed. A series of tests were performed to insure the accuracy of the approximations by comparing simulations to analytical solutions for contrived idealized conditions such as advective-dispersive-equilibrium transport and diffusive-only transport (Stright and Gierke, 1998).

MODEL PARAMETERS

Model parameters for this study are shown in Table 1. Unknown parameters, such as mass transfer rate, were found by calibrating the transport model (excluding biodegradation) to column studies of air-sparging performed by Wolff and Hein (unpublished data, 1998). These parameters were used as inputs for the O_2MT model such that a sensitivity analysis could be performed by changing channel geometry

RESULTS AND DISCUSSION

To illustrate the simulations on the same graph it is convenient to normalize the time based on the initial mass present. A throughput scale was used to normalize time. A throughput equal to 1 corresponds to the time required to volatilize all of the contamination assuming equilibrium and no dispersion (i.e., ideal removal conditions). The dimensionless concentration scale is the effluent gas concentration normalized by the initial gas concentration in equilibrium with the water.

For the case where the air channels would abut each other (r=w=0.2 cm), the removal is essentially an equilibrium transport condition for both the contaminant and the oxygen, so nearly all of the contamination is removed by volatilization and only a few percent of the initial contaminant mass was biodegraded (Figure 2).

When the air channels are spaced only 4 centimeters apart (w=2), the removal by volatilization is controlled by the contaminant's aqueous diffusion rate in the interchannel (Figure 2). If mass transfer limitations for oxygen are ignored, the fraction biodegraded is more than 45% (Figure 2), which is consistent with results published by Johnston et al. (1998) for a field sparging study of BTEX contamination where they surmised that almost 40% was biodegraded. The model predicts, however, that when oxygen transport within the interchannel is limited by its effective molecular diffusion rate, at most only 1% is biodegraded (Figure 2). This effect is due to the biodegradation reaction kinetics being slower for lower oxygen concentrations (equations 1-3), which occur because of the non-equilibrium conditions.

TABLE 1: Input oxygen properties for O_2MT (T=20°C)

Input Parameter (units)	Volatile Contaminant	Oxygen
Chemical Name	toluene	oxygen
Molecular Formula	C_7H_8	O_2
$C_{w,initial}$ (mg/L)[a]	10,000	--
D_L (cm^2/s)	0.08[b]	0.08[b]
D_x, D_z (cm^2/s)	10^{-6} [b]	10^{-5}[c]
F (mole O_2 / moles HC)	11	--
H [d]	0.23	--
K_H (L-atm/mg)[e]	--	0.024
K_ga (1/s)[a]	0.7	0.7
K_c (mg/L)[f]	0.37	--
K_o(mg/L)[a]	--	0.1
L (cm)	30	30
M_t (mg/L)[a]	1.0	--
P_{O2} (atm)[a]	--	0.205
$R_{d,m}$ [b]	4.3	--
$R_{d,im}$	1.8	1.0
v_z (cm/s)	0.06	0.06
μ_{max} (1/s)[f]	1.3 x 10^{-5}	--
θ_g	0.34	0.34
θ_w	0.34	0.34

[a]assumed

[b]value fit (Stright and Gierke, 1998) to data from column sparging of trichloroethlene in uniform sand (A. Wolfe and G. Hein, Michigan Technological University, unpublished data, June 1999). Molecular diffusion from Hayduk and Laudie (1974) correlation.

[c]value fit (Stright and Gierke, 1998) to data from column sparging of trichloroethlene in uniform sand (A. Wolfe and G. Hein, Michigan Technological University, unpublished data, June 1999) and adjusted for differences in molecular diffusion coefficients of oxygen and nonpolar organics in aqueous solution. Molecular diffusion from Holmen and Liss (1984).

[d]Ashworth et al. (1988)

[e]Montgomery Engineering (1988)

[f]Kelly et al. (1996)

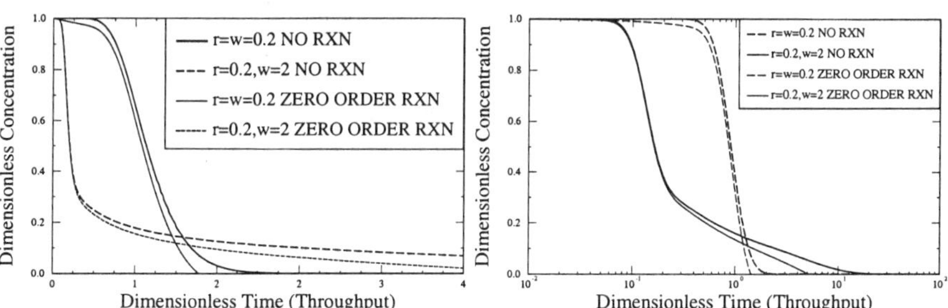

FIGURE 2: Volatilization and degradation, interchannel widths varying

Additionally model simulations have confirmed the general conceptualization that slower gas velocities yields a higher fraction of contamination that is biodegraded (Johnson, 1998). This is generally true because the volatilization rate decreases, however, since the oxygen delivery rate in sparge air is in such excess that a slower gas velocity has no impact on oxygen transport between air channels. However, our mass transfer model has shown that diffusion distance, represented here by a channel spacing, has a large impact on the volatilization rate and the effectiveness of the oxygen delivery. Therefore, a better understanding of the contaminant mass transfer mechanisms and the oxygen mass transfer mechanisms, as well as the biochemical reactions, that occur between air channels and how air channels are distributed is needed to develop appropriate predictive tools.

REFERENCES

Ashworth, R. A., G. B. Howe, M. E. Mullins, and T.N. Rogers. 1988. "Air-Water Partitioning Coefficients of organics in Dilute Aqueous Solutions," *Journal of Hazardous Materials. 18:* 25-36.

Hayduk, W and H. Laudie. 1974. "Prediction of Diffusion Coefficients for Non-electrolytes in Dilute Aqueous Solutions." *Journal of AIChE. 28:* 611.

Holmen, K. and P. Liss. 1984. "Models for Air-Water Gas Transfer: An Experimental Investigation." *Tellus. 36B:* 92-100.

Johnston, C. D., J. L. Rayner, B. M. Patterson and G. B. Davis. 1998. "Volatilization and biodegradation during air sparging of a petroleum hydrocarbon-contaminated sand aquifer." *Groundwater Quality: Remediation and Protection. 250.* 125-131.

Johnson, P. C. 1998. "Assessment of the Contributions of Volatilization and Biodegradation to In-Situ Air Sparging Performance." *Environmental Science and Technology. 32*(2)*:* 276-281.

Johnstone, T. L. 1996. *Modeling Solute Transport Under Fingered Flow Conditions: A Two Dimensional Approach.* MS Thesis, Michigan Technological University.

Kelly, W. R. 1996. "Kinetics of BTX biodegradation and mineralization in batch and column systems." *Journal of Contaminant Hydrology. 23*(1-2): 113-132.

Montgomery, J. M. 1985. *Water Treatment Principles and Design,* John Wiley and Sons. NY.

Rabideau, A. J. and J. M. Blayden. 1998. "Analytical Model for Contaminant Mass Removal by Air Sparging." *Ground Water Monitoring & Remediation. 181*(4)*:* 120-130.

Rifai, H. S. and P. B. Bedient. 1990. "Comparison of Biodegradation Kinetics With an Instantaneous Reaction Model for Groundwater." *Water Resources Research. 26*(4)*:* 637-645.

Stright, L. E. and J. S. Gierke. 1998. "Using Mathematical Moments to Quantify Mass Transfer Limitations in Air Sparging." *EOS, Transactions, American Geophysical Union. 70*(45): H31B-17, F334.

PREDICTION AND FIELD TESTING RADIUS OF INFLUENCE FOR BIOSPARGING

Ir. G.J. Mulder, ***Drs. C.C.D.F. van Ree***, (Delft Geotechnics, Delft, The Netherlands)
Dr. ir. C. Vreeken (BMS BodemMilieuSystemen, Delft, The Netherlands)

ABSTRACT: The "radius of influence" of a biosparging well is studied in a field scale project for intermittent sparging. Field tests included measuring the soil resistance and the development of oxygen concentrations. A conceptual model of air channeling and dead end pores is based on the field test results. Fast generation of air channels implies a heterogeneous modelling approach. The multi-phase flow model TOUGH2 is suitable for this approach because it can handle fractures.

Modelling results and measurements correspond with the increase in air saturation. Small permeability ratios between soil sediments are seen to be important. Simulation of the developing oxygen concentrations included convection, diffusion and oxygen exchange between the air phase and the water phase. The oxygen mass balance showed the importance of intermittent sparging for layered soil systems and the need to optimize the mode of operation.

INTRODUCTION

The capacity of a biosparging well is quantified by the "radius of influence" concept. The oxygen concentration inside this radius is high enough for biological degradation of the contaminants. In a pilot test the aeration radius and oxygen transfer are measured for intermittent sparging to provide more insight into the sparging mechanism. Field tests included the use of cone penetrometer equipment to measure the electrical resistance of the soil as a function of depth and at different times. A protocol for in situ respiration tests providing oxygen consumption rates is presented by "Ree et al. (this symposium)". Modelling is applied to simulate the measured aeration radius and to predict the oxygen response. The results facilitate the choice between continuous or intermittent sparging as the optimal mode of operation.

METHODOLOGY

Pilot Experiment. The test area is located at the EPON-site in Nijmegen, the Netherlands, where a strip of land measuring 150 m of length and 40 m wide on the bank of the river Waal is contaminated with light fuel oil. The soil profile consists of two sand layers separated by a clay layer. The clay layer is missing at some locations. The fluctuating level of the Waal caused a 6 m thick smear zone of oil in the sand layers. The upper sand layer is restored by bioventilation.

Eight sparging filters have been installed in the deepest sand formation. The clay layer acts as a barrier for vertical air transport. Air will spread out horizontally and escape through windows. The sparging process was studied by measuring both the electrical resistance and the oxygen concentration. Figure 1 shows the locations of the sparging wells and the oxygen monitoring points.

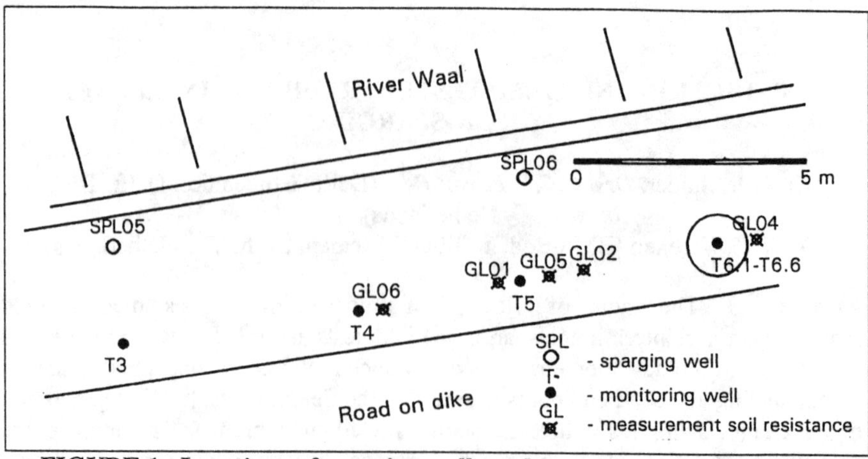

FIGURE 1. Locations of sparging wells and oxygen monitoring points.

Measuring the Electrical Resistance. An Archie-type relationship is applied to determine the relationship between electrical resistance and water saturation [Schima]:

$$S_w = \sqrt{R / R_0}$$

where: R specific resistance of the sparged soil
R_0 specific resistance of the saturated soil
S_w water saturation

The electrical resistance measurements were carried out by pushing an isolated cone-string fitted with electrodes into the soil using a CPT truck. One pair of electrodes is connected to a supply of electric current. The electrical resistance between two other pairs of electrodes is measured. The distance between the electrode pair A is 250 mm and between pair B is 30 mm (Figure 2).

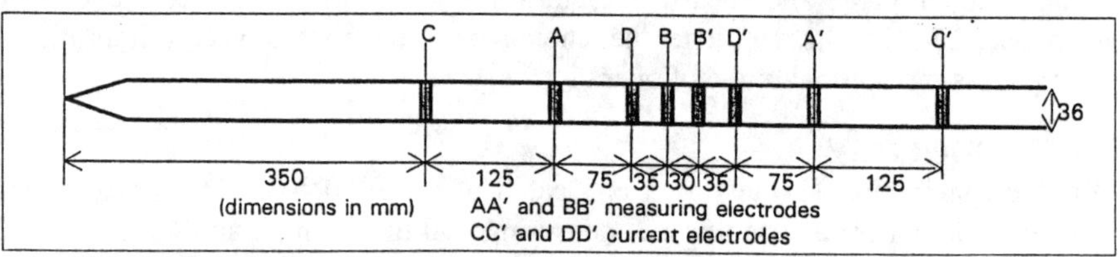

FIGURE 2. Cone-string with two pairs of electrodes.

Sparging Mechanism. Different stages can be distinguished for the aeration radius: an expansion stage, a collapse stage and a steady state. During the first stage, an air zone is formed in which the air pressure is high due to resistance against water displacement. When the pressure at the air/water interface exceeds the capillary pressure for air penetration, the air will displace the water resulting in a network of interconnected air-filled pores (see Figure 3A). At the collapse stage, the air zone reaches the water table and the air pressure reduces considerably. Water will partly replace the air so that only a limited part of the aerated zone contains an interconnected network of air-filled pores. The remaining part of the aerated zone consists of dead end pores (see Figure 3B). This mechanism has been proposed recently by "Clayton, 1998".

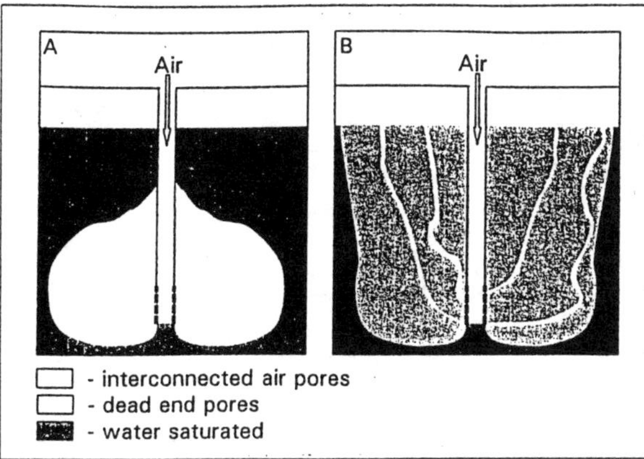

FIGURE 3. Air channeling and dead end pores.

For the layered soil system at the EPON site, the main transport mechanisms of air/oxygen due to air sparging are illustrated in Figure 4. These are:

- convection in air channels (soil air phase)
- diffusion in dead end pores (soil air phase)
- oxygen exchange between air phase and water phase
- convection due to mixing by intermittent sparging (soil water phase).

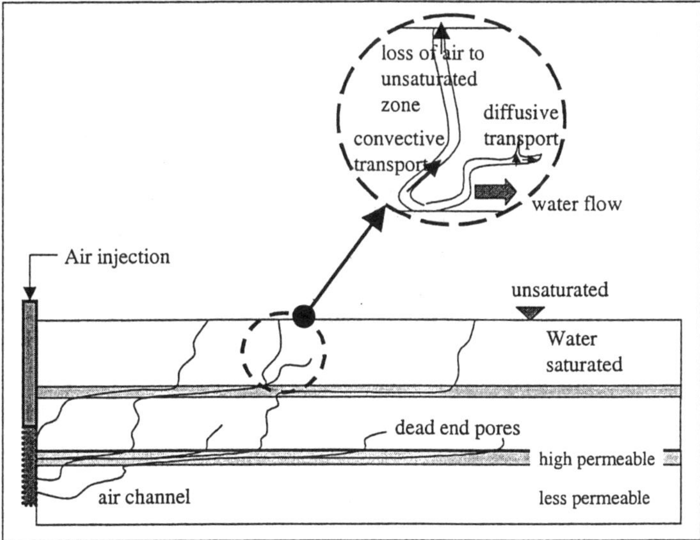

FIGURE 4. Transport mechanisms of air/oxygen.

RESULTS AND DISCUSSION

Measuring the soil resistance. At location GL02 (see Figure 1), the resistance of the soil is measured during air injection in SP05. Injection starts 20 minutes before the first measurements are taken and continues during the test. In Figure 5, the results are compared with the initial soil resistance (GL01). The results of electrode pair A at this location vary less than B, partly due to averaging of the soil resistance over a larger volume of soil.

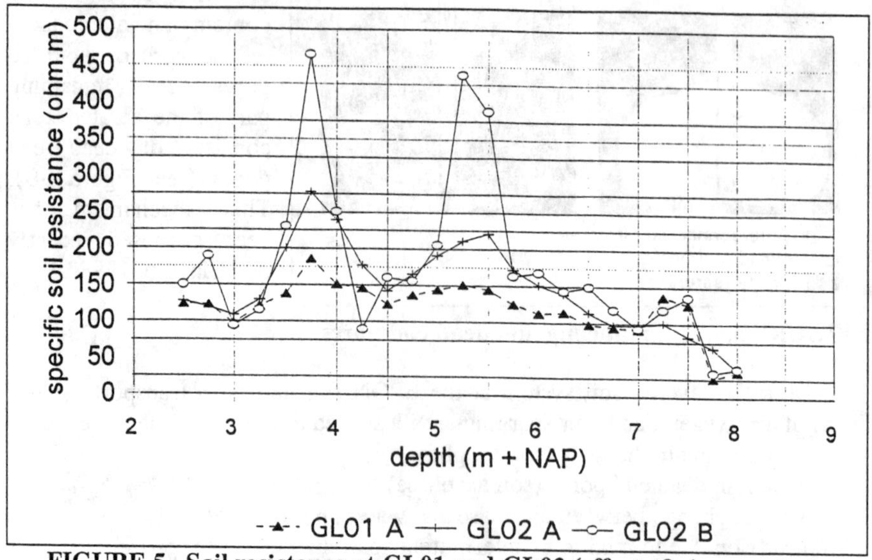

FIGURE 5. Soil resistance at GL01 and GL02 (effect of air sparging).

From the comparison of GL01A with GL02A and GL02B, it can be concluded that all investigated soil layers are aerated. From GL01A and GL02A, a mean reduction in water saturation of 11% is calculated. From the differences between GL02A and GL02B, it can be concluded that aeration is not homogeneous. Air saturations measured at a scale of 30 mm scatter more, but the qualitative agreement with measurements at a scale of 250 mm is mostly good. This contradicts with the idea of a few air channels in a saturated soil. At NAP + 3.75 m and NAP + 5.25 m (a gravel layer), the soil resistance of GL02B is 2-3 times larger than in the initial case. According to the Archie-type equation, water saturation at these depths is reduced to approximately 30-40%.

Simulation of Multi-phase Flow. Rapidly developing air saturations over large distances can only be understood by means of a fast generation of air channels, consisting of a limited volume of interconnected pores. The air will not penetrate so rapidly in dead end pores. This implies a heterogeneous modelling approach which includes the occurrence of air channels during sparging combined with dead end pores in the soil system. Therefore TOUGH2 the modelling tool, capable of handling dual permeability, is used to investigate the effect of air channeling and oxygen transfer on the radius of influence of a sparging well.

In the dual permeability model, both the matrix and the fractures are continua in which global flow occurs (see Figure 6). The matrix and fractures interact locally by means of "interporosity" flow. The highly permeable soil layers

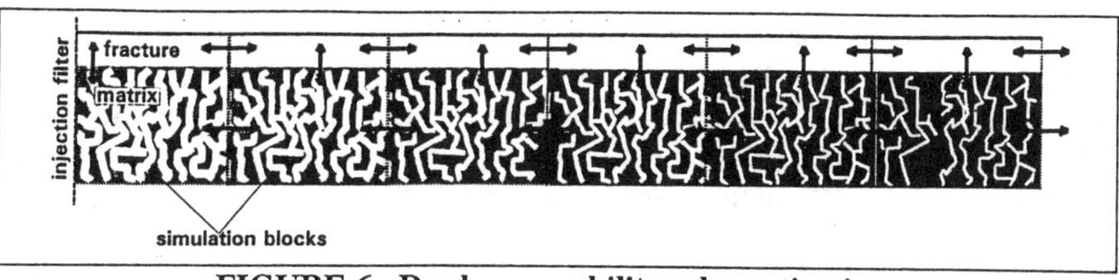

FIGURE 6. Dual permeability schematization.

in which air channeling develops much faster are modelled as fractures. The less permeable soil layers are modelled as the matrix, the main part of the sedimentary soil system. Parker relationships are used for the saturation/pressure/ relative permeabilities for these layer types. TOUGH2 simulations are carried out based on the soil properties of the test site and the sparging conditions as summarized in Table 1. Simulation of the oxygen concentrations include convection, diffusion and oxygen exchange (Henry's law).

Table 1. Input data for the TOUGH2 simulation.

Parameter	Value
filter depth sparging filters	2.5 – 4.5 m + NAP
outer diameter filters	40 mm
air injection rate	100 m^3/hr
injection regime	2 hr injection, 2.3 hr rest
permittivity (fractures + matrix)	4.65 x 10^{-11} m^2
permittivity ratio fractures/matrix	2.5
type of fracture	horizontal, 0.1 m apart
volume ratio fractures/matrix	0.05
ratio hor./vert. perm. fractures	5000
first order oxygen consumption rate	0.342 hr^{-1}

The simulated fast increase in air saturations corresponds with measurements for a small permittivity ratio of 2.5. The simulated less rapid increase in oxygen concentrations of 7.6 mg/l in water (14.5% in air) corresponds with the upper range of measured values at a distance of 5.4 m from the injection filter. For optimization purposes, intermittent sparging is simulated by two sparging cycles separated by a 2.3 hour interval. The effect of trapped air is taken into account (hysteresis) for the period between cycles. Figure 7 shows the oxygen mass balance for an injected amount of 52 kg oxygen (in two hours).

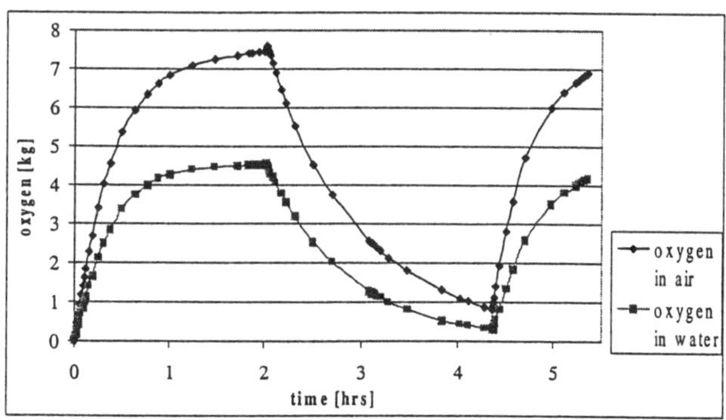

FIGURE 7. Oxygen mass balance, two sparging cycles.

The gaseous phase contains 7.5 kg and 4.5 kg oxygen dissolved in the water phase, implying a large loss to the unsaturated zone. During the 2.3 hour interval, most of the oxygen will be consumed due to the measured first order consumption. The simulations have been adjusted for both the increase in air saturation and the developing oxygen concentrations.

Continuous or intermittent sparging. Loss of air to the unsaturated zone in a layered soil system will be delayed depending on the vertical resistance to air flow. In such a system, the aeration radius and oxygen consumption determine whether intermittent air sparging or continuous air sparging will be the optimal method. Both the duration of the injection and the stand-still time need to be optimized to determine the most effective air sparging system. In a homogeneous (non-layered) soil, continuous air sparging will result in rapid loss of air to the unsaturated zone once the air has passed the freatic water level. Intermittent air sparging with a high frequency can therefore be an optimal operation mode.

The modelling results from the EPON site reflect an almost stationary aeration radius within 20 minutes, where oxygen concentrations are developed within a radius of 4 m. An injection time of 20 minutes and a frequency of two hours is the optimal mode of operation. Restrictions in oxygen availability mean that the outer radius will need more time for cleaning.

CONCLUSIONS

Electrical resistance measurements with cone penetrometer equipment can be applied to determine soil aeration by sparging. The measurements imply that air sparging provides aeration at a few centimetres scale, confirming the concept of air channeling and dead end pores.

The modelling approach, including the effect of air channeling and dead end pores, can simulate aeration and oxygen mass transfer near a sparging well. From the simulations, it is concluded that small contrasts in permeability (sedimentary layering) determine the aeration rate of individual soil layers. It is also concluded that intermittent sparging in layered soil systems will be the optimal operation mode for oxygen transfer. However the average oxygen concentration will be lower than for continuous sparging.

ACKNOWLEDGMENTS

This research has been sponsored by the Dutch NOBIS research programme.

REFERENCES

Schima, S., D.J. LaBrecque and P.D. Lundegard, "Using resistivity tomography to monitor air sparging." *Proc. Symp. On Application of Geophysics to Engineering and Environmental Problems, Environ. Eng. Geophys. Soc., pp 757-774.*

Clayton, W.S. 1999 "The Effects of Pore Scale Dead-End Air Fingers on Relative Permeabilities for Air Sparging in Soils." *Water Resour. Res., (under review).*

Ree, C.C.D.F. van, X.I. Kolkman and E.C.L. Marnette, In Situ Respiration, a Protocol, *this symposium 1999.*

PROTOCOL FOR THE ASSESSMENT OF IN SITU RESPIRATION RATES

Drs. C.C.D.F. van Ree, Ir. X.I. Kolkman (Delft Geotechnics, Delft, The Netherlands)
Dr. ir. E.C.L. Marnette (Tauw b.v., Deventer, The Netherlands)

ABSTRACT: In a field-scale study,a protocol has been developed and tested for two in situ respiration tests: the Push-Pull (PP) test and the Start-Stop (SS) test. These provide insight into in situ biodegradation rates at contaminated sites. A comparison has been made, based on technical aspects and their applicability in different stages of a remediation (feasibility, design and monitoring). Six monitoring wells have been tested over a twelve month period.

Results show that both tests can be used in feasibility investigations, for design and dimensioning purposes and process monitoring. The PP test can also be used as a diagnostic tool for optimization purposes. The SS test can only be used when a sparging system (a pilot plant) is present.

INTRODUCTION

Biosparging is used for enhancing aerobic biodegradation of both volatile and non-volatile contaminants by aeration of the soil. The efficiency is determined by the increase in oxygen consumption rates due to air injection. To design an in situ air sparging (IAS) system, information on the in situ biodegradation rate and the distribution of air/oxygen in the soil (Radius Of Influence [ROI] of the injection filters) are essential.

Several tests are used to determine biological activity. This study focuses on the use and comparability of two in situ respiration tests. The aim is to develop a protocol for the selection and application of these tests.

Site Description. Field tests are carried out at the EPON site at Nijmegen in the Netherlands. The site is situated near the river Waal and is contaminated with light fuel oil. The soil profile is shown schematically in Figure 1.

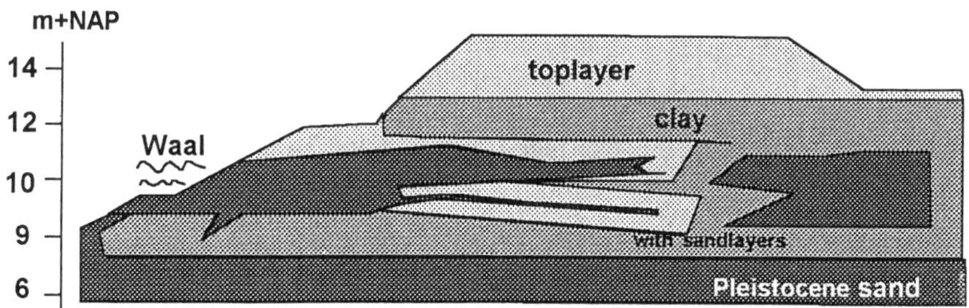

FIGURE 1. Soil profile at the EPON site.

Concentrations of up to 20,000 mg/kg dw are measured in the soil in smear zones of Light NAPL. Due to fluctuations in the river Waal's water level, a smear zone extends from approximately NAP + 11 m to NAP + 5.6 m Concentrations of several mg/l are measured in the groundwater. An IAS system and a bioventing system are installed to remediate the polluted soil.

MATERIALS AND METHODS

The in situ respiration test concept combines the original Push-Pull test of "Hinchee et al, 1992" and "Johnson et al 1995" for application in the saturated zone.

Push-Pull Test (PP Test). Oxygen is often depleted at the site prior to testing. The PP test is based on the principle that oxygenated water (and a tracer) is introduced into the soil ('push'), and samples are then periodically collected from the groundwater ('pull') to measure the decrease in oxygen concentration (respiration). Because of the introduction of well oxygenated water, the PP test gives an estimate of the *respiration rate at optimal aeration*. If there is no decrease in oxygen, biodegradation may be limited by nutrients. This can be investigated by adding nutrients to the injected mixture. If oxygen is again not consumed, biological activity may be substrate limited. In that case, there is either no contamination present at that sub area or the contamination is highly persistent. One test option is to add a C-source as substrate. Bromide tracer is used in all tests to determine the influence of dilution in groundwater by convection and dispersion.

The test is schematically depicted in Figure 2. Oxygenated water (50 l) is introduced into the soil by siphonation, the pump rate is 50 l per hour. Purging and sampling is performed by a peristaltic pump with a sampling rate of 100 ml per minute. Approximately 20% of the injection volume is withdrawn during a typical test.

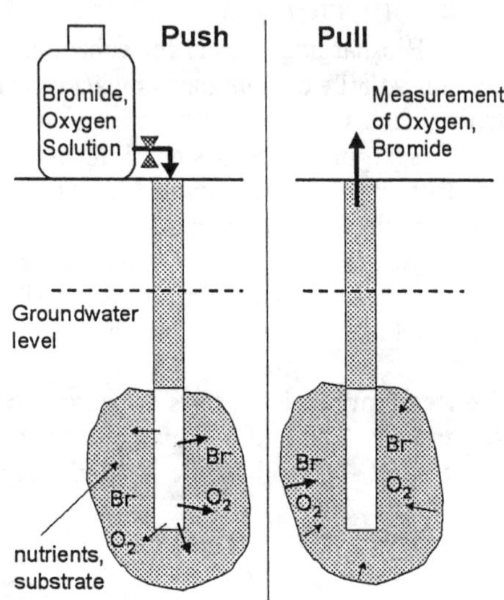

FIGURE 2. Push-Pull test.

The injected water contains 600 mg/l bromide. Samples are analyzed for oxygen in an online flow cell; and for nitrate, phosphate, bicarbonate and bromide using the anionograph (if possible on site). The rate of decrease of the oxygen concentration, corrected for dilution effects, allows the respiration rate to be estimated. Respiratory activity can be measured within two 2 hours of turning off the injection (push) for groundwater. Continuous extraction (pull) gives the most reliable maximum uptake rate.

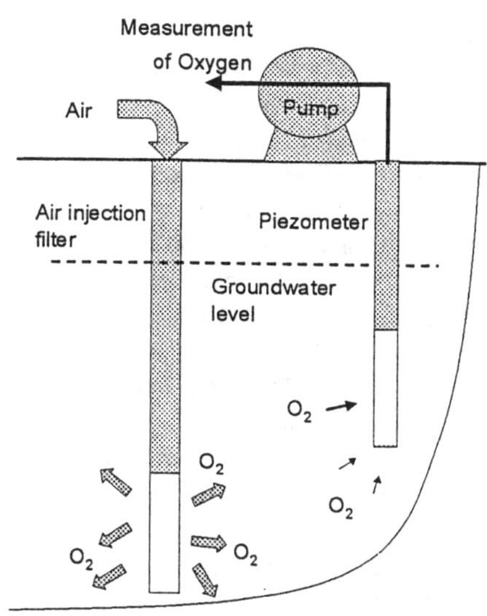

FIGURE 3. Start-Stop test.

Start-Stop Test (SS Test). The SS test is executed by oxygenating the groundwater by means of an existing IAS system. The test is schematically depicted in Figure 3. After the oxygen concentration has reached its maximum level, the system is turned off (Start-Stop) and groundwater samples are collected to measure the decrease in the oxygen concentration. Like the PP test, the rate of oxygen decrease is an estimate for in situ respiration rates.

Protocol. The applicability of PP and SS tests are related to different remediation phases (Table 1). The tests indicate whether biological degradation occurs and answers the question as to whether IAS will stimulate biological degradation at that location. The PP test gives an estimate of the respiration rate at optimal aeration, since well oxygenated water is added. The PP test can be carried out with or without an IAS system. The SS test gives an estimate of the actual respiration rate at the ambient oxygen concentration. An IAS system must be present (pilot test) to execute a SS test. With the decrease of the biodegradation coefficient over time due to a decrease in the biodegradability of the contaminants, in situ respiration tests can be used as a monitoring parameter. (e.g. at the end of remediation, to check whether there is enough biological activity left to ensure a significant bioremediation progress). Although both tests can be used for designing and dimensioning bioremediation plans, SS tests are preferable for monitoring purposes. PP tests are better suited for design and optimization purposes. The PP test can also be used as a diagnostic tool to investigate which factors limit biological degradation.

TABLE 1. Applicability PP and SS test.

Remediation phase		Push-Pull test	Start-Stop test
Feasibility	With pilot test (IAS present)	Yes	Yes
	Without pilot test (no IAS)	Yes	No
Design and dimensioning monitoring plan		Yes	Yes
Process monitoring		(No)	Yes
Optimization		Yes	No

When carrying out tests at different distances from the sparging filter, the tests can be considered as a tool for estimating the ROI and degradation rates within the ROI.

RESULTS AND DISCUSSION

In situ respiration tests have been performed in well screens situated 9 m below the surface level in the vicinity of a sparge well. There are two processes that may severely affect the determination of oxygen concentrations: diffusion and degassing. The effects of diffusion through the sampling tube have been studied. Residence time and tubing material are important, especially at low oxygen concentrations. Concentrations lower than 0.2 to 1 mg/l cannot be measured accurately under typical field conditions. Oxygen diffusion will not significantly affect the concentration measurements in this range, hence PE can be used as tubing material.

A peristaltic pump is used for sampling. Suction can result in degassing of the groundwater, which affects the outcome of the in situ respiration test. Investigation has resulted in a sampling protocol and the application of back pressure during sampling. A consistent recovery of 96% of the in situ value with a standard deviation within 10% of the average value was found. The back pressure set-up is shown in Figure 4 and consists of a column (height 10 cm, diameter 3 cm) filled with Wessem sand (grain size distribution 200-630 μm).

Interpretation of the SS test is more sensitive to measurement errors and the characteristics of the sparging system than the PP test, because the range of oxygen concentrations measured during the test are often smaller than for the PP test.

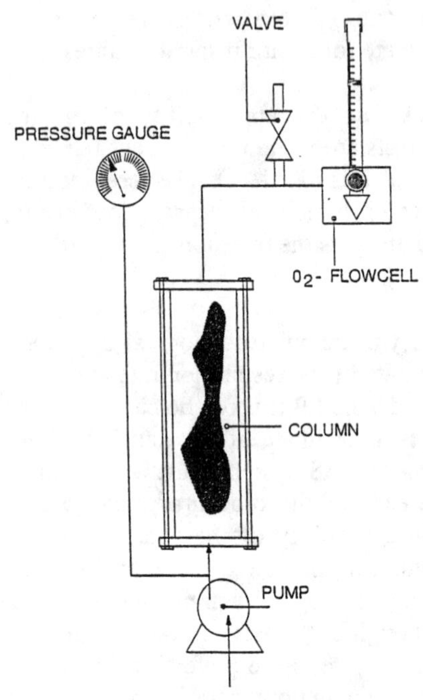

FIGURE 4. Back pressure set-up.

Calculation Procedure. The rate of oxygen decrease allows the degradation rate in the soil to be determined. Curve fitting is used for the field data to investigate whether the biodegradation is a zero order or a first order process. The optimal conditions in a PP test (oxygen is added in concentrations determined by the maximum solubility [12 mg/l]), means that a zero order degradation process is expected. For a SS test, limitations may occur due to the generally lower oxygen level (due to mass transfer limitations and oxygen consumption between the sparge well and the monitoring well). A first order degradation process is expected in this case. Field data indicate that this is true, with a first order degradation coefficient fitting the whole range of oxygen concentrations and correlation coefficients of 0.97. Table 2 contains the data for three series of tests

TABLE 2. Regression analysis test results (first order oxygen consumption).

Fil-ter	March 97 (PP-test) 1st order coeffi-cient	O2 start conc. (mg/l)	March 98 (PP-test) 1st order coeffi-cient	O2 start conc. (mg/l)	April 98 (PP-test) 1st order coeffi-cient	O2 start conc. (mg/l)	April 98 (PP- and SS-test) 1st order coeffi-cient	O2 start conc. (mg/l)
1	-0.07	9.67	-0.16	12.17	-0.34	10.75	impossible to fit	
2	-0.09	10.11	-0.07	12.98	-0.11	11.01	-0.08	10.74
3	-0.57*	9.37*	-0.51	10.00	-0.95	8.79	-0.52	6.88
4	-0.11	7.14	-0.13	10.69	-0.12	10.34	-0.10	10.15
5	-0.11	6.94	-0.16	9.72	-0.34	10.42	-0.13	8.59
6	-0.50*	7.93*	-0.41	9.26	-0.60	9.63	-0.48	8.75

* Regression analysis restricted to shorter period (higher values)

at six locations. The data from the SS tests fitted well in the data of the PP tests (last columns), carried out at the same locations. This indicates that both (different) methods used to estimate biodegradation are sound. A zero order fit through the PP test data does not combine well with SS test data. From further analysis, it is concluded that PP and SS tests can be correlated and fitted by a single first order degradation rate (see Figure 5).

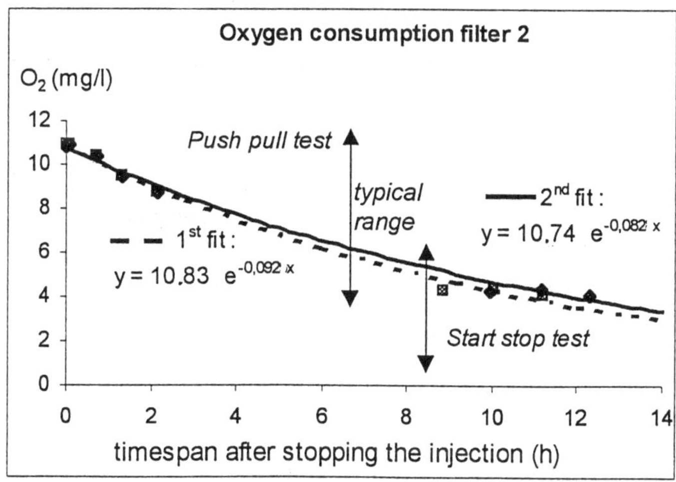

FIGURE 5. Combined results PP and SS test.

From the PP test data the timelag to reach the initial oxygen concentrations for the SS test is calculated. This timelag is used to correct the starting point of the SS test data, resulting in a first fit. This fit is used to improve on the initial estimate of the timelag, resulting in the second fit. The timelag results from the distance between the sparge well and the monitoring well and reflects amongst others the effects of oxygen transport phenomena.

Comparison of the PP test with the SS test gives the opportunity to assess the possible optimization ranges of the IAS system. This can be derived from the maximum oxygen consumption rates with optimal oxygen supply, and the actual oxygen consumption rates reached by the sparge-system at the location tested. Based on these differences in oxygen consumption rates, optimization of the IAS system is possible by:

- increasing the sparge density (by installing additional sparge wells)
- changing the sparge frequency (for intermittent sparging).

The optimization of oxygen transfer in an IAS system is addressed in an accompanying paper on the Radius Of Influence (Mulder et al., this symposium). The benefits of the additional capacity in terms of reduced remediation times and better aeration can thus be quantified.

CONCLUSIONS

In situ respiration tests provide data on oxygen consumption rates in the soil. Test results show high reproducibility in time. In situ respiration tests can be carried out using Push-Pull and Start-Stop tests. Oxygen consumption rates from both tests cover different oxygen concentration levels and can be described by a single first order consumption rate. Application of both tests in a remediation project provides added value, by helping determine the optimization range of the IAS system.

ACKNOWLEDGMENTS

This project has been sponsored by the Dutch Research Programme In Situ Bioremediation (NOBIS).

REFERENCES

Hinchee, R.E. and Ong, S.K., 1992. *A rapid in situ respiration test for measuring aerobic biodegradation rates of hydrocarbons in soil.* J. Air Waste Manage. Assoc. 42, 1305-1312.

Johnson, P.C., Johnson, R.L., and Kemblowski, M.W.,1995. *Diagnostic tools for the monitoring and optimization of in situ air sparging systems.* API Proposal 1-17.

Olie, J.J., 1997, *Technical report in situ respiration test, Final report, Phase I and II*, Delft Geotechnics CO-368790/189, p.p. 1-35.

BIOSPARGING UNDER A FUEL OIL PLUME TO EXPEDITE SITE CLOSURE

Raveendra Damera, P.E., Dev Murali, P.G., and
Rebecca Kinal (General Physics Corporation, Columbia, Maryland)
Raymond McDermott (Aberdeen Proving Ground, Maryland)

ABSTRACT: A ruptured fuel line released an estimated 46,000 gallons (172,500 L) of #2 fuel oil into the subsurface at a bulk fuel storage facility in APG. An extractive bioventing system and a free product recovery system augmented with water table depression has been operating at the site since July 1995. Although bioventing has been successful, free product recovery and bioventing of capillary zone were ineffective due to frequent interruptions of the pumping system resulting from high groundwater iron levels and geologic heterogeneity. A baseline respirometry test indicated the system is not effective in addressing the residual contamination. Pilot tests were conducted in hot spots to determine the feasibility of focused bioventing and biosparging. Air Force Center for Environmental Excellence (AFCEE) Bioventing test protocol was slightly modified to design the biosparging tests. The tests yielded encouraging results with biodegradation rates ranging from 1.1 to 2.5 mg/kg/day. Based on these test results, the existing system is currently being upgraded with biosparging to expedite site closure within 12-18 months.

INTRODUCTION

Building 345 site (Figure 1) is a bulk fuel storage facility in APG located in the northeastern portion of Maryland. Two 70,000-gallon (262,500-L) aboveground storage tanks (ASTs) were located at the site within an earthen dike. These ASTs (345-B and 345-C) were in operation since 1949 and always contained #2 fuel oil. A leak was suspected in the 345-C tank system when an extraordinarily large volume decrease (over 10,000 gallons) was recognized during a tank gauging event in 1993. A ruptured fuel line discovered during the initial excavation of the piping was believed to be the cause of the leak. Following the detection of the leak, the AST was taken out of service and dismantled. Subsequent excavation of all piping and demolition of tank revealed corrosion holes in the tank bottom plate and underground piping. The duration of the leaks and, therefore, the exact volume of release was uncertain. A further structural integrity evaluation indicated that the AST 345-B was intact. APG and General Physics (GP) Corporation performed 2 phases of site characterization, feasibility evaluation and pilot-scale tests, designed and installed the remediation system, and have been operating it since its activation in July 1995. The details of these activities and the results of the remediation are discussed in the following sections.

Site Characteristics. The site subsurface is primarily composed of three facies made up of silty sands, fine sands, and coarse to gravely sands. The site geology is complicated by the presence of silt-clay lenses in all three facies. These silt-clay lenses combine at some locations to form a semi-continuous fourth facies in the vicinity of the water table. The thickness of the silt-clay layer varies considerably across the site with the water table and/or capillary fringe located within the silt-clay layer at several locations. Groundwater at the site is located approximately 25 ft below ground surface (bgs) and flows in a southeasterly direction with a horizontal gradient of 0.001-0.003 ft/ft (0.001-0.003 m/m). Hydraulic conductivity of the water table aquifer was determined to be 30-40 ft/day (0.011-0.014 cm/sec).

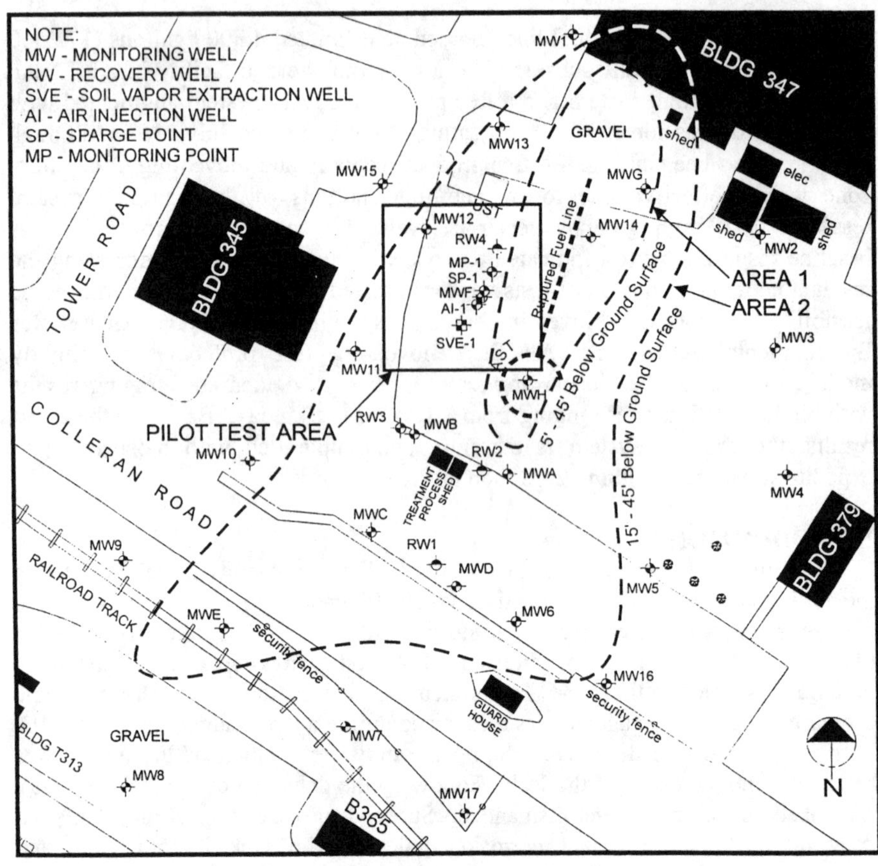

Figure 1. Site Plan, Extent of Spill and Test Area

Horizontal and vertical extent of the release is depicted in Figure 1. Areas 1 and 2 were delineated based on numerous data points and are indicative of the product migration pattern. Free product was detected in 10 monitoring wells (MW) with a maximum measured thickness of 6-10 ft (1.8-3.0 m). Elevated levels of petroleum hydrocarbons, exceeding the Maryland Department of the Environment (MDE) criteria, were detected at several locations with a maximum of 17,300 mg/kg. Majority of the contamination is in the free product and adsorbed phases. Based on the data, the mass of the spill was estimated to be approximately 325,000 lbs (147,400 kg) with approximately 4000 gal (15,000 L) present as free product.

Corrective Action. The existing remediation system consists of free product recovery and extractive bioventing with water table depression. Groundwater and free product are extracted from four recovery wells (RW-1 through RW-4) using a dual pumping system. Extracted groundwater is treated using granular activated carbon before it is discharged to the sanitary sewer. Water table depression through groundwater withdrawal was intended to increase the available vadose zone for bioventing and accelerate mobilization of free product by increasing the hydraulic gradient. Soil vapor is extracted from 11 of the existing monitoring wells and all four recovery wells to enhance aerobic biodegradation by supplying oxygen to the native bacteria. APG and GP have been operating the remediation system since its activation July 1995.

System Performance. To date a relatively small mass (7,000 lbs or 3175 kg) of hydrocarbons has been recovered through the physical processes of volatilization and free product recovery. Bioventing, on the other hand, proved to be a much more effective aspect of the existing system, with an estimated 211,000 lbs (95,690 kg) of hydrocarbons degraded. This quantity translates to 95% of the total hydrocarbons recovered by all processes (218,000 lbs or 98,870 kg) and 65% of the estimated spill.

Free product recovery has not achieved the anticipated results. To date the remediation system has recovered approximately 700 gal (2625 L) of free product, which is approximately 18% of the 4000 gal (15,000 L) estimated to be present before the remediation. This is attributed partly to geology and partly to the decreased pumping rates and fluctuating water levels due to iron fouling problems. The water table is generally located in the lower permeability silt-clay layer. Based on this finding and the understanding of the LNAPL behavior, it is concluded that migration of the free product to the recovery wells may have been retarded by the lower permeability material located at the water table. Although the pumping system was designed for 35-40 gpm (131-150 L/min), the pumping rate seldom exceeded 20 gpm (75 L/min) due to the iron fouling of pumps and well screens and often resulted in the pumping system shutdowns.

Soil sampling revealed that certain lower permeability zones within the vadose zone have not been addressed by bioventing probably due to preferential channeling of air flow through more permeable zones. Vacuum application on the existing wells also decreased the available vadose zone leading to ineffective remediation of the capillary fringe. In addition, water level fluctuations have increased the smear zone. It is estimated that over 75,000 lbs of hydrocarbons (34,000 kg) are still present in the capillary, smear and lower permeability vadose zones. Several options have been investigated to address these areas. Two options were found to be feasible: focused bioventing of the lower permeability vadose zone soils by injecting air into these zones; and biosparging in the capillary/smear zones and areas with significant free product.

BIOVENTING/BIOSPARGING PILOT TESTS

Prior testing at the site indicated air injection may be effective at stimulating biodegradation in the vadose and capillary zones, areas that are not addressed by the existing bioventing system. Previously collected soil samples contained bacterial populations ranging from 1.93 X 10^3 to 4.77 X 10^8 CFU/g. Groundwater quality parameters such as pH, temperature, dissolved oxygen (DO), oxidation/reduction potential (ORP) monitored throughout the remediation period indicated favorable conditions for the growth of aerobic bacteria. Nutrient samples indicated 16.8-44.9, 229-324, and 276-454 mg/kg of ammonia, total kjeldahl nitrogen and phosphorous, respectively. A comparison of these nutrient levels with the available data and research compiled by the U.S. Environmental Protection Agency (USEPA, 1995) indicated nutrient addition at this site would not be required.

Given the above, *in-situ* respirometry (ISR) tests were conducted at the site to further investigate and quantify biodegradation through air injection into the vadose zone (bioventing) and the saturated zone (biosparging). A secondary objective of the tests was to determine the effectiveness of small diameter wells installed by direct-push methods. Air permeability tests were also conducted to determine the pneumatic permeability of the vadose zone soils and the air-sparging zone of influence.

Test Well Installation. Based on the soil vapor data and free product measurements collected during the system operation and maintenance (O&M) the vicinity of MW-F was chosen as the test area (Figure 1). Current measurements indicate 1.0-1.7 ft (0.3-0.5 m) of free product within MW-F. SVE-1 was previously installed with a 3-ft (1.0 m) screen located entirely within the lower permeability zone. Monitoring well, MW-F was installed during the site characterization phase with screen spanning across the vadose zone and saturated zone. Newly-installed wells include AI-1, SP-1 and MP-1. AI-1 is an air injection well screened similar to SVE-1; SP-1 is an air sparging well screen approximately 5-7 ft (1.5-2.1 m) below water table and entirely within the saturated zone; and MP-1 is a monitoring point screened from 15-30 ft (4.6-9.1 m) bgs in both vadose and saturated zones. The specific locations of these wells were so chosen that some of the existing monitoring wells could be used as monitoring points during the tests. All new wells were constructed of ½-inch (12.5 mm) diameter PVC material and installed using direct-push methods.

Test Procedure. The pilot tests were conducted during October 13-28, 1997 and included: a baseline test; air permeability and zone of influence (ZOI) tests; ISR test with air injected into vadose zone (SVE-1 and AI-1); ISR test with air injected into saturated zone (SP-1). The baseline test was performed by turning of the existing extractive bioventing system and monitoring soil gas in select wells for a period of 4 days (October 13-16, 1997). Vapor samples collected from these wells were analyzed for oxygen (O_2), carbon dioxide (CO_2) and %LEL/PPM. Tracer gas was not used during this stage of testing. Pneumatic air permeability and sparge ZOI were measured based on pressure measurements in the surrounding wells. The vadose zone ISR test was conducted following the procedure outlined in the AFCEE bioventing test protocol (Hinchee et al., 1992) with helium tracer and soil gas was monitored for at least five days.

Unlike bioventing, information on biosparging test procedures is rather sparse. Therefore, biosparging tests were designed by slightly modifying the bioventing test protocol. Air mixed with 1-3% helium was injected into SP-1 over a period of 24 hours. Following the aeration period, soil gas was monitored for five consecutive days from various monitoring wells located within the sparge ZOI based on the assumption that the sparge air eventually travels upward into the vadose zone. Soil gas data from the test was analyzed as described in the AFCEE protocol.

RESULTS AND DISCUSSION

Air permeability test data was analyzed using the Hyperventilate computer program and yielded permeability values ranging from15.5 to 16.8 darcy. This range of values is typical of medium sands. An extraordinarily high value of 500.9 darcy observed during injection at SVE-1 is indicative of preferential channeling of air flow through a higher permeability zone. Vacuum/pressure influences during vadose zone and saturated zone testing were observed in monitoring points 30 and 14 ft (9.1 and 4.3 m) away from the test wells. Baseline and pilot respirometry test results for two representative wells are discussed in the following paragraphs.

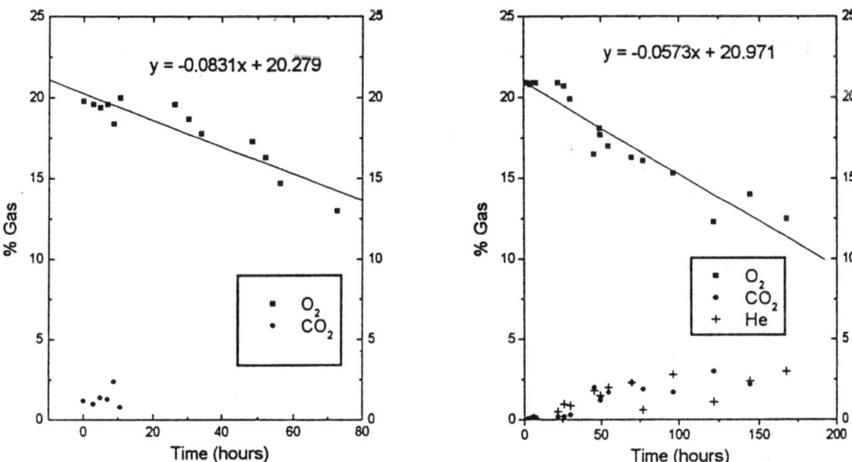

Figure 2. Pilot Test Results for SVE-1. (a) Baseline. (b) Injection Bioventing

Plots of percent gas (O_2, CO_2 and helium) against time for baseline and bioventing test performed at SVE-1 are presented in Figure 2(a) and 2(b), respectively. Equations of the lines of best fit determined by linear regression of the O_2 data are shown on the plots. The oxygen utilization rates (OUR) represented by the x-coefficient of the equation for the two cases are 0.083 and 0.057%/hr, respectively. Biodegradation rates corresponding to each OUR were calculated as described in Hinchee, et al. (1992). Given the short-duration nature of the test and the data scatter, the biodegradation rate at SVE-1 with air injection (1.10 mg/kg/day) is consistent with the baseline biodegradation rate with air extraction (1.6 mg/kg/day). This indicates that air can be effectively supplied to the contamination at SVE-1 through both air extraction and injection. This is most likely because both methods of air delivery are effective in the vadose zone, where SVE-1 is screened.

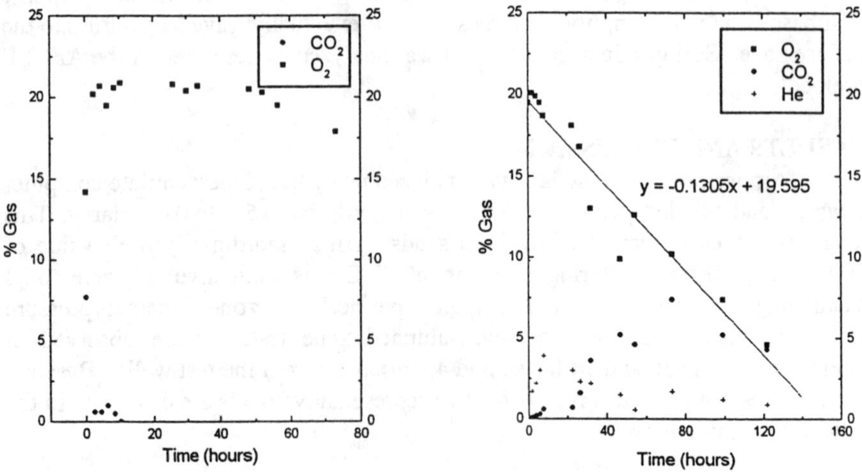

Figure 3. Pilot Test Results for MW-F. (a) Baseline. (b) Biosparging

Percent gas versus time plots for baseline and biosparging test performed at MW-F are presented in Figure 3(a) and 3(b). Baseline vapor monitoring did not show a significant decrease in O_2 levels within the first 80 hours indicating negligible oxygen utilization or bioactivity even though MW-F contained 1-1.7 ft (0.3-0.5 m) of free product. On the other hand, OUR during the biosparging test was 0.13%/hr corresponding to a biodegradation rate of 2.5 mg/kg/day.

The baseline test was performed by monitoring soil gas samples from MW-F after turning off the existing vapor extraction system to which MW-F is directly connected. Vacuum application in wells screened across the water table causes the water table to rise, thus increasing the saturated thickness of the aquifer at the extraction point and preventing air delivery to the capillary zone. As a result, extractive bioventing is ineffective at stimulating biodegradation in cases where the most contaminated zone is located within or just above the capillary fringe. This effect is clearly demonstrated by the baseline soil vapor monitoring results for MW-F. These results indicate that existing remediation system is addressing only a small portion of contaminants and that the majority of the contamination within the saturated/capillary zone is not receiving sufficient oxygen for biodegradation. Pilot test results indicated that oxygen can be more effectively delivered to the smear zone through air injection. When air was injected into the saturated zone at SP-1, measured OUR indicated that biodegradation was stimulated. Interpretation of OUR or biodegradation rates for biosparging alone is rather difficult due to the fact that the sparge air may result in some bioventing within the vadose zone as it moves upwards. Thus the biosparging rates invariably include some bioventing component and proper assumptions should be made in applying the test results to the design of full-scale remediation systems.

Figures 2(b) and 3(b) indicate that CO_2 concentrations increased during the testing providing additional evidence that biodegradation was stimulated by the air injection. The plots also indicate that helium was present in all of the soil gas samples collected, generally at concentrations ranging from 0.5-3.0%. Given that the helium was injected at a concentration of 1-3%, and that there is no decreasing trend in the soil gas helium concentrations, it can be assumed that significant dilution of the injected air did not occur, and that observed trends in O_2 and CO_2 concentrations can be attributed to bioactivity.

CONCLUSION

Biosparging tests have been successfully conducted using a slightly modified approach of the AFCEE bioventing protocol. Respirometry testing with biosparging under the #2 fuel oil plume yielded encouraging results. The small-diameter wells installed using the direct-push methods performed fairly well during the pilot testing. The tests clearly concluded that the existing remediation system is not effective in addressing the 75,000 lbs (34,000 kg) of residual contamination, which is primarily present as free product and within the smear and capillary zones. Upgrades to the existing system consist of 7 focused bioventing wells and 38 focused biosparge wells using small-diameter wells. The existing pumping system will be phased out. The system upgrades are expected to attain site closure in 12-18 months.

REFERENCES

Hinchee, R.E., Ong, S.K., Miller, R.N., Downey, D.C., and Frandt, R. 1992. *Test Plan and Protocol for a Field Treatability Test for Bioventing*. U.S. Air Force Center for Environmental Excellence (AFCEE), Brooks AFB, TX.

U.S. Environmental Protection Agency (USEPA). 1995. *Bioventing Principles and Practice, Vol. 1*. EPA/540/R-95/534a. Office of Research and Development, Washington, DC.

DIAGNOSTIC TOOLS FOR QUANTIFYING OXYGEN MASS TRANSFER DURING IN SITU AIR SPARGING

*Cristin L. Bruce**, *Illa L. Amerson***, Paul C. Johnson* and Richard L. Johnson**
* Arizona State University, Tempe, Arizona
** Oregon Graduate Institute, Beaverton, Oregon

ABSTRACT: In this study, two tracer-based diagnostic tools were used to determine rates of oxygen mass transfer during in situ air sparging (IAS). One approach involves a push-pull test of conservative and degradable tracer compounds, while the other measures the transfer of a volatile tracer from the injection air into groundwater. Results from these tests yield oxygen delivery rates for specific points and depths around the air injection system. These methods are unique in that oxygen transfer measurements can be made without pre-sparging data and in that they give a real-time measure of mass removal from biotransformation processes that is integrated over a local scale instead of a site-wide scale. Results from a test demonstration at the Port Hueneme National Test Site showed oxygen mass transfer rates to be in the range of 0-140 mg-oxygen/L-water/day for both tools. These results also suggest that 1) oxygen transfer may be occurring outside the zone of treatment defined by conventional air distribution measurements, 2) dissolved oxygen concentrations may not accurately reflect the oxygen transfer at a given point, and 3) the transfer of oxygen may continue well after air injection has been terminated.

INTRODUCTION

The performance of an IAS system is typically monitored by measuring dissolved oxygen and contaminant levels in groundwater monitoring wells, measuring aquifer pressure transducer responses, and measuring offgas concentrations from the soil vapor extraction system. Methods of measuring air saturation, such as capacitance or neutron probe measurements, or the use of electrical resistance tomagraphy, are less frequently applied. These measurements simply assess the distribution of air saturation in the aquifer, they do not quantify mass removal or degradation. Unfortunately, if an air channel happens to short circuit into a monitoring well, groundwater samples may not be representative of the conditions in the bulk fluid surrounding the well. Water pressure measurements can be instructive, giving a rough estimate of the amount of air entrapped in the aquifer. The only real measure of IAS mass removal comes from monitoring the contaminant concentrations in the effluent from a vapor extraction system. This gives an assessment of mass removal from volatilization integrated over the entire site. The objective of this study was to analyze the mass transfer of oxygen into groundwater, and to do it on a more local scale. This was achieved by the use of two tracer recovery tests, discussed below.

METHODS

Push-Pull Method Development. The multi-tracer push-pull test was developed and tested in a three dimensional physical model at Arizona State University. The physical model was 8 ft long x 4 ft wide x 4 ft deep. It was equipped with ¼ in stainless steel sampling points that terminated at depths ranging from 1 – 3 ft. below the soil surface. An air injection port was located along one side of the tank (Figure 1). The original tracer solution consisted of bromide (conservative), acetate (degradable, non-volatile), and sulfur hexafluoride (SF_6) (volatile, non-degradable). This combination of tracers was chosen to provide a comparison of the volatilization and enhanced degradation aspects of remediation via IAS.

The push-pull test was conducted with and without air injection in the physical model to a) assess the appropriateness of each component, b) assess each component's behavior under sparged and unsparged conditions and c) determine the appropriate in-situ holding time for the test. For each experiment, 800 mL of tracer solution was injected at one or more points in the physical model. The solution was left "in-situ" for holding times of 1 hr, 4 hrs, 12 hrs, 24 hrs, and 48 hrs. After the appropriate holding time, 7 to 10 L of water were extracted in one liter increments. The concentration and percent recovery of each component with extracted volume was evaluated for each experiment.

FIGURE 1. Cross-section of the ASU three-dimensional physical model.

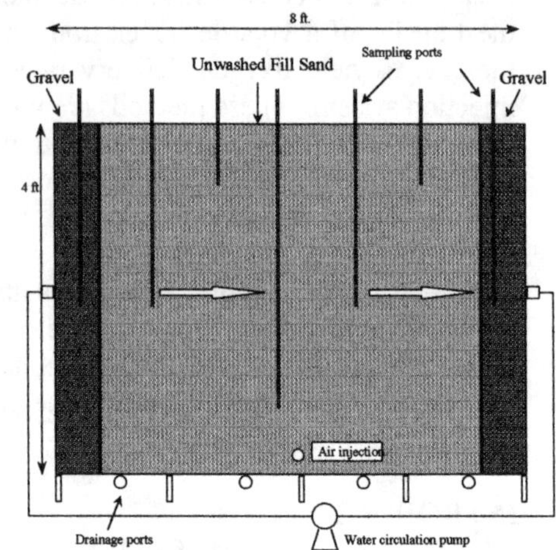

This phase of development resulted in the following conclusions:

- SF_6 recovery could not be firmly correlated with the experimental conditions. That is, only loose trends in recovery with increased holding time or IAS activity were observed.
- Acetate recovery correlated well with both the in-situ holding time and the presence or absence of air injection.
- Differences in acetate recovery at points affected and unaffected by air injection were observed for holding times between 4 and 24 hours, thus providing a large window for the time frame of the test.

The push-pull method was field tested at the Port Hueneme Naval Construction Battalion Center in June, August, and September 1997. The test was conducted at several depths and monitoring points surrounding the IAS well. All tests were conducted while the air sparging system was operating. The data indicated that dissolved oxygen measurements might not give an accurate assessment of oxygen transfer to the groundwater during IAS operations (Figure 2). This was most noticeable at points close to the IAS well where oxygen transfer as measured by acetate degradation was very high but dissolved oxygen was <1 mg/L, suggesting that oxygen is being consumed during degradation at least as quickly as it was being supplied.

FIGURE 2. Measured oxygen transfer compared to measured dissolved oxygen.

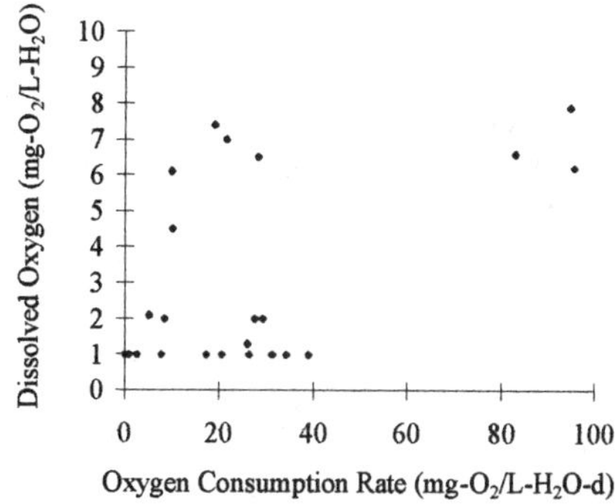

Field Implementation: Push-Pull Tracer Test. A modification of the above test was performed in situ to determine the amount of oxygen consumed by indigenous microorganisms after termination of air injection. This test involved injecting (push) a known quantity of a solution of conservative and degradable tracers into the subsurface, waiting 12-25 hours, and then extracting (pull) several times the injection volume (Figure 3). Assuming complete mineralization of the degradable compound, the relative concentrations of the degradable tracer to the conservative tracer allow a measurement of biological transformation, and by stoichiometry, a measure of oxygen consumption. For the tests performed at the Port Hueneme test site, 800 mL of a multi-tracer solution consisting of 50 mg/L acetate (CH_3COO^-) and 50 mg/L bromide (Br^-) were injected into each monitoring point. The recovered concentrations of the degradable acetate versus the non-degrading bromide were measured within 3 days of collection (samples were iced for ion preservation). Samples were analyzed via ion chromatography on a Dionex DX 500 Ion Chromatograph equipped with an Ionpac AS12A analytical column, Ionpac AG12A guard column, and electrochemical and conductivity detectors. The carrier eluent used in this setup was a 2.7 mM carbonate / 0.3 mM bicarbonate.

FIGURE 3. The "push-pull" tracer test setup.

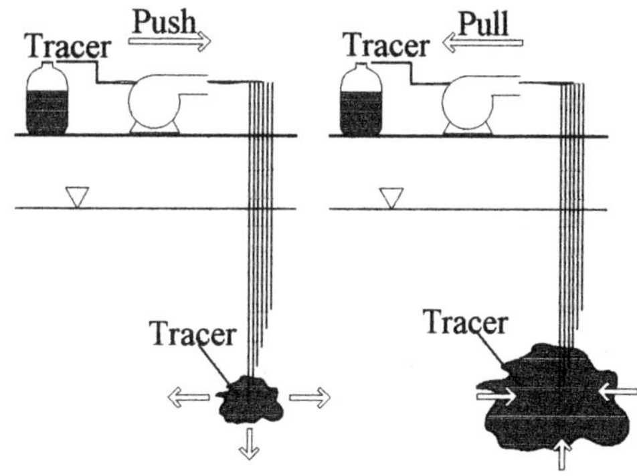

Field Implementation: Air-Injection Tracer Test. To determine the amount of oxygen being delivered to the subsurface, a continuous-injection tracer test was applied (Figure 4). This test requires that a conservative volatile tracer be metered in with the injection air. Groundwater concentrations of this tracer are then monitored to determine the mass of tracer transferred to the aquifer. The mass of tracer present in the groundwater should be proportional to the amount of oxygen transferred by the injected air. Oxygen's reactivity, both biological and chemical, masks the true amount of oxygen transferred to the aquifer. At the Port Hueneme test site, sulfur hexafluoride (SF_6) was the volatile conservative tracer. SF_6 is chemically stable, nontoxic, and can be detected in extremely low concentrations in air and water samples. SF_6 has about half the solubility of oxygen, and a Henry's Law coefficient of ~150 mg-SF_{6air} / mg-SF_{6water}. The fraction of oxygen in the influent air is ~0.20, while that of the tracer is ~10^{-6}.

FIGURE 4. The air-injection tracer test setup.

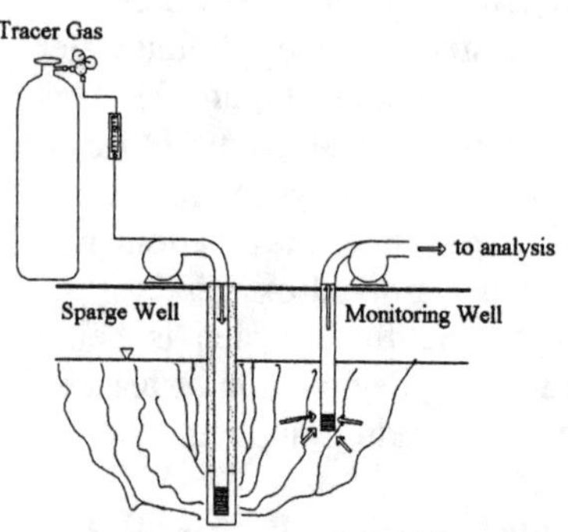

Site Description. The location used for this test was the National Test Site at the Port Hueneme Naval Construction Battalion Center. In 1985, a leak was found in pipes joining the underground storage tank to the Naval Exchange (NEX) Service Station fuel dispensers. This leak resulted in the release of several thousand gallons of fuel to the underlying semi-perched aquifer (Table 1). The test site is located approximately 400 yards from the NEX Station, and covers an area that is impacted with both immiscible product and dissolved contamination.

TABLE 1. Test Site Aquifer Properties

Property	Units	Value
Depth to groundwater	ft	8 – 9
Aquifer thickness	ft	~15
Hydraulic gradient	ft/ft	0.0029
Groundwater flow direction	NA	southwest
Porosity	%	30
Hydraulic conductivity	gpd/ft^2	1267 - 3000
Transmissivity	gpd/ft	286 – 45000
Storativity	NA	0.001 – 0.05
Avg. linear groundwater flow	ft/d	0.5 – 0.9

Air Sparging Test Description. In the summer of 1997 an experiment was performed at the Port Hueneme Test Site to determine the impact of varying operating conditions (Figure 5a) on air sparging performance. The system performance was monitored by conventional and experimental methods, including measuring dissolved oxygen and contaminant concentrations in the groundwater (Figure 5b), vapor contaminant concentrations, and aquifer pressure transducer response.

FIGURE 5. Test Site a) Conditions applied during the IAS test b) Well layout

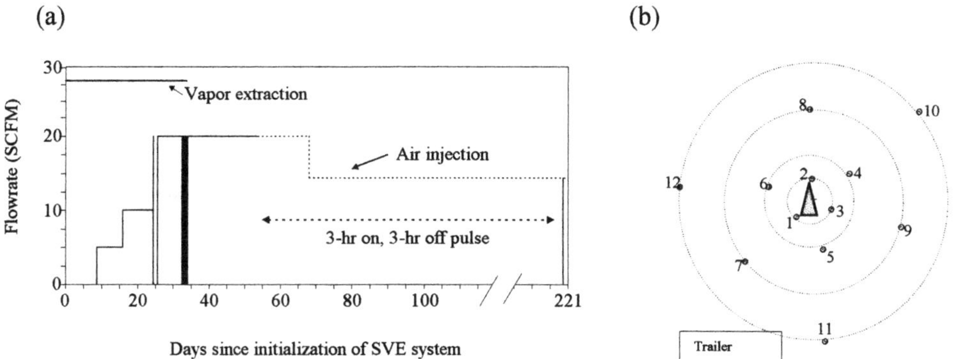

RESULTS AND DISCUSSION

Push-Pull Test. Total determined oxygen mass transfer in the treatment area ranged from 0 – 140 mg-oxygen/L-water/day after the termination of the air sparging test. These values did not show a significant change until several months after termination of air injection (Figure 6). This would suggest that the air transferred to the aquifer during air sparging was not immediately expended to the vadose zone. Instead, it would appear that a "bubble" (or series of bubbles) of air was somehow trapped in the aquifer – providing a continued source of oxygen to the indigenous subsurface populations. This effect may indicate that a more appropriate operation of air sparging in some field applications would be to utilize a pulsed air injection, to maximize the utilization of trapped air, and decreasing compressor run time. The oxygen mass transfer rate results determined from the push-pull test are in sharp contrast to the dissolved oxygen measurements, which dropped to below 2 mg-oxygen/L-water within days of termination of air injection. Values for dissolved oxygen and oxygen mass transfer are shown below (Figures 7a and 7b) for immediately after the air sparging system had been turned off. These contours show a disparity between the observed treatment zone and the actual zone of oxygen influence. This effect is also seen when comparing the results of the continuous tracer injection test with dissolved oxygen concentrations (Figures 8a and 8b).

Figure 6. Calculated oxygen mass transfer rates at monitoring point 3 (MP-3) from the "push-pull" diagnostic tool.

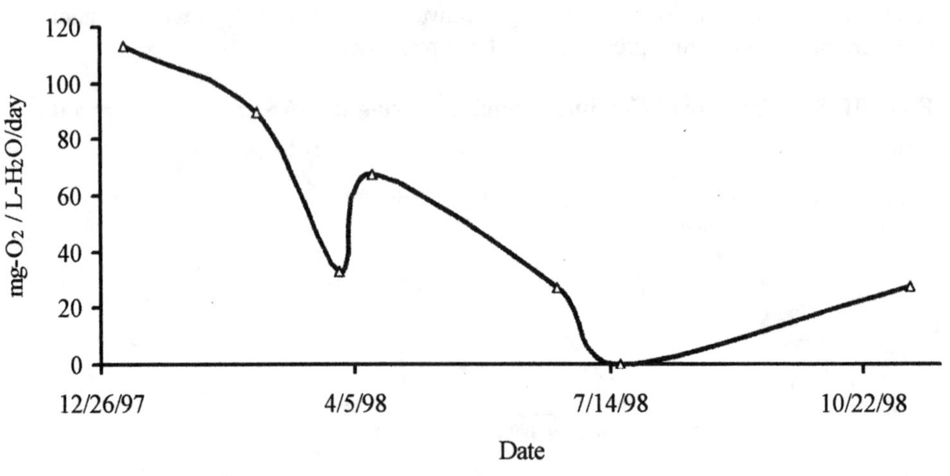

FIGURE 7. Contour maps for January 1998 (using the push-pull method).
(a) Dissolved oxygen concentrations [mg-oxygen/L-water],
(b) Calculated oxygen mass transfer rates [mg-oxygen/L-water/day]

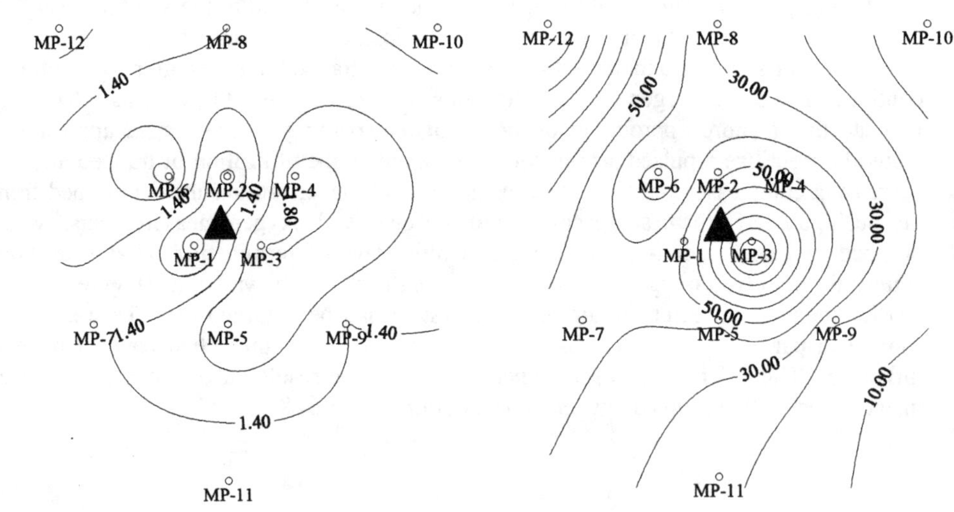

FIGURE 8. Contour maps for July 1997 (using the air injection method).
(a) Dissolved oxygen concentrations [mg-oxygen/L-water],
(b) Calculated oxygen mass transfer rates [mg-oxygen/L-water/day]

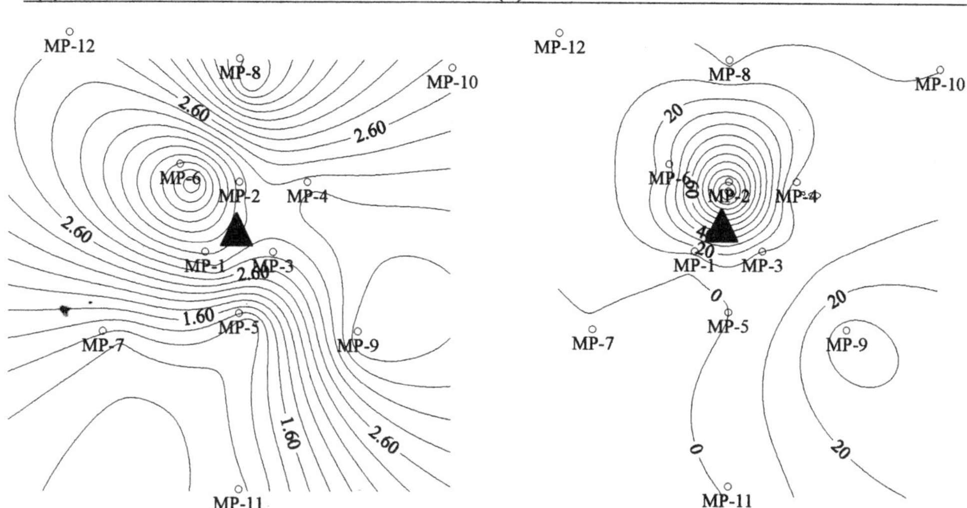

Air – Injection Tracer Test. Total determined oxygen mass transfer in the treatment area ranged from 0 – 130 mg-oxygen/L-water/day. The results from the measured dissolved oxygen and the calculated oxygen delivery rates from this test seem to agree more on the areas benefiting from injected air. Major inconsistencies are found in the well most likely to short-circuit (MP-6) in this test plot.

CONCLUSIONS

The two tracer-based methods of monitoring air sparging performance discussed above are unique in that oxygen transfer measurements can be made without pre-sparging data and in that they give a real-time measure of oxygen mass transfer that is integrated over a local scale instead of a site-wide scale. A demonstration of these diagnostic tools at the National Test Site in Port Hueneme, California, show oxygen mass transfer numbers ranging from 0 – 140 mg-oxygen/L-water/day for both tests. Results from these analyses suggest that 1) oxygen transfer may be occurring outside the zone of treatment defined by conventional air distribution measurements, 2) dissolved oxygen concentrations may not accurately reflect the oxygen transfer at a given point, and 3) the transfer of oxygen may continue well after air injection has been terminated.

REFERENCES

Amerson-Treat, Illa. 1997. Diagnostic tools for the monitoring and optimization of in situ air sparging systems. M.S. Thesis: Arizona State University.

Leeson, Andrea. 1996. Work Plan for D/NETDP Technology Demonstration. Battelle Memorial Institute, Columbus, OH.

MULTI-PHASE EXTRACTION PILOT TEST AT THE FORMER FIRE-FIGHTING TRAINING FACILITY, NAVAL STATION SAN DIEGO

Abram Eloskof, Jamshid Sadeghipour, Victor Owens, and Luis Gomez
(Foster Wheeler Environmental Corporation, Costa Mesa, CA)
Jim Reisinger (Integrated Science & Technology, Inc., Marietta, GA)
Darren Belton (Naval Facilities Engineering Command, SWDIV, San Diego, CA)
Theresa Morley (Naval Station, San Diego, CA)

ABSTRACT: A multi-phase extraction (MPE) pilot test was conducted at the former Fire Fighting Training Facility to evaluate the effectiveness of bioventing, free-hydrocarbon product recovery and *in-situ* bioremediation technologies in remediating two hydrocarbon plumes composed primarily of Jet Fuel No. 5 (JP-5) (approximately 10,000 to 15,000 gallons). The pilot test included a skimmer test, a stepped-up MPE test at varying well head vacuums and an *in-situ* respiration test. The tests were designed to generate the data needed for the design of a full-scale system. The radius of influence around a single extraction well was estimated to be approximately 40 ft (12.16 m). An average free hydrocarbon product recovery of approximately 8 gallons per hour (gph) was achieved from a single extraction point with an in-well vacuum of about 34 inches of water (inches-H_2O) and airflow of 31 standard cubic feet per minute (scfm). The results of the *in-situ* respiration test indicated that an average biodegradation rate of 7 milligrams of total petroleum hydrocarbon per kilogram of soil per day (mg-TPH/kg-soil/day) could be achieved at the site. Based on these results, a full-scale system was designed and is now in operation. The full-scale system operations started in November 1997. As of November 1998, approximately 11,500 gallons of hydrocarbons have been removed.

INTRODUCTION

MPE technology, also known as "bioslurping", is used to remove free-phase hydrocarbon product floating on groundwater. It combines the superior features of bioventing and vacuum enhanced free-hydrocarbon recovery to remediate hydrocarbon-contaminated sites. The vacuum enhanced recovery system increases product recovery rates over traditional skimming techniques by increasing the effective hydraulic gradient and the aquifer transmissivity. The effective hydraulic gradient is increased as a result of developing a cone of reduced pressure around the well, thus promoting a horizontal flow of fluids across the reduced pressure gradient. The increase in effective transmissivity is a result of increasing the saturated thickness near the well screen. The application of negative pressure to the vadose zone also results in air movement, which in turn promotes aerobic degradation of hydrocarbons.

Objective. The objective of the MPE pilot test was to generate the data needed to determine the site-specific effectiveness of MPE and *in-situ* bioremediation at the site and to develop the basis for the full-scale MPE system design. The required data to meet the MPE pilot test objective were as follows: (i) the radius of remediation and level of vacuum for a single extraction well; (ii) the rate (volume per unit time) of free-hydrocarbon product that could be extracted from a single extraction point; (iii) the ratio of hydrocarbons to water in the recovered fluids; (iv) the extent of tidal influence on free-hydrocarbon recovery; and (v) the rate of *in situ* biodegradation that might be occurring.

Site Description. The former Fire Fighting Training Facility is located along San Diego Bay in the central part of Naval Station San Diego (NAVSTA), San Diego, California. Structures, flight decks used for fire fighting exercises, and a system of tanks and piping that were used to store and convey fuel to the flight decks previously existed at the site. At the time of the pilot test, most of the structures had already been demolished and most of the subsurface structures had been removed. Since 1987, various site characterization activities have been performed at the site. Based on the results of these characterization activities, two large hydrocarbon plumes were identified at the facility as a result of leaking underground storage tanks and fuel lines. One plume is located on the northern half of the site and the other is located on the southern half (Figure 1). The major component of both subsurface fuel hydrocarbon plumes is JP-5. The MPE test was conducted at well PD-3 where the apparent thickness of JP-5 in the southern plume exceeded 2 ft (0.61m).

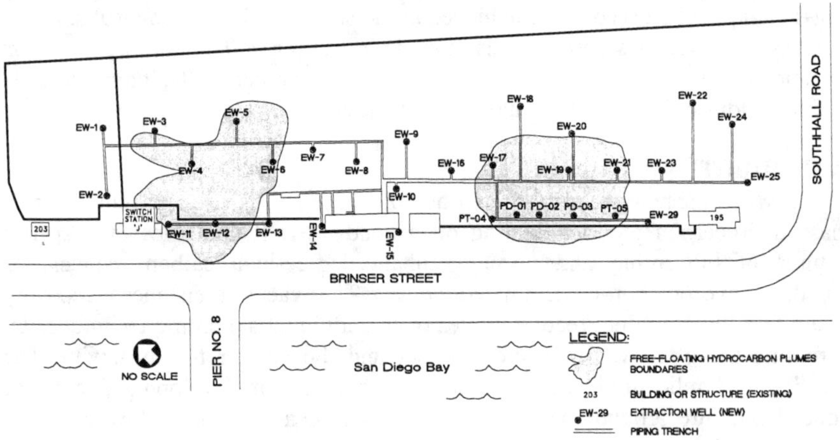

FIGURE 1. Free-Floating Hydrocarbon Plumes and Well Locations

MATERIALS AND METHODS

The pilot test system consisted of two medium-flow liquid-ring vacuum pumps (one standby), an air-liquid separator, a fluid transfer pump, a particulate bag filter, a 10 gallons per minute (gpm) oil/water separator, a second transfer pump, an integral liquid flow meter, and liquid- and vapor-sampling ports.

Figure 2 shows the schematic of the pilot test equipment. The vapor extracted during the pilot test was routed through vapor-phase carbon absorbers for treatment (in the full-scale system, a thermal oxidizer was used) prior to discharge to the atmosphere; the liquid was routed through liquid-phase carbon absorbers prior to discharge into a 21,000-gallon recovery tank. An existing monitoring well (PD-3) within the southern plume was used as the extraction well for the MPE pilot test.

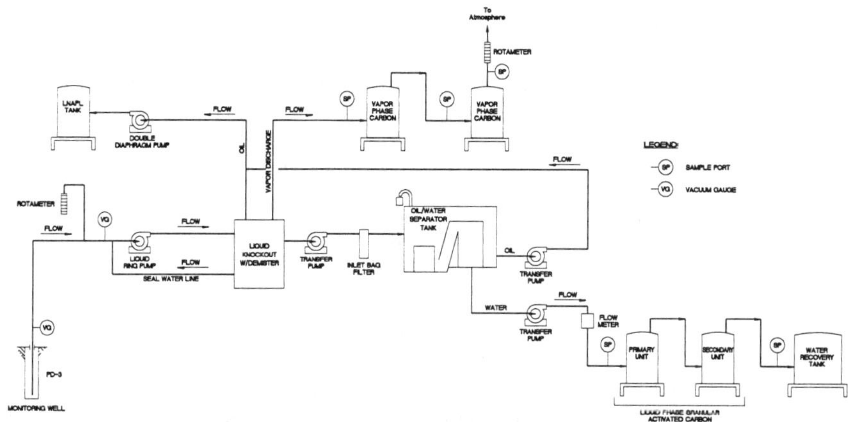

FIGURE 2. MPE Pilot Test Schematic

The pilot test was conducted in four stages: pre-test baseline monitoring; a liquid recoverability test with no vacuum enhancement (i.e., skimmer test); a stepped MPE recoverability test with concurrent vapor-monitoring data collection; and a long-term MPE test with vapor monitoring.

Pre-Test Baseline Monitoring. Four on-site wells, wells PD-1, PD-2, PT-4, and PT-5, located 73, 44, 88, and 56 ft (22.19, 13.37, 26.75, and 17.02 m) from the test well PD-3, respectively (Figure 1), were used to monitor fluid levels during the MPE pilot test. The depth to static groundwater and apparent hydrocarbon thickness were measured in the four wells before, during and after the pilot test using an oil/water interface probe. Baseline subsurface vapor concentrations (oxygen, carbon dioxide, and total hydrocarbon) were measured in surrounding vapor-monitoring points located in the northern and southern plumes.

Skimmer Test. The skimmer test was conducted on the test well (PD-3) for 18 hours. The test was conducted by lowering a 1-inch drop tube inside the 4-inch well casing and placing it within the free-hydrocarbon product zone. The well was left open to the atmosphere by opening the gate valve attached to the well head assembly. Vacuum was applied through the drop tube to recover the hydrocarbon product from the well. The rates and volumes of hydrocarbons and water recovery were quantified during and after the test. Water and hydrocarbon thicknesses in the wells were measured at regular intervals during and after the test.

MPE Test. The MPE test was conducted with stepped vacuum at estimated extraction flow rates of 20, 28, and 31 cfm at well-head vacuums of 8, 19.5, and 34 inches-H_2O, respectively. The well head was isolated from the atmosphere while vacuum was applied through the drop tube. Throughout the stepped vacuum testing, subsurface vadose-zone pressure measurements, water levels and hydrocarbon product thickness were measured in the monitoring wells to monitor possible changes in the radius of vacuum influence due to tidal fluctuations. The extracted water and free product volumes and rates, and vacuum pressure at the extraction well were recorded periodically during each step.

At the conclusion of the stepped vacuum MPE tests, the optimal steady-state vapor flow rate, well vacuum, and drop-tube setting were determined for the test well configuration which would produce the greatest hydrocarbon product recovery rate and the highest vacuum in the formation. The MPE test continued for 3 additional days at these optimal parameters. Total volatile hydrocarbons, vapor temperatures, liquid and free product as well as several process parameters were recorded during and after the 3-day long-term MPE test.

***In-Situ* Respiration Test.** An *in-situ* respiration test was conducted within the northern plume area to reduce the potential interference from the MPE test. The test was conducted via four vapor-monitoring probes or stations. Baseline concentrations of oxygen, carbon dioxide, and total hydrocarbons were measured in each of the vapor probes. Following baseline characterization, ambient air with 1.0 percent helium was injected into each monitoring point. The helium was added as a check on oxygen decay resulting from diffusion. After 24 hours, the injection was terminated and oxygen, carbon dioxide, hydrocarbon, and helium concentrations were measured with decreasing frequency in each monitoring point over a period of 4 days.

The oxygen utilization rate from the *in-situ* respiration test was used to calculate the biodegradation rate according to equation 1.

$$K_B = K_O A D_O C/100 \quad (1)$$

Where: K_B = Biodegradation rate (mg / kg - day)
K_O = Oxygen utilization rate (%O_2 / day)
A = Volume of air / kg soil (L / kg)
D_O = Density of oxygen gas (mg / L)
C = Mass ratio of hydrocarbon : oxygen required for mineralization

Substituting a porosity value of 0.03, soil bulk density of 1,440 kg/m^3, an oxygen density of 1,330 mg/L, and a hydrocarbon to oxygen ratio of 1 to 3.5 into equation 1 results in equation 2.

$$K_B = \frac{(K_O)(0.21)(1{,}330)(1/3.5)}{100} = 0.798 K_O \quad (2)$$

RESULTS AND DISCUSSION

Skimmer Test. The skimmer test was conducted at an average vapor flow rate of 70 scfm. The total volume of hydrocarbons and water recovered during the test was 14.2 gallons (approximately 0.8 gph) and 152 gallons (approximately 8 gph), respectively. The ratio of hydrocarbons recovered to water recovered was approximately 1:10.

MPE Steps #1, #2, and #3. These tests were conducted at three separate vacuums, 8, 19.5 and 34 inches-H_2O. Results indicated that Step #3 resulted in the most efficient recovery as well as the highest response. During Step #3, the total volume of hydrocarbons and water recovered was 69.1 gallons (approximately 17.3 gph) and 436.2 gallons (approximately 109 gph), respectively. The ratio of hydrocarbons recovered to water recovered was approximately 1:6.

Long-Term Test. The long-term MPE test was conducted at a measured well-head vacuum of 34 inches-H_2O. The drop tube was set at the identical depth used for the skimmer test. A total of 418 gallons of hydrocarbon product and 3,201 gallons of water (ratio of 1:8) were recovered during this step test over 56 hours (approximately 7.5 gph overall). Vacuum levels measured in the monitoring wells at the end of this step were 1.60 (PD-2), 0.46 (PT-5), and 0.47 (PD-1), and 0.00 (PT-4) inches-H_2O, respectively, which indicate a strong tendency for subsurface air movement induced by vacuum applied at the test well.

Radius of Influence. Radius of vacuum influence was measured in the vadose zone using monitoring wells PD-1, PD-2, PT-4, and PT-5 during both the MPE step tests and the 3-day steady-state test. Normally the estimated limit of vacuum influence to induce air flow is set at 0.1 inches-H_2O negative pressure. To be conservative, a second limit of vacuum influence at 0.5 inches-H_2O was considered to account for the site heterogeneity. The results are listed in Table 1.

TABLE 1. Radius of Influence at Various Vacuum Levels

MPE Test Designation	Test Well (PD-3) Vacuum (inches-water)	Est. Radius of Vacuum Influence (ft) at 0.1-inch Water Limit (less conservative)	Est. Radius of Vacuum Influence (ft) at 0.5-inch Water Limit (more conservative)
Step #1	8	81 (24.62 m)	14.2 (4.32 m)
Step #2	19.5	91 (27.66 m)	40 (12.16 m)
Step #3	34	108 (32.83 m)	49 (14.89 m)
Steady State	34	101 (30.7 m)	49 (14.89 m)

Tidal Effects. Significant tidal effects were observed during the MPE test as the depth to water and free hydrocarbons, and apparent hydrocarbon thickness vary in response to the tides. Well PT-5 displayed the greatest fluctuations in water level, hydrocarbon level, and apparent hydrocarbon thickness. The other wells monitored (PD-1, PD-2, PT-4) also displayed variations in hydrocarbon level and thickness, but not of the same magnitude as PT-5. The large magnitude of fluctuations in PT-5 are likely related to its proximity to a large storm sewer that discharges into San Diego Bay. It is speculated that the pipe and/or the backfill for the sewer may be hydraulically connected to the Bay water. Therefore, areas along the trend of the sewer would be expected to reflect the magnitude of the tidal fluctuation to a greater extent than areas that are more distant from the sewer.

Extracted Vapor Analysis. The results of the laboratory analysis of the vapor samples collected during the long-term test indicated a shift from anaerobic to aerobic environment in the soil media. Trends showed an increase in the concentration of oxygen during the 56-hour test period, which indicates that substantial subsurface oxygenation was occurring as a result of the soil-vapor extraction aspect of the MPE. In addition, the concentrations of methane and hydrocarbons in the extracted air stream decreased.

***In-Situ* Respiration Test.** Significant oxygen depletion was measured throughout the *in-situ* respiration test in most of the monitoring points. The rate of carbon dioxide production was much less than the corresponding oxygen decay rate. This likely results from biological uptake of carbon by soil microorganisms. Estimated biodegradation rates, based on the oxygen depletion measured at the sample points, ranged from 0.85 mg-TPH/kg-soil/d to 11.05 mg-TPH/kg-soil/d. The arithmetic average of the rates at the eight sample points was 7.05 mg-TPH/kg-soil/d. The lowest rates were observed at the sample points which displayed the lowest vapor-phase hydrocarbon concentrations [VMP-3 at 4 ft (1.22 m) and VMP-4 at 4 ft (1.22 m)]. The low biodegradation rates estimated for these sample points are likely a result of low hydrocarbon availability that limits aerobic respiration during the test.

CONCLUSIONS

Analysis of data gathered during the MPE pilot test led to the following conclusions: (i) successful demonstration of vacuum enhanced hydrocarbon recovery and MPE was achieved during the pilot test; (ii) an average of approximately 8 gph of hydrocarbons was recovered during the step and long-term aspects of the MPE test, while only 0.8 gph was achieved with non-vacuum enhanced skimming; (iii) the tidal fluctuations clearly influenced the rates of hydrocarbon and water recovery such that during the periods of low groundwater levels in the southern plume, the rate of hydrocarbons recovered appeared to be greater than the periods of high groundwater levels; (iv) soil-vapor extraction during the course of the MPE pilot test led to enhanced oxygen concentrations in the vadose zone; (v) estimated biodegradation rates (K_B) ranged from 0.85 to 11.05 mg-TPH/kg-soil/d, with an average K_B of 7.05 mg-TPH/kg-soil/d; and (vi) the site appears to be a very good candidate for aerobic bioremediation.

DUAL-PHASE EXTRACTION AND BIOVENTING OF PETROLEUM-CONTAMINATED SOILS AND GROUND WATER AT GASOLINE STATION NO. 63-232-0049, SEATTLE, WASHINGTON

Christian E. Houck, P.E. (ThermoRetec, Seattle, Washington)
Thomas C. Morin (Environmental Partners, Inc., Bellevue, Washington)
Alexander P. Jones (Environmental Partners, Inc., Bellevue, Washington)
Tony Palagyi (Equiva Services, LLC, Bellevue, Washington)

ABSTRACT: Equiva Services, LLC initiated an aggressive soil and ground water remediation in the fall of 1994, at a retail gasoline station in glacial till soils, following the discovery of light non-aqueous phase liquid (LNAPL) on the water table aquifer and in the neighboring basement sump of a down gradient fast food restaurant. The system utilized dual-phase extraction (DPE) as the primary remediation technique beginning in August 1995. During the first two years of operation, two additional petroleum releases occurred at the gasoline station property due to a faulty petroleum distribution system. In the third year, soil vapor extraction concentrations asymptotically declined, however, petroleum constituents in the ground water remained at elevated levels. It appeared the system was not working effectively to remediate residual non-volatile smear zone contaminants. Therefore, modifications were made to take advantage of the existing system components to initiate a more effective in-situ remediation approach to address the residual contaminants.

In March of 1998, the remediation system was modified to include eight bioventing wells to passively supply ambient air using the high vacuum (i.e., 16" Hg) induced by the system's liquid ring pump (LRP). The bioventing wells were placed within the overlapping radius of vacuum influence surrounding each of the four extraction wells. The bioventing wells were screened across the smear zone and supply atmospheric oxygen to facilitate aerobic biodegradation. It was believed that the smear zone was acting as an ongoing source of petroleum hydrocarbon dissolution to ground water and that the existing system configuration was only having limited success with these contaminants. Aerobic conditions created by the addition of atmospheric oxygen creates a "bioreactor" degrading the smear zone contaminants in situ. As part of this modification, it is possible to re-direct a portion of the system off-gases to the subsurface via a manifold to the bioventing wells. This configuration would use the in situ "bioreactor" to degrade extracted petroleum hydrocarbon and saves on vapor phase carbon polishing costs. This portion of the system is currently being tested and will be reported in the future.

INTRODUCTION

In November 1993, employees at a Seattle fast-food restaurant noted petroleum odors emanating from a basement sump room. Upon further inspection, light, non-aqueous phase liquid (LNAPL) and dissolved total petroleum hydrocarbons (TPH) where identified in the basement sump. Following initial cleanup and mitigative responses, a series of subsurface investigations where conducted to identify potential TPH sources.

The site configuration consists of a retail gasoline station covering the western 0.25 acres and a fast-food restaurant and associated parking and thoroughfares occupying the eastern 0.80 acres. Retail gasoline services have been present at the site for more than fifty years.

A series of subsurface investigations showed two TPH source areas at the site and the leading edge of a third TPH plume originating from an off-site source. The main source (TPH-gasoline, diesel) had originated from a leaking fitting in the UST cavity. The original plume was supplemented by two diesel releases of less than 50 gallons each from leaking valves in the fuel dispensing system within the UST cavity following the installation of the active remediation system. A second source (TPH-gasoline) was traced to an abandoned sump associated with a former station located between the current dispenser islands. A third source, containing LNAPL (TPH-oil) appears to have originated off-site and moved onto the restaurant property and into the basement sump from the north.

The objectives of this project were to: mitigate impacts inside the restaurant; remove LNAPL from the basement sump and UST cavity plume; gain hydraulic control of the plumes; perform the remediation in a cost effective manner; and ultimately cleanup the soils and ground water to levels acceptable to the State of Washington for obtaining a no further action designation for the site.

The remediation system was designed to incorporate two techniques; 1) a total fluids pumping system in the restaurant basement sump and 2) dual-phase extraction (DPE) with the subsequent addition of bioventing.

DPE was selected as the primary remedial technology because it could establish hydraulic control of the dissolved-phase contaminants, enhance volatilization of sorbed contaminants, and facilitate oxygen transport to the vadose zone.

SYSTEM CONFIGURATION AND PERFORMANCE

DPE (also referred to as bioslurping) was implemented at this site using a liquid ring pump (LRP). Given the low permeability of the glacial till soils, DPE with an LRP would produce high vacuums with moderate air flow rates while maximizing ground water extraction rates.

For this site, four extraction well were installed (a fifth [MW-16] was added in added in August 1998) within and down-gradient of the source areas. LRP drop tubes were set in each of the four extraction wells at between 17 and 20 feet below grade (i.e., about 7 to 10 feet below the static water table) to expose a smear zone present in the vadose zone and subject it to vapor extraction. The DPE system was started in August 1995 and over the next 3.5 years operated at an average vacuum of 16" Hg, an average extracted air flow rate of 45 cfm, and an average ground water extraction rate of 2.5 gpm.

During this period of operation the DPE system removed a total of about 8,200 pounds of contaminant and was very effective at maintaining hydraulic control of the dissolved-phase plume. However, since late 1996 and early 1997, the contaminant removal rates had become asymptotic and ground water quality had ceased to improve.

This condition strongly indicated a low-volatility source of residual contaminant was still present in the smear zone. Without directly addressing this dissolution source there was no foreseeable end to the remedial effort and Equiva would be required to expend significant expense to keep the system in operation without remedial benefit.

Bioventing Wells. The remediation system was modified in March of 1998 to include bioventing wells as a means to address the smear zone. The addition of the bioventing wells would use the high vacuum of the LRP to draw in atmospheric oxygen with the goal of creating strong aerobic conditions in large areas surrounding each extraction wells. The switch from anaerobic conditions to aerobic conditions would facilitate the growth of a population of aerobic petroleum hydrocarbon degrading bacteria and significantly increase the degradation rate of the sorbed smear zone contaminants.

The bioventing system consisted of eight 2-inch diameter vent wells screened from 4 to 20 feet below grade. These wells were installed at distances of 10 to 25 feet from the extraction wells in locations that would create the largest possible area of aerobic conditions. Two 2-inch diameter wells were also installed to allow the potential future re-injection, and below grade treatment, of system off-gases. The system process flow diagram in shown in Figure 1.

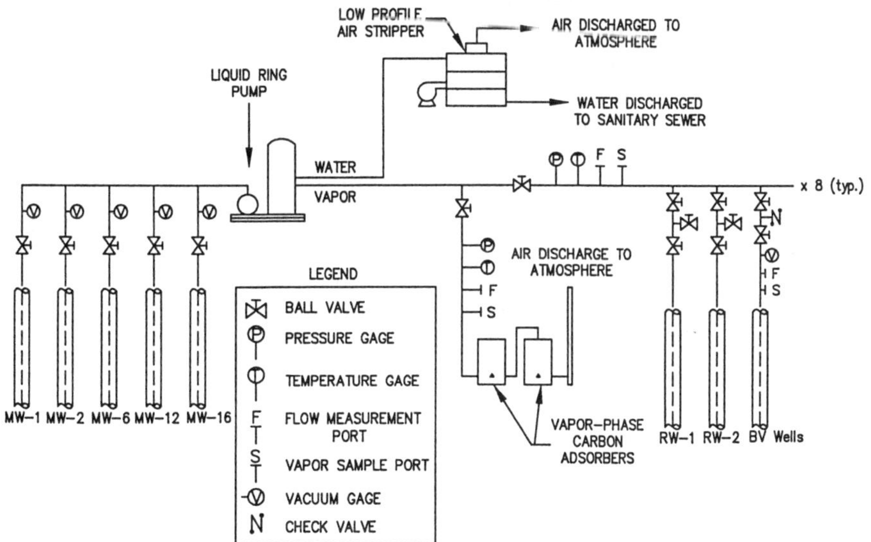

FIGURE 1. Schematic of the dual-phase extraction and bioventing system

Initial airflow rates into the individual bioventing wells ranged from 0.25 to 2.18 cfm and totaled 6.11 cfm. After seven months these rates had stabilized and ranged from 0.05 to 1.64 cfm and totaling 3.73 cfm. This decrease in flow rates was largely associated with a seasonal rise in the water table and thinning of the vadose zone.

Measurements indicative of aerobic respiration have been collected each month since the bioventing wells went on-line. Measurements of oxygen use and carbon dioxide generation were collected at each of the extraction wells.

Mass destruction through biodegradation was calculated by measuring the oxygen concentration in the extracted air system at each extraction well. It was assumed that the only oxygen demand in the subsurface was biological and related to the TPH and that all subsurface vapor originated from the atmosphere with an oxygen concentration of 20.9 percent. The difference between the concentration of oxygen in the extracted vapor and the atmospheric concentration was attributed solely to biodegradation of TPH. A stoichiometric relationship of 3 pounds of oxygen to 1 pound of hydrocarbon was used as the basis for calculating mass destruction and all vapor flow rates were converted to standard conditions. The oxygen concentrations extracted from the subsurface have ranged between 13.4 percent to 20.9 percent depending upon the location of the well. The average biological destruction rate calculated from these measurements has ranged from about 1 pound/day at MW-1 to about 0.2 pounds/day at MW-6.

RESULTS AND OBSERVATIONS

Ground water potentiometric maps have been generated quarterly since startup and have consistently indicated hydraulic capture of the dissolved-phase plume. The addition of the bioventing wells did not perceptibly affect ground water extraction rates or hydraulic capture.

The operational data for the DPE system indicate that, initially, the system worked very well for contaminant mass removal. However, as the remediation progressed the DPE system became increasingly ineffective at removing the lower volatility contaminants and the residual smear zone contaminants within the glacial till soils. About 90 percent of the mass removed by the DPE system was removed in the first 2 years of system operation. DPE removal rates had become asymptotic to about 2.5 pounds/day 16 months after system start-up and 0.9 pounds/day 28 months after start-up.

The addition of bioventing wells, which utilized the LRP vacuum to draw in atmospheric oxygen, provided a method for inducing aerobic biodegradation and targeting that biodegradation to the smear zone. Operational data suggest that the biological mass destruction resulting from the bioventing totaled about 360 pounds of TPH and averaged about 1.4 pounds/day. The addition of the bioventing wells increased the mass removal/destruction of the DPE system by about 2.5 times. Figure 2 presents cumulative mass removal rates for the DPE and bioventing systems.

The most dramatic demonstration of the effectiveness of the bioventing system is the large drop in dissolved-phase contaminant concentrations observed since the start of bioventing. The combined TPH and benzene concentration in well MW-1 dropped from about 83,000 μg/L to about 3,477 μg/L over a period of 7 months. The combined concentration in this well had been above 60,000 μg/L for the previous 60 months. Figures 3A and 3B represent TPH and benzene concentration contours over the life of the system.

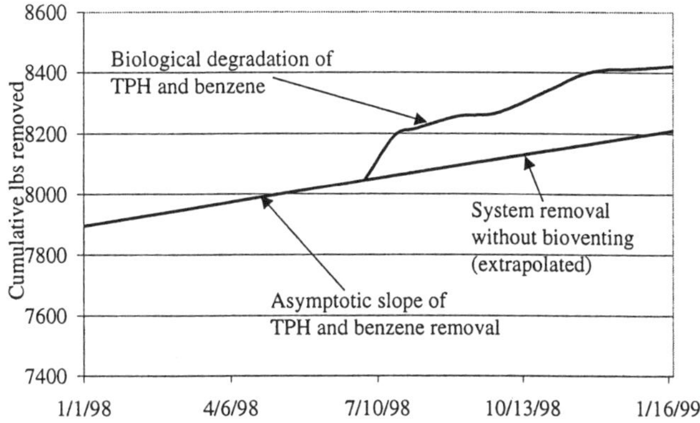

FIGURE 2. Bioventing contributions to TPH and benzene removal

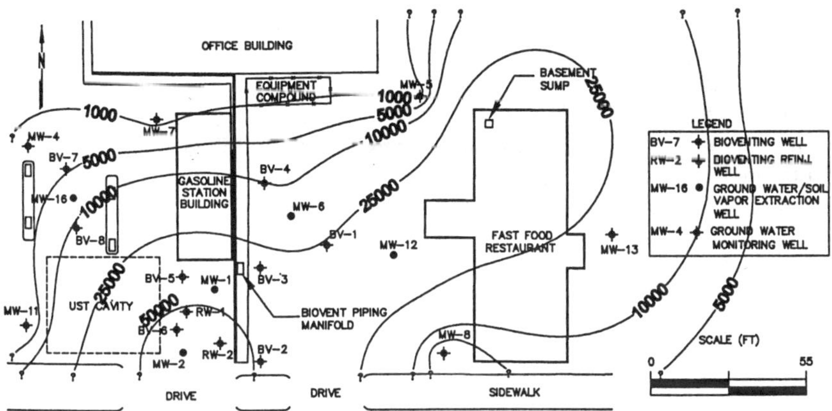

FIGURE 3A. TPH and benzene concentration contours (μg/L) – April/May 1994

The establishment of a large population of aerobic bacteria was demonstrated by the volume of bioslime observed in each of the extraction wells and treatment equipment. This slime was not observed prior to installation of the bioventing wells.

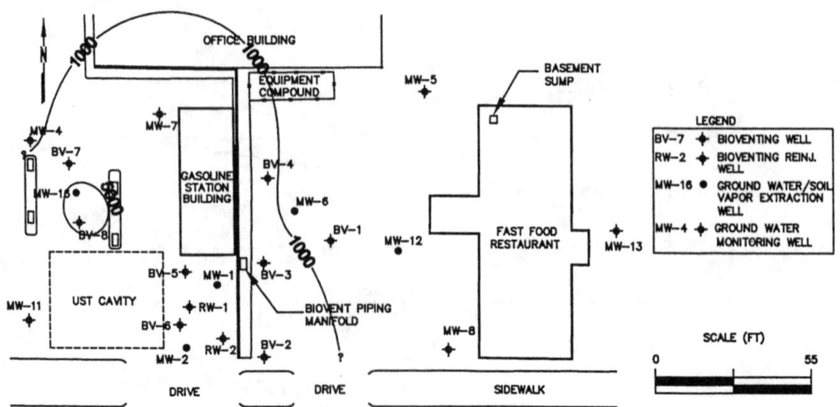

FIGURE 3B. TPH and benzene concentration contours (µg/L) – Nov/Dec 1998

CONCLUSIONS

The primary conclusion of this field study of bioventing as an augmentation to DPE with an LRP is that in low permeability soils such as glacial tills, DPE can be significantly enhanced by the addition of bioventing wells. The DPE system will quickly remove volatile and readily mobile contaminants but is less effective at removing residual contaminants that act as diffusion limited sources of release. By placing the bioventing wells in locations surrounding the DPE extraction wells it is possible to create a large zone of active biodegradation.

The addition of bioventing to the DPE system did not adversely affect the systems ability to maintain hydraulic control of the dissolved contaminants and the biological slime that developed within the extraction wells also did not adversely affect the LRP operation or increase operation and maintenance requirements.

The effectiveness of the bioventing system is demonstrated by the operational data of oxygen utilization and, most directly, by the steep declines in contaminant concentration in ground water.

For future DPE systems the installation of bioventing wells at the start of the project should be strongly considered. If these wells had been installed at the startup of the DPE system it appears that, based on the current biological destruction rates, the overall remedial effectiveness of the system could have been significantly increased.

ACKNOWLEDGEMENTS

We wish to thank Equiva Services, LLC for their assistance with this project.

BIBLIOGRAPHY

Kuo, J. 1999. *Practical Design Calculations for Groundwater and Soil Remediation.* Lewis Publishers, Boca Raton, FL.

U.S. Army Corps of Engineers. 1995. *Engineering and Design: Soil Vapor Extraction and Bioventing.* Department of the Army, Washington, DC.

AN EVALUATION OF ALTERNATIVE VACUUM SYSTEMS FOR IMPLEMENTING BIOSLURPING

David S. Woodward; Michael S. Niederreither, P.G.; Thomas L. McMonagle
Earth Tech, Inc., Harrisburg, Pennsylvania, USA

ABSTRACT: A successful bioslurping project requires expertise in the areas of hydrogeology, biology and chemistry, but it also requires significant mechanical process knowledge. Based on the authors' collective experience, many bioslurping problems result from a lack of experience and knowledge with the mechanical process equipment that is employed. This paper focuses on the mechanical process (vacuum pump, slurp tube, etc.) aspects of bioslurping systems. This paper also presents the results of a Bioslurping Demonstration Project in which alternative vacuum systems were utilized and evaluated on four large LNAPL plumes within varying geological settings located throughout the United States. A number of vacuum source alternatives were utilized and are discussed and evaluated, including: liquid ring vacuum pumps (LRVPs); a dry vacuum blower (DVB); and, an internal combustion engine (ICE). Inherent problems with bioslurping are discussed including LNAPL emulsification and ferric hydrate deposition and fouling caused by iron bacteria. A method developed for quantifying the LNAPL recovery rate from individual wells is also discussed. Experience gained at a number of additional sites and from numerous pilot tests was also used to develop this paper. These experiences have collectively been utilized to evaluate alternative vacuum sources and other mechanical process equipment utilized for bioslurping.

INTRODUCTION

Bioslurping is an innovative in situ remediation technology that utilizes an applied vacuum to recover Light Non-aqueous Phase Liquids (LNAPL) through small diameter recovery piping (the "slurp tube") while simultaneously biodegrading adsorbed phase hydrocarbons in the vadose zone due to the inherent replenishment of oxygen that occurs (Woodward and Niederreither, 1998). Haas et al (1997), Gibbs et al (1997) and others have previously demonstrated that bioslurping can be a very effective and efficient remedial technology for recovering LNAPL. However, despite the successful application and implementation of this technology, many bioslurping systems continue to be misapplied, poorly designed, and haphazardly implemented. Based on the authors' collective experience, many bioslurping problems result from a lack of experience and knowledge with the mechanical process equipment that is employed. Often, inexperienced practitioners are not familiar with available alternative vacuum systems. A successful bioslurping project requires expertise in the areas of hydrogeology, biology and chemistry, but also requires significant mechanical process knowledge.

This paper presents the results of a Bioslurping Demonstration Project conducted on four large LNAPL plumes within varying geological settings located throughout the United States. Experience gained at a number of additional sites and from numerous pilot tests was also used to develop this paper. These experiences have collectively been utilized to evaluate alternative vacuum sources and other mechanical process equipment utilized for bioslurping.

BIOSLURPING SYSTEM COMPONENTS

Primary bioslurping system components typically include: a recovery well(s); the slurp tube (and integral wellhead); a vapor/liquid separator; a vacuum source; and water and vapor treatment systems, as required. Most recovery well configurations can be adapted for implementing bioslurping so recovery well construction is not discussed. Typical vapor treatment systems can be utilized including granular activated carbon and thermal/catalytic oxidizers and as such, treatment systems are also not discussed.

Slurp Tube Configuration and Materials. The slurp tube is used to simultaneously convey LNAPL, groundwater and vapors to the recovery piping. The slurp tube height must be adjustable so that it can be quickly and easily moved up and down the well to maximize LNAPL recovery. Short sections of flush threaded piping (two to five feet) supported by an expandable well seal can be used to accommodate rapid adjustments to the slurp tube height. Quick couplings and flexible vacuum hose can be used to connect the slurp tube to the recovery piping.

The wellhead must be constructed with a bleed valve to allow ambient air to enter the well, if necessary. The "bleed" air prevents the well from vapor locking during periods of higher groundwater levels. Bleed valves also provide a means of enhancing the LNAPL recovery rate by increasing the velocity at the slurp tube inlet. The increased velocity entrains LNAPL in the vapor stream and enables the system to recover LNAPL from greater depths without requiring a deeper vacuum. The slurp tube velocity (and LNAPL recovery rate) can be increased by cutting a notch or V-notch in the bottom of the slurp tube and/or using incrementally smaller piping. The bleed valve also provides a means of adjusting and balancing the system so that the wells that yield the highest LNAPL recovery rates generate the most flow. Although it is more costly, individual recovery piping from each well should be installed back to a central manifold near the vacuum source to reduce operation, maintenance and monitoring (OM&M) costs and provide greater operational flexibility.

There is no optimal slurp tube diameter since the most productive size depends on the permeability of the well, system flow rate, well diameter, amount of bleed air, system vacuum capacity and several other factors. The design should accommodate use of several different sizes of slurp tube. Clear Polyvinyl Chloride (PVC) piping is readily available for use as a slurp tube and serves as a flow indicator for monitoring the LNAPL and groundwater flow from individual wells.

Another inherent problem with a typical bioslurping configuration is that there is no way to quantify the volume of LNAPL recovered from individual wells. The individual LNAPL recovery rate will change over time in response to precipitation events and as the site is remediated. It is beneficial to monitor the individual well LNAPL recovery rate so that system adjustments can be made to maximize recovery volumes. The authors developed a method of quantifying the

LNAPL recovery rate during the completion of the subject bioslurping demonstration project. The method allows you to measure the ratio of vapors, groundwater and LNAPL in individual recovery piping at any time. The recovery well manifold is constructed with a calibrated (known volume) section of clear PVC piping equipped with a ball valve on both ends and a drain valve on the bottom. During system operation, both ball valves are simultaneously closed to trap the vapors, groundwater and LNAPL within the closed section of piping. The liquids are drained from the piping section into a graduated cylinder or other measurement device to quantify the vapor/groundwater/LNAPL ratios. The ratios can be determined several times for each well and averaged to increase the accuracy of the measurements.

Vapor/Liquid Separators. A vapor/liquid separator (VLS) is typically a tank or vessel designed to separate the vapors and liquids that have been recovered simultaneously in the slurp tube. The position, type, size, and complexity of the VLS is a function of the vacuum source utilized. The VLS can consist of a simple tank designed with a centrally located tangential inlet, a demister pad at the outlet, and a manual drain valve. However, more sophisticated and efficient centrifugal separators consist of a tank with a conical tuyere and agglomerator blades which impart centrifugal motion and move the liquids to the periphery of the tank. The liquids drain to a sump equipped with a level switch activated pump that pumps the LNAPL and water from the VLS. The VLS should also be equipped with a large diameter cleanout port and a sight tube. Our experience has shown that typical design and operational problems with a bioslurping VLS include:

- the drain pump cavitates because the operational vacuum is higher than expected or because the pump was not designed to overcome the VLS vacuum;
- the drain pump is undersized because the groundwater yield is higher than expected;
- mineralization from hard groundwater accumulates on level switches and prevents them from activating the drain pump and/or high-high shutdown;
- ferric hydrate is produced in the VLS (and piping/water treatment equipment) by iron bacteria due to the presence of iron rich (and/or manganese rich) and oxygenated groundwater.

Vacuum Source Alternatives. Bioslurping systems have typically been designed using a water-cooled LRVP. Single stage LRVPs produce up to 25 inches of mercury ("Hg.) and two stage designs can achieve greater than 29 "Hg. Water cooled LRVPs have been utilized because they produce relatively deep vacuums and can be configured to utilize the recovered groundwater (and LNAPL) to provide the necessary liquid coolant and sealant for the pump, thereby eliminating the need for a fresh water supply. Partial recovery LRVPs have generally been utilized because they come equipped with a VLS for recycling a portion of the coolant. Based on our experience, this configuration can cause several problems because the LNAPL/water/vapor mixture directly enters the pump and the VLS is positioned after the vacuum source. Problems with this configuration can include:

- premature seal failure due to the exposure of LNAPL to the pump seals;
- scaling on the internal pump surfaces originating from minerals deposited from hard groundwater;
- excessive emulsification of the LNAPL which results from the inherent water/LNAPL mixing that occurs inside the pump;
- solids deposition inside the pump and failure of the pump rotors due to abrasion when the groundwater contains excess solids; and,
- variable well yields and groundwater surges provide a variable coolant flow rate and may cause the pump to overheat or cavitate.

Emulsification of recovered LNAPL is an inherent problem with bioslurping systems because water, LNAPL and vapors simultaneously travel up the slurp tube together. However, based on our experience, using the recovered groundwater/LNAPL as a coolant in an LRVP can exacerbate this problem. Alternative configurations and vacuum sources can be used to resolve these problems and provide other benefits.

Alternative vacuum sources include: oil-cooled LRVPs; a specially staged dry vacuum rotary lobe blower (DVB); a rotary vane blower (RVB); and, a specially configured internal combustion engine (ICE). High vacuum (up to 16 "Hg.) "2nd generation" regenerative blowers are also available, but are non-positive displacement blowers and cannot operate as efficiently as other alternatives for bioslurping applications. An oil cooled LRVP provides some of the same benefits as a water cooled LRVP, but can still be operated without a continuous supply of fresh coolant. An oil cooled LRVP contains an oil reservoir and fan cooled heat exchanger to cool the recycled oil in the system. A DVB produces sufficient vacuum levels for bioslurping systems and can be operated without a continuous supply of liquid coolant. The DVB is designed with a unique jet plenum (or intercooler) that allows ambient air to be used to cool the blower. The ambient air does not enter the vapor stream. The authors have experience utilizing Roots model DVJ Whispair® DVBs. RVB's generally produce 18-20 "Hg. vacuum although two stage designs can achieve 28"Hg. RVBs are best suited when high flow rates are expected in more permeable geologic formations. An ICE can provide the necessary vacuum levels and simultaneously serve as the vapor treatment unit. ICE's are essentially automobile engines that have been customized with a special carburetor system that allows them to run on high BTU vapor; natural gas or propane as the primary fuel and recovered vapors are used as a supplemental fuel source. The authors have experience utilizing a Remediation Service, International (RSI) ICE capable of up to 18 "Hg and air flow rates of up to 80 cubic feet/minute (CFM). The oil-cooled LRVP, DVB, RVB, and ICE alternatives necessitate installing a VLS between the blower and the recovery wells to separate the groundwater/LNAPL from the vapor stream before the vacuum source.

BIOSLURPING DEMONSTRATION PROJECT

A bioslurping demonstration project was conducted at four active military bases that each contained laterally extensive diesel and/or jet fuel LNAPL plumes. The depth to groundwater ranged from six (6) to eighteen (18) feet below ground surface and the apparent LNAPL thickness ranged from one (1) to nine (9) feet.

The geologic conditions varied from fine sands and gravel to silty-clay overlying limestone bedrock.

Some sites contained existing equipment from prior projects and some sites had no equipment. Upon initiating the project, the existing designs were evaluated and the nature of the OM&M problems reviewed. The existing systems had been utilized for a number of pilot tests to confirm the feasibility of using bioslurping to recover LNAPL and remediate the sites. Although the systems had proven effective for short-term pilot testing, OM&M of the systems were considered excessive and system shutdowns were frequent. The objective of the demonstration project was to evaluate several process configurations to reduce the OM&M requirements and maintain a high runtime percentage. The existing equipment consisted of trailer mounted systems equipped with partial recovery water-cooled LRVPs that used recovered groundwater/LNAPL as a coolant.

The following mechanical/process configurations were evaluated for the demonstration project:

- an existing trailer mounted water-cooled LRVP with significant process modifications;
- an existing trailer mounted water-cooled LRVP without the process modifications;
- an oil cooled LRVP equipped with a VLS;
- a trailer mounted DVB equipped with a VLS; and,
- a trailer mounted ICE with a detached skid mounted VLS.

The DVB, ICE and oil cooled LRVP were each configured with a VLS between the vacuum source and the recovery wells. Each of these alternative vacuum sources were selected because they produce sufficient vacuum levels for bioslurping systems, but can still be operated without a continuous supply of liquid coolant. The ICE was also utilized because it can simultaneously serve as the vapor treatment unit. One existing trailer mounted LRVP was utilized to serve as a baseline and one existing LRVP system was modified to determine if significant process modifications could enhance the original system configuration.

PROJECT RESULTS

Each of the vacuum sources proved to be effective for implementing bioslurping. As expected, certain configurations operated more efficiently and were less OM&M intensive. Configurations that included a VLS between the vacuum source and the recovery wells generally performed better and provided several benefits including:

- the degree of LNAPL emulsification is less;
- negative effects associated with surges of groundwater were eliminated;
- the chance for mineralization to develop on the internal surfaces of the vacuum source was reduced; and,
- more efficient separation of vapors, groundwater and LNAPL

Although the ICE provides the benefit of serving as the vacuum source and vapor treatment system, it cannot produce as much vacuum as other sources, the flow capacity is relatively limited, and at sites with low BTU vapors, the primary fuel costs can be high. The DVB provides another good alternative to water cooled

LRVPs, but cannot operate in areas with sustained high ambient temperatures because it uses ambient air as a coolant for the blower. The oil cooled LRVP also worked effectively. However, systems equipped with explosion-proof motors are not as readily available, they may also be difficult to operate in areas with sustained high ambient temperatures (due to overheating of the oil), and, a more costly and sophisticated VLS may be required to eliminate the chance for groundwater to contaminate the oil coolant.

As discussed above, LNAPL emulsification is an inherent problem with bioslurping systems. However, problems described previously as emulsification of LNAPL in Pilot Test Reports for the sites actually originated from ferric hydrate deposition caused by iron bacteria. This problem occurs inherently with bioslurping systems due to the presence of dissolved iron concentrations in the recovered groundwater and the oxygenated water that results from the mixture of vapors and liquids in the slurp tube and VLS (and pump with LRVP systems). Iron bacteria oxidize ferrous (dissolved) iron to ferric iron, which is precipitated as ferric hydrate. Dissolved iron concentrations as low as 1 to 3 mg/l can cause this problem. Ferric hydrate is best described as a "pumpkin pie" like substance that can plug piping and foul pumps and valves. Some system designs limit the growth of iron bacteria (by minimizing turbulence and oxygenation) or can be configured to minimize the rate of ferric hydrate production. Chlorination can be used to control iron bacteria, but nonoxidizing microbicides such as ammonium halides, organo-sulfur compounds and phenolic-amine compounds are also effective.

CONCLUSIONS

The results of this demonstration project along with the authors' collective experiences confirm that practical experience, knowledge of vacuum systems and attention to detail during the OM&M phase of the project are key factors to successfully implementing bioslurping technology. Each of the different process configurations and vacuum sources can be successfully applied, however, alternatives to LRVPs are available and may prove to be more effective and less OM&M intensive for implementing bioslurping. LNAPL emulsification and ferric hydrate deposition/fouling are inherent problems with bioslurping. LNAPL emulsification can be minimized by using alternative vacuum systems, however, ferric hydrate deposition and fouling also occurs when using alternative vacuum sources and must be addressed through chemical addition.

REFERENCES

Gibbs, J. T., Leeson, A., Peyton, B., Mills, J., Kuch, D., and J. Spivey. 1997, *"Bioslurping/Biovenitng in Tight Soils With Dewatering to Remediate Large Smear Zone."* Batelle Memorial Institute, Volume 1, Pages 287-294.

Haas, P. E., and J. A. Kittel. 1997. *Determining the Feasibility of LNAPL Free Product Recovery: Bioslurping Versus Traditional Technologies.* Batelle Memorial Institute, Volume 1, Page 273.

Woodward, D. S. and M. S. Niederreither. 1998. *Practical Considerations for Optimizing Bisolurping Systems.* Proceeding of the 1998 Petroleum Hydrocarbons and Organic Chemicals in Groundwater Conference.

FULL-SCALE BIOSLURPING DESIGN, INSTALLATION, OPERATION AND COMPLETION

Ralph S. Baker, Gregory Smith (ENSR Corp.: Acton, MA, Warrenville, IL) and Fred M. Blechinger (Lucent Technologies Inc., Morristown, NJ)

ABSTRACT: Soil investigations during the decommissioning of an industrial tank farm consisting of eleven 50,000-gallon (189,000-liter) underground fuel oil storage tanks indicated that a release had occurred. Soils were silty-sand saprolite over fine silty-sand saprolite, with mean hydraulic conductivity equal to 4 x 10^{-4} cm/sec. The zone of residual fuel oil contamination was believed to extend from approximately 15 to 30 ft (4.6 to 9.1 m) below ground surface (bgs). In situ bioventing (BV) was initially selected for site remediation based on the results of laboratory biotreatability testing. A system was designed that incorporated groundwater extraction, reinfiltration of treated groundwater to optimize moisture content, and installation of BV wells. The BV system included three groundwater recovery wells and 68 air extraction/air induction wells, along with a vacuum pump, instrumentation, controls, piping and treatment facilities.

At the time of BV system installation and start-up early in 1995, light non-aqueous phase liquid (LNAPL) was observed for the first time in newly installed wells. Planning then commenced to convert some of the BV wells to bioslurping (BSL) to enhance free-product recovery and to initiate BV. Multiphase flow modeling (TIMES) was performed to optimize the BSL well spacing and design. Due to ease of conversion, a total of 29 BV wells were converted to BSL and operated over a period of three months during 1997, during which time over 250 kg of free product was recovered. The BSL system was then turned off, and no LNAPL reappeared after 15 months. BV to address residual NAPL has commenced and is ongoing. Reliance on in situ remediation has eliminated the need for excavation and off-site disposal of affected soils.

INTRODUCTION

In 1994, Lucent Technologies Inc. (then AT&T Network Systems, Inc.), with ENSR Corp. as its contractor, began decommissioning a tank farm consisting of eleven 50,000-gallon (189,000-liter) underground fuel oil storage tanks at its former manufacturing facility in Winston-Salem, NC. Soil investigations performed prior to and during tank removal indicated that a release of petroleum hydrocarbons (PHC) had occurred, with soil concentrations above the remedial goals of 40 mg/kg Total PHC. No LNAPL was evident at that time. Given the depth to which the oil had migrated, an in situ remediation method was desired. Consequently, a BV system was installed in the Tank Farm area to remediate the PHC-affected soils. The BV system included three groundwater recovery wells and 68 air extraction and air induction wells installed in a triangular grid at 6.1 m horizontal spacings, covering a 40 x 55 m surface area. Additional BV system components included a vacuum pump, instrumentation, controls, piping and treatment facilities. The purpose of the groundwater extraction wells was to lower the water table and thereby expose the deeper portion of the smear zone (i.e., below the pre-pumping water table) to in situ bioremediation.

In 1995, following start-up of the BV system, LNAPL consisting of free-phase No. 2 fuel oil was for the first time observed and recovered from the groundwater extraction wells. Consideration turned to methods of removing the LNAPL that might be compatible with the newly installed system. Conventional free product recovery techniques such as skimming were not viewed as attractive because of the relatively small horizontal and vertical extent of free-phase product, the limited available drawdown in the wells, coupled with the moderate permeability of the silty-sand saprolitic soil in the LNAPL-affected zone.

BIOSLURPING CONCEPTS

Although the exertion of vacuum on wells to enhance LNAPL recovery was a common oil production technique in the 1800s (Lindsley, 1926), it was only within the past fifteen years adopted by the remediation industry. The relatively recent term "bioslurping" incorporates into vacuum-enhanced free product recovery the benefits of stimulating, through the processes of BV, aerobic biodegradation of PHCs within the unsaturated zone (AFCEE, 1994).

A bioslurper system employs a suction tube positioned near the LNAPL-water interface to induce a pressure gradient causing water, LNAPL and/or air to flow into the extraction well. Typically, water and/or LNAPL that is drawn into the well is lifted and conveyed to an air-water separator. The liquid phase is subsequently conveyed to an oil-water separator. BSL systems are designed and operated in a manner that attempts to maximize LNAPL recovery while minimizing groundwater and air-phase recovery. The BV aspect of BSL is initially less important than the primary objective of enhancing free-product recovery. Since BV is integrated with BSL, however, treatment of residual LNAPL and PHCs in the unsaturated zone can begin during product recovery.

SITE-SPECIFIC APPLICATION OF BIOSLURPING

At the Winston-Salem Tank Farm area, the pre-pumping water table was at approximately 7.6 m bgs, and the LNAPL-affected smear zone ranged from approximately 4.6 to 9.1 m bgs. Numerous BV wells screened across that zone were already in place at the time that the LNAPL was discovered. Pump test results indicate that the silty-sand saprolitic soil within the LNAPL-affected zone had a geometric mean hydraulic conductivity of 4×10^{-4} cm/sec.

The overall objective of the BSL system was to remove recoverable free product within two years. To this end, there were three primary components of the development of BSL design specifications:

- *Subsurface design.* The number and spacing of BSL wells needed to be established. This task was initially accomplished with the aid of multiphase flow modeling.

- *Extraction well design.* Conversion of wellhead assemblies was necessary to accommodate insertion and adjustment of suction tubes.
- *Aboveground design.* Modification of the aboveground BV equipment was necessary to ensure that the required airflow and vacuum could be exerted, to incorporate oil/water separation, and to enable connection with the existing facility water treatment system.

BSL DESIGN MODELING

Analytical and numerical models that can simulate multiphase hydrostatics and flow may be used in the assessment and design of BSL systems. A comprehensive Engineer Manual on Multi-Phase Extraction (MPE) recently prepared by ENSR for the U.S. Army Corps of Engineers includes discussion of the application of models for MPE and BSL design (USACE, 1999).

Modeling objectives. Specific objectives of the modeling employed for the subject site (ENSR, 1996) included: a) to evaluate various well configurations to optimize number and location of BSL wells and achieve removal of recoverable free product within two years; and, b) to predict airflow rate and determine the maximum vacuum to be applied, for determination of required blower size.

Model Selection. The numerical model TIMES (1996) was used to facilitate this analysis. TIMES is a finite-element model that simulates the two-dimensional horizontal or vertical flow of water, air and LNAPL in an unconfined aquifer, as well as dissolved-phase transport. Capable of simulating heterogeneous and anisotropic hydraulic properties, TIMES simulates areal movement of water and free phase hydrocarbon under natural gradients as well as under conditions involving hydrocarbon skimming and vacuum-enhanced pumping.

Input Parameters and Modeling Scenarios. Oil specific volume may be computed from well product thickness data, if vertical equilibrium pressure distributions are assumed and soil capillary pressure-saturation relations are measured or can be estimated. Site-specific groundwater elevation data were employed. Literature values for soil parameters were used, except for saturated hydraulic conductivity, which was derived from site-specific data (ENSR, 1996). LNAPL properties were obtained from literature values for No. 2 fuel oil.

Eleven simulations were conducted to optimize the number and location of BSL wells. The numbers of wells considered were 10, 7 and 3, with applied vacuums ranging from 5 to 15" Hg. The simulations assumed no physical drawdown (i.e., suction tube positioned at the LNAPL/water interface).

RESULTS AND DISCUSSION

Model Results. Free oil specific volume computed from the well product thickness data for the assumed soil properties was approximately 600 gal (2300 l).

Simulations of product recovery were carried out for each case for 1000 days or until asymptotic recovery was achieved. Modeling results (ENSR, 1996) indicated that asymptotic recovery could be achieved in ≤1000 days with either 10, 7, or 3 wells, provided applied vacuums were ≥10" Hg (34 kPa). Associated airflow and water flow rates were predicted to fall within the ranges of 20-40 scfm (0.6-1.1 std. m^3/min) and 20-40 gpm (75-150 l/min), respectively. The results also suggested that although about 5-10% of the initially free oil would eventually become trapped due to water table mounding away from BSL wells, it could be addressed through removal by BV processes and dissolution. Most importantly, the model results predicted that even if only three BSL wells were operated (at an applied vacuum of 15" Hg [51 kPa] and steady-state airflow and water flow rates of 24 scfm [0.68 std. m^3/min] and 25 gpm [95 l/min]), approx. 60% of the recoverable product would be extracted within the first 100 days. This represented an average product recovery rate over that period of 3 gpd (11 l/d).

Field BSL Results. Initially three BV wells were converted to BSL, by means of inserting 1" dia. (2.54 cm) suction tubes through wellhead compression fittings. A Roots-Dresser 412 DVJ rotary lobe positive displacement blower was used to exert a vacuum of 15" Hg (51 kPa) and achieve a system-wide steady-state airflow of 500 scfm (14 std. m^3/min). Although LNAPL recovery rates were judged as adequate, they were too low to enable daily measurement. Because facility-supplied labor was available and moving the suction tube assemblies from well to well proved easy, the decision was made to cycle BSL among 29 BV wells, with approximately 3-6 BSL wells being operated at a given time. Apparent free product thicknesses of as much as 25" (51cm) were measured on 3/7/97, during the onset of BSL (Figure 1). By 4/25/97, ≤ 2" (5cm) of apparent free product was evident (Figure 2), and by 5/28/97, ≤ 1" (2.5cm) was evident. Over the three-month period of BSL operation in 1997, over 750 l of product was recovered, at a rapidly diminishing rate. The BSL system was then turned off. No LNAPL has reappeared during the ensuing 15 months, during which BV to address residual LNAPL has continued. In addition, although the single groundwater extraction well (TF-3) that continues to be operated is located within the area that formerly yielded free product, no sheen has been evident in the oil-water separator to which it discharges.

CONCLUSION

The full-scale conversion of an ENSR-designed BV system at the Lucent Winston-Salem site to BSL proved to be inexpensive and effective. The design work, including modeling, was carried out by ENSR for a fixed price of <$15,000. Operation of the BSL system for three months was sufficient to inexpensively solve the free product problem, and operation of BV for treatment of residual PHC is advancing towards closure.

ACKNOWLEDGEMENTS

The authors thank David K. Ramsden, Ph.D., author of the bioremediation operating manual for the site and its revision to accommodate BSL; Ram Pemmireddy, modeler; Mark McGlathery, design engineer; Roland B. Norris II, P.E., senior reviewer; and Henry Campbell, facility engineer.

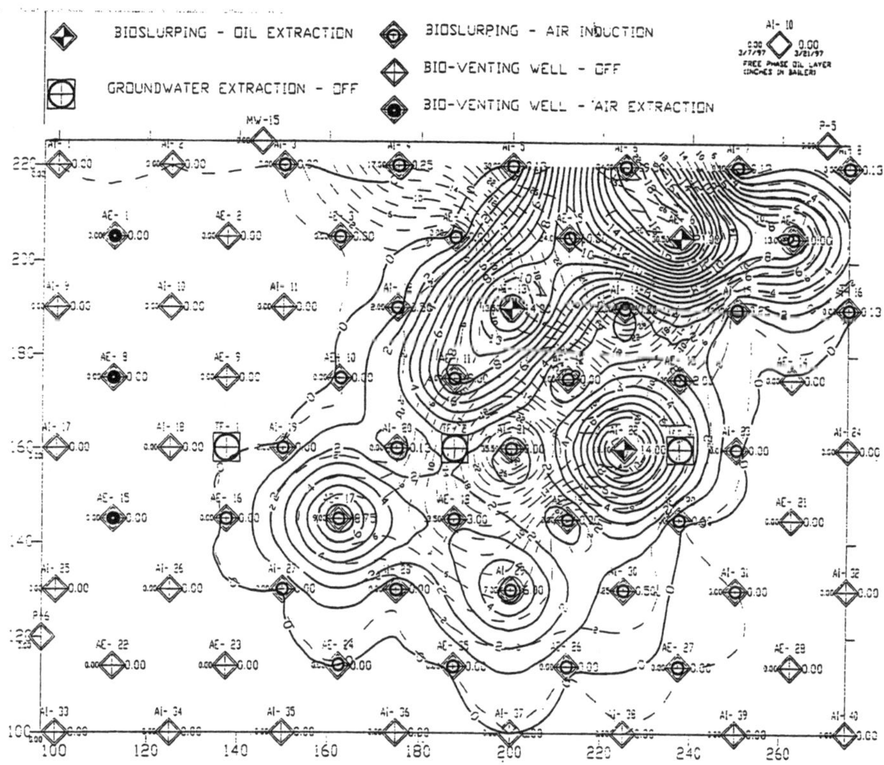

FIGURE 1. Apparent free product thickness (inches) during onset of bioslurping. Dashed and solid contours are for measurements made on 3/7/97 and 3/21/97, respectively.

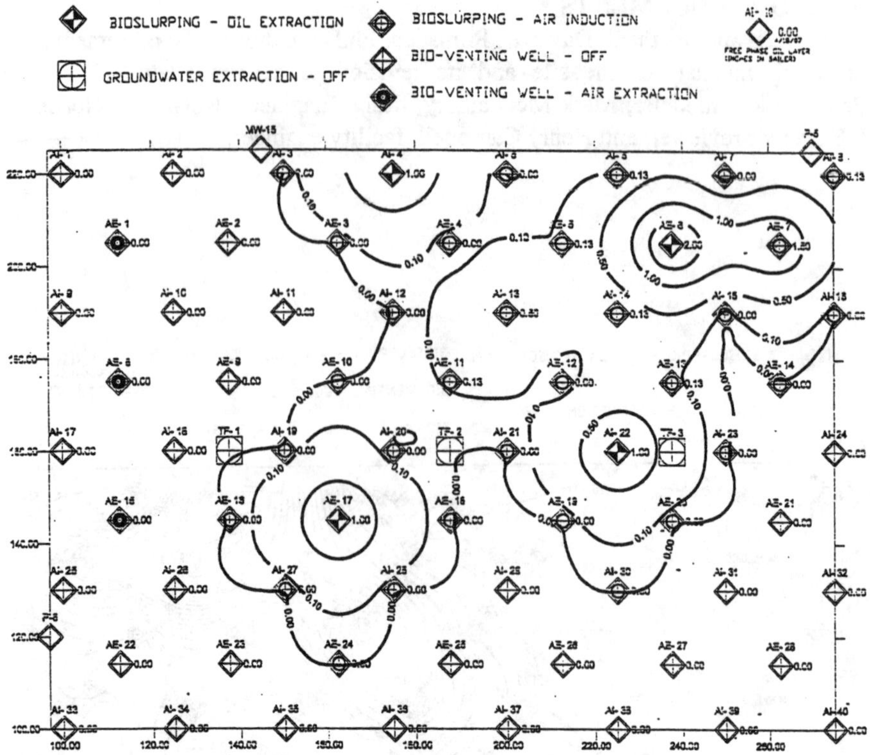

FIGURE 2. Apparent free product thickness (inches) measured on 4/25/97, after 7 weeks of bioslurping.

REFERENCES

Air Force Center for Environmental Excellence. 1994. *Technology Profile: Vacuum-Mediated LNAPL Free Product Recovery/Bioremediation(Bioslurper).* Issue 1. March, 1994.

ENSR, 1996. *Report on Modeling Conducted to Support Development of the Technical Specifications – Lucent, Winston-Salem, NC.* ENSR Consulting and Engineering, Acton, MA.

TriHydro Integrated Models for Environmental Solutions (TIMES). 1996. v. 1.0. TRIHYDRO Corporation, Laramie, WY.

USACE. 1999. *Multi-Phase Extraction: Engineering and Design*. Engineer Manual EM 1110-1-4010. Prepared by ENSR for U.S. Army Corps of Engineers.

INTRODUCING USACE'S MULTIPHASE EXTRACTION ENGINEER MANUAL

Ralph S. Baker, ENSR Corp., Acton, Massachusetts USA
David J. Becker, U.S. Army Corps of Engineers, Omaha, Nebraska USA

ABSTRACT: The U.S. Army Corps of Engineers (USACE) Hazardous, Toxic and Radioactive Waste Center of Expertise (HTRW CX) in Omaha, NE conducts an ongoing program to develop peer-reviewed guidance and guide specifications pertaining to technologies which it and its customers utilize. Multi-phase extraction (MPE) is a rapidly emerging, in situ remediation technology for simultaneous extraction of vapor phase, dissolved phase and separate phase contaminants from vadose zone, capillary fringe, and saturated zone soils and groundwater. It is a modification of soil vapor extraction (SVE), and is most commonly applied in moderately permeable soils. It is applicable to both light and dense non-aqueous phase liquids, as well as to both dissolved and sorbed volatile and semi-volatile organic compounds.

The HTRW CX recognized that due to the rapidly expanding application of MPE technologies including two-phase extraction, dual-phase extraction, and bioslurping, standardized guidance was needed to ensure that MPE is used only where it is appropriate. In October 1997, a multidisciplinary team of in situ remediation experts at ENSR began a 15-month long effort to prepare an MPE Engineer Manual for the HTRW CX. The EM (USACE, 1999) presents the principles underlying MPE and contains specific guidance for critically evaluating the feasibility of performing MPE, including tools such as checklists and flow charts. The EM identifies the physical, chemical, and biological site characterization data needed for feasibility assessment and design of MPE systems; suggests means to better screen the applicability of MPE during feasibility studies; provides guidance on the performance and monitoring of MPE pilot tests; and presents detailed design considerations for MPE systems. In addition, the EM addresses operations and maintenance, system shutdown, confirmation of cleanup; and maintains consistency with previous guidance developed by ENSR for USACE regarding two technologies often used in conjunction with MPE: SVE and bioventing (USACE, 1995). In a few months, anyone will be able to access an Adobe Acrobat® version of the MPE EM at http://www.usace.army.mil/inet/usace-docs/eng-manuals/em1110-1-4010/toc.htm. Note that the USACE SVE and Bioventing EM can be obtained at http://www.usace.army.mil/inet/usace-docs/eng-manuals/em1110-1-4001/toc.htm.

MULTI-PHASE EXTRACTION BACKGROUND

In-situ soil and groundwater remediation techniques are increasingly being relied on as methods that are less expensive than excavation and that do not simply move the contamination to another location. However, the limitations of many solitary in-situ technologies are becoming more apparent, especially longer-than-expected remediation times. In addition, solitary technologies may only treat one phase of the contamination when, in fact, the contamination is often spread through multiple phases and zones. For example, SVE and bioventing treat only the vadose zone, and groundwater pump-and-treat removes dissolved material only from the saturated zone. Most recovery systems for removal of light non-aqueous phase liquid (LNAPL) rely on gravity alone to collect and pump the LNAPL. In contrast, MPE can extract: groundwater containing dissolved constituents from the saturated zone; soil moisture containing dissolved constituents from the unsaturated zone; LNAPL floating on the groundwater; non-residual LNAPL in soil; perched or pooled dense non-aqueous phase liquid (DNAPL), under some conditions; and, soil gas containing volatile contaminants. MPE is therefore a technology that finds its widest use in source areas.

MPE Application Strategies. One generally chooses MPE to enhance the extraction of one or more of the following phases:

- NAPL, to accomplish free product recovery.
- Soil gas, to accomplish mass reduction through SVE or BV in soils having low air permeabilities.
- Groundwater, to improve pump-and-treat yields. (This objective is the least common of the three.)

These application strategies may be pursued separately or in combination. For example, a reason for implementing MPE may be to accomplish contaminant mass removal from saturated zones via both gas- and liquid-phase extraction; another may be to improve mass removal from the vadose zone primarily via gas-phase extraction.

In general, MPE works by applying a high vacuum (relative to SVE systems) to a well or trench that intersects the vadose zone, capillary fringe and saturated zone. Because the resulting subsurface pressure is less than atmospheric, groundwater rises and, if drawn into the well, is extracted and treated aboveground before discharge or reinjection. If liquid and gas are extracted within the same conduit (often called a suction pipe or drop tube), this form of MPE is termed bioslurping (BSL) when used for vacuum-enhanced LNAPL recovery, or two-phase extraction (TPE), often when used to address chlorinated solvents. If separate conduits for vapor and liquids are used, we term the technology dual-phase extraction (DPE). The terms TPE and DPE more commonly refer to situations where LNAPL is absent. If LNAPL is present floating on the water table, it will also flow into the well screen and be removed. Due to the imposed vacuum, soil moisture and/or NAPL retained by capillary forces within the soil can, to some degree, also move to the well for collection and removal. Over time, the groundwater level may be lowered; thereby creating a larger vadose zone that can be treated by the SVE aspect of MPE. The soil gas that is extracted is, if necessary, conveyed to a vapor-phase treatment system (i.e., activated carbon, catalytic oxidation, etc.), prior to its discharge.

Because air movement through the unsaturated zone is induced during MPE, oxygen can stimulate the activity of indigenous aerobic microbes, thereby increasing the rate of natural aerobic biodegradation of both volatile and non-volatile hydrocarbon contamination.

IMPORTANT CONDITIONS AND PROCESSES AFFECTING MPE

MPE systems impose a vacuum on the subsurface via extraction wells that are screened across the stratigraphic interval from which water, air, volatile hydrocarbons, and/or NAPL is to be extracted. MPE systems also provide the means for removing each of the fluid phases that arrive at extraction wells. As vacuum propagates from such a well out into the formation, a pressure gradient is established that is the driving force for fluid flow toward the well. Whether flow of NAPL, water and/or air is induced through the formation and into the well depends on a number of factors: the vacuum imposed, the saturation of each fluid and the history of saturation, the pore sizes occupied by each fluid, the associated permeabilities of the various available pathways, and the fluid properties (e.g., density, viscosity).

Since vacuum is propagated most readily within the subsurface if it can be applied through contiguous air phase pathways, MPE effectiveness generally depends on there being unsaturated soil or soil that can be dewatered enough to transmit airflow during MPE. It also depends on the distribution of such air-filled pathways within the soil, relative to the distribution of contaminants. Early MPE practitioners claimed that MPE could be applied to both low and moderate permeability soil. Recent evidence, however, suggests that although high vacuums applied during MPE can dewater the largest pores (i.e., preferential pathways), they may not produce conditions that can dewater matrix blocks within low permeability soil (Baker and Groher, 1998). Furthermore if matrix blocks are not dewatered, diffusion-limited mass transfer from such blocks to advective pathways can be excessively prolonged (McWhorter, 1995). Therefore, soil structure and stratigraphy are important determinants of the effectiveness of MPE, as well as other fluid-based in situ remediation technologies. In addition, the ability of soils to become dewatered enough to transmit airflow depends on their capillary pressure-saturation relationship. The associated air permeability varies strongly as a function of both saturation and capillary pressure and differs greatly for various soil types.

NAPL recovery is a process of managing conditions within the NAPL flow paths to optimize NAPL flow. The EM emphasizes that adequate characterization of NAPL extent is necessary prior to selection of free product recovery method. As is the case with MPE to dewater soils for the purpose of accomplishing SVE, the effectiveness of MPE to enhance NAPL recovery depends on the distribution of vacuum influence exerted on the formation relative to the distribution and connectedness of NAPL-filled pores. These and related factors are discussed in the EM, and guidance is provided as to how the relevant soil properties can be determined, as well as how to use such data in technology screening decisions and in MPE design.

HISTORY OF THE USACE MPE ENGINEER MANUAL EFFORT

The USACE HTRW CX noted in 1996 and 1997 that there was a significant increase in the consideration of MPE, especially for product removal, on USACE projects. In many cases, the controlling factors affecting the applicability of MPE were clearly not incorporated in the analyses or designs. In order to improve the designers' understanding of the important considerations in evaluating the applicability of MPE and in designing MPE systems, work on a comprehensive Engineer Manual (EM) was begun in late 1997.

The HTRW CX contracted with ENSR to prepare the EM in September of 1997 and directed the ENSR team to provide guidance on the appropriate site characterization, pilot testing, design approach, start-up, operation and maintenance, and site closure in support of MPE projects. The EM was to address the various configurations of MPE, including DPE, TPE, bioslurping, and DNAPL recovery. The guidance was to be as specific as possible, and meant for the engineers and scientists actually performing or overseeing the design work or operations.

ENSR assembled a team to prepare various sections of the EM. These include: Ralph Baker, Rob Bukowski, Liz Denly, Jim Galligan, Dan Groher, Josh Lieberman, Alan Moore and Ellen Moyer. The EM received the benefit of review comments from Dave McWhorter of Colorado State University, George Mickelson of Wisconsin Department of Natural Resources, and Tom Peargin of Chevron Research and Technology. Numerous USACE personnel provided comments as well, including Chuck Coyle, Bill Crawford, Rod Dolton, Phil Durgin, Bernie Gagnon, Dee Ginter, Cheryl Groenjes, Jack Keeton, Tom Liefer, Tomiann McDaniel, Okan Nalbant, Rex Ostrander, Doug Pendrell, Clif Rope, Don Schlack, Johnette Shockley, Ted Streckfuss and Laura Tate. Currently the EM is complete and is in publication.

MPE EM CONTENTS AND EMPHASIS

The EM provides information and guidance on all phases of a MPE project, from site characterization through design to operations and closure. It emphasizes both pore-scale and large-scale processes that affect the success of MPE and describes appropriate sampling and characterization methods to gain a site-specific understanding of these processes.

The EM provides flow charts to assess the applicability of MPE to two general classes of application, product recovery and enhanced SVE/bioventing/vacuum dewatering (Figures 1 and 2). Key parameters include soil permeability, apparent NAPL thickness, volatility, and biodegradation parameters. The EM recommends pilot testing methods and data analysis techniques. It encourages the consideration of monitoring liquid saturation levels, particularly where MPE is meant to apply SVE to a dewatered zone. This may involve the use of neutron probes or other repeatable non-destructive measurement techniques.

The EM presents a strategy for subsurface system design for both NAPL recovery and enhanced SVE/BV that considers the interrelationship between air, water and liquid production. This strategy is a general "road map" for selecting

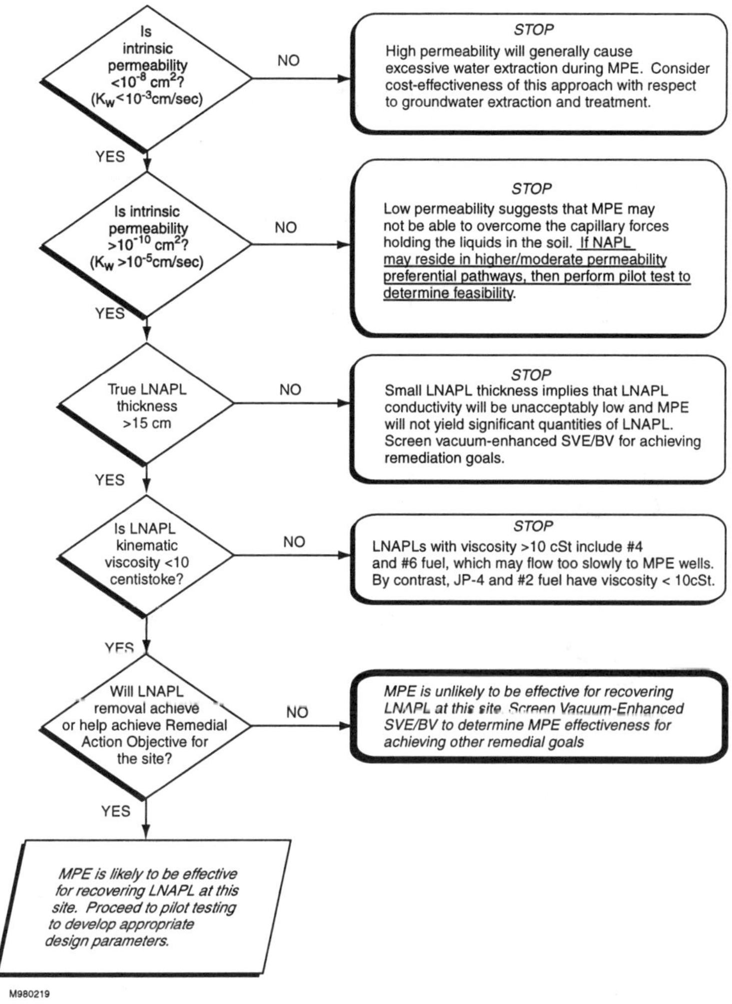

Figure 1. Technology screening matrix: vacuum-enhanced LNAPL (free product) recovery.

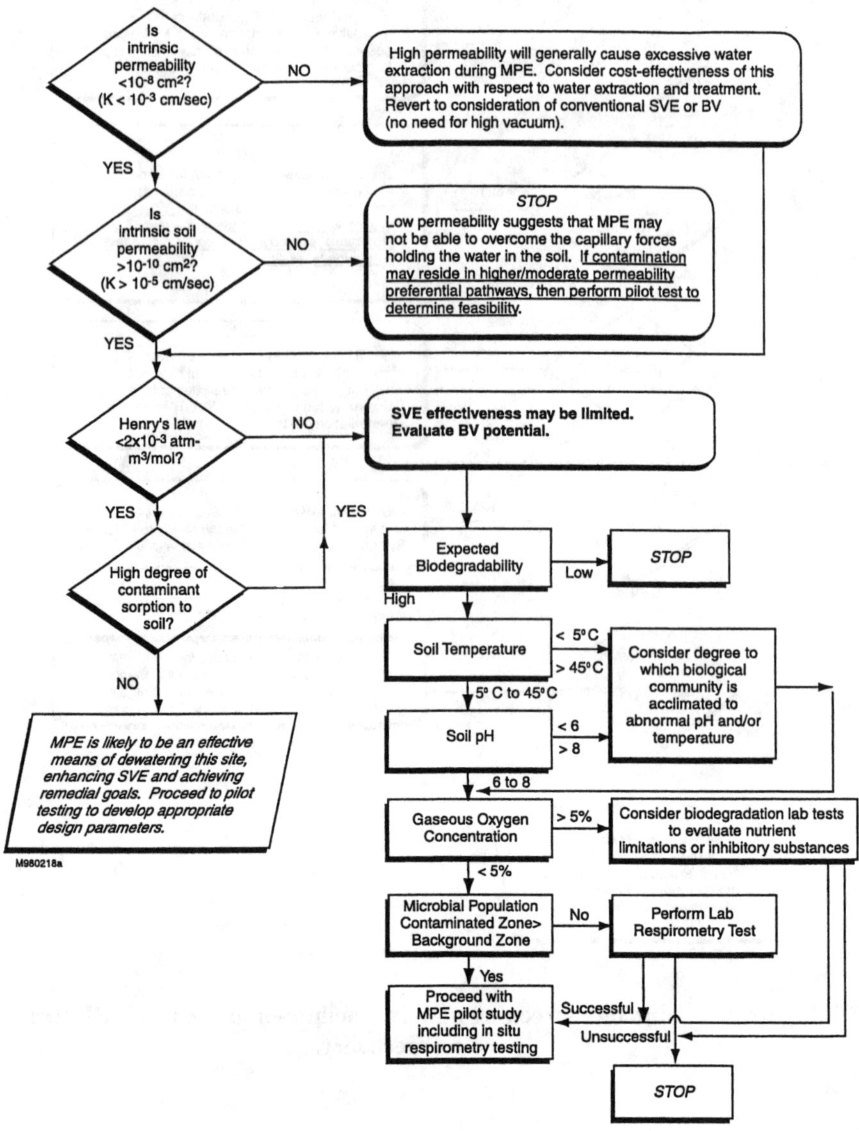

Figure 2. Technology screening matrix: Vacuum-enhanced SVE/BV, incl. vacuum dewatering.

well layout and predicting flows of air and liquids. The EM recommends the use of computer models for simulating the flow of air and liquids, and summarizes the commercial or public domain models currently available. The EM also provides detailed design guidance, including aboveground equipment selection, well construction, and piping design. Although brief discussions are provided on air and water treatment systems, the EM refers to other guidance for detailed design of those components.

The EM provides guidance on the start-up, operations, and maintenance of MPE systems. Several checklists are provided, including ones addressing: 1) baseline data to collect prior to startup, 2) pre-commissioning/shakedown checks, 3) parameters to monitor during operation,
4) operational performance considerations, 5) a system troubleshooting guide, 6) an operational strategy guide, and 7) typical maintenance activities.

The EM also summarizes regulatory, patent, and health and safety issues related to MPE.

SUMMARY

The effects of the simultaneous application of a vacuum and extraction of fluids, including immiscible liquids are not well understood by the many designers of MPE systems. The strong dependence of flow on saturation is the controlling process. In order to improve both MPE technology screening and design, the EM provides guidance on site characterization, pilot testing, design, and operation that considers this interrelationship.

REFERENCES

Baker, R.S. and Groher, D.M. 1998. "Does Multiphase Extraction Require Soil Desaturation to Remediate Chlorinated Sites?" pp. 175-180. In Wickramanayake, G.B., and Hinchee, R.B. (Eds.) *Physical, Chemical and Thermal Technologies: Remediation of Chlorinated and Recalcitrant Compounds*. Battelle Press, Columbus, OH.

McWhorter, D.B. 1995. "Relevant Processes Concerning Hydrocarbon Contamination in Low Permeability Soils." In T. Walden (Ed.), *Petroleum Contaminated Low Permeability Soil,* pp. A1-34. Pub. No. 4631. American Petroleum Institute, Washington, DC.

USACE. 1995. *Soil Vapor Extraction and Bioventing: Engineering and Design.* Engineer Manual EM 1110-1-4001. Prepared by ENSR for U.S. Army Corps of Engineers.

USACE. 1999. *Multi-Phase Extraction: Engineering and Design.* Engineer Manual EM 1110-1-4010. Prepared by ENSR for U.S. Army Corps of Engineers.

OPTIMIZING LNAPL RECOVERIES USING COMBINED EXTRACTION TECHNOLOGIES

Drew F. Lessard, U.S. Air Force, Holloman Air Force Base, New Mexico
David L. Rizzuto, Indian Environmental, Alamogordo, New Mexico
Dino Bonaldo II, Mission Research Corporation, Albuquerque, New Mexico

ABSTRACT: A 3-month pilot study was performed to determine if down-hole hydrocarbon pumps, equipped with hydrophobic filters, can be used as an adjunct to vacuum recovery systems for the efficient recovery of light non-aqueous phase liquid (LNAPL) hydrocarbons. Since onsite groundwater is classified as non-potable, it not necessary to remediate it beyond the removal of any LNAPL thickness. The study goal was to achieve a higher rate of LNAPL removal without an increase in costs (i.e., additional water treatment), to insure the greatest volume of LNAPL recovery from the subsurface per dollar spent. Application of various stepped vacuums and pumping intervals resulted in an average LNAPL recharge rate increase by an approximate factor of five (5X). Increased product recoveries from 1.5 gallons (gal)/day/well in Phase I of the study to greater than 10 gal/day/well in Phase II were recorded. The dynamic wellfield management techniques tested increased pre-study LNAPL product recovery by 59 percent when compared to similar time intervals for the previous 12-month operational period of high-vacuum dual-phase extraction (HVDPE) operation alone.

INTRODUCTION

Vacuum-enhanced technologies have been demonstrated to successfully remediate hydrocarbon-impacted soil in low permeability formations. Combining technologies and realizing combined benefits, which are greater than additive without each technology's limitations (synergistic effect), are essential to moving sites toward closure sooner. In the pilot study performed, an existing HVDPE system was combined with down-hole free-product skimmer pumps.

A significant feature of vacuum-enhanced recovery is the induced airflow, which in turn induces LNAPL flow toward the extraction well. The pressure gradient created in the air phase results in a driving force on the LNAPL that is greater than that which can be induced by pumping the LNAPL with no induced airflow (gravity). Also of importance is the fact that airflow created by the vacuum enhances the LNAPL content surrounding the extraction well. This accumulation around the well ensures that the permeability controlling the conductivity of LNAPL is maximized. For these reasons, in addition to mass removal via soil vapor extraction, vacuum-enhanced recovery has the potential for removing more LNAPL and at greater rates than do other source removal mechanisms. However, the increased vacuums required to bring the recovered LNAPL to the surface from greater depths brings with it significant volumes of groundwater. Groundwater extracted during vacuum-enhanced recovery must be handled, treated, and discharged, which increases costs in necessary equipment and long-term operations.

Alternatively, skimmer systems withdraw little to no groundwater. The selected skimmer has a polyethylene LNAPL collection filter, a 2.5-ft (0.76 m) self-adjusting depth range capable of compensating for rising/falling groundwater effects, and weighted to float at the groundwater/LNAPL interface. The skimmer filter is hydrophobic and has a programmable timer, which periodically applies air pressure to the collection chamber to push accumulated LNAPL to the surface. In general, the resulting limited pressure head (actual free-product-saturated formation thickness) provides minimal LNAPL drive toward the extraction well. This passive movement typically yields relatively low hydrocarbon recovery rates.

Objective. Determining the effect of using skimmer pumps as an adjunct to vacuum-enhanced recovery of LNAPL was the objective of the pilot study. The study evaluated the efficacy and feasibility of combining two existing technologies in an effort to increase LNAPL recovery, while minimizing groundwater removal, thus decreasing equipment and long-term operational costs.

Site Description. In 1991, approximately 2,000 gal (7,571 liters [L]) of JP-8 jet fuel leaked from underground lines leading from an aboveground storage tank at the T-38 Test Cell (TC) on Holloman Air Force Base (HAFB). The base is located in southern New Mexico, 7 miles west of the city of Alamogordo. In subsequent investigations 450,000 to 485,000 gal of JP-8 LNAPL were detected over an 11-acre area. HVDPE was selected to remediate the site. An interim system consisting of 11 extraction wells was constructed in 1994. A full-scale system, consisting of 122 additional wells capable of remediating the entire 11-acre site, was constructed in 1996.

Shallow soil beneath the TC extending to approximately 15 ft (4.6 m) consists of intercalated layers of alluvial sediment. Upper soil is typically tan to light brown, fine-grained, silty sand/sandy silt or silty clay/clayey silt. Underlying this layer to approximately 30 ft (9.1 m) is reddish-brown silty clay/clayey silt with interbedded fine-grained silt and sand. Discontinuous caliche layers are also present across the site. Moisture contents of vadose zone soil ranges from 2 to 11 percent, increasing with depth. The heterogeneic character of HAFB soil, combined with the high percentage of fine-grained silt present throughout the shallow soil, results in low permeabilities. These characteristics limit the migration of the LNAPL plume, but also hinder hydrocarbon recoveries.

Groundwater occurs at 6 to 20 ft (1.8 to 6.1 m) below ground surface depending on topography surrounding the TC. Hydraulic conductivities, based on slug tests, range from 7.9×10^{-5} to 2.9×10^{-4} centimeters/second. Porosities averaged 33 percent. However, effective porosities are interpreted to be significantly less than 33 percent in most cases (Woodward-Clyde, 1994).

The gypsiferous nature of the site soil has resulted in a total dissolved solids content in excess of 10,000 milligrams/liter in most of the basin.

Groundwater beneath HAFB is classified as non-potable under the New Mexico Water Quality Control Commission Regulations. Therefore, it does not require remediation beyond the removal of any LNAPL contamination that might exist.

SCOPE OF PILOT STUDY

As with most remediation sites, extensive operational costs are associated with necessary equipment and ongoing operation and maintenance efforts to keep systems operating as close to optimal as possible. Altering the TC-recovery system to become more flexible (efficient) would be extremely beneficial, financially and temporally, in the long run. Two tests were performed to assess proposed modifications to increase system performance while reducing operation and maintenance costs. Phase I involved baseline data collection and the application of down-hole hydrophobic skimmer pumps to remove the free (mobile) phase product. Phase II included vacuum-enhanced recovery and the concurrent use of the hydrophobic pumps.

Phase I: Baseline Data Collection. Pilot study preparation (equipment procurement, system configuration, etc.) and collection of baseline field data including fluid-level measurements (depth to groundwater and LNAPL thickness) and baildown tests were performed in the first 4 weeks. Ten existing 4-inch recovery wells were selected for the study: five wells with the greatest observed LNAPL thickness (D-Section, best recovery potential) and five wells with the smallest observed LNAPL thickness (B-Section). Baildown tests were performed and static (no vacuum enhancement) LNAPL recovery rates were recorded.

Phase II: Vacuum Step Tests and Optimization. Incremental (low to high) vacuum pressures were applied to the test wells and LNAPL recovery rates were recorded for the remaining 8 weeks of the study period (system was not operated over the weekend). Recovered free product was collected in a 55-gal drum at each well head and was measured daily. Based on site soil and previous operational knowledge, vacuum steps were 10, 20, and 30 inches of water column (IWC). Various skimmer-pump cycle times (both interval and duration) were tested in an effort to observe the effect on recovery and/or any hysterisis effects.

DISCUSSION OF RESULTS

During the 3-month study period, the skimmers removed an approximate total of 7,812 pounds (lbs) (3,543 kg) or 1,260 gal (4,770 L) of free phase LNAPL from the subsurface and only 31 gal (117 L) of water. Hydrocarbon mass removed in the aqueous phase was estimated based on the minimal groundwater extracted and the average concentration of total hydrocarbons detected in the monthly regulatory samples. The vapor extraction portion of the study removed approximately 9,000 lbs (4,082 kg) or 1,452 gal (5,495 L) based on laboratory analytical results. Overall, the investigation removed approximately 16,819 lbs (7,629 kg) of total hydrocarbons or 2,713 gal (10,270 L), based on a laboratory tested specific gravity of 0.83 for the recovered fuel, almost half in the free phase. Table 1 summarizes the hydrocarbon recovery data.

Table 1. Summary of Hydrocarbon Recovery Data

Hydrocarbon Phase	Volume Removed gal [L]	Mass Removed lbs [kg]	Percent of Mass Removed
Liquid (Free Phase)	1,260 [4,770]	7,812 [3,543]	46.5
Aqueous	1.2 [4.5]	7.4 [3.4]	< 0.1
Gaseous	1,452 [5,495]	9,000 [4,082]	53.5
Total	2,713 [10,270]	16,819 [7,629]	100.0

In situ respirometry tests were performed at the conclusion of the pilot study. The tests confirmed *in situ* microbial activity was occurring, was oxygen-limited, and was enhanced by the combined extraction methodology via the induced air flow. Although beyond the scope of this study and not included in the cumulative product-removal results, the *in situ* respirometry test will be used to estimate an additional hydrocarbon removal (biodegradation) rate.

Increased Recovery Rates. As illustrated in Figure 1, product recovery rates increased following vacuum application, increasing the initial recovery rates from 1.5 gal/well/day to greater than 10 gal/well/day. Approximate recharge rate increase factors ranged from 2.5 for the poorest recovery well to 10 for the greatest recovery well, averaging 5.6.

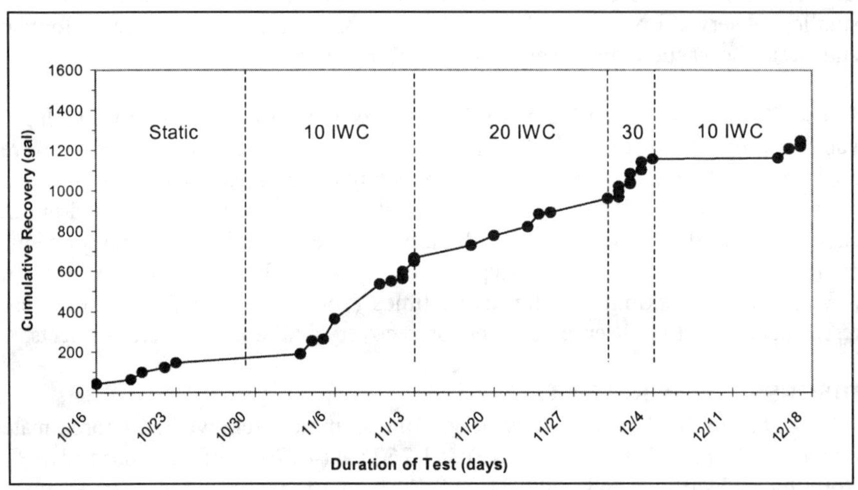

Figure 1. Free Product Recovery Versus Time Throughout Pilot Study.

The first vacuum step was performed at 10 IWC. Vacuum-enhanced hydrocarbon recoveries increased by a factor of four (4X) and the free product was recovered with no groundwater. Properly positioned skimmer pumps (adjusted for product thickness groundwater depression) reduced the required maintenance necessary for efficient operations. At 20 IWC vacuum, additional maintenance activities were required. Recoveries were not only less than at 10 IWC, but emulsification of the LNAPL caused water to be entrained in the recovered fuel and increased required filter replacements. Similarly at 30 IWC vacuum, although recovery rates were temporarily higher, maintenance was required more frequently and recoveries greater than one or two well volumes were not sustainable. Product saturation in the soil surrounding each well was successfully reduced, however, long-term recovery diminished to volumes less than recoveries obtained at 10 IWC. Figure 2 illustrates the step tests and hydrocarbon recovery rates at well D-10.

Additionally, given the low permeability of the site soil, groundwater up-welling at both of the higher applied vacuums reduced the accessible LNAPL saturated formation thickness. This is of particular interest in thin product-bearing formations; since the limiting factor for product yield is that drawdown cannot exceed the saturated thickness of the product-saturated zone (assuming no induced gradient). So a balance must be obtained in that the highest achievable gradient is desired, with the least amount of induced up-welling reducing the product-bearing formation, and limiting recovery to the well.

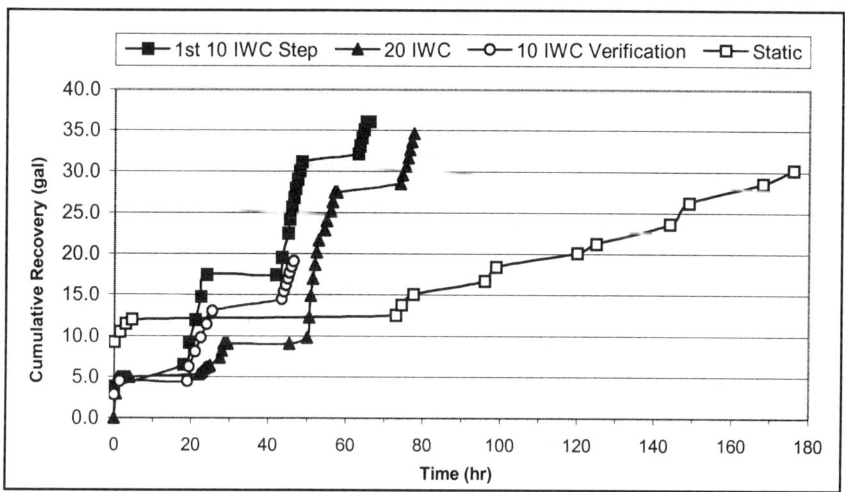

Figure 2. Vacuum-Enhanced Recovery Relative to Static Recovery.

For verification and repeatability, the final study vacuum step was repeated at 10 IWC. Repeating the step test provides a more accurate basis for comparing sustainable LNAPL recovery rates with other removal technologies. Similar to the first step test at 10 IWC, identical recoveries were observed in the last step.

Pump interval and duration were varied to find the effect on recovery. Pump intervals were initiated at 1.5 hours and durations of more than 10 minutes. Well recoveries were not great enough to sustain these durations, therefore, durations were decreased as recoveries were increasing ultimately using a 30-second duration at an interval of every 30 minutes. This combination of interval and duration, applying 10 IWC vacuum, was found to be the most productive in terms of the greatest LNAPL recovery rate.

Engineering Comparison. HVDPE is the voluntary corrective action selected technology to remediate the site. The two remedial objectives are removal of any LNAPL thickness and a total petroleum hydrocarbon soil concentration below 1,000 parts per million in site vadose zone soil. Historical HVDPE free-product recovery rates were reviewed and compared to pilot study recoveries. Results indicated the combined methodology recovered 59 percent more free-product.

HVDPE, on a larger scale, concurrently achieved progress toward both cleanup goals, though not as efficiently or as cost effectively as the demonstrated combined methodology. The increase in source-removal rates (decrease in LNAPL thickness), with minimal increase in costs (i.e., initial skimmer capital, equipment, and filters, etc.), was substantial.

The demonstrated combined methodology proved a great adjunct to the existing system. The goal satisfied by skimming systems is the collection of free product with little to no groundwater. Concurrent realization of both remedial objectives will save costs, opposed to vadose zone remediation completed prior to LNAPL thickness removal due to higher mass-transfer rates in the vapor phase during HVDPE. This sequential approach is inefficient and will result in prolonging the remedial timeframe, increasing long-term operation and maintenance costs.

Concerns/Limitations. The following concerns/limitations warrant mentioning:

- Application of higher vacuums would be expected to produce proportionately higher yields of free product. However, groundwater up-welling (in lower permeability soil) decreases the accessible product-bearing saturated formation and, therefore, limits recovery to the well.
- Hydrophobic filters used in this study were density selective. Higher vacuums also resulted in limiting recoveries as a result of water entrainment, as the free product emulsified to the eventual point of being pumped with water.
- The increased mass-removal rates and the effectiveness of the combined technology are directly dependent on site characteristics (geologic, hydrogeologic, and contaminant characteristics, etc).
- Vacuumed-enhanced increased removal rates in higher conductivity soil must be weighed against reduced control of free-phase plume migration.

CONCLUSION

Increased mass-removal rates decrease total removal costs. The investigation results demonstrated the combined extraction methodology ensured effective source removal, which will ultimately decrease long-term operation and maintenance costs. Combining existing technologies and realizing the combined benefits without each technology's limitations are essential to moving sites toward closure sooner.

REFERENCES

Woodward-Clyde. 1994. *RCRA Facility Investigation Report for SWMUs 229 and 230.* Prepared for Holloman AFB, Environmental Restoration.

ANALYSIS OF BIOVENTING AT EIELSON AIR FORCE BASE, ALASKA

Daniel McKay (U.S. Army Corps of Engineers, Hanover, NH)

ABSTRACT: Source zone removal is a necessary precursor to effective ground water treatment through natural attenuation. Two sites undergoing active vadose zone remediation at Eielson Air Force Base, Alaska, were studied to compare conventional bioventing of unsaturated soils with bioventing accompanied by air sparging. Here, an investigation of injected air distribution is described. The shallow water table at each site has a typical seasonal fluctuation of 0.6 m, normally limiting the available range of air distribution for soil oxygenation and subsequent aerobic respiration. At site ST20 E-9, air is injected in wells screened above the seasonal high water table. The system installed at ST10 introduces air through screens that are completely submerged below the seasonal low water table. Air sparging tests were conducted at both sites to assess the benefit or potential benefit of sparging to enhance biodegradation in the seasonally saturated smear zone. Methods to evaluate airflow distribution included neutron probe air saturation tests, helium tracer tests, and measurements of soil pressures and oxygen levels near the water table. At ST20 E-9, saturated zone airflow followed lateral preferential pathways below the targeted smear zone, yielding desultory effects near the water table. It was thus concluded that bioventing at this site would not receive value added from the introduction of air below the water table. The data support the use of sparging at ST10 however, due in large part to the macroscale uniformity of soil properties.

INTRODUCTION

Sites contaminated with light non-aqueous phase fuel hydrocarbons (LNAPLs) are typically characterized by a smear zone containing residual-phase petroleum that remains following free (mobile) product recovery. Because fuel hydrocarbons are lighter than water, non-aqueous product is suspended in soil above whatever may be the current water table. Climatic or seasonal fluctuation of the water table on the scale of tens of centimeters or even meters is typical of most sites. With the rise and fall of groundwater elevation, coupled with rain or snow melt events, the smear zone provides a persistent source of groundwater contamination. The U.S. Air Force has adopted a policy of source removal in combination with natural attenuation of groundwater to treat most petroleum-contaminated sites. Active attenuation of contaminated soils by the injection of air to deliver oxygen to aerobic soil bacteria has been termed bioventing. Less restrictive capillary effects and a higher diffusion coefficient for soil gas relative to pore water allow significantly more uniform and efficient mixing of oxygen in unsaturated soils as compared to saturated soils. Consequently, air is generally supplied only to the region above the current table water.

Bioventing to remedy a smear zone may be accelerated in some cases by injecting air beneath the seasonally fluctuating water table (air sparging). To be effective, the site must be conducive to uniform air flow in the saturated soils in a way that does not adversely affect the distribution of oxygen in the unsaturated smear zone. Mechanisms that increase the rate of hydrocarbon reduction are volatilization from groundwater and oxygenation with subsequent biodegradation. Chemical equilibrium drivers as defined by constituent sorption coefficients, K_d, may then reduce the level of sorbed hydrocarbon in saturated smear zone soils.

Objectives. Two sites at Eielson Air Force Base, Alaska, were studied to evaluate the distribution of air above and below the water table when air is injected into saturated soils. At site ST10, a full-scale bioventing system had been constructed for air injection into the saturated zone. At site ST20 E-9, a full-scale system was constructed to inject air above the water table, limiting active treatment to the unsaturated soils. As the nature and distribution of stratigraphy, soil properties, and contaminants are unique to each site, the objective was to assess the limitations or benefits of saturated zone aeration specific to these sites. Limitations of two-phase flow are largely attributed to zones of contrasting permeabilities and wetting properties that typically exist within a targeted region. This study sought to establish the pattern of saturated zone and unsaturated zone air flow with regard to the observed stratigraphic delineation. The tools and approaches utilized can be applied at other locations undergoing design considerations.

Site Description. Eielson Air Force Base is located approximately 39 kilometers southeast of Fairbanks, Alaska, within the Tanana-Kuskokwim Lowland physiographic province. Climate is subarctic, characterized by large diurnal and seasonal variations and low precipitation. Average winter temperatures range between -26 and -13°C while average summer temperatures are between 7 and 16°C.

The unconfined aquifer is composed of 60 to 90 m of loose alluvial sands and gravel overlying bedrock and is characterized by high transmissivity. Static water is 2-3 m below the ground surface (BGS) with about 0.6 m of seasonal fluctuation owing to precipitation and recharge from the Tanana River. Discontinuous permafrost is present while the depth of active frost may be two or more meters.

Site ST10, a source area within Operable Unit 2 (OU2), is the location of the E-2 tank farm consisting of six 2500-m^3 aboveground tanks for storage of JP-4 jet fuel and aviation gasoline. Boring logs describe the soils at as poorly graded sand with silt from 0 to 2 or 3 m BGS, overlying well-graded gravel with sand. A full-scale air sparging / bioventing system has been operating since November 1995. Contaminants of concern are benzene, ethylbenzene, toluene, xylenes (BTEX) and napthalene.

Site ST20 E-9 is a refueling loop complex consisting of a fuel pump house and three 190,000-L underground tanks (USTs) for storage of JP-4 and one 95,000-L defuel UST. The shallow vadose and aquifer material is characterized as sandy gravel and silty sand with layering of aquifer materials on a scale of several centimeters to meters. A full-scale bioventing system has been operating

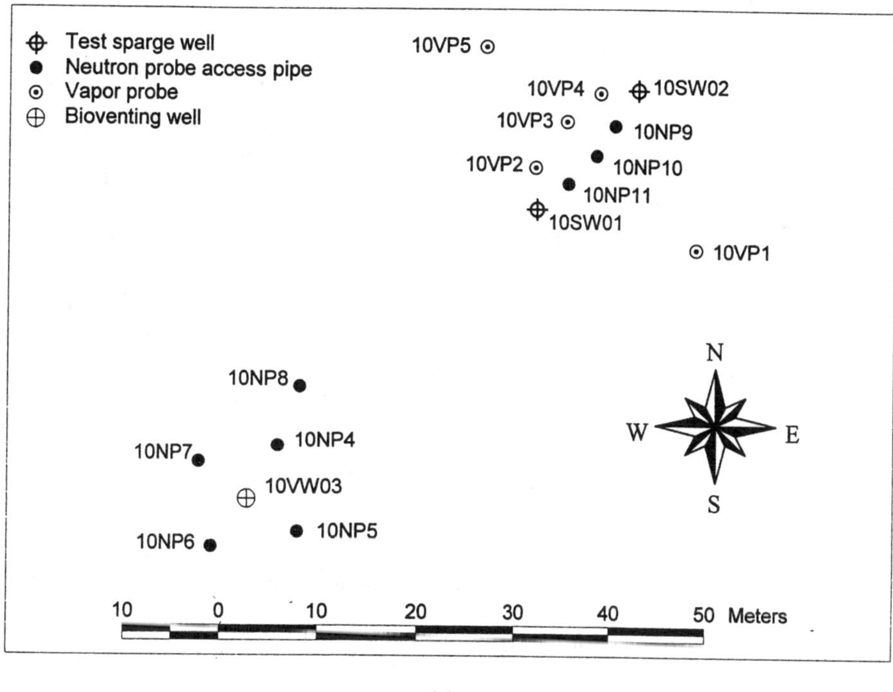

(a)

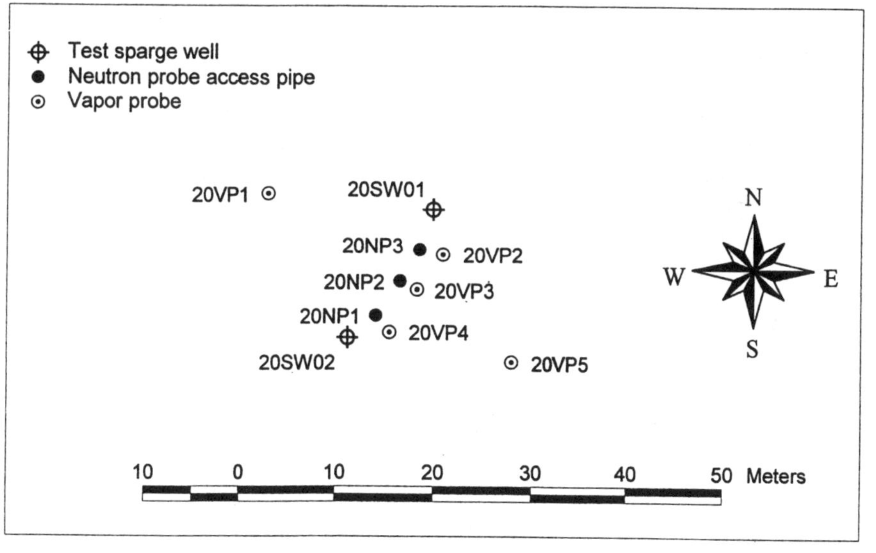

(b)

FIGURE 1. Test site layout. (a) Site ST10. (b) Site ST20-E9.

since 1993. Contaminants of concern are benzene, ethylbenzene, toluene, xylenes (BTEX) and napthalene.

MATERIALS AND METHODS

Saturated zone air flow during sparging was characterized by changes in soil moisture as measured with a model 501 borehole neutron moisture/density probe (Campbell Pacific Nuclear, Martinez, CA). Two air injection wells were each installed at ST10 and ST20 E-9 to assess the impact of flow rate and injection depth on flow uniformity and radius of influence (Figs. 1-a and b). SW01 and SW02 were placed at overall depths of 5 m and 5.8 m BGS respectively. The wells were constructed of 5.1-cm PVC riser extending above a 60-cm long slotted (0.025-cm) screen. The upper screen elevations of 10SW01 and 20SW01 were 1.89 and 2.49 m below the water table during the period of measurements. The neutron probe access pipes were constructed of 5.1-cm (39-mm wall thickness) black iron pipe with a welded drive point. These were driven into the ground to eliminate cuttings and minimize air flow between the pipe and the formation below the water table.

Additional investigation was performed at site ST10 since a full-scale sparging system was already operating. Four neutron probe access pipes were uniformly spaced a distance of 1.8 m from injection well 10VW03 to assess flow uniformity. A fifth access pipe was installed at a distance of 3.6 m to evaluate the radius of influence. The upper screen elevation of 10VW03 was 1.91 m below the water table during the period of measurements.

Soil gas distribution was characterized by measurements of injected helium, oxygen and pressure in the subsurface. Soil gas probes were placed at various distances near the injection wells and at midpoint elevations between 33-53 cm above the water table. The stainless steel wire screen implants, connected to low density polyethylene tubing, were embedded in an average of 33 cm of sand, then overlaid by bentonite to seal the borehole.

Air was injected into each well, SW01 and SW02, for two hours at a target flow rate of 600 L/min. Following each two hour period, helium was added to the air stream at 5 std. L/min. Helium in soil gas was measured with a model 9822 helium detector (Mark Products, Sunnyvale CA) calibrated for 0.02, 1.0, and 100% concentrations. Oxygen was measured with a model GT201 portable gas monitor (GasTech, Newark, CA) calibrated in atmospheric air.

RESULTS AND DISCUSSION

Soil Vapor. Prior to any air injection, helium, oxygen, and pressure were measured at the soil gas probes to confirm proper detector and monitor operation and establish initial conditions. All values were zero except for 1.4 and 2.4 % oxygen levels at 20VP1 and 20VP5. Measurements were recorded again after approximately two hours of air injection at 11 std. L/s. Injection pressure ranged from 30-37 kPa among all the wells. The results are summarized in Table 1. Maximum helium concentrations are slightly higher than predicted per the measured flow rate (1.1%), indicating bias in either the mass flow controller or the helium detector. However, the data are useful for observing consistent conditions throughout the Site 10 test area and non-uniform flow behavior at site 20. Both oxygen levels and pressure attained notably higher and consistent values at site 10 compared to site 20.

TABLE 1. Summary of soil conditions during air injection test.

10SW01	10VP1	10VP2	10VP3	10VP4	10VP5
Pressure (cm H_2O)	14.73	15.24	19.05	15.75	9.65
Oxygen (%)	20.2	20.7	20.5	20.3	20.4
Helium (%)	1.2	1.3	1.4	1.4	1.2
10SW02					
Pressure (cm H_2O)	19.05	8.89	15.49	33.02	5.59
Oxygen (%)	19.0	19.9	19.9	20.2	19.3
Helium (%)	1.0	1.2	1.4	1.5	1.2
20SW01	**20VP1**	**20VP2**	**20VP3**	**20VP4**	**20VP5**
Pressure (cm H_2O)	5.33	17.53	9.65	6.86	23.11
Oxygen (%)	19.8	20.9	20.2	19.6	20.9
Helium (%)	0.0	2.3	1.6	0.2	2.3
20SW02					
Pressure (cm H_2O)	3.05	2.79	4.32	6.86	14.48
Oxygen (%)	16.2	15.0	16.4	19.2	15.4
Helium (%)	0.00	0.51	0.25	0.23	0.06

Air Saturation. Changes in water content were measured with the neutron probe to determine the percentage of voids containing air after approximately two hours of air injection at 10-12 std. L/s. The results are presented in Figures 2 (a)-(c). Single sample uncertainty attributed to the combined random error of measurements (initial and test conditions) were determined to range from 0.7 to 1.5 %. Actual uncertainty due to all possible factors was not considered.

A satisfactory degree of homogeneity is exhibited at site 10, as similar air flow behavior is observed in all directions. At sparge well 10SW01, air saturation near the water table (i.e. smear zone) is about 5% at all three locations. At Site 20, air saturation approaches zero near the water table, suggesting excessive horizontal flow beneath a lower permeability layer. The relatively high air saturation (28%) about 1 m beneath the water table indicates a likely location of this layer.

CONCLUSION

The non-uniform distribution of injected air at ST20E-9 as evidenced by distributed measurements of saturated zone air content, soil gas pressure, oxygen and helium tracer indicate that smear zone remediation would not receive added benefit sufficient to justify air sparging. Thus the vadose zone bioventing system that is currently operating is appropriate.

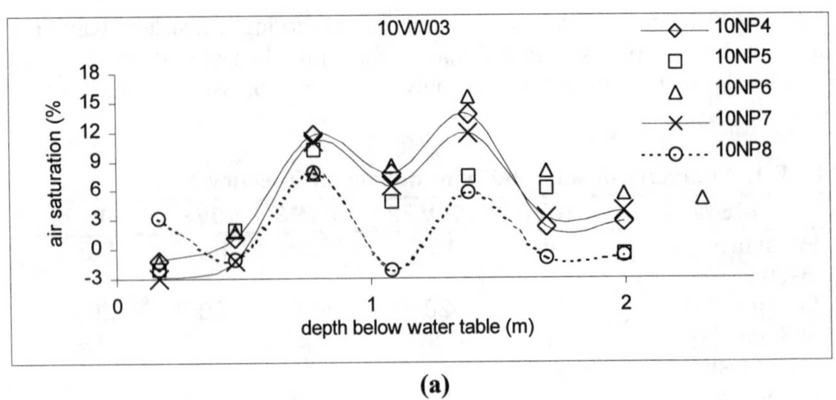

(a)

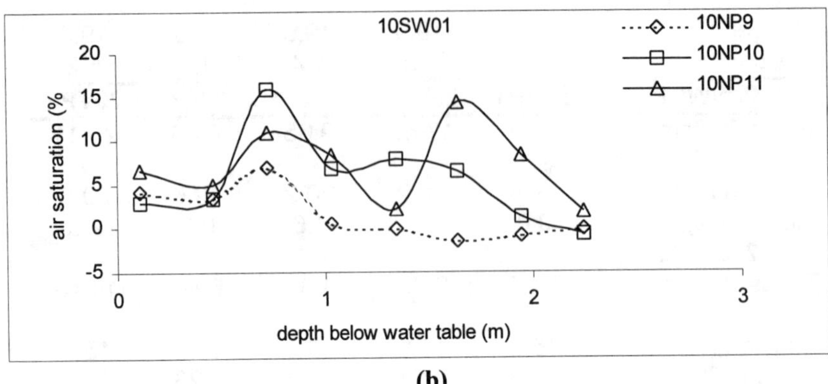

(b)

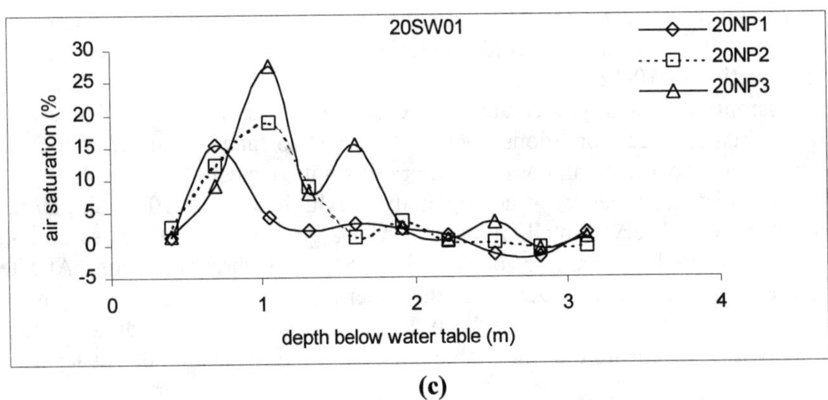

(c)

FIGURE 2. Measurements of air saturation below water table during sparging. (a) Site 10 bioventing well 10VW03. Air flow rate is 11.9 std. L/s. (b) Site 10 test well 10SW01. Air flow rate is 11.6 std. L/s. (c) Site 20 test well. Air flow rate is 10.8 std. L/s.

Conversely, uniform air distribution was observed out to 3.6 m horizontally from the air injection wells at site 10 (measurements at greater distances were not recorded). The 5% air saturation observed in the smear zone is sufficient to warrant consideration of air sparging as a supplement to vadose zone bioventing.

ACKNOWLEDGEMENTS

Research funds were provided to the Army Corps of Engineers through Eielson Air Force Base. Special thanks are extended to Conrad Christianson and Mike Raabe for logistical support.

DEVELOPMENT OF A REAL-TIME CONTROL SYSTEM FOR THE BIOVENTING PROCESS

Philippe Vanderby, Josée Gagnon, Michel Perrier and Réjean Samson (NSERC Industrial Chair in Site Bioremediation, École Polytechnique de Montréal, Montreal, Quebec, Canada)
Denis Millette (Biotechnology Research Institute, National Research Council, Montreal, Quebec, Canada)

ABSTRACT: In this work improvement of the bioventing technique is considered by using a real-time control system. The real-time control system provides the appropriate air flow rate to maintain the oxygen gas phase concentration at a desired set point. Two packed soil columns of 22 L are used to simulate the bioventing process. The soil used in the experiments was mostly contaminated with heating oil with an average concentration of 11 400 mg/kg of total petroleum hydrocarbons (TPH). While performing dynamic tests, air flow rates and oxygen concentrations in the gas phase are monitored on-line, with mass flow controllers and an electrochemical sensor, respectively. A dynamic input-output (I/O) model is identified from experimental data while forcing transient states by varying inlet air flow rates. Results from the transient tests reveal that the dynamic input/output model of the process is well represented by a first order plus time delay transfer function and the model parameters are then used to obtain the tuning constants of a proportional-integral (PI) controller. The controller allows for a robust dynamic performance in closed-loop. Set points ranging from 13 to 16,5 % are attained in satisfactory response times.

INTRODUCTION

Soil venting methods, used for decontamination of the vadose zone, induce air flow in the subsurface using an above-ground vacuum blower/pump system. Off-gas usually requires treatment. Soil venting techniques include soil vapor extraction (SVE) and bioventing, which have two very different objectives. The primary goal of SVE is to volatilize hydrocarbon constituents *in situ* to the gas phase. The principal objective of bioventing is to allow for sufficient oxygen transfer to stimulate aerobic biodegradation of petroleum hydrocarbons by the indigenous microorganisms. However both processes occur due to air circulation in the vadose zone.

Relatively high air flow rates typically ranging from 1 to 15 pore volumes per day are common in SVE operation (Dupont, 1993). For bioventing, design guidelines to evaluate air flow rate are based on gas phase oxygen concentration (U.S.EPA, 1995a). To maintain aerobic conditions in the contaminated soil area, minimal gas phase oxygen concentrations of 2% (Dupont, 1993), 5% (U.S.EPA, 1995a) or 10% (U.S.EPA, 1991) are given. Typical air flow rates usually range than from 0.1 to 0.5 pore volumes a day (Dupont, 1993) or 0.25 to 1 pore volumes per day (U.S.EPA, 1995b). Lower air flow rates hence increase the percentage of hydrocarbon removal by biodegradation and decrease the percentage of removal by volatilization (U.S.EPA, 1995b).

Oxygen soil gas concentrations depend on oxygen utilization rates of the indigenous microorganisms. These oxygen utilization rates are influenced by several factors: biomass activity and concentration, soil moisture, pH and temperature as well as nutrient and contaminant concentration. These factors can vary with time and according to spatial coordinates of the contaminated soil area. Moreover, field scale respiration test, first described by Hinchee and Ong (1992), show that oxygen utilization rates of the indigenous microorganisms generally decline as bioventing treatment progresses with time (U.S.EPA, 1995b). Widrig and Manning (1995), in a study of no. 2 diesel fuel biodegradation in soil columns, observed a steady decline in bacterial respiration over 240 days. The most likely explanation for this phenomenon is that the more easily degraded constituents are biodegraded in the first phases of treatment.

Objective. The goal of this study is to develop a control system where the appropriate air flow rate is adjusted in real-time to maintain the oxygen concentration in the gas phase at a desired set point. This paper presents experimental results of dynamic tests where the gas phase oxygen concentration is studied in relation with the air flow rate. The dynamic behavior of this system is modeled using system identification techniques and subsequently the appropriate process model is used for the evaluation of the controller tuning constants. Controller performance is then verified in closed-loop by changing set point values.

Rationale. Implementation of a control system to bioventing is beneficial in three ways. First, the required oxygen is provided in real-time to maintain aerobic conditions essential to the microorganisms. Consequently, air flow rates are not unnecessarily high, which means that volatilization of petroleum hydrocarbons and the need for off-gas treatment is reduced. Secondly, soil contamination levels can be estimated in real-time using a stoichiometric oxygen mass balance and prediction of cleanup time is evaluated more accurately. Finally, the electrical energy consumption of the blower/pump system decreases.

MATERIALS AND METHODS

Two stainless steel (grade 316) columns of approximately 22 L (1.1 m long and 16 cm inner diameter) were used to contain the soil. Stainless steel ends with Teflon™ o-rings were used to close up the columns at the ends. Glass beads (particle diameter ranging from 0.52 to 0.59 mm) were filled at the base of the column to a height of 16 cm to insure adequate air distribution to the soil. The soil was packed manually in the remainder of the columns with a metal cylinder rod. Before entering the column the air was humidified by passing through a 2 L Erlenmeyer flask containing water.

Air flow rates were regulated with mass flow controllers (model 1179a, MKS instruments, Nepean, Ontario). Gas phase oxygen concentrations were measured at the column outlets with an electrochemical sensor (model 308 BWP,

Nova, Hamilton, Ontario). Solenoid three way stainless steel valves (Asco, Brantford, Ontario) automatically send the selected outlet air from the appropriate column towards the oxygen sensor. Tygon® tubing with internal Teflon™ coating was used to connect the apparatus mentioned above with the columns inlet and outlet. Air flow rates and gas phase oxygen concentrations values were monitored on-line using the LabVIEW software for data acquisition.

The soil used in the experiments was contaminated with heating oil with an average concentration of 11400 mg/kg of total petroleum hydrocarbons (TPH).

RESULTS AND DISCUSSION

System identification techniques were used to identify the appropriate dynamic model of the process (Ljung, 1987). To insure that aerobic conditions were maintained throughout the column experiments were realized with oxygen concentration of about 10%. Calculations of the parameters of the models were performed with the MATLAB® System Identification Toolbox. Because experimental results from columns 1 and 2 were reproducible, illustrative data will be presented and discussed. Results in open loop will be from column 1 and in closed-loop from column 2.

A pseudorandom binary sequence (Ljung, 1987) (PRBS) input signal with a minimal switch period of 4 h and for air flow rates fixed between 13.3 and 17.6 ml/min for column 1 and 14 and 18.3 ml/min for column 2 was used for model identification.

Figure 1 shows the dynamic response of the oxygen gas phase concentration for the PRBS on air flow rate for column 1(with y, experimental data and y_m, model output). The dynamic behavior of the gas phase oxygen concentration is quite reproducible from day 1 to 11. However, an unsought drop of almost 10 °C in ambient temperature occurred for a few hours around day 11, and the oxygen uptake rate (for both columns) decreased provoking the upwards deviation of the gas phase oxygen concentration. Consequently, only the experimental data from days 1 to 11 are used to model the process dynamics. An auto-regressive with exogenous input (ARX) model adequately represented the process dynamics for both columns and the model parameters for the continuous transfer function, which are of first order with a time delay, are (for columns 1 and 2 respectively):

- steady state gain ($\%O_{2g}$/ml/min) of 1.086 and 1.019
- dominant time constant (hours) of 11.71 and 12.93
- time delay (hours) of 1.00 and 0.00

The results from the PRBS indicate that process dynamics are well represented by a first order plus time delay transfer function. A proportional-integral (PI) control algorithm is therefore appropriate and usually gives good results in closed-loop (Marlin, 1995; Ogunnaike and Ray, 1994). The controller parameters were evaluated with the direct synthesis method (Ogunnaike and Ray, 1994) with a closed-loop time constant of half the dominant time constant to insure robust dynamic performance.

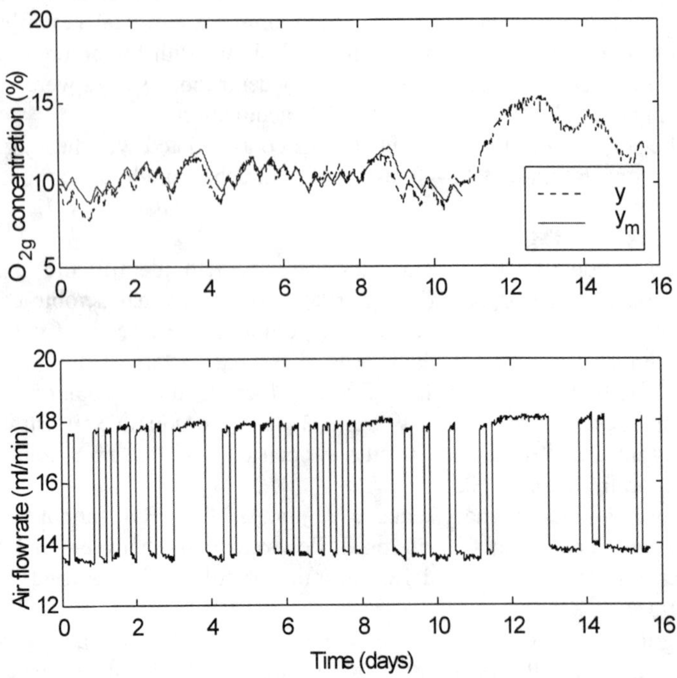

FIGURE 1. Dynamic behavior of oxygen concentration in the gas phase during PRBS on air flow rate.

Using the model found for column 2 from the PRBS test, a controller gain of 1.96 (ml/min / %O_{2g}) and integral time constant of 12.93 (h) were obtained. Results presented in closed-loop were obtained 4 months after initial testing began. As expected, a decrease in bacterial respiration was observed with time. Controller performance had to be evaluated for air flow rates according to the mass flow controller specifications. Consequently, operating conditions were different from the identification experiments, and the corresponding oxygen concentrations became higher than 10%.

Figure 2, shows the dynamic response of oxygen concentration in the gas phase and air flow rate for step changes on the set point (y_{sp}) ranging 13 to 15% (fig. 2a) and 16.5 to 13% (fig. 2b). It can be seen that the set point was reached in a short period, considering that the dominant time constant was evaluated as 12.93 hours. Comparison between air flow rates in figs. 2a and 2b shows that the response time in closed-loop system was slower for lower air flow rates. Since residence time of oxygen is inversely proportional to air flow rate, it seems reasonable to think that the time of response will be slower for lower air flow rates.

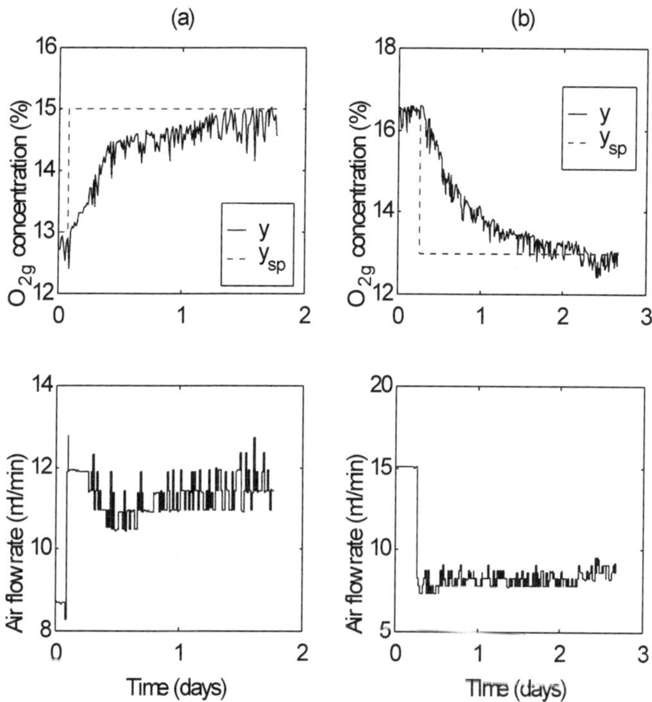

FIGURE 2. Step changes on set point. (a) First change. (b) Second change.

CONCLUSIONS

A control system for the bioventing process is proposed and a dynamic model relating oxygen concentration in the gas phase with air flow rate was identified through system identification experiments. Results from the transient tests reveal that process dynamic are well represented by a first order plus time delay transfer function. Tuning constants of a PI controller were determined from the parameters of the transfer function. The controller allowed for a robust dynamic performance in closed-loop. Set points ranging from 13 to 16,5 % were attained in satisfactory response times.

Having established the technical feasibility of this control system at a laboratory scale, the next step is to evaluate the economic advantages of automating the bioventing process. A study will be performed to determine the economic benefits relating to: off-gas treatment due to reduction of volatilization of petroleum hydrocarbons, better monitoring of the process, and decreased electrical energy consumption by the blower/pump system.

ACKNOWLEDGEMENTS

The authors acknowledge the financial support from the industrial Chair partners: Alcan, Bodycote/Analex, Bell Canada, Browning-Ferris

Industries, Cambior, Centre québécois de valorisation de la biomasse et des biotechnologies (CQVB), Hydro-Québec, Natural Science and Engineering Research Council (NSERC), Petro-Canada and SNC-Lavalin.

REFERENCES

Dupont, R. R. 1993. "Fundamentals of bioventing applied to fuel contaminated sites." *Environmental Progress.* 12(1): 45-53.

Hinchee, R. E. and Ong, S. K. 1992. "A rapid in situ respiration test for measuring aerobic biodegradation rates of hydrocarbons in soil." *Journal of Air and Waste Management Association.* 42(10): 1305-1312.

Ljung, L. 1987. *System identification: theory for the user.* Prentice-hall, Englewood Cliff.

Marlin, T. E. 1995. *Process control: designing processes and control systems for dynamic performance.* McGraw Hill, New York.

Ogunnaike, B. A. and Ray, W. H. 1994. *Process dynamics, modeling, and control.* Oxford University Press, New York.

U.S.EPA. 1991. *Innovative treatment technologies: overview and guide to information sources*, U.S. Environmental Protection Agency, Washington D.C.

U.S.EPA. 1995a. *Bioventing principles and practice volume II: bioventing design*, U.S. Environmental Protection Agency, Washington D.C.

U.S.EPA. 1995b. *Bioventing principles and practice volume I: bioventing principles*, U.S. Environmental Protection Agency, Washington D.C.

Widrig, D. L. and Manning, J. F. 1995. "Biodegradation of no. 2 diesel fuel in the vadose zone: a soil column study." *Environmental Toxicology and Chemistry.* 14(11): 1813-1822.

OPERATING MODE IMPACTS ON BIOVENTING SYSTEM PERFORMANCE

R. Ryan Dupont, Thimappa Lakshmiprasad
(Utah State University, Logan, UT, U.S.A)

ABSTRACT: A series of 6-in diameter 6-ft long soil columns was used to evaluate the effect of soil gas purge rate (0.5 and 1.0 pore volume/d) and purge scheduling (pulse venting at 0.5 pore volume/purge) on the degradation, volatilization and overall mass removal of JP-4 constituents from a native Utah soil. Pulse-venting reduced the overall volatilization of JP-4 constituents from the column by 42%, while increasing the initial removal rate of JP-4 constituents attributed to biodegradation by a factor of three to four. The pulse operating mode improved overall JP-4 removal efficiency as compared to continuous flow systems while using 35 to 70% less air during treatment. Degradation accounted for 89 to ≈100% of the mass removal observed in the columns for toluene, ethylbenzene, p-xylene, n-heptane, n-octane, n-nonane, and naphthalene. Benzene removal via biodegradation was maximized within the pulse-vented column, however, a total of 72% of the benzene mass was still attributed to volatilization losses under this modified flow mode.

INTRODUCTION

The evaluation of air flow management for optimizing bioventing system performance has been carried out by a number of investigators. Increased biodegradation and decreased volatilization with decreasing air flow rates was documented in a field scale study of JP-4 degradation conducted at Tyndall AFB, Florida (Miller et al., 1991). Pulse cycle venting has been applied at a gasoline-contaminated site to reduce blower operating time by more than 90% without compromising contaminant removal efficiency (Downey et al., 1995). Methods to reduce vapor extraction rates and maximize vapor retention times in the soil result in minimizing volatilization, potentially eliminate the need for off gas treatment, maximize the utilization of oxygen in situ.

The objective of this study was to evaluate the relationships among air flow rates, biodegradation rates, and volatilization rates under simulated bioventing conditions to develop an optimal operating mode that maximized biodegradation and minimized volatilization of JP-4 jet fuel constituents.

MATERIALS AND METHODS

Experimental Apparatus. The laboratory soil microcosms were made of glass columns of 200-cm length and 15.25-cm I.D. They were provided with 12 side sampling ports spaced at approximately 15-cm intervals. The sampling ports were made of glass O-ring joints through which 1-mm I.D. metal tubes were passed and extended to the center of the soil column. The other end was fitted with a Swage Lok fitting with Teflon ferrules and a septum to facilitate sampling of the soil gas with gas tight syringes. Glass caps on the top and bottom of the column were provided with fittings for atmospheric air inlet and for the application of vacuum. Vacuum was applied through the fitting in the bottom cap via the in-house vacuum that was regulated to the required vacuum with a vacuum regulator (Spectrum®, Houston). The atmospheric air was passed through an activated carbon bed to remove hydrocarbons prior to entering the column. An Optiflow 650 digital flow meter (Humonics®, California) was connected at the inlet to measure the soil gas flow rate.

The ends of each column were packed with gravel to assure uniform flow through the soil column. A total of 54.8 kg (120.9 lb) of soil were added to each column to achieve a bulk density of 1.5 g/cc. The soil was acclimated by adding 10 ppm of JP-4 jet fuel to soil maintained at 75% field capacity, and was then loaded into the columns and allowed to sit for 15 days. The required vacuum condition was initiated and the %O_2 depletion across the column was measured. This was quantified as background respiration and was subtracted from the subsequent oxygen depletion measured after introducing the contaminant to the columns. The top 1/3 of the soil in each column was contaminated by injecting 68.7 mL of JP-4 through the top two sampling ports to yield an initial contaminant concentration of 1,000 ppm. The soil column was then sealed and allowed to equilibrate within the column for approximately 3 days.

Uncontaminated soil from Hill AFB, Utah, was used in the experiments. Prior to use the soil was air dried and sieved, and material passing a 2 mm sieve was retained for use in this study. The soil has a sandy texture, with a pH of 7.2, 0.1% organic carbon, a CEC of 3.0 meq/100 g, a field capacity of 6.6%, a total porosity of 30%, and a packed soil air permeability was in the range of 80 to 90 Darcys.

The JP-4 jet fuel used in this study was supplied by Hill AFB and was weathered by bubbling compressed air through it at 10 mL/min overnight to simulate weathered fuel conditions. A diluted aliquot of his weathered product was analyzed via gas chromatography to determine its distribution of 19 specific straight-chain (2,4-dimethylpentane, C-6 through C-15) and aromatic hydrocarbon constituents (BTEX, propylbenzene, butylbenzene, naphthalene, and methylnaphthalene). These compounds of interest accounted for 52 wt% of the total fuel added to the test soils, resulting in an average weathered fuel molecular weight estimated at 128 g/gmol.

Column Venting Modes. After the equilibration period, initial soil vapor total petroleum hydrocarbon, and microbial respiration product/reactant gas (CO_2 and O_2) composition were monitored by taking samples at alternate sampling port locations. Two columns were operated in the continuous venting mode. One column was maintained at a flow rate to remove 1 pore volume (10 L) of soil gas per day (1 pv/day) and another to remove 1/2 pore volume of soil gas per day (1/2 pv/day). The moisture content in the columns was maintained throughout the duration of the experiment by adding 10 mL of DDW every day from an opening at the top of the column. The off-gas was passed through an activated carbon absorber prior to release to the atmosphere. Soil gas samples were collected every 24 hours over the duration of the study from alternate sampling port locations using gas tight syringes and were analyzed by direct injection into a GC. Samples were collected for O_2/CO_2 and hydrocarbon analysis in separate syringes at the outlet port.

In the pulse venting mode samples were collected at zero time for hydrocarbon and O_2/CO_2 analyses and monitored continuously thereafter. As soon as the O_2 level in any of the ports dropped to less than or equal to 2 vol%, the soil column was purged from the top with 1/2 a pore volume (approximately 5 L) of humidified, compressed air at a rate of 250 mL/min. The purged gas was collected in a 5 L Tedlar bag connected to the opening in the bottom cap. Duplicate samples were taken from each Tedlar bag and analyzed for O_2/CO_2 and total hydrocarbon concentrations by direct injection into a GC.

Extraction and Analysis Methods. At the end of each experimental run, soil samples were taken at each sampling port location to quantify the residual concentration of the compounds throughout each column for mass balance calculations. Purge and trap extraction procedures (3 to 5 g of a soil sample in a 40 mL screw cap vial containing a known mass of methanol) were used to recover volatiles, and soxhlet extraction (10 g of the soil + 10 g of anhydrous sodium sulfate extracted with 100 to 150 mL of methylene chloride) was used to quantify the semi- and non-volatiles from column soils prior to analysis by gas chromatography.

Hydrocarbon concentrations were determined using gas chromatography/ flame ionization detection. Total petroleum hydrocarbon analyses were calibrated using a standard containing C-6 to C-15 hydrocarbons. A gas chromatograph equipped with a thermal conductivity detector was used for O_2/CO_2 analyses.

O_2 uptake and CO_2 production data were evaluated for their fit to a zero-order and first-order regression model based on 95% confidence intervals of their regression slopes being significantly greater than zero, and by inspection of regression residuals.

Mass balance calculations were carried out for the test compounds based on the soil vapor, off-gas and residual soil compound concentrations, and respiration product measurements. At the end of each run, soil samples were collected from the columns and analyzed for hydrocarbon concentrations. The amount of contaminant loss due to biodegradation was calculated from the difference between the amount of contaminant added to the soil at the beginning of the run and the sum of residual contaminant concentration at the end of the run plus the contaminant removed during the venting operation and in gas sampling. An independent determination of contaminant biodegradation was made based on O_2 depletion across each column assuming an O_2 to hexane equivalent of 3.5 g O_2/g hexane.

RESULTS AND DISCUSSION

Contaminant Removal Results. The cumulative amount of JP-4 volatilized in both columns operated in the continuous venting mode were identical (6% of the initial applied JP-4 mass). As might be expected, complete removal of volatiles from the 1/2 pv/d reactor took approximately twice as long as in the 1 pv/day column (750 hrs versus 300 hrs). The cumulative volatilized mass from the pulse vented column was only 3.5% of the applied mass, representing more than a 40% reduction in JP-4 volatilization in this column.

The initial rate of biodegradation was higher in the 1 pv/d column than in the 1/2 pv/d column (Figure 1), however, once long-term rates were achieved in these continuous flow columns, both columns yielded statistically identical rates (6.1 versus 6.9 mg/hr in the 1 and 1/2 pv/d columns, respectively). The initial biodegradation rate in the pulse vented column was three to four times higher than the continuous flow columns, and persisted at these high rates until 400 hr into the run. At 400 hr, when volatiles were completely removed from this column, respiration rates decreased significantly, falling from 56 to less than 12 mg/hr. Final rates in this pulse vented column were only 1.6 mg/hr, well below long-term rates in the continuous columns.

Total hexane equivalent JP-4 removal achieved during approximately 4000 hours of operation of the columns due to both volatilization and degradation under each flow management scheme were 68%, 62% and 72% for the 1 pv/d, 1/2 pv/d, and pulse vented columns, respectively. The total amount of air required to achieve these overall JP-4 removal rates are summarized in Table 1. Not only did the pulse venting operation reduce volatilization, it also yielded more rapid overall removal of contaminant with significantly smaller volumes of air moved than either continuous flow system evaluated.

Mass Balance Determinations. Mass balance calculations were carried out for eight specific compounds in JP-4 of interest in this study. The fraction of the initial mass loaded to each column that was not accounted for (i.e., apparent loss) was attributed to biodegradation. Biodegradation was found to be the major loss mechanism for all specific compounds except benzene, accounting for 86 to nearly 100% of their observed mass removal during the study (Table 2). Only 27% of the benzene's mass removal could be attributed to biodegradation, with this occurring in the pulse vented column. Pulse venting was found to be more effective than continuous flow systems in the degradation of a highly volatile, slowly degradable

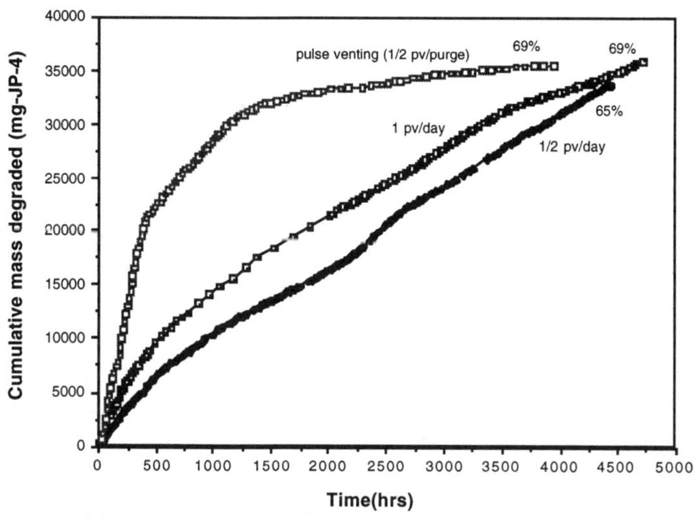

FIGURE 1. Cumulative mass degraded versus time for various flow management options.

TABLE 1. Comparison of air requirements for overall JP-4 removal observed at 4000 hr for each air flow management scheme.

Column #	Flow mgt. Option	Amount of air moved (L)	Time (hrs) operation	Volatilized (%)	Biodegraded (%)	Total (%) removed
1	1 pv/day	1830	4000	6.0	63.0	69.0
2	1/2 pv/day	805	4000	6.0	60.0	66.0
3	1/2 pv/purge	520	4000	3.5	68.5	72.0

TABLE 2. Specific compound mass balance calculations for the various flow management options.

Compound	Column #1			Column #2			Column #3		
Name	%Voln	% Resi	%Biodg	%Voln	% Resi	%Biodg	%Voln	% Resi	%Biodg
Benzene	93.73	0.00	6.16	103.00	0.00	-3.17	72.20	0.35	27.38
Toluene	5.11	0.15	94.62	5.13	0.00	94.70	2.61	0.06	97.27
Ethylbenzene	2.13	0.74	97.02	9.07	0.00	90.76	5.56	0.00	94.38
p-Xylene	0.14	0.41	99.33	1.01	0.00	98.82	0.67	0.00	99.27
n-Heptane	0.97	0.00	98.91	6.56	0.00	93.26	13.86	0.00	86.09
n-Octane	1.25	0.00	98.64	0.55	0.00	99.28	0.38	0.24	99.32
Nonane	0.14	0.56	99.19	1.45	0.00	98.37	0.10	0.04	99.81
Naphthalene	0.00	0.00	99.88	0.00	1.82	98.18	0.41	10.22	89.31

Note: Column #1 = 1 pv/d, Column #2 = 1/2 pv/d, Column #3 = 1/2 pv/purge

substrate, reducing benzene volatility by 20 to 30%, and increasing its removal by biodegradation by more than a factor of four.

CONCLUSIONS

Based on the results of this laboratory study, the following findings and conclusions can be stated:

1. Pulse venting operation (1/2 pv/purge, with purging contingent upon less than 2 vol% O_2 in column soil gas) reduced contaminant volatilization rates by more than 40% as compared to continuous flow columns.
2. This pulse venting mode resulted in initial respiration rates that were three to four times that of the continuous flow columns, while providing greater overall JP-4 removal efficiency with significantly smaller volumes of air moved.
3. Mass balance results showed that degradation accounted for 89 to nearly 100% of the mass lost for toluene, ethylbenzene, p-xylene, n-heptane, n-octane, n-nonane, and naphthalene, while only 27% benzene removal could be attributed to biodegradation. Benzene biodegradation was maximized within the pulse vented column.

Due to the enhanced performance of the pulse venting versus continuous flow systems in terms of volatilization rates, biodegradation rates, and required gas volume per mass of contaminant treated, it would appear that pulse venting systems are generally preferable to more conventional bioventing system designs. Significant cost savings result for the pulse vent design due to significantly lower off-gas volumes. Off-gas treatment my even be eliminated if regulated specific contaminant emission rates can be met by the reduced volatilization rates provided by pulsed venting.

REFERENCES

Downey, D. C., R. A. Frishmuth, S. R. Archabal, C. S. Pluhar, P. G. Blystone, and R. N. Miller. 1995. "Using In Situ Bioventing to Minimize Soil Vapor Extraction Costs." In R. E. Hinchee, R. M. Miller, and P.C. Johnson (Eds.), *In Situ Aeration: Air Sparging, Bioventing, and Related Remediation Processes,* pp. 247-266. Battelle Press, Columbus, OH.

Miller, R. N., Vogel, C. C., and Hinchee, R. E. 1991. "A Field-Scale Investigation of Petroleum Hydrocarbon Biodegradation in the Vadose Zone Enhanced by Soil Venting at Tyndall AFB, Florida." In R. E. Hinchee and R. F. Olfenbuttel (Eds.), *In Situ Bioreclamation: Applications and Investigations for Hydrocarbon and Contaminated Site Remediation*, pp. 283-301. Butterworth-Heinemann, Stoneham, MA.

REMOTE MONITORING OF A BUNKER C FUEL OIL BIOVENTING SYSTEM

Scott Adamek, Patrick J. Evans and Dave W. Ashcom (AGI Technologies, Bellevue, Washington)
Joseph A. Kurrus (AVISTA Corporation, Spokane, Washington)

ABSTRACT: A bioventing system has been in operation since November 1997 remediating hydrocarbons from an underground Bunker C fuel oil spill. Air injection and extraction systems were installed to enhance microbial degradation of petroleum hydrocarbons in the capillary fringe. Instrumentation allows remote monitoring of carbon dioxide and oxygen concentrations and relative humidity from individual extraction wells. Monitoring data are automatically transferred to a database to assess performance. To date, data indicate greater biodegradation rates near the hydrocarbon source and during late summer and early fall.

INTRODUCTION

Approximately 70,000 gal (265 m^3) of Bunker C fuel oil were released from underground storage tanks at a former steam generating facility in downtown Spokane, Washington. The current extent of the oil plume is approximately 550 ft (150 m) long and 15 to 40 feet (4.6 to 12 m) below ground surface. The oil is present in the capillary fringe and the upper portion of the saturated zone. Concentrations of petroleum hydrocarbons in the capillary fringe range up to 50,000 mg/kg. Remediation efforts included UST closure, stormwater diversion and control, installation of a subsurface jet-grouted barrier wall, groundwater hydraulic control, free product recovery, and bioventing to stimulate in situ Bunker C biodegradation.

Biodegradation of Bunker C fuel oil is expected to be relatively slow due to its content of long-chain hydrocarbons. Longer chain hydrocarbons are degraded at lower rates during bioventing (Leeson and Hinchee, 1996). Treatability studies (Dames and Moore, 1995) indicated a maximum specific mineralization rate of 10.7 mg-C/kg-day based on carbon dioxide (CO_2) production from Bunker C biodegradation. Ideal moisture and nutrient concentrations appeared to promote the rate of biodegradation.

Since in situ biodegradation rates were an important criterion for bioventing performance, the design incorporated on-line instrumentation to enable biodegradation rate monitoring. A critical component was acquisition of real-time data that could be used to monitor and optimize biodegradation rates across the site and over time.

MATERIALS AND METHODS.

Ten extraction wells (BE 1 to 10) were installed at the site to optimize airflow through the vadose zone within the Bunker C fuel oil plume. Seven air injection wells (BI 1 to 7) were installed to ensure sufficient air flow and to reduce competition (dead zones) between multiple extraction wells. Buildings and paving cover the entire site. Extraction and injection well screen intervals were designed to correspond to hydrocarbon-contaminated soil thickness as determined during the remedial investigation. The intent was to localize oxygen delivery in regions of Bunker C oil rather than throughout the entire vadose zone. The design air flow for each extraction well was based on microbial oxygen requirements calculated from treatability study data. A total extraction airflow rate of 200 ft^3/min (5.7 m^3/min) was designed based on the above calculations and the likelihood that biodegradation rates will decrease with time.

Figure 1 shows a screen capture from the Wonderware InTouch Human Machine Interface (HMI) and shows the bioventing system, associated instrumentation, and real-time data collected at 12:01:26 PM on December 24, 1998. The extraction system is shown at the top of the figure and the injection system at the bottom. A summary table for the individual extraction wells is shown on the right side of the figure. At that time, CO_2 concentrations in extracted air ranged from 0.03 to 0.82% and air flows ranged from 1.6 to 35.6 ft^3/min (0.045 to 1.01 m^3/min) in individual extraction wells. Total extracted air flow was 138.3 ft^3/min (3.9 m^3/min). Parameters measured for each extraction well include flow rate, vacuum, oxygen (O_2) and CO_2 concentrations, temperature, and relative humidity. O_2 and CO_2 concentrations, relative humidity, and temperature are measured in slipstreams from each extraction well line routed through a vacuum booster pump. Solenoid valves automatically direct extracted air from selected wells to analytical instrumentation in a programmed sequence. Subsurface temperatures are measured at four locations at the site. Data collected in calendar year 1998 are presented in this paper.

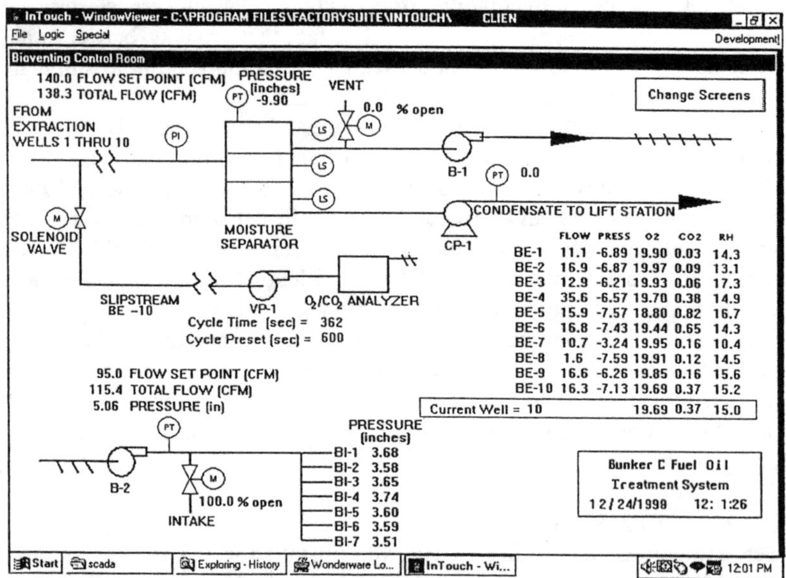

FIGURE 1. HMI screen of the instrumented bioventing system.

Monitoring data are continuously recorded and automatically downloaded to a database. These data are used in periodic reports to comply with regulatory requirements and to provide system performance criteria such as calculation of hydrocarbon degradation rates in the subsurface.

The entire remediation system is automated with a supervisory control and data acquisition (SCADA) system including instrumentation and controls, a programmable logic controller (PLC), a personal computer running HMI software, and modem access. This design allows system monitoring and control from anywhere with telephone access. Air flow rates to both extraction and injection blowers are automatically controlled using motor-operated valves that are controlled through proportional integral derivative loops in the PLC logic software.

RESULTS AND DISCUSSION

Figure 1 shows that wells BE 4, 5, 6, and 10 produced air containing the greatest concentrations of CO_2. These wells are located nearest the source of the released Bunker C fuel oil. The lowest concentration of CO_2 was observed in well BE 1 which is furthest from the source and adjacent to the subsurface barrier wall containing the plume. Well BE 5 had the highest CO_2 concentration and, as expected, the lowest O_2 concentration.

Figure 2 shows the CO_2 response in four extraction wells following a 106-hour unplanned shutdown for system maintenance. CO_2 concentrations were observed to transiently increase and then exponentially decay to steady-state values upon reinstatement of air flow at zero hours. During the shutdown hydrocarbons were being biodegraded and CO_2 was accumulating in the vadose zone. Steady-state was attained for all four wells after approximately 20 hours.

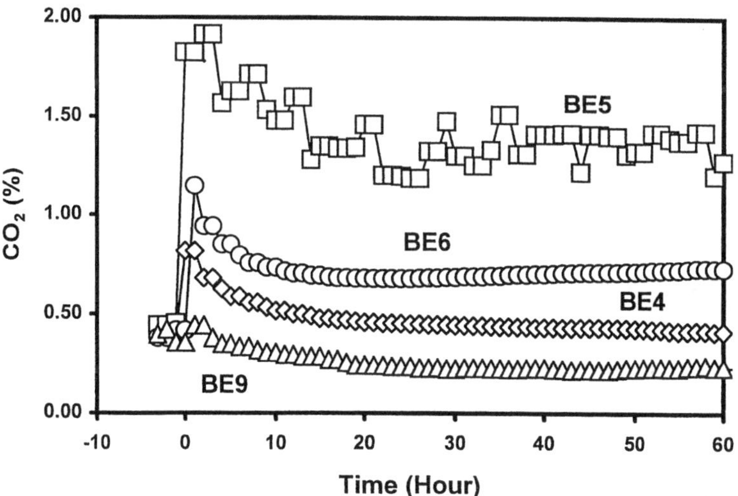

FIGURE 2. Transient response of CO_2 concentration following shutdown.

Spatial Variation of Biodegradation. The rate of hydrocarbon mineralization was observed to be greatest near the source of the hydrocarbon release (Figure 3). Mineralization rates were calculated for each well by multiplying the CO_2 concentration by the air flow rate. Other parameters were investigated as a function of distance from the source to determine the cause of this correlation. These parameters included hydrocarbon-contaminated soil thickness, hydrocarbon concentration and composition, relative humidity, and Total Kjeldahl Nitrogen (TKN) concentration as an indicator of nutrients. Greater TKN was observed in groundwater near the source (7 mg/L) compared to 480 ft (150 m) downgradient (1 mg/L). The source of nitrogen is not known but may be the Bunker C fuel oil itself. Since the treatability study indicated a stimulatory effect of nutrients, the greater TKN may, in part, explain the trend in Figure 3. Hydrocarbon concentration or composition would be expected to decrease farther from the source and thus affect the rate. Limited data collected to date do not support this hypothesis.

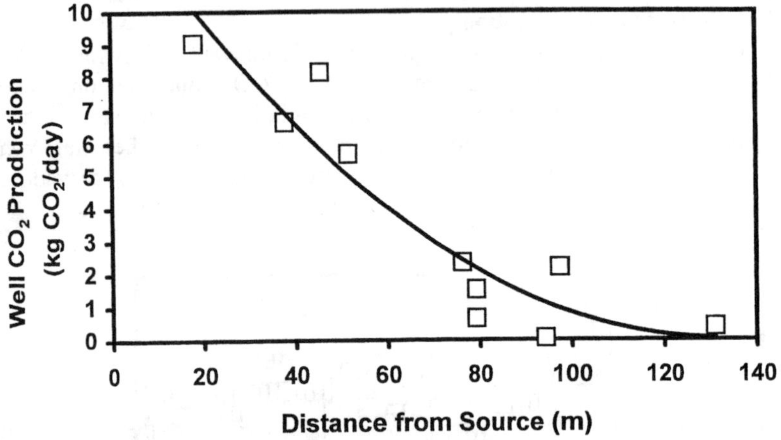

FIGURE 3. Spatial variation of hydrocarbon mineralization rate.

Temporal Variation of Biodegradation. Figure 4 shows trends in CO_2 production (i.e., hydrocarbon mineralization) during 1998. Also shown are trends of ambient surface temperature (i.e., atmospheric), average subsurface temperature, and absolute humidity in extracted air. Subsurface temperature data from January through April were not collected. The CO_2 and humidity data are for the entire system including all 10 extraction wells. The CO_2 production rate ranged from 16 to 67 kg/day which translates approximately to 1.4 to 6.0 gal/day (0.0055 to 0.023 m^3/day) of Bunker C fuel oil being mineralized. Aerobic biodegradation of hydrocarbons typically results in 50% of the carbon being assimilated into biomass. Taking assimilation into account theoretically increases the range of biodegradation rates to 2.8 to 12 gal/d (0.011 to 0.046 m^3/day).

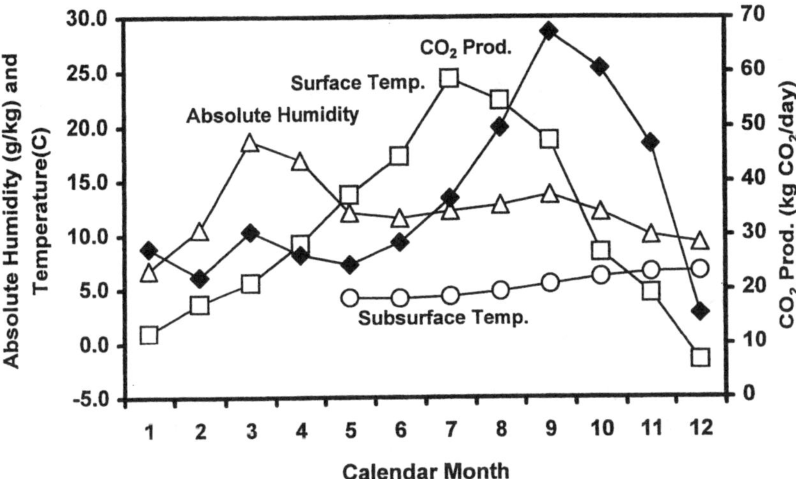

FIGURE 4. Temporal variation of biodegradation and associated factors.

Biodegradation rates can be affected by temperature and moisture. Figure 4 shows that the maximum combined CO_2 production in September (Month 9) was observed following a maximum surface temperature in July and prior to a maximum subsurface temperature (recorded about 30 feet below ground surface) recorded in December. This maximum CO_2 production also corresponded to the highest absolute humidity recorded in the second half of the year.

Several interrelated phenomena appear to have resulted in a maximum biodegradation rate in September. The increased surface temperature resulted in warmer air being injected into the subsurface. This likely caused more water evaporation from the capillary fringe and increased oxygen availability to microorganisms. Injection of warmer air also caused, in part, increased subsurface temperatures recorded during later months. Biodegradation rates generally double for every 10°C increase in soil temperature (Leeson et al., 1996). The reason for continued increase in subsurface temperature during the final months of the year (Figure 4) is uncertain, but may be attributable to slow heat transfer from the surface to the subsurface. Heavy rain events in June contributed to increased subsurface humidity in subsequent months which was shown to stimulate biological activity in the treatability study. Even though absolute humidity was greatest in March and April, probable low subsurface temperatures prevented realization of high CO_2 production rates. Thus, the observed increase in hydrocarbon mineralization appears to be attributable to a number of factors associated with temperature and moisture.

Comparison to Treatability Data. The greatest mineralization rate for the site (i.e., CO_2 production rate) observed was 9.1 kg-CO_2/day for BE 5 (Figure 3). The greatest specific mineralization rate reported in the treatability study was 10.7 mg-C/kg-day. Comparison of these data required assuming a zone of influence for well BE 5. This zone was assumed to be equal the 60-ft (18-m) design radius of influence and the well screen interval of 5 ft (1.5 m). The calculated specific mineralization rate for BE 5 was 0.83 mg-C/kg-day which is an order of magnitude less than 10.7 mg-C/kg-day from the treatability study. This latter number was observed in the presence of nitrogen and phosphorous nutrients, moisture and in a laboratory-controlled environment. The treatability result in the presence of moisture alone was 0.74 mg-C/kg-day which is comparable to the result for BE 5. These data suggest that nutrients may be limiting biodegradation of hydrocarbons at the site.

CONCLUSIONS

An instrumented field bioventing system is used to monitor in situ biodegradation of Bunker C fuel oil at a site in Spokane, Washington. This system provides the capability of on-line measurement of biodegradation rates and the ability to identify the correlation between operating and environmental parameters. Biodegradation rates were observed to be greatest nearest the source of the release, possibly due to greater nutrient levels. Biodegradation rates showed a trend over the course of a year that appeared to be related to temperature and moisture. Overall hydrocarbon biodegradation at the site may be limited by nutrients.

Additional studies planned include in situ respiration testing where the injection and extraction blowers will be stopped but gas sampling and analysis will continue. These studies will allow further investigation of the CO_2 production transients similar to those shown in Figure 2.

REFERENCES

Dames and Moore. 1995. *Report – Soil Column Bioventing and Surfactant Flushing Treatability Study, WWP Central Steam Plant, Spokane, Washington.* Prepared for Washington Water Power (now AVISTA Corporation).

Leeson, A. and R.E Hinchee. 1996. *Principles and Practices of Bioventing, Volumes I and II: Bioventing Design.* Prepared for the U.S. Air Force, Center for Environmental Excellence.

COST-EFFECTIVE BIOVENTING OF KEROSENE CONTAMINATED SOIL

G. Malina (Technical University of Czestochowa, Poland)
J.T.C. Grotenhuis and W.H. Rulkens
(Wageningen Agricultural University, The Netherlands)

ABSTRACT: Bioventing efficiency was compared for a continuous and intermittent airflow in bench-scale columns, packed with 5 kg of kerosene contaminated soil (the initial concentration of 2660 mg·kg^{-1}), at controlled temperature of 20 °C (± 2.5 °C). The continuous flow of air was at constant rate of 1L/h. The intermittent flow was controlled by the O_2 content in the soil gas, which at 4% did not limit biodegradation. To reach similar efficiency of 17-29% and 11-30% within the same remediation time of 28-29 days, almost 8 times less air was used during the intermittent flow, as compared to the continuous flow. The intermittent flow resulted in 5.3-7.7 times more effective use of air and in an increase by factor of 1.5-2.3 the contribution of biodegradation to the overall removal of kerosene from soil, as compared to the continuous flow.

INTRODUCTION

Reduction of the amount of air used to achieve the minimum soil aeration during soil bioventing (SBV) that provides for high biodegradation rates may decrease the total remediation costs.

To find out the effective air flow regime to combine optimum evaporation-to-biodegradation ratio, acceptable remediation time and residual kerosene concentrations, simulation of SBV was carried out in bench-scale columns with soil from a site contaminated with kerosene.

Previous SBV experiments with soil artificially contaminated with model hydrocarbons (Malina et al., 1997) showed that to reach comparable removal efficiency at comparable times, the intermittent flow regime (IFR) required 2.8 and 1.8 times less air, as compared to the continuous flow regime (CFR) at a constant and a 2-phase flow rate, respectively.

Based on these results, the IFR was tested as the cost-effective SBV strategy, in comparison to the CFR at the constant rate.

MATERIALS AND METHODS

Site and Contaminants Characteristics. Contaminated and reference (not contaminated) soil was collected from a military base in the Netherlands at depths of 0.2-1.2 m. Soil properties were determined using disturbed samples, homogenized, air dried, sieved (fraction < 4 mm) and stored in dark at 4 °C. Sieving was done at 0 °C to minimize evaporation losses of kerosene from soil. For the same reason, columns were packed at 0 °C.

Jet fuel (kerosene) contamination from a leaking fuel pipeline consisted mainly of the aliphatic hydrocarbon mixture C_{10} - C_{16}, BTEX and naphthalene derivatives. Quantitatively, the average composition comprised 17.2 % (v/v) of BTEX and 82.8% (v/v) of alkanes.

Column Experiments. SBV was simulated in the installation (Figure 1) with 4.3 L glass columns (60 cm x 9.6 cm i.d.), at controlled temperature of 20 °C (± 2.5 °C), (Malina and Grotenhuis - submitted).

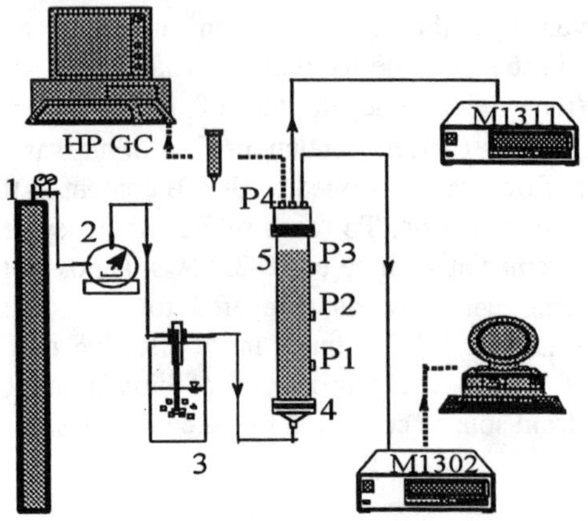

Figure 1. Installation for simulation of SBV and SVE.
Notation: 1 - CO_2-free air; 2 - wet gas meter; 3 - gas washer; 4 - glass filter; 5 - soil column (glass); P1-P4 sampling ports; M1302, M1311 -gas monitors.

Kerosene contaminated soil (5 kg) at the initial concentration of 2660 $mg{\cdot}kg^{-1}$, as determined by extraction with CS_2, and water content of 6% (w/w), was packed to 45 cm height. CO_2-free air was applied at the bottom of the column at different flow regimes. CFR was done at the constant upflow rate of 40 $cm^3{\cdot}cm^{-2}{\cdot}h^{-1}$ (i.e. 1 $L{\cdot}h^{-1}$). The IFR was realized at the same rate for 7-8 h, then the flushing was stopped until the O_2 content in the soil gas dropped to ca. 4-5%, which was found in previous experiments not to limit biodegradation (Malina et al., 1997), and resumed at the same rate until ca. 21% of O_2 was reached again.

Soil vapors in ports 2, 3 and 4 (headspace) were sampled for total volatile organic compounds (TVOCs) once a day, and CO_2 and O_2 were monitored continuously in the outlet gas. A column with the reference soil (not contaminated) was used as a control for indigenous respiration. Wet gas meters were used to determine the air volume that passed through the columns. To minimize adsorption all pipe connections having contact with kerosene were either from glass or Teflon™, and sampling ports were sealed with Viton™ caps. After termination, soil was analyzed for total extractable organic compounds (TEOCs), water content and pH.

Soil vapors and headspace samples were transferred by 100μL gastight syringes and analyzed for TVOCs immediately after collection by HP GC 5890

Series II with flame ionization detector (FID), and for CO_2 and O_2 on GC 8000 Series (Fisons Instruments) with thermal conductivity detector (TCD).
The Bruel&Kjaer Multi-gas Monitor Type 1302, and Industrial Emission Monitor Type 1311, associated with the PC software package BZ5156, were used for continuous measurements of CO_2 and O_2 in the outlet gas.
TEOCs were determined by extraction of 25 g of soil with 50 mL of CS_2.

RESULTS

Simulation of Bioventing. Mass balance of kerosene was based on GC measurements of TVOCs, and CO_2 and O_2 contents, with the assumptions:
(1) chemical formula (weighted average) of kerosene: $CH_{1.8}$ (Damera and Hill 1997);
(2) general chemical formula of biomass: CH_2O;
(3) calculations based on the mineralization reactions at biomass production yields:

$$\textit{for } Y = 0.35 \quad CH_{1.8} + 1.1 \cdot O_2 = 0.65 \cdot CO_2 + 0.55 \cdot H_2O + 0.35 \cdot CH_2O \tag{1}$$
$$\textit{for } Y = 0.5 \quad CH_{1.8} + 0.95 \cdot O_2 = 0.5 \cdot CO_2 + 0.4 \cdot H_2O + 0.5 \cdot CH_2O \tag{2}$$

(4) biodegradation modeled by first order kinetics, with the rates determined as (Leeson and Hinchee 1997; Malina et al., 1998):

for CFR:

$$r_{deg} = 0.5\left([CO_2]_1 + [CO_2]_2\right)Q_a d_{CO2} MW_{norm} / 100 m_s Y \tag{3}$$
$$r_{deg} = 0.5\left([O_2]_1 + [O_2]_2\right)Q_a d_{O2} / 100 m_s R_{O2/ker} \tag{4}$$

for IFR

$$r_{deg} = k_{CO2} V_a d_{CO2} MW_{norm} / m_s Y \tag{5}$$
$$r_{deg} = k_{O2} V_a d_{O2} / m_s R_{O2/ker} \tag{6}$$

(5) evaporation rates calculated as:

$$r_{evap} = 0.5\left(C_1 + C_2\right)Q_a / m_s \tag{7}$$

where: r_{deg}, r_{evap} - kerosene biodegradation and evaporation rates (mg/kg·d); $[CO_2]$, $[O_2]$ - CO_2 and O_2 contents (%) in the soil gas at time t_1 and t_2; C_1, C_2 - concentrations of kerosene as TVOC, or as CH_4 in the outlet gas (mg/L) at t_1 and t_2; Q_a - air volume passing the column (L/d); k_{CO2}, k_{O2} - CO_2 production and O_2 utilization rates (%CO_2/d, %O_2/d); V_a - gas phase volume in the column, estimated as 2.5 L; d_{CO2}, d_{O2} - densities of CO_2 and O_2 (mg/L); Y - biomass production yield (mass of C in biomass per mass of C in substrate); m_s - mass of soil (kg); MW_{norm} - molecular weight of kerosene normalized on carbon (mg/mg), for kerosene MW_{norm} = 1.15 mg kerosene/mg C; $R_{O2/ker}$ - equivalent of O_2 required for biodegradation of kerosene (mass of O_2 per mass of substrate degraded), $R_{O2/ker}$ = 2.21 (at Y=0.5) and $R_{O2/ker}$ = 2.55 (at Y=0.35).

Continuous Flow Regime. CFR at the constant flowrate of 40 $cm^3/cm^2{\cdot}h$ resulted in evaporation of ca. 1392 mg, i.e. 11% of kerosene initially present in soil, at the rate of 10.3 mg/kg·d (Figure 2). Simultaneous biodegradation at the rate of 5.1-16.3 mg/kg·d, for Y = 0.35 (Figure 3) and 5.8-11.4 mg/kg·d at Y = 0.5 (results not shown), depending on the estimation method, contributed to subsequent removal of 6-18% of kerosene.

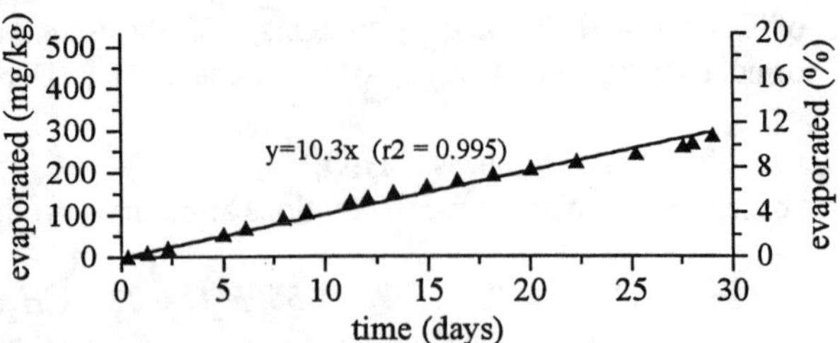

FIGURE 2. Evaporation of kerosene at CFR (as TVOC measured by GC).

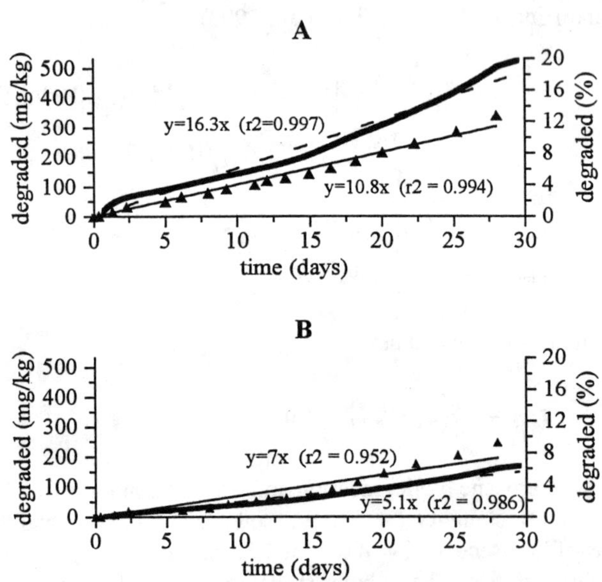

FIGURE 3. Biodegradation of kerosene at CFR based on (A) CO_2, and (B) O_2 concentrations in the outlet gas (at biomass production yield Y=0.35). Legend: ■ measured by M1311, ▲ measured by GC.

The overall SBV efficiency was within the range of 17-29%, for which 29 days and 713 L of air was required (Table 1).

TABLE 1. Results of SBV at CFR vs. IFR.

Flow Regime	Remediation Time (d)	Air Used (L)	Effectiveness of Air Used (L/mg)	Removal Efficiency (%)	Evaporation-to Biodegradation Ratio (%)
CFR	29	713	0.23-0.31	17-29	11:6 - 11:18
IFR	28	90	0.03-0.06	11-30	2:9 2:28

As evaporation and biodegradation rates are comparable, SBV at the IFR should give the more effective use of air, at a comparable time scale and without decreasing the removal efficiency.

Intermittent Flow Regime. IFR was realized at the rate of 40 $cm^3/cm^2 \cdot h$ in 10 flushing intervals of duration depending on O_2 contents in the soil gas (Figure 4).

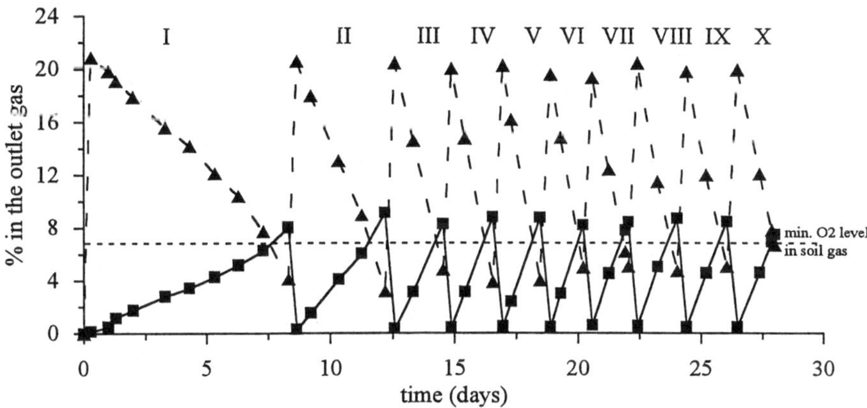

FIGURE 4. O_2 and CO_2 concentrations in the outlet gas during SBV at IFR. Legend: ▲O_2; ■ CO_2; I-X - non-flushing periods.

Some 245-257 mg of kerosene (i.e. 2%) was removed by evaporation after 28 days (total flushing time of 82 hours), at the rate of 12.2-16.8 mg/kg·d as determined from GC measurements, and 14.6-17.2 mg/kg·d (CH_4 signal measured by M1311).

On the other hand, about 1458-2292 mg and 1162-2326 mg of kerosene was degraded, as calculated from CO_2 production and O_2 uptake, respectively, during periods when the gas flow was off, assuming the yield Y = 0.5. At the yield of 0.35, amounts degraded where 2083-3273 mg (CO_2 production) and 1003-2011 mg (O_2 uptake). Biodegradation rates varied depending on the estimation methods and flushing periods from 3.2 to 33.4 mg/kg·d (Y=0.5), and from 2.6 to 47.8 mg/kg·d (Y=0.35).

The amount of kerosene degraded during flushing periods was estimated as 185-461 mg, based on the averaged biodegradation rates of non-flushing periods.

CONCLUSIONS

To reach similar overall SBV efficiency within the same remediation time of 28-29 days, almost 8 times less air was used during the IFR, as compared to the CFR.

The effectiveness of air used, defined as volume of air per mass of kerosene removed from soil, was 5.3-7.7 times higher at IFR than CFR.

The shift of the evaporation-to-biodegradation ratio led to an increase from 35-62% to 82-93% the contribution of biodegradation to the overall kerosene removal from soil.

Most probably the advantage of the IFR would be even more spectacular at lower concentrations of kerosene in soil during the "tailing" phase of SBV, with diffusion limited desorption, evaporation and bioavailability of kerosene. However, long term column and large-scale experiments under controlled conditions are required to confirm this hypothesis, as well as to fully verify suggested cost-effective SBV strategy for different soil and oil contamination.

ACKNOWLEDGMENTS

The research was founded mainly by the Department of Environmental Technology, Wageningen Agricultural University, The Netherlands, and Heidemij-Arcadis b.v., Research and Development, Waalwijk, The Netherlands.

REFERENCES

Damera, R., and M.R. Hill M.R. 1997. "Rapid clean-up of a multiple fuel spill". In B.C. Alleman and A. Leeson (Eds.), *In Situ and On site Bioremediation: Volume 1*, pp. 199-204. Battelle Press, USA.

Leeson, A., R.E. Hinchee. 1997. *Soil Bioventing Principles and Practice*. CRC, Lewis Publishers.

Malina, G., and J.T.C Grotenhuis. "The Role of Biodegradation during Bioventing of Soil Contaminated with Jet Fuel". Submitted to *Appl. Biochem. Biotech.*

Malina, G., J.T.C. Grotenhuis, W.H. Rulkens, S.L.J. Mous, and J.C.M. de Wit. 1998. "Soil vapour extraction versus bioventing of toluene and decane in bench-scale soil columns". *Environ. Technol., 19*: 977-991.

Malina, G., J.T.C. Grotenhuis, and W.H. Rulkens. 1997. "Effective Use of Air During Bioventing of Model Hydrocarbons". In B.C. Alleman and A. Leeson (Eds.), *In Situ and On site Bioremediation: Volume 1*, pp. 367-372. Battelle Press, USA.

EFFECTS OF NUTRIENT ADDITION ON BIODEGRADATION OF FUEL-CONTAMINATED SOILS

James E. Abbott, Bruce C. Alleman, Melody J. Drescher, Albert J. Pollack, and Thomas C. Zwick (Battelle, Columbus, Ohio)
George Watson (NFESC, Port Hueneme, California)
Leon Bowling (MCAGCC, Twentynine Palms, California)

Abstract: A field study was conducted to determine if biodegradation rates in arid soils, where fuel contamination exists to approximately 40 ft below ground surface (bgs), were either moisture and/or nutrient limited. The study was conducted in three phases and utilized three multilevel soil-gas monitoring points (MP-A, MP-B, and MP-C) for the introduction of amendments and sampling. During the first phase, MP-A served as a nonamended control. An agricultural starter fertilizer solution was injected at MP-B, and MP-C was amended only with water. For the second phase of the study, nutrients were added to MP-C, and water was introduced to MP-A. The final phase of the study involved the introduction of nutrients at MP-A. In situ respiration tests were performed at each monitoring point prior to any amendment injections to establish baseline hydrocarbon biodegradation rates. Subsequently, respiration tests were conducted following each phase of the study to evaluate effects of amendments on biodegradation rates. The data from MP-A showed that biodegradation rates remained constant under nonamended conditions and following the water-only amendment. However, rates increased substantially following nutrient amendment. Nutrient amendment at MP-B caused a significant and sustained increase in biodegradation rates. The results at MP-C showed little change following water amendment and only a slight increase in microbial activity following nutrient addition. The results from this study indicate that biodegradation of the fuel at this site is generally nutrient-limited, and that the addition of nutrients will substantially reduce the time of treatment.

INTRODUCTION

The study site (Site 1573), located at the Marine Corps Air Ground Combat Center (MCAGCC), is a former underground storage tank (UST) site where soils are primarily contaminated with bunker fuel to a depth of approximately 40 ft (12.2 m) bgs as a result of leaking tanks. Total petroleum hydrocarbon (TPH) concentrations in soils 15 to 40 ft (4.6 to 12.2 m) bgs range from 5,000 to 29,000 mg/kg. Due to the depth of contamination, existing underground utilities, and surrounding structures at the site, in situ bioventing was determined to be the most cost-effective and feasible remediation technology for the site. Bioventing was initiated at the site during 1995, with in situ respiration tests performed approximately every 6 months. The results from these tests indicated that biodegradation rates ranged from 0.3 to 11 mg TPH/kg soil/day. However, a more representative average in situ biodegradation rate following bioventing would be <3.0 mg/kg/day. Previous research conducted at other hydrocarbon-contaminated sites at MCAGCC showed that the background soil moisture and nutrient levels were low and limiting the biodegradation process. One research study in particular showed that following the addition of water and nutrients, biodegradation rates increased by an order of magnitude relative to baseline conditions (Zwick et al., 1995; Zwick et al., 1997). To further evaluate the effects of subsurface amendments on biodegradation of hydrocarbon-contaminated soils, Battelle, in consultation with the Naval Facilities Engineering Services Center (NFESC) and MCAGCC, initiated a pilot study at Site 1573. Water alone and aqueous-based nutrients were injected into the subsurface during three separate phases. The nutrient material selected for the study was a commercial-grade, agricultural starter fertilizer (5:15:10; 5% N: 15% P_2O_5: 10% K_2O). An on-line environmental monitoring system (OEMS) developed by Battelle was used to perform automated soil-gas monitoring during this study.

Site Description. The MCAGCC is an active military installation located in south-central San Bernardino County, California. The installation covers ~932 mi^2 of remote desert terrain used primarily for live-fire training exercises. The study site is located at Building 1573 of the Mainside Facility, a developed portion of the Base located approximately 5 mi north of the city of Twentynine Palms, California. Currently, Building 1573 serves as an Officer's Club for the Marine Corps. According to facility records, four 10,000-gallon USTs storing waste oil, diesel fuel, and bunker fuel for heating existed near the northeast corner of the building. Prior to removal of the USTs in 1987, bunker fuel and some waste oil leaked into the vadose zone, resulting in vertical migration of hydrocarbon constituents to a depth of 40 ft (12.2 m) bgs. The soil at the site consists of medium- to coarse-grained sands, silts and clays. Depth to groundwater is approximately 336 ft (102.4 m) bgs.

Objective. The objective of this study was to evaluate the effect of nutrient and water amendments on biodegradation rates of fuel-contaminated soils during bioventing.

MATERIALS AND METHODS

System Design. The bioventing system utilized for this study consisted of the following: four vent wells, three multilevel soil-gas monitoring points (MP-A, MP-B, and MP-C), and a 2-hp regenerative blower (Figure 1). The three soil-gas monitoring points also served as water and/or nutrient solution injection points.

The OEMS was installed inside Building 1573 and connected to soil-gas probes at MP-A at depths of 20, 30, 35, and 40 ft (6.1, 9.1, 10.7, and 12.2 m) bgs; at MP-B at 15, 30, 35, and 50 ft (4.6, 9.1, 10.7, and 15.2 m) bgs; and at MP-C at 15, 25, 35, and 40 ft (4.6, 7.6, 10.7, and 12.2 m) bgs (Figure 1). Monitoring point sampling modules were deployed at each of the three multilevel soil-gas monitoring point locations so that soil gas could be selectively sampled at each discrete depth. The OEMS was connected to a commercial telephone line and soil-gas data were downloaded via a modem from the remote monitoring device to Battelle's Columbus Laboratories for data reduction.

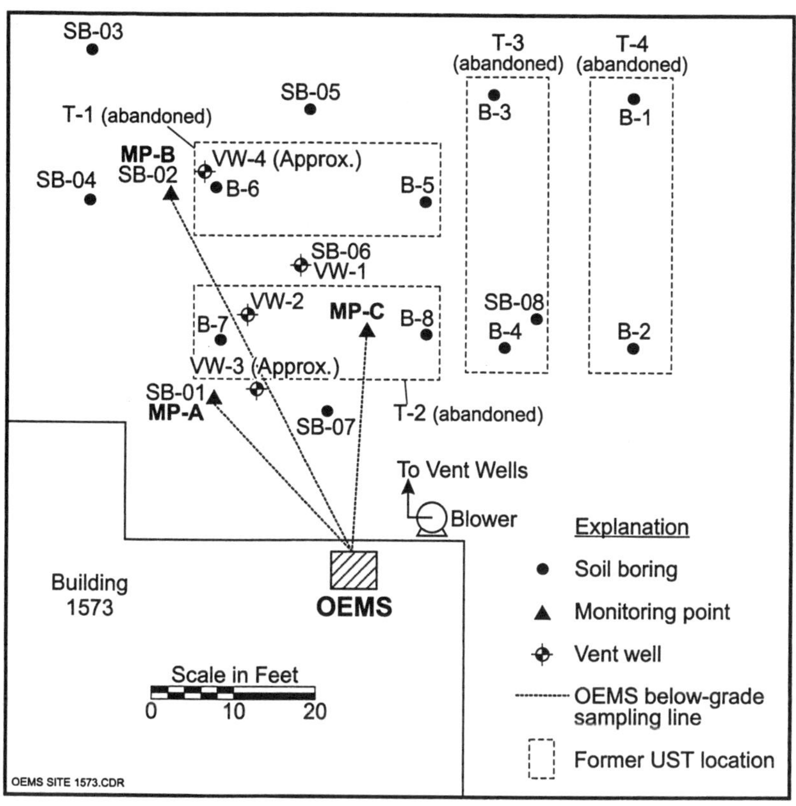

FIGURE 1. Bioventing system at Site 1573.

TABLE 1. Phases 1, 2, and 3.

Monitoring Point-Depth (ft bgs)	Phase 1 Initiated March 1997	Phase 2 Initiated November 1997	Phase 3 Initiated February 1998
A-20	Control	Water	Nutrient Solution
A-30	Control	Water	Nutrient Solution
A-35	Control	Water	Nutrient Solution
A-40	Control	Water	Nutrient Solution
B-15	Nutrient Solution	None	None
B-30	Nutrient Solution	None	None
B-35	Nutrient Solution	None	None
B-50	Nutrient Solution	None	None
C-15	Water	Nutrient Solution	None
C-25	Water	Nutrient Solution	None
C-35	Water	Nutrient Solution	None
C-40	Water	Nutrient Solution	None

Field Testing Methodology. The study was conducted over a 2-year period and included three phases (Table 1). Phase 1 was initiated in March 1997 and began with the addition of nutrients to MP-B and water amendment to MP-C. MP-A served as a control during this phase. During phase 2 (initiated in November 1997), nutrient solution was injected at MP-C, where water had been previously applied, and soil hydration at MP-A was investigated to determine if low soil moisture was a limiting factor for bioremediation. During phase 3 (initiated in February 1998), nutrients were introduced to MP-A for further evaluation of nutrient addition at the site. MP-B was amended with nutrient solution in phase 1 and received no further amendment in subsequent phases to evaluate the long-term effects of nutrient addition.

For soil hydration, approximately 360 gal (1,363 L) of tap water were loaded into a graduated 500-gal polyethylene holding tank. A total of 90 gal (340 L) of water was pumped from the tank at ~0.6 gal/min (2.3 L/min) into the monitoring points. Assuming a soil retention capacity of 2 gal/yd^3, each 90-gal (340 L) volume of water would hydrate 4.5 yd^3 (3.44 m) of soil around each of the monitoring-point depths.

To add nutrients, 5 gal (18.9 L) of fertilizer (5:15:10) were poured into the tank and uniformly mixed with 360 gal (1,363 L) of tap water. This nutrient solution was injected into the monitoring points in the same manner as the water for soil hydration. The nutrient addition rate was determined by assuming a soil mass of 5,500 kg per 4.5 yd^3 (3.44 m) of soil and a nitrogen requirement of 56 mg of elemental nitrogen per kg of soil. Based on these assumptions, each of the four monitoring points was amended with approximately 308 grams of elemental nitrogen. Under laboratory conditions, this application rate produced enhanced biodegradation rates of hydrocarbon-contaminated soils collected from another site at the MCAGCC (Zwick et al., 1997).

Upon completion of water and/or nutrient solution injection, ambient air was pumped into the sampling lines from each monitoring probe to purge liquid from the line, probe, and the sand pack around each sampling probe. A vacuum was then pulled on each sampling probe to confirm that the liquid had been displaced and that soil gas was available for monitoring.

In Situ Respiration Tests. Throughout all three phases, in situ respiration tests were performed prior to any amendment to establish baseline or "pre-amendment" biodegradation rates. Soil-gas samples were automatically collected and analyzed using the OEMS every 4 to 6 hours while the blower was turned off. Subsequently, the vadose zone was aerated by operating the blower at approximately 65 cubic feet per minute (cfm), until oxygen (O_2) levels were increased to >15%. During reaeration periods, O_2 soil-gas concentrations were tracked by the OEMS and the ratio of air injected into the four vent wells was adjusted so that the site was uniformly aerated.

To quantify any enhancement, in situ respiration tests were performed following amendment and aeration. Again, samples were automatically collected for O_2, carbon dioxide (CO_2), and TPH measurement every 4 to 6 hours. The O_2 data were used to determine O_2 utilization rates, and the CO_2 data were used to verify biodegradation. TPH biodegradation rates were calculated as mg of TPH per kg of soil per day using the O_2 utilization rates (% change in O_2 per day) and based on the stoichiometric oxidation of hexane (Leeson and Hinchee, 1997).

RESULTS AND DISCUSSION

Table 2 provides biodegradation rates calculated for each soil-gas probe location over the course of this study. Figure 2 illustrates the biodegradation rate trends observed at MP-A at 35 ft (10.7 m), MP-B at 35 ft (10.7 m) and MP-C at

TABLE 2. Estimated biodegradation rates during phases 1, 2, and 3.

MP-Depth (ft)	Phase 1					Phase 2				Phase 3			
	Mar '97	Apr '97	Nutrient Amendment to MP-B and Water Amendment to MP-C	Apr '97	Apr '97	Nov '97	Nutrient Amendment to MP-C and Water Amendment to MP-A	Dec '97	Jan '98	Feb '98	Nutrient Amendment to MP-A	Mar '98	Apr '98
A-20	8.3	7.5		8.0	6.0	5.2		2.1	0.5	2.1		2.2	17.2
A-30	10.8	11.7		12.0	11.6	6.0		8.1	6.3	5.1		NA	23.6
A-35	5.7	5.4		6.3	6.3	4.8		6.9	4.2	3.7		6.3	26.7
A-40	1.4	1.9		1.5	1.3	1.0		1.6	3.8	1.7		1.1	11.2
B-15	2.0	2.2		0.5	4.0	2.4		3.6	1.3	1.7		2.6	1.8
B-30	2.2	2.4		6.3	19.8	10.4		17.2	9.0	6.7		10.3	7.4
B-35	2.3	3.0		12.3	24.7	15.1		19.3	15.4	11.3		12.4	12.4
B-50	0.1	0.1		0.3	0.4	0.3		0.1	0.8	0.6		0.5	4.4
C-15	4.3	2.9		1.3	0.8	7.9		NA	11.9	4.5		9.7	ND
C-25	1.3	1.4		2.0	1.5	1.6		4.4	3.6	2.0		3.4	6.7
C-35	0.5	0.4		0.9	0.9	0.9		3.9	2.5	2.8		1.4	3.2
C-45	0.5	0.3		0.7	0.6	0.6		1.1	1.2	1.1		0.9	4.2

NA = Point not adequately aerated for biodegradation field measurement; O_2 concentration <10 %
ND = Biodegradation not detectable.

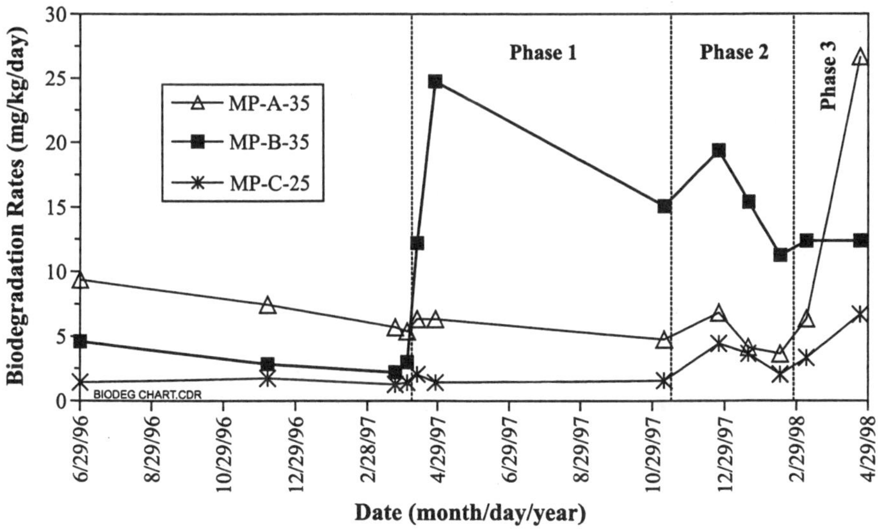

FIGURE 2. Biodegradation rate trends at MP-A-35, MP-B-35, and MP-C-25 ft bgs.

25 ft (7.6 m) bgs prior to and following amendment. During phase 1, MP-A was selected as a control based on the higher respiration rates observed at this location prior to amendment compared to MP-B and MP-C. Overall, the biodegradation rates at MP-A remained fairly constant between pre- and post-amendment phase 1 respiration tests conducted between March and April 1997.

Before injecting the nutrient solution into MP-B, two respiration tests were performed to establish nonamended rates. In the upper contaminated soils, approximately 15 to 35 ft (4.6 to 10.7 m) bgs, biodegradation rates ranged from 2.0 to 3.0 mg/kg/day with a mean rate of 2.4 mg/kg/day. At the 50-ft (15.2 m)-bgs depth, where there are nondetectable levels of soil contamination, baseline rates were 0.01 mg/kg/day. Immediately following nutrient addition, the third respiration test showed that biodegradation decreased at the shallowest depth, but increased at 30 and 35 ft (9.1 and 10.7 m) to an initial post-amendment level of 6.3 and 12.3 mg/kg/day, respectively. At these depths, respiration tests 1 and 2 show that O_2, measured by the OEMS, was depleted over a mean of 28 sampling events or 112 hours after the blower was shut down. Following the injection of nutrients, an O_2-limited condition was achieved in nine sampling events or 36 hours. Biodegradation rates subsequently increased during respiration test 4, where all depths above 35 ft (10.7 m) bgs displayed further biodegradation rate enhancements ranging from 4.0 to 24.7 mg/kg/day.

At 30 and 35-ft bgs, O_2 concentrations of ~15% was essentially consumed in two sampling events or 8 hours after the start of the last respiration test. Biodegradation rates at the 50-ft (15.2 m)-bgs depth showed increases above the baseline measurements, but, as expected, continued to be low because of the limited carbon source outside of the highly contaminated soil region.

The injection of water into MP-C during phase 1 resulted in limited increases in biodegradation rates. When compared to baseline biodegradation rates, the introduction of water to the shallow depth actually resulted in a decrease of biological activity. This could reflect some displacement of limited hydrocarbon contamination from the region around this monitoring depth. At the 25-ft (7.6 m)-bgs depth just below where 29,000 mg/kg TPH as diesel was measured in the soil, biodegradation rates, although increasing slightly, were similar to previous test results. This was also the case for the 35 and 40-ft (10.7 and 12.2 m)-bgs depths.

During phases 2 and 3, six additional respiration tests were conducted at MP-B from November 1997 to April 1998. At 30 and 35 ft (9.1 and 10.7 m) bgs, biodegradation rates ranged from 6.7 to 17.2 and from 11.3 to 19.0 mg/kg/day, respectively. These data would indicate first, that the microbial population had acclimated to the available nutrients, and second, that very high biological activity was sustained over an extended period of time. After 1 year following nutrient addition, the respiration rates at depths characterized with higher TPH concentrations remained 3 to 5 times higher than the pre-amendment rates.

The injection of nutrients into MP-C during phase 2 resulted in an increase of biodegradation rates, on an order of magnitude, at all four depths. Continued testing during phase 3 confirmed these results. The most notable increase was at the 25-ft (7.6 m)-bgs depth, where biodegradation rates ranged from 2.0 to 6.7 mg/kg/day.

MP-A served as a control during phase 1 of this study. During phase 2, water addition caused only slight changes in respiration rates. These results agreed well with those observed at MP-C during phase 1. During phase 3, MP-A was amended with the nutrient solution. The results from the respiration test conducted immediately following amendment (March 1998) showed no change in biological activity. Data collected during the last respiration test conducted 7 weeks after amendment (April 1998) showed a significant increase in O_2 utilization and CO_2 production at all four depths of MP-A. Compared to the data from the respiration test conducted just prior to nutrient addition (February 1998), the data from April 1998 indicated that bioactivity increased severalfold. Biodegradation rates at MP-A at 20, 30, 35, and 40 ft (6.1, 9.1, 10.7, and 12.2 m) bgs increased from 2.1 to 17.1, 5.1 to 23.6, 3.7 to 26.7, and 1.7 to 11.2 mg/kg/day, respectively.

CONCLUSIONS

The data generated during this study defined conditions limiting biological activity at Site 1573 at the MCAGCC. Based on the results of this study, it was concluded that the microbial activity was not moisture-limited, as the addition of water to MP-C in phase 1 and MP-A in phase 2 had little effect on respiration rates. Based on the results following nutrient addition at all 3 monitoring points over the course of the study, it was concluded that the microbial activity was nutrient limited. It was also concluded that the nutrient addition provided long-term benefit, based on the data from MP-B, and nutrient amendment would significantly reduce the time required for bioventing the fuel at Site 1573. The challenge would be to devise a method for effective delivery and distribution of the nutrient solution throughout the contaminant volume of the vadose zone. Vapor-phase nutrients may prove useful for accomplishing this goal.

REFERENCES

Leeson, A. and R.E. Hinchee. 1997. *Soil Bioventing Principles and Practice.* CRC, Lewis Publishers, Boca Raton, FL.

Zwick, T.C., A. Leeson, R.E. Hinchee, R.E. Hoeppel, and L. Bowling. 1995. “Soil Moisture Effects During Bioventing in Fuel-Contaminated Arid Soils.” In R.E. Hinchee, R.N. Miller, and P.C. Johnson (Eds.), *In Situ Aeration: Air Sparging, Bioventing, and Related Remediation Processes.* Battelle Press, Columbus, OH.

Zwick, T.C., E.A. Foote, A.J. Pollack, J.L. Boone, B.C. Alleman, R.E. Hoeppel, and L. Bowling. 1997. “Effects of Nutrient Addition During Bioventing of Fuel Contaminated Soils in an Arid Environment”. In B.C. Alleman and A. Leeson (Eds.), *Proceedings of the Fourth International Symposium on In Situ and On-Site Bioremediation*, Volume I, pp. 403-409. Battelle Press, Columbus, OH.

DEGRADATION OF HYDROCARBONS IN CRUDE OIL BY PURE BACTERIAL CULTURES UNDER DIFFERENT REDOX CONDITIONS

Vladimir G. Grishchenkov (IBPhM (RAS), Pushchino, Russia)
Richard T. Townsend, Thomas J. McDonald, Robin L. Autenrieth, James S. Bonner (Texas A&M University, College Station, TX) and Alex M. Boronin (IBPhM (RAS), Pushchino, Russia)

ABSTRACT: Nitrate-reducing bacterial strains (*Pseudomonas sp.* BS2201, BS2203 and *Brevibacillus sp.* BS2202) isolated from petroleum-contaminated soil were capable of degrading petroleum hydrocarbons under aerobic and anaerobic conditions. Under aerobic conditions (a 10 day experiment in liquid media) the strains degraded 20-25% of the total extractable material (TEM), including up to 90-95% of all alkanes analyzed (n-C_{10}-C_{35}). Under anaerobic conditions (a 50 day experiment) these organisms degraded 15-18% of the TEM, 20-25% of some alkanes, and 15-18% of selected polycyclic aromatic hydrocarbons (PAHs). The strains also degraded saturated hydrocarbons under anaerobic conditions in the absence of nitrates as electron acceptors.

INTRODUCTION

Biodegradation of hydrocarbons can occur under aerobic and anaerobic conditions. Under anaerobic conditions, however, the rate and extent of hydrocarbon biodegradation decreases and the variety of substrates degraded is typically narrower (Coates et al.,1997; Bertrand et al.,1989). Microorganisms that degrade hydrocarbons anaerobically can be divided into at least two groups relative to molecular oxygen. The first group includes facultative anaerobes, in particular nitrate-, iron- and manganese-reducing microorganisms. The second group comprises strict anaerobes, such as sulfate-reducers. Reported studies on the anaerobic oxidation of hydrocarbons by pure bacterial cultures have emphasized the bacterial ability to conduct this process under particular, pre-determined conditions such as sulfate-, nitrate-, iron- or manganese-reducing conditions (Coates et al.,1997). At the same time, the ability of facultative bacteria to exist under both aerobic and anaerobic conditions makes it possible to achieve more stable population numbers and metabolic (in this case, catabolic) activity within a wide oxidation-reduction potential range. This is true of such facultative bacteria as nitrate-reducers. These organisms can oxidize organic substrates releasing large quantities of energy, comparable to the oxidative system in anaerobes (Valiela,1984).

Objective. This work was undertaken to determine the potential of nitrate-reducing microorganisms to degrade hydrocarbons under different oxidation-reduction potentials in suspended cultures.

MATERIALS AND METHODS

Microorganisms. Nitrate-reducing bacterial strains (*Pseudomonas sp.* BS2201, BS2203 and *Brevibacillus sp.* BS2202) capable of growing on petroleum hydrocarbons under anaerobic conditions were used for these studies.

Media and reagents. Lauria broth (LB), agarized lauria broth and a minimal salts liquid medium comprised of the following (g/L): Na_2CO_3 - 0.1; $CaCl_2$ x $6H_2O$ 0.01; $MnSO_4$ x $7H_2O$ - 0.02; $FeSO_4$ - 0.01; Na_2HPO_4 - 1.5; KH_2PO_4 -1.0; $MgSO_4$ x $7H_2O$ - 0.02; NH_4Cl - 2.0; NaCl -1.0, were used to culture the microorganisms. Media pH was adjusted to 7.0. Arabian light crude oil subjected to atmospheric effects (weathered to achieve 25% volume reduction) was used as the sole source of carbon and energy in each bioreactor. This crude oil was supplemented to the minimal salts medium of each bioreactor using a syringe. Anoxic conditions in the culture medium were generated according to Kaspar and Tiedje (1982).

Experiments and chemical analysis. Bacterial cultures were grown in the LB media under aerobic conditions. Cells were centrifuged and re-suspended to achieve a cell concentration of $5x10^9$ cells/ml in anoxic minimal salts medium. Aliquots (0.2 ml) were dispensed into 100 ml anaerobic vials with medium (40 ml in each) using a syringe. For aerobic incubation, the volume of medium per vial was 10 ml. Reactors were incubated at 28°C on a rotary shaker at 200 rpm for 10 and 50 days for the aerobic and anaerobic experiment, respectively. In the latter case, nitrate (1.20-1.4 g/L of medium) was used as an alternative electron acceptor. Nitrate removal was estimated quantitatively three times: at the zero point, after 15 days, and after 40 days using an ion chromatograph. Petroleum hydrocarbon degradation was determined gravimetrically and with GC-MS analysis to monitor concentrations of selected *n*-alkanes and PAHs. A total of 27 *n*-alkane compounds, including n-C_{10} – C_{35}, pristane, and phytane, and 37 PAHs, including the parent and substituted homologs of naphthalene, dibenzothiophene, fluorene and phenanthrene, were analyzed. Hopane (17α(H), 21β(H)-hopane) was also quantified. Further details concerning the GC/MS analysis can be found elsewhere (Mills, 1997).

RESULTS AND DISCUSSION

As was expected, nitrate-reducing strains BS2201, BS2202 and BS2203 proved to be facultative anaerobes and were able to grow under aerobic conditions as well. Taking this into consideration, the characteristics of petroleum hydrocarbon degradation were evaluated by the above strains under both aerobic and anaerobic conditions. As seen in Figure 1, under aerobic conditions the cultures degraded 90-95% of the *n*-alkanes examined (n-C_{10}-C_{35}). However, no appreciable or statistically reliable degradation of any of the examined PAHs was detected for the indicated period of time.

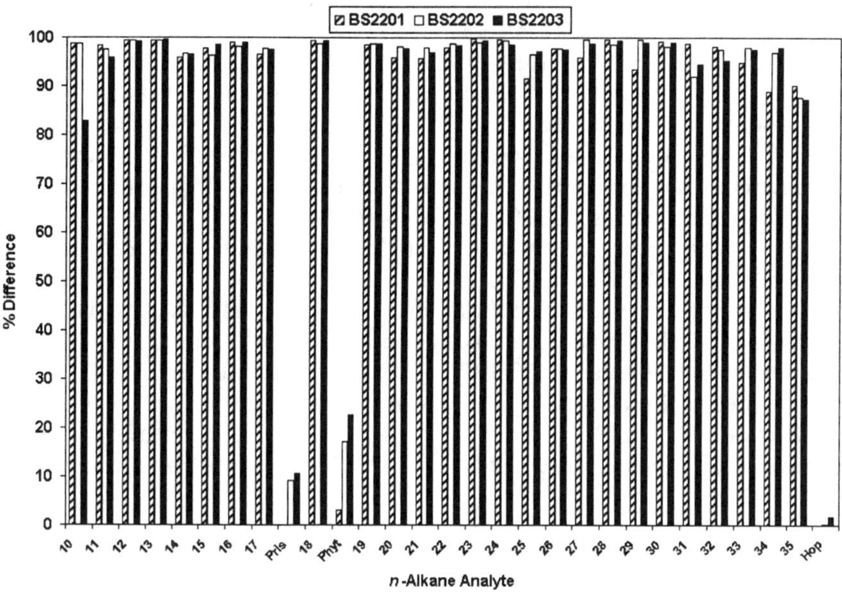

FIGURE 1. Percent differences in *n*-alkane concentrations (from control) in aerobic reactors with selected strains after 10 day incubation on crude oil.

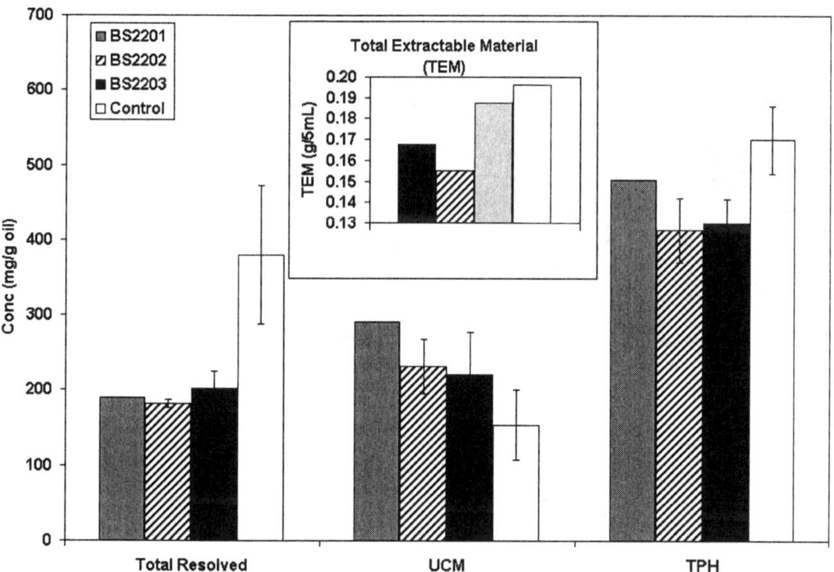

FIGURE 2. Total extractable materials (TEM), total resolved hydrocarbons, unresolved complex mixture (UCM) & total petroleum hydrocarbons (TPH) in aerobic reactors with selected strains after 10 day incubation on crude oil.

As for the degradation of petroleum as a whole, i.e. its degradation as determined by total extractable material (TEM) weighing analysis, the picture was as follows: in the bioreactors with strains BS2202 and BS2203, TEM decreased by 21% and 15%, respectively. In addition, the concentration of the total resolved hydrocarbons (those monitored using GC/MS) decreased by approximately 50% and that of total petroleum hydrocarbons (TPH) by an average of 18% (Figure 2). The concentration of the unresolved complex mixture (UCM) for all strains under aerobic conditions increased relative to the control.

Anaerobic petroleum hydrocarbon degradation was evaluated for the individual strains BS2202, BS2203 and a mix of strains BS2201, BS2202, and BS2203. Figure 3 illustrates the decrease in concentrations of *n*-alkanes, though to a lesser extent than under aerobic conditions. Thus, with strain BS2202 concentrations of the alkanes *n*-C_{11}-C_{14} and *n*-C_{35} decreased only 25-30%, while concentrations of the alkanes *n*-C_{15}-C_{18} decreased by 20-25%. The mix of strains degraded alkanes slightly better in the *n*-C_{18}-C_{30} range compared with the performance of strains BS2202 and BS2203 which decreased this fraction by 15-20%.

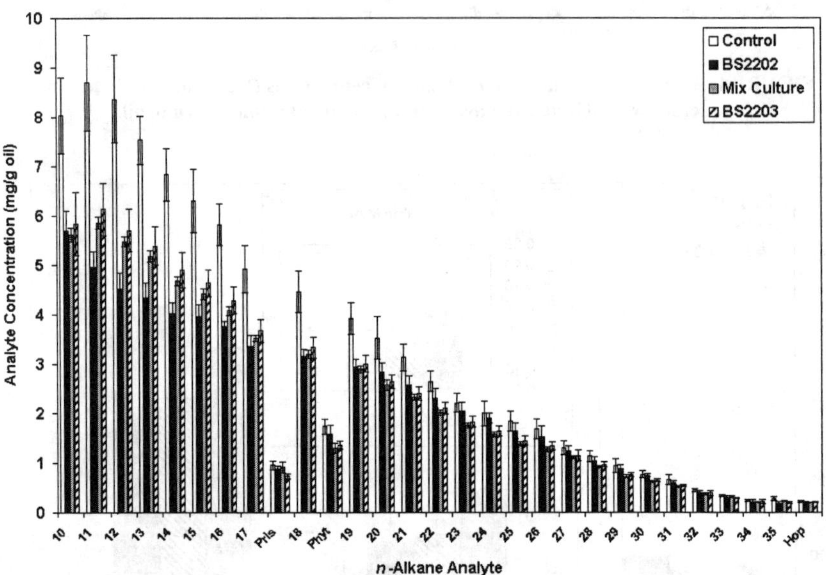

FIGURE 3. Concentrations of *n*-alkanes in anaerobic reactors with selected strains after 50 day incubation on crude oil.

As for PAHs, a statistically reliable decrease in the concentrations was noted for strain BS2202 which degraded C1-naphthalene, C4-naphthalene and C3-fluorene by 13%, 18% and 23%, respectively. The mixture of strains decreased the naphthalene concentration by 13% (Figure 4).

An 18% decrease in TEM was also observed in the anaerobic bioreactor containing the mixture of strains as well as a 15% decrease in the concentration of the total resolved hydrocarbons and a 28% decrease in the UCM. Though anaerobic incubation was carried out for 50 days, nitrate reduction was observed only during the first two weeks, during which nitrate was reduced to nearly 25% of its initial concentration. No further decrease in nitrate concentrations was observed. This result indicates that for this period of time the microorganisms accomplished the process, but the cultures were not sampled until day 50.

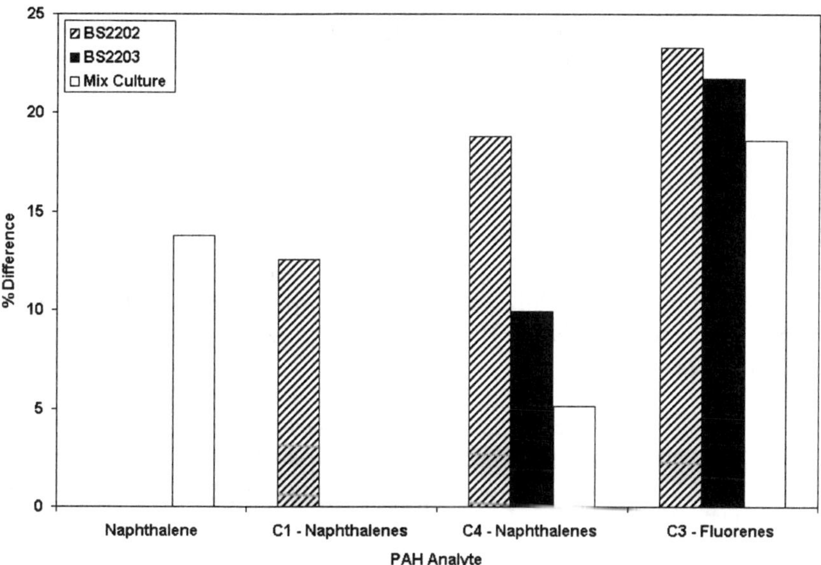

FIGURE 4. Percent differences in PAH concentrations (from control) in anaerobic reactors with selected strains after 50 day incubation on crude oil.

It is noteworthy that in anaerobic bioreactors without nitrate the consumption of *n*-alkanes also occurred and at a level comparable to that of nitrate-reducing conditions (data not shown). In other words, the microorganisms degraded the *n*-alkanes at a redox potential lower than -100 mV, thereby, coupling this process with fermentation. Thus, it may be stated that the isolated pure cultures were capable of biotransformating petroleum and its components, i. e. both *n*-alkanes and selected PAHs, under anaerobic, nitrate-reducing conditions.

It was also demonstrated that the selected strains were able to degrade hydrocarbons in a wide range of redox potentials: aerobic conditions at Eh > +400 mV, anaerobic, nitrate-reducing conditions - at Eh within +50 - +150 mV and at Eh < -100 mV, fermentation. It can be concluded that these microorganisms may be capable of catabolizing petroleum hydrocarbons at different redox potentials not only under model laboratory conditions, but in nature as well.

ACKNOWLEDGEMENTS

This research was supported by the Fulbright Senior Scholar Program grant N22232 awarded to V. Grishchenkov.

The authors are grateful to Merin Brodsky for assistance in chemical analysis, to Madhu Kodikanti for help with nitrate reduction analysis, to Jim Sweeney and Frank Stephens for fruitful discussions and to Dr. E. Ariskina (All-Russia Culture Collection of Microorganisms, Pushchino, Russia) for identification of the bacterial strains.

REFERENCES

Bertrand J. C., P. Caumette, G. Mille, M. Gilewicz and M. Denis. 1989. "Anaerobic Biodegradation of Hydrocarbons." *Science Progress* 73: 333-350.

Coates J. D., J. Woodward, J. Allen, P. Philp, and D. R. Lovley. 1997. "Anaerobic Degradation of Polycyclic Aromatic Hydrocarbons and Alkanes in Petroleum-Contaminated Marine Harbor Sediments." *Appl. Environ. Microbiol. 63*(9): 3589-3593.

Kaspar, H.F. and J.M. Tiedje. 1982. "Anaerobic Bacteria and Processes." In A.L. Page, R.H. Miller and D.R. Keeney (Eds.), Methods of Soil Analysis, Part 2: Chemical and Microbiological Properties, Second Edition, pp. 989-1009. ASA-SSSA, Madison, WI.

Mills, M. A. 1997. Bioremediation of petroleum hydrocarbons in aqueous and sediment environments. Ph.D. Dissertation, Texas A&M University, Civil Engineering Department, August 1997.

Valiela I. 1984. *Marine Ecological Processes*. Springer-Verlag, New-York.

IN SITU BIOREMEDIATION OF PITS FROM PUERTO LA CRUZ REFINERY

Carmen Infante (PDVSA Intevep, Los Teques, Venezuela)
Mariela Romero (PDVSA Intevep, Los Teques, Venezuela)
Alfonso Arrocha (PDVSA Manufactura y Mercadeo, Puerto La Cruz, Venezuela)
Diana Gilbert (PDVSA Manufactura y Mercadeo, Puerto La Cruz, Venezuela)
Francisco Brito (PDVSA Manufactura y Mercadeo, Puerto La Cruz, Venezuela)

ABSTRACT: The Puerto La Cruz refinery is located to the northeast of Venezuela. Presently, plans for the management and treatment of wastes for their transformation in an environmentally safe manner are conducted. In order to remove environmental liabilities, *in situ* bioremediation of 14 pits were performed. These pits were located in the following areas of the refinery: Los Nisperitos, Portón 27 and Pamatacual. A total of 52.000 m^3 of contaminated soil with oily sludges were cleaned up.

The applied technique was based on the biostimulation of indigenous microorganisms from the soil by means of nutrients, pH adjustments, and humidity control, and by appropriate mixing of the soil to allow proper aeration conditions. The key of the process consisted in obtaining an optimal mixture with bulking agents to improve the biodegradation of the hazardous material.

To ensure biodegradation feasibility, studies were first conducted at laboratory scale, prior to field practices. Afterward, the process was applied in field where a section of soil, six meters deep, was removed in some pits while in other sections of two to three meters were removed.

Previously to the application of the *in situ* biotreatment technique to the pits, the crude oil present at upper layers was removed and transported to a center for oil recovery. Subsequently, the remaining oily sludges, present at the bottom of the pits, were biotreated. However, in some cases the crude oil layer was not removed as it was highly weathered. Consequently, the waste was first crumbled and placed around the pits, before the application of the bioremediation technique. Then, the crumbled material was incorporated into the pit. Bulking agents and nutrients were then added to promote the biodegradation process.

The efficiency of the bioremediation technique was established by monitoring the following parameters: decrease in biodegradable fractions of crude oil which are saturates and aromatics, through 3-4 months of treatment; toxicity characteristic leaching procedure (TCLP) analyses before and after the biotreatment; microbial activity by respirometer methodology, and the bacterial density expressed as the CFU/g of soil.

The results showed 70% of depletion of the biodegradable fractions after the application of the technique; TCLP levels for the Venezuelan legislation were met; waste resulted non toxic; the concentration of total crude oil in the waste after biotreatment were less than 3%. On the other hand, microbial activity was found to be higher in the soil samples, contaminated with oily sludges, than in the control sample (without oil). The bacterial density was found to be around 10^5-10^6 CFU/g of soil during the biotreatment period, which indicates that the bacterial number was enough to promote the biodegradation process.

These results showed that bioremediation is one of the most promising and safest technologies to be used for the restoration of crude oil contaminated sites.

INTRODUCTION

PDVSA Intevep, together with Puerto La Cruz Refinery, carried out an intensive cleanup program for recovering more than 52,000 m^3 of hydrocarbon-contaminated soils as a result of oilspills and oily sludge accumulation in pits. In this program, bioremediation technology called INTEBIOS™ was applied. This is based on the stimulation of soil indigenous microorganisms through the application of nutrients, bulking agents, as well as effective control and monitoring to obtain appropriate conditions for the crude oil biodegradation in tropical soils. This technology has been successfully applied for cleaning up soils impacted by several organic compounds from the Venezuelan oil industry (Infante et al 1997, Infante 1998).

Objective. The purpose of this study is to show the results of the application of INTEBIOS™ technology for cleaning up 52,000 m^3 of soils contaminated by hydrocarbons coming from 14 pits and an oilspill area, located at Puerto La Cruz Refinery. The cleanup was aiming at reducing in 70% the biodegradable fractions of crude oil, obtaining concentrations below 3% of oil and waxes, meeting TCLP maximum levels established by the Venezuelan legislation, and reaching a significant decrease of toxicity measured by means of bioassays. Meeting these parameters constitutes the basis or criteria for the cleanup of pits with crude oil or highly weathered oily sludges, as in the case of Puerto La Cruz Refinery.

Site description. Puerto La Cruz Refinery is located to the northeast of Venezuela, Anzoátegui State. A total annual rainfall of 608 mm and the mean temperature of 26.5 °C per year characterize climatic conditions.

During the last 20 years, oily sludges coming from oilspills, tank bottom, API separators and other sources have been disposed of in pits located in 3 sectors: Los Nisperitos, Portón 27 and Pamatacual. Los Nisperitos sector had 11 pits (14,246 m^3); Portón 27 sector accounted for 2 pits and an oilspill area of 3,044 m^3; and Pamatacual sector had a macropit of 33,000 m^3 with contaminated soils.

MATERIALS AND METHODS

Feasibility of Biotreatment. Treatability studies were conducted at laboratory scale in 1996, using representative samples of oily sludges for determining their crude oil biodegradation rate. Initially evaluated parameters in samples were content of crude oil, saturates (S) and aromatics (A), total metals and TCLP. Test took 90 days, and CO_2 production or respirometry, and microbial content were evaluated. Decrease of crude oil and S + A (w/w) fractions in 90 days determined the biodegradation rate. The results of this treatability study are not reported in this document, however they indicate the biotreatment application was feasible for cleaning up the Refinery soils impacted with crude oil.

In situ Biotreatment Application (INTEBIOS™). The crude oil present at upper layers was removed and transported to a center for crude oil recovery. This storage is aiming at facilitating the bioremediation process and reducing project' costs.

Pit depths varied significantly: for instance Pamatacual macropit accounted for 6 m of contaminated soils, whereas Los Nisperitos sector's pits were 2-3 m deep. The biotreatment period was also different for each pit, and depended on the weathering rate of oily sludges. In some cases, it was necessary to remove the hardened layers of up to 1 m deep, which were dispersed around the pits and then incorporated into it.

The applied technology consisted in adding bulking agents together with organic fertilizers for meeting the microbial nutritional demand. Mixing was carried out with a large-extent digger, building walls inside the pits. The mixture and irrigation frequency (aeration) was established in field based on the humidity content and sludge/soil mix texture.

Efficiency of the bioremediation technique was established by monitoring the following parameters: decrease in crude oil biodegradable fractions (S+A), during 3-4 months of treatment; TCLP analyses before and after biotreatment; microbial activity by respirometric methods; and bacterial density expressed as CFU/g of soil. In the case of Pamatacual sector, the biotreatment process took 6 months, due to the large volume and to the difficulty in homogenizing the material present in the macropit. At the end of the biotreatment, toxicity studies with bioassays were carried out to evaluate innocuousness of the remaining crude oil in soils. Treatment in all pits was conducted during 1997.

RESULTS AND DISCUSSION

Los Nisperitos Sector. In all pits, the upper layers crude oil was highly weathered which not only restrained its extraction but also made the clean-up process very complex. In this sense, a rigorously breaking up of the layers had to be carried out, in order to increase the surface area and the contact between microorganisms and S+A fractions in sludge. Table 1 shows the results of the biotreatment conducted in some of the pits from this sector. It was observed that in a 90-120 day period, there was a removal above 70% of biodegradable fractions (S+A). Likewise, the values of metals, oils and total waxes, and TCLP demanded by the Venezuelan legislation (Decree 2,289) were met. Toxicity tests using earthworms showed non-toxic levels, at the end of the biotreatment.

Portón 27 Sector. Table 2 reports the results from one of the pits from this sector. Also, there was a removal above 70% of biodegradable fractions (S+A), and Decree 2,289 resolutions for TCLP were met. None of the cases reported toxicity in the remaining material in soil after cleaning up.

The results obtained from the CO_2 production as measured in field inside the pit showed an increase with respect to control or non-contaminated soil contiguous to pits (data not registered in this paper). These results reflect increase of the biological activity in cleanup sites.

Pamatacual Sector. Cleaning up of this area was very complex and took much longer time (6 months) than in other sectors. This was due to the high volume of soil to be treated (33,000 m^3). So, the treatment had to be conducted in two phases. In the first one, biotreatment was applied to the first 1.5 m of soil, from an initial concentration of 0.23 gram of crude oil per gram of sample (gC/gM) to 0.075 gC/gM after 120 days. In the second phase, the contaminated soil was treated up to 6 m deep. Soils were placed in piles that reached up to 4 m high. The initial crude oil content was 0.040 gC/gM and values of 0.018 gC/gM were reached in two months.

TABLE 1. Crude oil and biodegradable fraction contents in function of time during treatment in Los Nisperitos pits.

	T0	T2 (30 days)	T3 (45 days)	T4 (90-120 days)
		Pit 1		
gC/gM*	0,14	0,13	0,11	0,03
gS/gM**	0,04	0,03	0,03	0,01
gA/gM***	0,06	0,04	0,04	0,01
% crude oil removal	-	7	21	79
% S+A removal	-	29	29	79
		Pit 4		
gC/gM	0,16	0,08	0,09	0,04
gS/gM	0,05	0,02	0,03	0,01
gA/gM	0,07	0,03	0,03	0,01
% crude oil removal	-	50	44	75
% S+A removal	-	59	49	83
		Herradura Pit		
gC/gM	0,18	0,09	0,08	0,04
gS/gM	0,04	0,02	0,02	0,01
gA/gM	0,08	0,04	0,03	0,01
% crude oil removal	-	50	56	78
% S+A removal	-	50	57	82

*gC/gM= gram of crude oil per gram of sample
**gS/gM= gram of saturates per gram of sample
***gA/gM= gram of aromatics per gram of sample

TABLE 2. Crude oil and biodegradable fraction contents in function of time during treatment in Pit 1, Portón 27 Sector.

	T0	T2 (30 days)	T3 (45 days)	T4 (90 days)
gC/gM*	0,12	0,06	0,05	0,04
gS/gM**	0,05	0,02	0,02	0,01
gA/gM***	0,05	0,02	0,02	0,01
% crude oil removal	-	50	58	67
% S+A removal	-	60	60	80

*gC/gM= gram of crude oil per gram of sample
**gS/gM= gram of saturates per gram of sample
***gA/gM= gram of aromatics per gram of sample

CONCLUSIONS

Cleaning up of these sectors allowed:

- To restore the soil ecology
- To remove environmental liabilities, meeting the Venezuelan legislation
- To reduce risks for health and environment
- To recover a total area of 10 ha, with a high economic value, which will be employed for the refinery in its facility expansion plans and in the construction of a biotreatment center to treat soils contaminated with crude oil inside the refinery.

ACKNOWLEDGMENTS

This project was funded by PDVSA M&M Puerto La Cruz refinery.

REFERENCES

1. Infante, C. 1998. "Bioremediation of oil spills". *Revista CODICID*. 42-45.

2. Infante, C.; Viale Rigo, M.E.; Salcedo, M. 1997. "In Situ Bioremediation of Oil Sludges. In Fourth International In Situ and On-Site Bioremediation Symposium. 4:409-411.

CONTAMINANT REDISTRIBUTION CAN CONFOUND INTERPRETATION OF OIL-SPILL BIOREMEDIATION STUDIES

Brian A. Wrenn (Washington University, St. Louis, MO), ***Albert D. Venosa*** (U.S. Environmental Protection Agency, Cincinnati, OH), and Makram T. Suidan (University of Cincinnati, Cincinnati, OH)

ABSTRACT: The physical redistribution of oil between the inside and outside of experimental plots can affect the results of bioremediation field studies that are conducted on shorelines contaminated by real oil spills. Because untreated oil from the surrounding beach will enter the plot, and treated oil will leave it, the observed rate of target analyte removal will be lower than the actual rate, and the final extent of remediation will be less extensive than would be achievable if the entire shoreline were treated. Normalization of the analyte concentration to the concentration of recalcitrant biomarkers cannot correct for this error.

INTRODUCTION

Crude oil is a complex mixture containing thousands of different organic compounds. Aliphatic hydrocarbons are the most abundant components, but the aromatic compounds are usually of more concern due to their greater toxicity and their tendency to bioaccumulate. Both aliphatic and aromatic hydrocarbons, especially the more toxic and mobile low molecular weight compounds, are biodegradable by a wide variety of microorganisms (Atlas, 1981; Cerniglia, 1992). As a result, bioremediation may be an effective and economical way to restore oil-contaminated coastal areas to acceptable conditions. Unfortunately, very few well-designed, large-scale field studies of coastal oil-spill bioremediation have been performed (Venosa, 1998). Such studies are needed to elucidate the fundamental processes that control bioremediation effectiveness. Field studies are also needed to develop guidelines that can be used by spill responders who wish to use bioremediation to clean up oiled shorelines. Guidelines that can be used by spill responders to determine whether bioremediation can restore contaminated areas to acceptable conditions are particularly important, as are guidelines for optimizing bioremediation operations.

Since it can be very difficult to obtain permission to intentionally contaminate otherwise clean beaches for the purpose of conducting bioremediation field studies, this type of research will usually be conducted after major oil spills on beaches that are set aside (i.e., not subjected to conventional shoreline cleaning operations) by the federal on-scene coordinator. The purpose of this paper is to describe a phenomenon that confounds the interpretation of bioremediation data and increases the difficulty of demonstrating complete ecosystem restoration during field studies conducted on set-aside beaches following these "spills of opportunity". This phenomenon, the transport of oil or oiled sand between relatively small treated areas (i.e., experimental plots) and large untreated areas, can cause remediation to (apparently) cease before the cleanup goals are achieved, even when bioremediation would be capable of effecting complete cleanup if it were applied to the entire beach. Awareness of this potential limitation is as essential to proper interpretation of the results of field studies using spills of opportunity as is proper experimental design. Furthermore, the transport of contaminant between treated and untreated zones is not limited to beach remediation studies but can affect the interpretation of experimental data from any field study where either the soil/sediment matrix or the contaminant are mobilized as a result of treatment or natural processes. For example, the transport and mobilization of contaminants during bioventing or biosparging experiments may redistribute contaminants between treatment units and controls.

Under the best of conditions, it is difficult to demonstrate the effectiveness of oil-spill bioremediation technologies in the field, because the experimental conditions cannot be controlled as well as is possible in the lab, and it is more difficult to quantify the rates of physical and chemical removal mechanisms. Nevertheless, well-designed field studies can provide strong evidence for the success of a particular technology if one can convincingly show that (1) oil disappears faster in treated areas than in untreated areas and (2) biodegradation is the main reason for the increased rate of disappearance. Biomarkers – compounds that are known to be resistant to biodegradation – are sometimes used to facilitate both of these demonstrations (Bragg et al., 1994; Venosa et al., 1996; Lee et al., 1997). The ideal biomarker would be subject to the same physical and chemical removal mechanisms that affect the biodegradable compounds. Unfortunately, this is rarely the case in practice. Although biomarkers are very useful for quantifying the rate at which oil is transported out of treated plots into a large unoiled area when bioremediation field studies are conducted with intentionally contaminated plots (Venosa et al., 1996; Venosa et al., 1997), correction for oil transport in field studies conducted on spills of opportunity is much more difficult. Nevertheless, its potential effects must be recognized to properly interpret the results.

MODEL:

Bioremediation field studies conducted on spills of opportunity often involve treatment of a relatively small portion of an oil-contaminated beach. Even if an entire beach is dedicated to the field study, a large fraction of the oiled area will remain untreated, because buffer zones must be maintained between the plots to prevent migration of amendments from treated to untreated or differently treated plots. The size of the buffer zone will depend on the characteristics of the beach, such as wave energy, long-shore currents, and the direction of overland or subsurface flow of water, but a distance at least equal to the length of the plot is a reasonable estimate. This constraint will tend to limit the size of the plots so that an adequate number of replicates can be used and will result in a large ratio of untreated to treated surface area. Therefore, transport of oil between the inside and outside of the plots, either as floating oil or as oil-coated sand, will tend to remove oil that is more highly biodegraded from inside the treated plots and replace it with oil that is less degraded from untreated areas outside of the plots. The net result of this transport is that the average concentrations of analytes inside of the plots will be higher than they would have been if the treated areas could be completely isolated from the surrounding beach. In practice, most oil transport will occur as a result of the movement of oil-coated sand, because oil that has been in contact with beach sediments for several days is relatively resistant to refloating (Hayes et al., 1979; Owens et al., 1987).

Normalization of target analyte concentrations to the concentration of a recalcitrant biomarker is an effective way to correct for the physical exchange of oil between the inside and outside of experimental plots when all of the oil is initially present inside of the plots and most of the beach is clean (Venosa et al., 1996). It will not work, however, for studies that involve spills of opportunity. Since the biomarker concentration is characteristic of the source oil, its concentration will be the same inside and outside of the plots at the start of the study (assuming that the initial oil distribution is uniform). By definition, biomarkers are conservative tracers, and the biomarker concentration at any position on the beach will reflect the initial extent to which that particular lot of sand was contaminated but not its subsequent weathering history. Therefore, the biomarker concentration of oil coming into the plots from untreated areas of the beach will be the same as the oil leaving them, but the concentrations of target analytes will be higher because less biodegradation will have occurred. Since the hopane concentration of oiled sand inside the treated plots will be unaffected by oil transport, it cannot be used to quantify the rate of oil exchange across the plot boundaries. Nevertheless, transport of oil and oiled sand between the experimental plots and untreated areas outside of them can result in significant redistribution of the oil (Owens and Robson, 1987; Venosa et al., 1996).

A simple model illustrates how physical exchange of oiled sand between treated and untreated areas of the beach affects the observed biodegradation rate of target analytes. For the nutrient-treated plots, the rate of change in the analyte concentrations, A_{treat}, is given by:

$$\frac{dA_{treat}}{dt} = \frac{A_{treat}}{T}\frac{dT}{dt} - k_{treat}A_{treat} + \left(\frac{1}{B}\frac{dB}{dt} - \frac{1}{T}\frac{dT}{dt}\right)A_{con} \tag{1}$$

where A_{con} is the analyte concentration in the control plots; k_{treat} is the first-order biodegradation rate coefficient for the treated plots; B is the concentration of the biomarker; and T is the concentration of a hypothetical nonbiodegradable tracer that is present in the oil inside but not outside of the plots (such as, for example, if a hydrophobic fluorescent dye were added to the oiled sand inside the plots at the start of treatment). For the untreated control plots and in the untreated areas of the beach, the rate of analyte disappearance is:

$$\frac{dA_{con}}{dt} = \frac{A_{con}}{B}\left(\frac{dB}{dt}\right) - k_{con}A_{con} \tag{2}$$

where k_{con} is the first-order biodegradation rate coefficient for the control plots. Equation (1) describes the rate of change of the analyte concentration inside the treated plots due to transport of treated oiled sand out of the plots, biodegradation, and transport of untreated oiled sand into the plots. The rates of physical loss of treated oiled sand from inside the plots (Eq. 3) and loss of oiled sand from the beach (Eq. 4) are assumed to be first-order processes:

$$\frac{dT}{dt} = -k_T T \qquad (3)$$

$$\frac{dB}{dt} = -k_B B \qquad (4)$$

where k_T and k_B are the first-order loss coefficients for the nonbiodegradable tracer and the biomarker, respectively. Solving these equations for the biomarker-normalized analyte concentration inside of the plots at any time gives:

$$\left(\frac{A_{treat}}{B}\right) = \left(\frac{A_o}{B_o}\right) e^{(k_B - k_T - k_{treat})t} \left[\frac{(k_{treat} - k_{con}) + (k_T - k_B)e^{(k_T + k_{treat} - k_{con} - k_B)t}}{k_T + k_{treat} - k_{con} - k_B}\right] \qquad (5)$$

RESULTS AND DISCUSSION:

Although suitable tracers are available, it probably is not practical or desirable to use them in a study of this type (e.g., adding a tracer to the oiled sand in such a way that the oil becomes uniformly labeled throughout the contaminated depth would be very difficult). Nevertheless, the model described by Eq. (5) is useful for illustrating the effects of exchange of oiled sand between the inside (i.e., treated) and outside (i.e., untreated) of the experimental plots. Since biomarkers serve the same role in field studies involving intentional contamination that we propose for the hypothetical hydrophobic tracer, T, in this model (i.e., they are initially present inside the experimental plots but not outside of them), we can estimate k_T from the hopane loss rate observed in a large-scale field study involving intentionally oiled experimental plots (Venosa *et al.*, 1996). In that study the rate coefficient for exchange of oiled sand across the plot boundaries was 0.025 day^{-1}. For the purposes of this illustration, we will assume that k_B is very small relative to k_T, because the area of the beach is large relative to its length, whereas the area of the plots is small relative to their perimeter. (Although we have no way to verify this assumption at this time, it makes intuitive sense, because if the half-life of oil on a contaminated beach were only 28 days, we probably would not be concerned about finding ways to speed the cleanup.)

The results of this model, are shown in Figure 1. The effect of exchange of oiled sand between the inside of treated plots and the untreated beach that surrounds them is to reduce the observed degradation rate relative to the true rate.

Sand exchange has no effect on the observed biodegradation rate in the control plots, because biodegradation is assumed to occur at the same rates inside and outside of those plots. Obviously, the magnitude of the effect of sand transport on the observed biodegradation rate in the nutrient-amended plots will depend on the rate at which the exchange occurs and the background biodegradation rate. We do not know how wave energy and sand particle size affect the sand exchange rates, but the rates are probably much higher on exposed beaches than on the sheltered beaches. Therefore, even if the same true biodegradation rates can be achieved during bioremediation on high- and low-energy beaches, complete remediation is more likely to be *observed* in low-energy environments when only a small portion of the contaminated beach is treated.

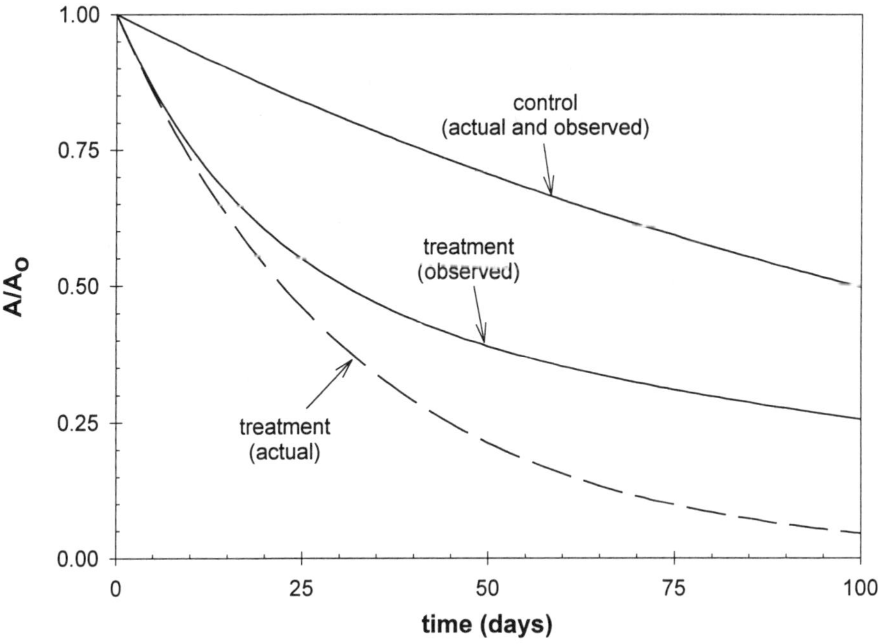

Figure 1: Reduction in the observed biodegradation rate for petroleum hydrocarbons due to exchange of oiled sand between the inside of treated plots and the untreated surrounding beach.

An important effect of oil transport is that the apparent rate of remediation can decrease, or stop altogether, while relatively large amounts of oil remain. This could lead to the incorrect conclusion that bioremediation cannot restore the contaminated shorelines to acceptable conditions, even though its effects on the initial cleanup rate are clear. Our analysis shows that behavior of this type might be an artifact of treating a small fraction of the total contaminated area and that more complete remediation could be expected if the entire beach were treated. Nevertheless, there are many possible explanations for incomplete remediation, including the inability to maintain sufficient nutrients in the bioremediation zone and, especially on high-energy beaches, a high loss rate of bacteria from the oiled surfaces (e.g., due to scouring by waves). A thorough performance monitoring program is a very important component of oil-spill bioremediation research, because ancillary data, such as nutrient concentrations, the size of the hydrocarbon-degrading microbial population, and biodegradation activity, can help determine whether problems are due to treatment inadequacies or are artifacts of natural transport phenomena when cleanup goals are not achieved during small-scale field studies.

REFERENCES:

Atlas, R. M. 1981. "Microbial Degradation of Petroleum Hydrocarbons: An Environmental Perspective." *Microbiol. Rev. 45*: 180-209.

Bragg, J. R., R. C. Prince, E. J. Harner, and R. M. Atlas. 1994. "Effectiveness of Bioremediation for the *Exxon Valdez* Oil Spill." *Nature 368*: 413-418.

Cerniglia, C. E. 1992. "Biodegradation of Polycyclic Aromatic Hydrocarbons." *Biodegradation 3*: 351-368.

Hayes, M. O., E. R. Gundlach, and L. D'Ozouville. 1979. "Role of Dynamic Coastal Processes in the Impact and Dispersal of the *Amoco Cadiz* Oil Spill (March, 1978) Brittany, France." In *Proceedings, 1979 Oil Spill Conference*, pp. 192-198. American Petroleum Institute, Washington, D.C.

Lee, K., G. H. Tremblay, J. Gauthier, S. E. Cobanli, and M. Griffin. 1997. "Bioaugmentation and Biostimulation: A Paradox Between Laboratory and Field Results." In *Proceedings, 1997 International Oil Spill Conference*, pp. 697-705. American Petroleum Institute, Washington, DC.

Owens, E. H. and W. Robson. 1987. "Experimental Design and the Retention of Oil on Arctic Test Beaches." *Arctic 40*: 230-243.

Owens, E. H., J. R. Harper, W. Robson, and P. D. Boehm. 1987. "Fate and Persistence of Crude Oil Stranded on a Sheltered Beach." *Arctic 40*: 109-123.

Venosa, A. D., M. T. Suidan, B. A. Wrenn, K. L. Strohmeier, J. R. Haines, B. L. Eberhart, D. King, and E. Holder. 1996. "Bioremediation of an Experimental Oil Spill on the Shoreline of Delaware Bay." *Environ. Sci. Technol. 30*: 1764-1775.

Venosa, A. D., M. T. Suidan, D. King, and B. A. Wrenn. 1997. "Use of Hopane as a Conservative Biomarker for Monitoring the Bioremediation Effectiveness of Crude Oil Contaminating a Sandy Beach." *J. Ind. Microbiol. Biotechnol. 18*: 131-139.

Venosa, A. D. 1998. "Oil Spill Bioremediation on Coastal Shorelines: A Critique." In S. K. Sikdar and R. L. Irvine (Eds.), *Bioremediation: Principles and Practices, Vol. III*, pp. 259-301. Technomic Publishing, Lancaster, PA.

BIOREMEDIATION OF A MARINE OIL SPILL IN THE ARCTIC

R. C. Prince, R. E. Bare, R. M. Garrett, M. J. Grossman, C. E. Haith and L. G. Keim (Exxon Research and Engineering Co., Annandale, New Jersey)
K. Lee (Maurice Lamontagne Inst., Mont-Joli, Quebec, Canada)
G. J. Holtom (AEA Technology, Abingdon , Oxfordshire, U.K.)
P. Lambert (Environment Canada, Ottawa, Ontario, Canada)
G. A. Sergy (Environment Canada, Edmonton, Alberta, Canada)
E. H. Owens (Polaris Applied Sciences, Bainbridge Island, Washington)
C. C. Guénette (SINTEF Applied Chemistry, Trondheim, Norway)

ABSTRACT: An international consortium carried out a field trial of bioremediation on an Arctic beach on Spitsbergen (78°N, 17'E), using an intermediate fuel oil and soluble and slow release fertilizers. While the primary mechanism of oil removal from the shoreline was physical, in association with fine sediment particles, there was good evidence that biodegradation was stimulated (approximately 8-10 fold in the early stages) by the bioremediation treatment. Fertilizer application was successful at delivering nutrients to the oiled sediment, and oxygen consumption and carbon dioxide evolution were substantially stimulated by the treatment. Microbial biomass, quantified by phospholipid fatty acids, was also increased. By 399 days after the application of fertilizer (about 195 ice-free days), changes in the chemical composition of the residual oil indicated that biodegradation was a significant fate for the oil.

An important secondary goal was to provide guidelines for monitoring bioremediation strategies in the field. Simple colorimetric test-kits proved adequate to measure ammonium, nitrate, and dissolved oxygen levels in the beach interstitial water, and a portable IR spectrometer was able to measure CO_2 evolution from the beach. These tools will make it significantly easier (and less expensive) to monitor and optimize bioremediation strategies for oil spills in aquatic environments.

INTRODUCTION

Bioremediation has proven to be an environmentally acceptable and cost-effective treatment of oiled shorelines in temperate climates (Prince, 1993, Bragg *et al.*, 1994, Lee *et al.*, 1995, Swannell *et al.*, 1996, Venosa *et al.*, 1996), and there have been strong indications that similar results can be expected in the Arctic (e.g. Sendstad *et al.*, 1982, Sveum and Ladousse, 1989). Hydrocarbons are generally quite biodegradable and oil-degrading microbes are ubiquitous, but oil biodegradation is limited in most marine environments by sub-optimal levels of nutrients such as biologically available nitrogen and phosphorus. Bioremediation of oil spills has thus focused on alleviating this limitation by adding fertilizers to oiled shorelines. As part of a project entitled *In Situ Treatment Of Oiled Shoreline Sediments* (Sergy *et al.*, 1998), we have addressed the potential that bioremediation by adding fertilizer stimulates oil biodegradation in Arctic climates.

Objective. The primary objective was to demonstrate that bioremediation would be an appropriate part of a response to an oil spill that impacted an Arctic shoreline. Thus we set out to demonstrate that fertilizer application resulted in the delivery of nutrients to the interstitial water surrounding the oiled sediment, and that this was followed by an increase in microbial activity (oxygen consumption, carbon dioxide evolution, and an increase in biomass), and that this in turn was followed by biodegradation of oil, revealed by changes in oil composition that are known to be the result of biodegradation.

A secondary objective was to determine whether nutrient levels could be measured in the field with sufficient precision and accuracy that fertilizer applications could be monitored, and adjusted, in the field.

Both objectives were achieved.

Site description. The experiment took place on shorelines near Sveagruva, Spitsbergen (approximately 78°N, 17'E) in the summer of 1997. Air and water temperatures in August, when oil and initial treatments were applied, were 3-7°C, although interstitial water in the shoreline was typically slightly warmer (4-9°C). The beach was relatively sheltered gravel above a shallow subtidal environment. It faced approximately south-east, with an approximately 10 km fetch to the Paulabreen glacier. Winds from the south-east generated 10-30 cm waves during the summer, and on occasion deposited small icebergs near the test beaches. Winds from other directions left the beach rather sheltered.

Oil The oil was an IF-30, an intermediate fuel grade made by diluting relatively heavy distillate with lighter fractions to obtain the desired viscosity. It was applied at approximately 5 L m^{-2} directly onto the shoreline and penetrated to a depth of approximately 15 cm. A total of 140 m of shoreline was oiled, and this was subsequently divided into four plots that were treated on a low tide one week after the oil was applied. One plot was left untreated, and two plots were tilled to a depth of approximately 20 cm by drawing tines through the plot both down the beach and across it (once in each direction) with a Bobcat front-end loader. This tilling extended to just beyond the depth of oil penetration. Fertilizer was applied to one of these tilled plots after the first tilling, and also to an untilled plot.

Fertilizer The first application of fertilizer, 7 days after oiling, was of 100 g m^{-2} prilled NH_4NO_3, 10 g m^{-2} $Ca(H_2PO_4)_2$, 1 g m^{-2} 89% $FeSO_4.H_2O$, 5% $MgSO_4$, 0.2% $MnSO_4$, and 0.1 g m^{-2} yeast extract. The second application of fertilizer, 7 days after the first, was of 140 g m^{-2} Inipol SP1 (CECA, Paris La Defense, France), a slow-release formulation containing 18% N as NH_4, and 1% P (as P_2O_5), $FeSO_4$ and yeast extract as before. A third application, of 100 g m^{-2} Inipol

SP1 was made 16 days later, before the site was left for six weeks. A fourth application of 50 g m^{-2} NH_4NO_3, 5 g m^{-2} $Ca(H_2PO_4)_2$, 70 g m^{-2} Inipol SP1 plus $FeSO_4$ and yeast extract was made 58 days after the initial treatment

RESULTS AND DISCUSSION

Interstitial water from within the oiled sediment layer was collected from perforated wells with a syringe. Nitrate, ammonium and phosphate levels were measured within one hour of collection using Chemetrics K-6902, K-1510 and K-8510 kits. Laboratory studies confirmed that these tests accurately measured the nutrients down to a detection limit of approximately 1μM phosphate, 6μM ammonium and 2μM nitrate in distilled water. A representative set of field samples was analyzed using a Technicon Autoanalyzer, and a comparison of the two methods for ammonium and nitrate is presented in Figure 1. The slope of the ammonium data is very close to 1, but it is clear that the field nitrate tests kit underestimates the amount of nitrate by about 40%. This is in accord with the manufacturer's indication that there is a systematic underestimate of nitrate levels in seawater. The test kits are clearly adequate for monitoring fertilizer application. Phosphate levels were rather lower, and the correlation between field and laboratory measurements was poorer.

Fertilizer applications were successful in delivering nutrient to the oiled zone, even in October, maintaining levels of total available nitrogen above 100μM for prolonged periods (data not shown). This was followed by a decrease in dissolved oxygen in the interstitial water (measured with a Chemetrics K-7512 kit), and an increase in the rate of CO_2-evolution from the beach (measured with a portable IR spectrometer, Swannell *et al.*, 1994) (Figure 2). This figure shows all the data collected on Days 8 to 11 after fertilizer application (n = 80 for CO_2, 61

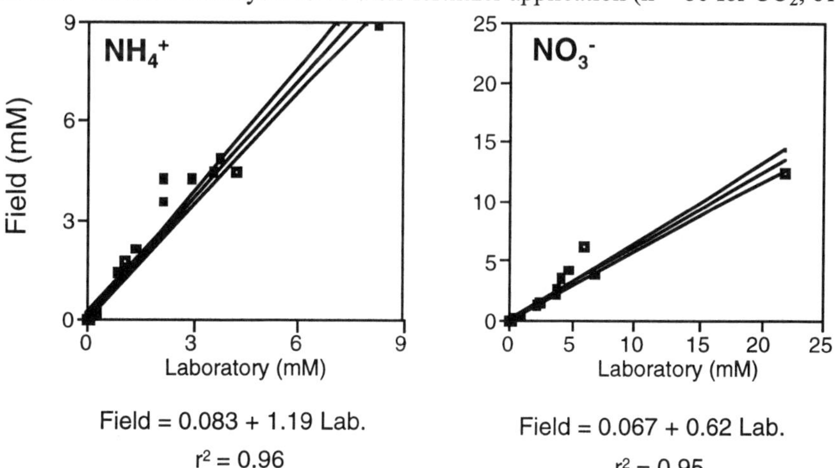

FIGURE 1. Comparison of field and laboratory measurements of fertilizer nutrients.

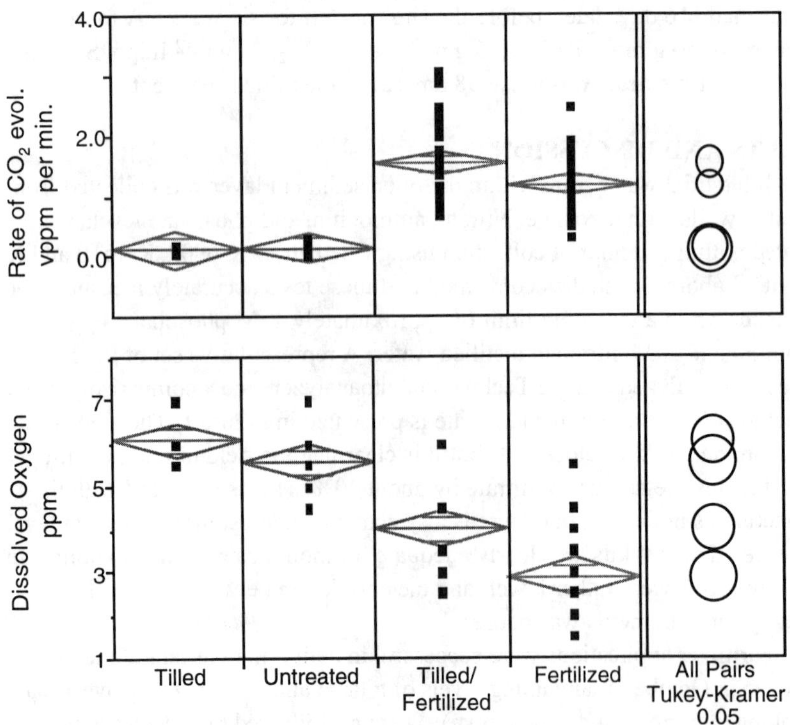

FIGURE 2. A comparison of *in situ* rates of CO_2 evolution and interstitial dissolved oxygen concentrations.

for dissolved oxygen). The heights of the diamonds encompass the 95% confidence limits of the estimates of the mean (center of diamond), using a pooled estimate of variance. For both analyses, the Tukey-Kramer test indicates that the untreated and tilled plots are statistically indistinguishable, while the fertilized and tilled/fertilized are different from the untreated (at $p<0.001$) and from each other (at $p<0.05$ for dissolved oxygen, $p<0.10$ for CO_2).

Samples collected 59 days after fertilizer application showed that this was in turn followed by an increase in microbial biomass (Figure 3), assessed by the amount of phospholipid fatty acids (White *et al.* 1996).

Despite purchasing all the oil for this experiment in one batch, there were significant differences in the composition of the oil applied to the different plots. Fortunately the ratio of phenanthrene to the dimethyl and ethyl phenanthrenes in the oils was similar. Aerobic biodegradation is expected to degrade phenanthrene before its alkylated forms, and Figure 4 shows that this expectation was met in samples collected 399 days after fertilizer application (about 195 ice-free days).

Taken together, our data provide convincing evidence that fertilizer applications were successful at delivering nutrients to the oiled sediment, resulting in increased microbial activity (O_2 consumption, CO_2-evolution) and growth

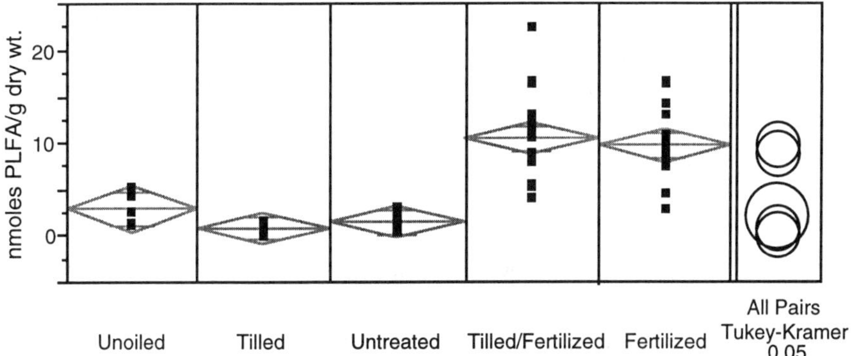

FIGURE 3. Nanomoles of phospholipid fatty acid per gram of sediment (dry weight) 59 days after treatment. The amounts of phospholipid fatty acid in oiled/untreated and oiled/tilled plots were statistically indistinguishable from an unoiled remote site, while the fertilized and tilled/fertilized plots had significantly more (at p<0.005).

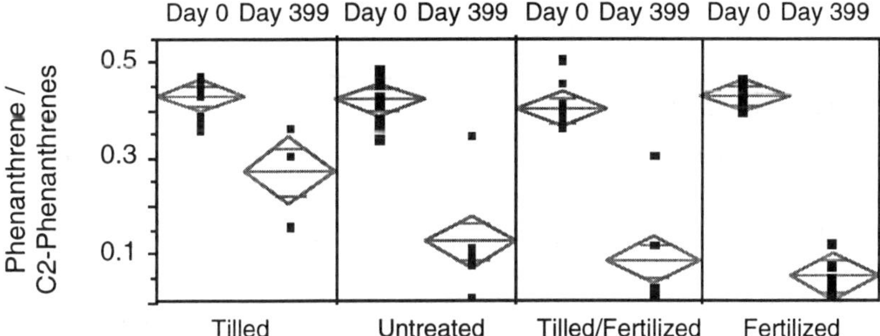

FIGURE 4. Changes in the ratio of phenanthrene to the dimethyl and ethyl phenanthrenes in oils collected on Days 0 and 399. The differences are statistically significant (p<0.05) on each plot.

(increased phospholipid fatty acids) and oil biodegradation (preferential removal of phenanthrene). Bioremediation is thus a potentially important tool in such environments. In addition, our data indicate that the success of fertilizer applications can be measured with simple test kits in the field, which should allow faster and cheaper optimization of fertilizer application strategies in all aquatic environments.

ACKNOWLEDGMENTS

The *In situ* Treatment of Oiled Sediment Shorelines Program was sponsored by an international partnership of spill response and research agencies composed of; Canadian Coast Guard, Environment Canada, Exxon Research and

Engineering Co., Fisheries and Oceans Canada, Imperial Oil Canada, Marine Pollution Control Unit (UK), Minerals Management Service (USA), Norwegian Pollution Control Authority, Swedish Rescue Service Agency and the Texas General Land Office. We are indebted to these organizations, and the field crews who helped apply the oil.

REFERENCES

Bragg, J. R., Prince, R. C., Harner, E. J. and Atlas, R. M. 1994. "Effectiveness of bioremediation for the *Exxon Valdez* oil spill." *Nature 368*, 413-418.

Lee, K., Tremblay, G. H. and Cobanli, S. E. 1995. "Bioremediation of oiled beach sediments: assessment of inorganic and organic fertilizers." In *Proceedings of the 1995 International Oil Spill Conference*, American Petroleum Institute, Washington DC, pp. 107-113.

Prince, R. C. 1993. "Petroleum spill bioremediation in marine environments." *Critical Reviews Microbiology 19*, 217-242.

Sendstad, E., Hoddo, T., Sveum, P., Eimhjellen, K., Josefsen, K., Nilsen, O. and Sommer, T. 1982. "Enhanced oil degradation on an arctic shoreline." In *Proceedings of the Fifth Arctic Marine Oilspill Program Technical Seminar*, Edmonton, Alberta, pp. 331-340.

Sergy, G. A., Guénette, C. C., Owens, E. H., Prince, R. C. and Lee, K. 1998. "The Svalbard shoreline oilspill field trials." In *Proc. 21st Arctic and Marine Oil Spill Program Seminar*. Environment Canada, pp. 873-889.

Sveum, P. and Ladousse, A. 1989. "Biodegradation of oil in the Arctic: enhancement by oil-soluble fertilizer application." In *Proceedings of the 1989 International Oil Spill Conference*, American Petroleum Institute, Washington DC, pp. 439-446.

Swannell, R. P. J., Lee, K. and McDonagh, M. 1996. "Field evaluations of marine oil spill bioremediation." *Microbiol. Rev. 60*, 342-365.

Swannell, R. P. J., Lee, K., Basseres, A. and Merlin, F. X. 1994. "A direct respirometric method for in-situ determination of bioremediation efficiency." In *Proceedings of the Seventeenth Arctic Marine Oilspill Program Technical Seminar*, Environment Canada, Ottawa, pp. 1273-1286.

Venosa, A. D., Suidan, M. T., Wrenn, B. A., Strohmeier, K. L., Haines, J. R., Eberhart, B. L., King, D. and Holder, E. 1996. "Bioremediation of an experimental oil spill on the shoreline of Delaware Bay." *Environ. Sci. Technol. 30*, 1764-1775

White, D. C., Stair, J. O. and Ringelberg, D. B. 1996. "Quantitative comparisons of *in situ* microbial biodiversity by signature biomarker analysis." *J. Ind. Microbiol. 17*, 185-196.

BIODEGRADATION OF SPILLED OIL OF NAKHODKA ACCIDENT IN JAPAN, 1997

Takashi Ohashi, Satoshi Imano
(Asano Engineering Co. Ltd., Chuo-ku, Tokyo, JAPAN),
Kazuhiro Iwasaki, Osami Yagi
(National Institute for Environmental Studies and CREST,
Tsukuba, Ibaraki, JAPAN)

ABSTRACT: The Nakhodka oil spill accident of January 2, 1997 was the most serious ever in Japan. Since the accident, bioremediation has received a great deal of attention. To study the application of bioremediation in cleaning up the Nakhodka oil spill, we determined the components of the spilled oil and studied their biodegradability. We also evaluated the effects of the spilled oil on the microbial population of the beach sediments.

We collected samples of beach sediments from 2 coastal areas near the spill and extracted them with hexane. The extracts were fractionated and analyzed by gas chromatography-mass spectrometry (GC–MS). The chromatograms of oil from samples collected both several days and a year after the accident were similar. We determined the biodegradation of the components of the oil in a liquid medium with and without nitrogen, phosphorus and exogenous oil-degrading bacteria that had been isolated from soil. We observed significant biodegradation of the n-alkane fractions in samples both with and without added oil-degrading bacteria. Microbial studies were conducted over a 1-year period on the sediments of one of the polluted beaches. Both the number of bacteria and the diversity index were very high soon after the accident. It appeared that the spilled oil was a carbon source for the oil-degrading bacteria in the beach sediments.

INTRODUCTION

A serious oil spill occurred off the coast of Japan in the early hours of January 2, 1997. The Russian tanker, Nakhodka, spilled 6240 kl of heavy oil (class C), contaminating a broad area of coastline bordering the Sea of Japan.

Early attempts to clean up the spilled oil were hampered by bad weather and the remoteness of the location. Many conventional clean-up techniques were employed to remove oil from the surface of rocks and beaches; these included the use of booms, skimmers and manual scrubbing. However, these techniques were unable to remove all of the oil from the beach sediments. As a consequence, public attention in Japan was directed towards alternative clean-up technologies, particularly bioremediation (US EPA 1990; Pritchard and Costa, 1991).

In our study we evaluated the weathering of hydrocarbons in sandy sediments of the oil-polluted zone between January 1997 and January 1998. We also evaluated the biodegradation of the components of the spilled oil in a liquid medium by both the indigenous sediment microorganisms and exogenous microorganisms. Further, over a 1-year period we studied the effect of the oil spill on the number and diversity of indigenous bacteria in the sediments.

MATERIALS AND METHODS

Samples. We collected samples of oil-contaminated surface sediments at a depth of 5 to 15 cm for microbial and hydrocarbon analysis. The samples were taken from 2 sites on the coast of Mikuni, Fukui Prefecture. All the samples were kept at 10℃ until they could be processed for chemical and microbiological analysis.

Measurement of Hydrocarbon Content. Hydrocarbons were extracted from both the sediment samples and the liquid cultures by vigorous shaking for 10 min with 10 ml of S-316 (polychlorotrifluoroethane, Horiba Co., Kyoto). The hydrocarbon-containing solvent layers were then separated off and dehydrated with Na_2SO_4. Quantitative analysis was performed using an oil-content analyzer (OCMA-350, Horiba Co., Kyoto).

Chemical Analysis. One gram of sample was extracted 3 times for 10 min with 20 ml of hexane. The extract was fractionated with 5% hydrous-silica-gel-column chromatography. It was then injected into a gas chromatograph (model GC-17A, Shimadzu Co., Tokyo) equipped with a mass spectrometer (model QP5050). Each fraction was analyzed using a capillary column TC-1 (0.32 mm × 30 m, GL Science Co., Tokyo). The column temperature was programmed to go from 50℃ (5-min hold) to 300℃ (15-min hold) at 10℃/min for the aliphatic fraction, and to go from 50℃ (1-min hold) to 120℃ at 5℃/min, then from 120℃ to 300℃ (5-min hold) at 12℃/min for the aromatic fraction.

Bacterial Strains and Culture. The bacterial strains used in the study were isolated from uncontaminated soil by enrichment culture. Dibenzo-p-dioxine-degrading bacteria were grown in a minimum medium containing dibenzo-p-dioxine as the sole carbon source. The composition of the minimum medium was as follows (mg/l): NH_4Cl, 2140; K_2HPO_4, 1170; KH_2PO_4, 450; $MgSO_4{\cdot}7H_2O$, 120; $FeSO_4{\cdot}7H_2O$, 28; $Ca(NO_3)_2{\cdot}4H_2O$, 4.8; H_3BO_3, 0.005; $MnSO_4{\cdot}4\text{–}6H_2O$, 0.06; $Na_2MoO_4{\cdot}2H_2O$, 0.001; $Co(NO_3)_2{\cdot}6H2O$, 0.06; $ZnSO_4{\cdot}7H_2O$, 0.01; $NiSO_4{\cdot}7H_2O$, 0.006; $CuSO_4{\cdot}5H_2O$, 0.006; H_2SeO_4, 0.004. The final pH was 6.8. The culture was reciprocally shaken at 30℃. Dibenzofuran- dibenzothiophene-, hexadecane- and benzene-degrading bacteria were also used in this study.

Biodegradation Experiment. The biodegradation experiment on the contaminated beach sediments was done in screw-capped test tubes, each containing 5 ml of minimum medium. We added 50 mg of contaminated beach sand and 100 μl of the mixed culture to each test tube. These tubes were incubated at 30℃.

Cluster Analysis. We used a plate-count method with a spiral plater (model EDDY JET, IUL, S.A., Barcelona) to count the microorganisms in the sediment samples and perform cluster analyses. The plates (Marine Agar, Difco) were incubated at 25°C. Colonies were toothpicked from the plate into a marine broth (Difco) and grown at 25°C for 24 to 48 h. Each culture was diluted with distilled water, and the cell suspension was tested for utilization of 95 carbon sources by being placed on a GN microplate (Biolog Inc., Hayward, Calif.) at 25°C for 24 h. For cluster analysis of strains we used the UPGMA method. The diversity index (Yokoyama 1993) was calculated as follows: Diversity Index (DI) = sum of distances between each cluster × average distance between the clusters.

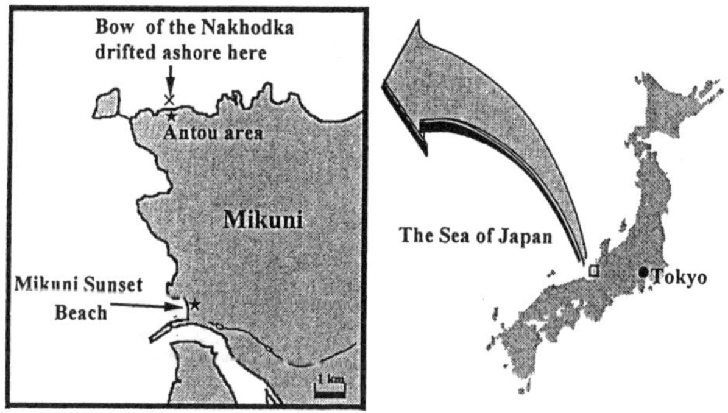

FIGURE 1. Locations of the 2 sampling sites

RESULT AND DISCUSSION

Persistence of Spilled Oil. Figure 1 shows the 2 sites from which we collected the samples. By several days after the accident large amounts of spilled oil had been deposited on Mikuni Sunset Beach and in the Antou area. A year after the Nakhodka oil spill, no oil remained on Mikuni Sunset Beach, but in the Antou area oil remained protected under rocks and in areas not constantly affected by wave action.

Comparison of Chromatograms of Spilled Oil. Figure 2 shows the chromatograms of the spilled oil analyzed by GC-MS. In the aromatic fraction there was phenanthrene, plus several compounds composed of methyl side-chains added to phenanthrene and naphthalene. The chromatograms performed a year after the accident were similar to those performed several days after; the composition of the spilled oil had not changed dramatically. This result indicated that the heavy oil in our 1-year samples—oil that had settled near the high tide mark, where it was not constantly affected by wave action—had barely been degraded. In an attempt to find a possible solution to this problem we studied the biodegradability of the spilled oil.

Biodegradation of Spilled Oil in the Presence of Added Nutrients With and Without Mixed Culture. Figure 3 shows the effects of hydrocarbon-degrading bacteria on biodegradation of the spilled oil. The initial oil concentration in the samples was about 1000 mg/l. After 4 weeks 40% of the oil had been degraded in the presence of the mixed culture; without the mixed culture only 25% of the oil had been degraded.

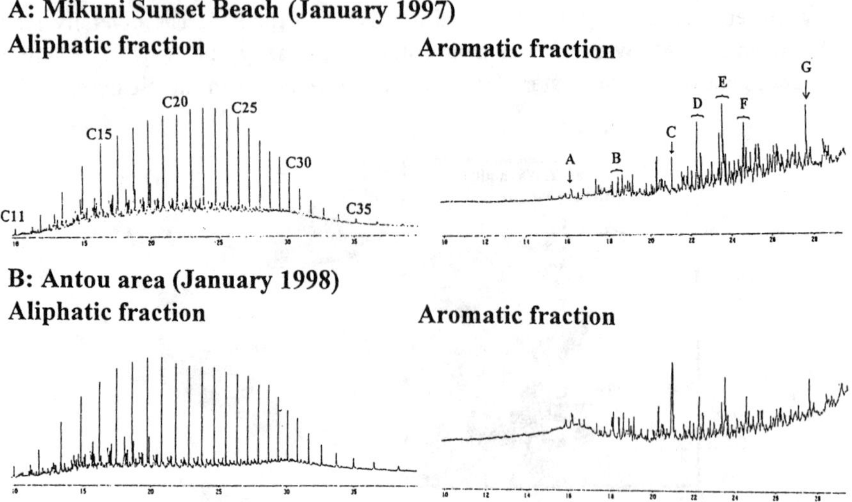

FIGURE 2. Chromatograms of the aliphatic and aromatic fractions of spilled oil extracted from samples collected several days (A) and a year (B) after the spill. A: dimethylnaphthalene; B: trimethylnaphthalene; C:phenanthrene;D: methylphenanthrene; E:dimethylphenanthrene; F:trimethylphenanthrene; G: methylchrysene

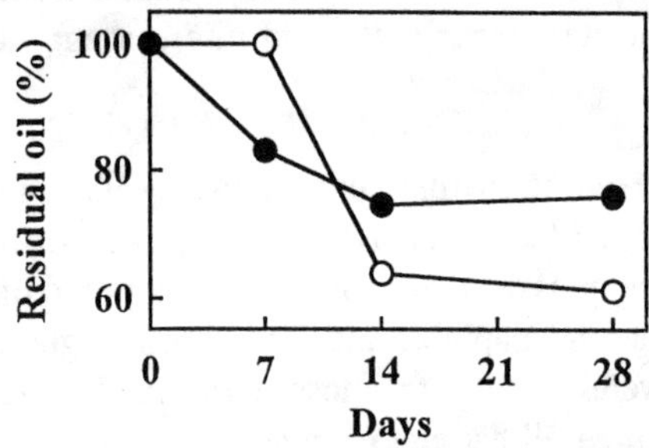

FIGURE 3. Effect of bioaugmentation on biodegradation of spilled oil after the addition of nutrients. ○ mixed culture added ● no mixed culture

Figure 4(A) shows changes in the chromatograms of the aliphatic fraction with no added mixed culture. The linear-alkane level decreased considerably after

1 week, and few linear-alkanes remained after 4 weeks. Figure 4(B) shows changes in the chromatograms of the aromatic fraction with no added mixed culture. There was a rapid decrease in the levels of light compounds, such as aromatics with 3 or fewer rings; the decline in the level of mono-methylphenanthrene was rapid. By contrast, aromatics with 4 or more rings (such as chrysene and di- tri- or tetramethylphenanthrene) were degraded slowly. In summary, adding nutrients only (nitrogen and phosphorus) to the samples decreased the levels of linear-alkanes and light compounds by encouraging the growth of the indigenous bacteria. Adding nutrients and exogenous hydrocarbon-degrading bacteria to the samples also decreased the levels of linear-alkanes and light compounds, and degraded the oil further.

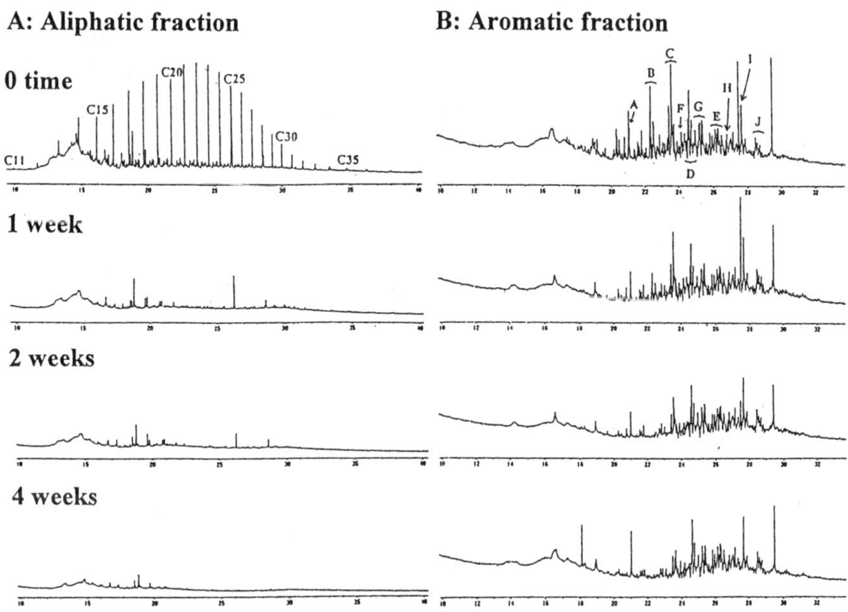

FIGURE 4. Changes in the chromatograms of the aliphatic fraction (A) and the aromatic fraction (B) of spilled oil from samples with added nutrients but no added mixed culture. A: phenanthrene; B, C, D, E: mono-, di-, tri-, tetramethylphenanthrene, respectively; F: pyrene; G: methylpyrene; H: chrysene; I: methylchrysene; J: dimethylbenzoanthracene

Effect of the Spill on the Microbial Population of the Coastal Sediments. There have been few trials of bioremediation in Japan because the appropriate risk-assessment technologies are still being developed. To evaluate the influence of the spilled oil on the microbial population in the sediments of Mikuni Sunset Beach, we determined bacterial number and DI.

Table 1 shows the number and DI of the bacteria in samples collected at Mikuni Sunset Beach. In January 1997, soon after the spill, both the number of bacteria and the DI were very high. A sample taken from an uncontaminated section of coastline was found to contain 10^3 to 10^4 cells/g (data not shown). At the polluted beach the number of heavy-oil-degrading bacteria and the DI appeared to have increased as the bacteria began utilizing the spilled oil. However, although the number of bacteria in the sediments at this study site increased soon after the spill, the increase was insufficient to degrade all of the spilled oil. From the results of the biodegradation experiment, it is likely that adding nutrients would have been an effective way to accelerate biodegradation of the oil at this site. In January 1998 there was no spilled oil at this site, and the numbers of bacteria and the DI had accordingly dropped.

TABLE 1. Numbers and diversity indices of bacteria in samples of sediments taken from Mikuni Sunset Beach over a 1-year period following the spill.

Sampling date	Cell number (CFU / g)	Diversity index (DI)
1997.1.19	6.3×10^6	2339
1997.8.15	3.3×10^5	—
1997.10.4	1.3×10^4	1230
1998.1.17	1.8×10^4	333
1998.8.15	1.7×10^5	1067

CONCLUSIONS

We found that, in the wake of the Nakhodka spill, oil remained in those beach sediments that were not constantly affected by wave action; 1 year later, this oil had scarcely been degraded. Adding nutrients to samples of these oil-contaminated sediments decreased the levels of linear-alkanes and light aromatic compounds. In samples taken from Mikuni Sunset Beach, both the number of bacteria and the DI were very high soon after the spill, indicating that the spilled oil was a carbon source for the indigenous bacteria.

REFERENCES

Alaskan Oil Spill Bioremediation Project. 1990. Prepared for the U.S. Environmental Protection Agency, office of Research and Development. EPA 600/8-89/073.

Pritchard, P. H., and C.F. Costa. 1991. "EPA's Alaska oil Spill Bioremediation Project." *Environ. Sci. Technol.* 25(3): 372-379.

Yokoyama, K. 1993. "Evaluation of Biodiversity of Soil Microbial Community." *Biol. Intern.* 29: 74-78.

[1]THE USE OF BIOREMEDIATION TO TREAT AN OILED SHORLINE FOLLOWING THE *SEA EMPRESS* INCIDENT.

Richard Swannell & David Mitchell (AEA Technology, Culham, UK)
Gordon Lethbridge & David Jones (Shell Research Ltd, Chester, UK)
David Heath & Michelle Hagley (Texaco, Pembroke Refinery, UK)
Martin Jones & Stuart Petch (University of Newcastle upon Tyne, UK)
Roger Milne & Rosie Croxford (Environment Agency, Cardiff, UK)
Kenneth Lee (Fisheries and Oceans, Canada, Mont-Joli, Canada)

ABSTRACT: Bioremediation was investigated as a method of treating a mixture of Forties Crude Oil and Heavy Fuel Oil stranded on Bullwell Bay, Milford Haven, UK after the grounding of the *Sea Empress* in 1996. A randomised block design in triplicate was used to test the efficacy of two bioremediation treatments: a weekly application of inorganic mineral nutrients and a single application of a slow-release fertilizer. Concentrations of residual hydrocarbons normalised to the biomarker 17α(H),21β(H)-hopane showed that after two months the oil was significantly ($p<0.001$) more biodegraded in the treated plots than in the controls. The results confirm that bioremediation can be used to treat a mixture of crude and heavy fuel oil on a pebble beach. In particular, the data suggest that the application of a slow-release fertilizer alone may be a cost-effective method of treating low-energy, contaminated shorelines after a spill incident.

INTRODUCTION

The *Sea Empress* ran aground in Milford Haven on 15 February 1996, releasing 72,000 tonnes of Forties Blend crude oil and 480 tonnes of heavy fuel oil over the next 7 days. A wide range of clean-up techniques were used to treat shorelines oiled after the incident (MPCU 1996, Lunel *et al* 1995) and for the first time after an oil spill in western Europe, a survey was conducted to determine whether any of the contaminated shorelines could be treated with bioremediation. Bioremediation has been shown to have potential to treat oiled shorelines (Prince 1993, Swannell *et al* 1996). Field experiments in the UK (Swannell *et al* 1997a), Canada (Lee *et al* 1995) and the USA (Venosa *et al* 1995) have demonstrated that bioremediation can stimulate *in situ* oil biodegradation. Work following oil spill incidents (Bragg *et al* 1994, Swannell *et al* 1996) has also shown that bioremediation could stimulate oil biodegradation on oil-contaminated shorelines. However, although the scientific understanding of the use of bioremediation has improved considerably, there have been few trials which tackle the operational use of bioremediation after spill incidents (Prince 1993, Swannell *et al* 1996). It was recognised that a well-constructed test of the use of bioremediation to treat oil

---

[1] This paper is shortened version of a paper accepted for publication in the journal *Environmental Technology*. The full paper is expected to be published in the summer of 1999.

contamination arising from the *Sea Empress* might improve the operational understanding of bioremediation, clarifying more precisely the role of bioremediation as an oil spill treatment technology. This paper describes the field evaluation of two bioremediation strategies in triplicate on a shingle and pebble beach, and their effect on the shoreline and nearshore environments using a similar methodology to that employed by Venosa *et al.* (1995).

MATERIALS AND METHODS

Site Selection On the basis of the data obtained from a site survey (Lunel *et al* 1995), Bullwell Bay (5.04°W 51.7°N), a beach of limited amenity use, was considered the most suitable site for a demonstration project. The substrate was a mixture of free-draining shingle and pebble (mean density = 1.6 kg/l) overlaying clay (depth to clay layer = 10-40 cm), and had a gradient of 10-12.5% (n= 4). The beach faced north, had a short fetch of <1 km, and was normally subject to low wave and tidal energy. As such, physical removal of oil was thought to be slow, and oil biodegradation by indigenous micro-organisms was thought likely to be a significant mechanism for oil removal. Preliminary analyses showed that the beach sediment possessed a competent hydrocarbon-degrading population ($>10^5$ cells.g^{-1}) and that the low levels of N and P (<0.18 mg.l^{-1}) may limit oil biodegradation.

Experimental Design and Monitoring. A randomised block design was utilised with 3 blocks of 3 plots (each 9 m by 0.9 m wide) placed perpendicular to the sea 1.25 m apart; a distance proven by lithium tracer experiments to be adequate to prevent migration of mineral nutrients between plots. In each block one plot was treated with inorganic fertiliser (1.15 kg $NaNO_3$ and 0.08 kg KH_2PO_4 in 9 litres of seawater/plot/week), the second with a pelleted slow release fertiliser based ammonium nitrate phosphate. These pellets were secured in mesh bags (0.9m by 1m) positioned at 1 m intervals along the length of the plot) to provide an equivalent amount of N & P cumulative for the experimental duration as for the plots treated with liquid fertiliser. The third plot was left untreated as a control.

The plots were monitored on a monthly basis. They were divided into three equal zones of 2.7 m^2 in area. Three sub-samples were taken at random in each of these zones and mixed together to give one sample per zone. These samples were analysed for residual hydrocarbons using a method described by Swannell *et al* (1997b). A sub-sample (250 g) was washed with 120 ml 0.5M $NaHCO_3$ adjusted to pH 8.5 and analysed for mineral nutrient content. Toxicity of the water soluble fraction (obtained by mixing oiled sediment in filtered seawater for 12 hours) was analysed using an oyster embryo bioassay (using *Crassostrea gigas*). Microtox™ tests were also carried out using the Organic Solvent Extract - Basic Test in accordance with the manufacturer's testing manual (Microbics Corporation 1994). Seawater samples were also taken at stations within Bullwell Bay and in Milford Haven to determine mineral nutrient levels in the nearshore environment.

RESULTS

Residual Hydrocarbon Analysis. At the start of the experiment no significant difference was found in the mean TPH levels, whereas at the end of the experiment the levels were significantly lower on the plots treated with liquid fertiliser (p=0.024) and on the plots treated with the slow release fertiliser (p=0.074). The TPH/hopane and Total Resolvable Hydrocarbons (TRH)/hopane ratios were also reduced on the treated plots (Figure 1). Given the weathered state of the oil seven months after the spill, any changes in this ratio could be attributed to biodegradation rather than to any other weathering process (Bragg *et al* 1994). After 2 months TPH/hopane and Total GC Resolvable Hydrocarbon (TRH)/hopane ratios were significantly reduced (p<0.001) on the bioremediated plots relative to the controls (using ANOVA). There was no difference in the average ratios at Day 0, and the magnitude of the difference between controls and treated plots increased with time (Figure 1). At the end of the experiment, the oil on the control plots was on average 13% more biodegraded than at the beginning. In contrast the oil on the treated plots was on average between 35 and 43% more biodegraded. There was no statistical difference in the degree of oil biodegradation on the bioremediated plots.

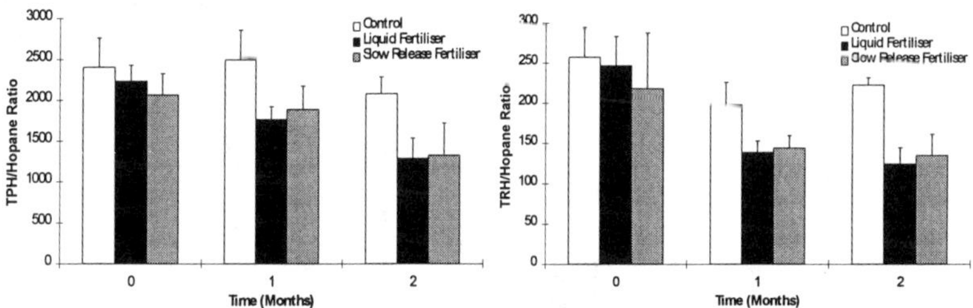

FIGURE 1. Ratios of total petroleum hydrocarbons (TPH) and total resolvable hydrocarbons (TRH) to hopane over time

The concentration of hopane did decrease slightly over the course of the experiment suggesting that some of the reduction in TPH could have been attributed to physical oil processes (data not shown). However, these changes were not statistically significant, and were not influenced by treatment. Thus, the oil chemistry results strongly suggests that the bioremediation treatments stimulated oil biodegradation thereby reducing the TPH levels and the TPH/hopane ratio.

Mineral Nutrient Levels. Data from the 2-month sample point clearly shows the efficacy of liquid fertiliser applications in elevating N and P concentrations within the beach sediment (Figure 2). By contrast the plots treated with the slow-release agent show only marginally higher levels of nitrate than are observed in the control.

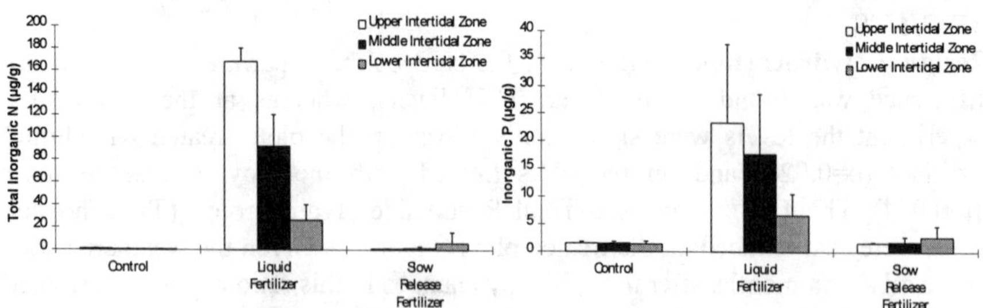

FIGURE 2. N and P distribution on bioremediation plots at 2 months

Toxicity Assessment. Acute toxicity (analysed using Microtox™) within the extracted sediment samples contaminated with the 6-month weathered oil was low and could only be found in 3 of 9 samples taken at Day 0 and after 2 months (mean = 53.0 mg & 52.8 mg eq/mL). These samples are considered to be only marginally toxic as "uncontaminated" control sediments have been reported to have a mg eq/ml value of 49.5 (Johnson, 1997). With the restricted number of positive samples, variation in toxicity between samples could not be correlated to treatment. Remaining samples were below the EC_{50} detection limit of the Microtox™ Basic Test. Using the oyster larvae assay a more complex picture emerged as illustrated in Figure 3.

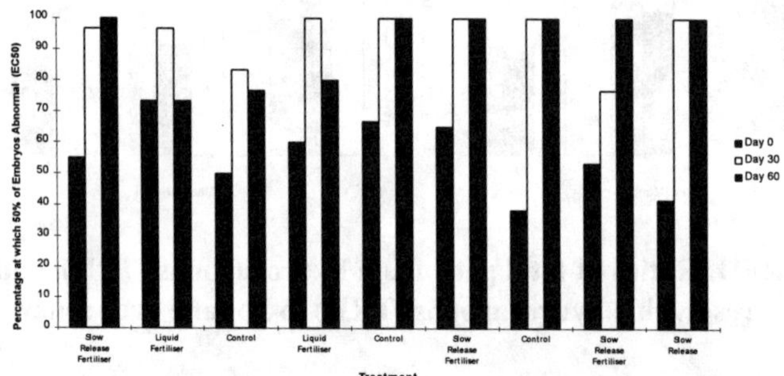

FIGURE 3. EC_{50} Values for Samples from Bioremediation Plots Tested Using the Oyster Larvae Assay

As can be seen all plots showed residual toxicity in the samples taken on Day 0. However on Days 30 and 60 only samples from certain plots gave a toxic response and this appears not to be related to treatment.

Background Nutrient Concentrations. Figure 4 shows the results from the analysis of samples taken in Bullwell Bay for background nutrient concentrations. Results for a site in the Milford Haven waterway are shown for comparison. Looking at the totality of the data and comparing with the data for the Milford

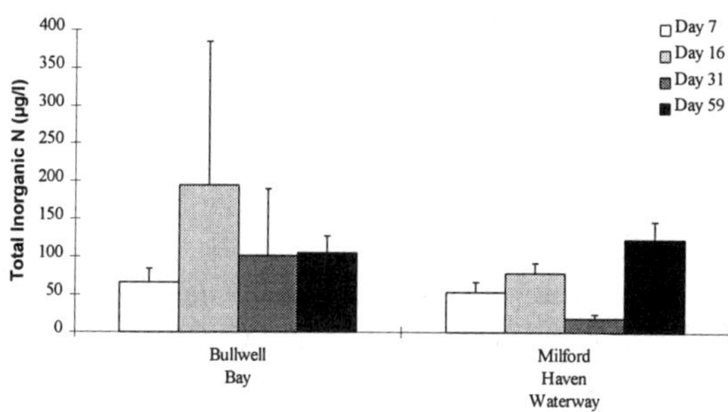

FIGURE 4. Total inorganic nitrogen concentrations in seawater sampled in Bullwell Bay and Milford Haven.

Haven waterway, it is clear however that the bioremediation treatments had minimal impact on nutrient levels in the nearshore environment.

CONCLUSIONS

The work clearly shows that both liquid inorganic fertiliser and slow release mineral fertiliser were capable of stimulating the biodegradation of oil on Bullwell Bay. It should also be noted that Bullwell Bay was contaminated with a mixture cargo oil (Forties Blend) and fuel oil (HFO) spilled during the grounding of the *Sea Empress*. Whilst there may be differences in the rates at which these oils degraded *in situ*, the results clearly show that both treatments were effective in stimulating the biodegradation of the oil mixture.

When looking at the mode of action of the two treatments there is, however, a striking difference. Levels of mineral nutrients detected in the plots treated with liquid fertiliser were much higher than those treated with the slow release agent. We suggest this difference is due both to microbial utilisation and physical removal of nutrients released from the pelleted fertiliser at a slow rate on the beach. The liquid agent was deposited weekly on the plots in bulk in a single application, thus leading to a spike in N and P concentrations for shortly afterwards. These differences had no effect on the overall result. The slow release agent is however much less labour intensive to apply and requires application only every 4 months. For this reason we see the use of such slow release mineral fertilisers as the most promising way forward from the standpoint of cost, convenience and efficacy. This trial has also demonstrated that bioremediation has a minimal impact on the surrounding environment in line with the observations made after the *Exxon Valdez* incident (Bragg *et al* 1994, Pritchard *et al* 1991). The data shows that there was little, if any, impact on nutrients levels within Bullwell Bay and that bioremediation treatments did not change the toxicity of the oiled sediment.

Bioremediation has been the subject of study for many years. Numerous field trials have now been conducted around the world both experimentally and after spill incidents (Venosa *et al* 1995, Oudot *et al* 1998, Lee *et al* 1995, Swannell *et al* 1996, 1997a, Bragg *et al* 1994). The authors believe that this trial should aid the implementation of this technique alongside the other oil spill treatments available to responders at the time of an incident in a timely and appropriate manner.

REFERENCES

Bragg, J. R., R C. Prince, E. C. Harner, and R. M. Atlas. 1994 "Effectiveness of Bioremediation for the *Exxon Valdez* Oil Spill" *Nature 368*: 413-418.

Johnson, B.T. 1998 "Microtox™ Toxicity Test System – New Developments and Applications". In P. G. Wells, K. Lee and C. Blaise (Eds.). *Microscale Testing in Aquatic Toxicology – Advances, Techniques, and Practice,* pp. 201-219. CRC Press, Boca Raton, FL, USA.

Lee, K., G. H.Tremblay, and E. M. Levy. 1995. "Bioremediation: Application of Slow Release Fertilisers on Low Energy Shorelines". In: *Proceedings of 1995 Oil Spill Conference,* pp. 107-113. American Petroleum Institute, Washington, DC, USA.

Lunel, T., R. Swannell, J. Rusin, N. Bailey, C. Halliwell, L. Davies, M. Sommerville, A. Dobie, D. Mitchell, M. McDonagh and K.Lee. 1995. "Monitoring of the effectiveness of response options during the Sea Empress incident: A key component of the successful counter-pollution response". *Spill Science & Technology Bulletin 2* (2/3): 99-112.

Merlin, F.X. (1995) "Assessment of the Efficiency of Bioremediation Strategies through Field Experiment" In: *Proceedings of the 2nd International Oil Spill Research and Development Forum,* pp 37-44. IMO, London, UK.

Microbics Corporation. 1994. *Microtox® M500 Manual. A Toxicity testing handbook. Version 3*. Carlsbad, CA.

MPCU. 1996. "The *Sea Empress* Incident." Published by the Coastguard Agency, Spring Place, 105 Commercial Road, Southampton, SO15 1EG. ISBN 1-901518-00-0

Prince, R. 1993. "Petroleum Spill Bioremediation in Marine Environments" *Critical Reviews in Microbiology 19*: 217-242.

Pritchard, P.H., C. F. Costa, and L. Suit. 1991. *Alaska Oil Spill Bioremediation Project Science Advisory Board Draft Report.* Report EPA/600/9-91/046a Environmental Research Laboratory, USEPA, Gulf Breeze, FL, USA.

Swannell, R. P. J., D. J. Mitchell, D. M. Jones, A. Willis, K. Lee, and J. Lepo. 1997a "Field Evaluation of Bioremediation to Treat Crude Oil on a Mudflat." In: *In situ and On-Site Bioremediation Conference* 4(4), pp. 401-406. Battelle Press, Columbus, Ohio, USA.

Swannell, R. P. J., D. J. Mitchell, D. I. Little and J. Smith 1997b *The Fate of Oil from Cleaned and Uncleaned Beaches Following the Sea Empress Incident.* Published by SEEEC, Cardiff, UK, ISBN No. 0-7058-1753-9.

Swannell, R.P.J., K. Lee and M. McDonagh. 1996. "Field Evaluations of Marine Oil Spill Bioremediation." *Microbiological Reviews 60*:342-365.

Venosa A.D., M.T. Suidan, B.A. Wrenn, K.L. Strohmeier, J.R. Haines, B.L. Eberhardt, D. King and E. Holder. 1996. "Bioremediation of an Experimental Spill on the Shoreline of Delaware Bay". *Environmental Science and Technology 30*:1764-1775.

CRUDE OIL COMPONENT BIODEGRADATION KINETICS BY MARINE AND FRESHWATER CONSORTIA

E. L. Holder and K. M. Miller (Univ. Of Cincinnati, Cincinnati, OH)
J. R. Haines (U. S. EPA, Cincinnati, OH)

Abstract: Sediments from locations throughout the U.S. were enriched for hydrocarbon degraders using artificially weathered Prudhoe Bay crude oil. Biodegradation rates were measured for 28 days in triplicate or quadruplicate shake flasks with either artificial sea water or freshwater. Flasks were sacrificed periodically for analysis of residual crude oil constituents by gas chromatography / mass spectrometry (GC/MS). Rates of alkane and aromatic hydrocarbon biodegradation were calculated by nonlinear regression analysis of hopane normalized analyte depletion assuming a first order relationship. Hydrocarbon degrading bacteria were enumerated using a most probable number (MPN) technique.

All culture consortia were competent alkane degraders with 7 of 8 marine and 5 of 6 freshwater consortia removing greater than 98% of measured alkanes. Biodegradation rates for summed n-alkanes ranged from 0.14 to 0.61 day $^{-1}$. The rates for pristane and phytane were always lower. The biodegradation rates for n-alkanes tended to decrease as chain length increased especially for the freshwater consortia. The removal of measured aromatic compounds by marine consortia ranged from 63 to 92 % and from 36 to 74% by the freshwater consortia. The biodegradation rate coefficients for the summed aromatic compounds were greater for 7 of the marine cultures than for 5 of the freshwater cultures. Lower rates were found with increasing ring number and alkyl substitution.

INTRODUCTION

Biodegradation of complex mixtures such as crude oil has been subject to study for many years. One of the key problems to be addressed in this field is the determination of the reaction kinetics of any single compound in a complex mixture. A common assumption is that the presence of many different possible substrates for the microorganisms will affect the metabolism of any one substrate compound. The problem is further compounded when biodegradation is carried out by a consortium of microorganisms rather than by a single species or very simple group of organisms. Work presented by Rogers et al., 1997 showed the kinetics of degradation of two substrates in the presence of two species of microorganisms. Results of field work have shown the kinetics of crude oil degradation in a marine environment with some effort directed toward microbial population behavior (Venosa et al., 1998).

The National Risk Management Research Laboratory carried out this study to determine the biodegradation kinetics of individual hydrocarbons in a crude oil matrix by mixed cultures of organisms. This approach perhaps more closely follows natural systems because few pollution events occur with single compounds. The kinetics of degradation of any one compound or group of compounds in crude oil were developed in the presence of the complete matrix. The 14 microbial populations used in this study were originally enriched from different sources. Each enrichment was a mixture of about fifteen different species of organisms. The results of the study permit the derivation of the kinetics of degradation of n-alkanes, branched alkanes, two, three, and four membered ring compounds, and their alkyl substituted homologs.

MATERIALS AND METHODS

Oil degrading enrichment cultures were developed by inoculating 10g of sediment from different areas of the United States into flasks containing 100mL mineral salts medium and North Slope 521 oil (NS521, Prudhoe Bay oil artificially weathered at 521°F under vacuum). Sediments from oceanic beaches were inoculated into artificial seawater and sediments or soils from terrestrial sources were inoculated into Bushnell Haas medium. The flasks were shaken on a rotary shaker at 200 rpm for three weeks at 20°C. The cultures that had grown were then transferred to fresh medium and incubated as before. Transfers were done until a stable, competent oil degrading consortium had been established. Of the 8 marine consortia, 3 were isolated from Alaskan sites impacted by the Exxon Valdez oil spill (4Org, Kn, and Snug), 3 from Maine (HI, Cv, and SS), 1 from Delaware (FB) and 1 from Texas (BS). The freshwater consortia were all from local sites (AF - Ohio River downstream from petroleum storage tanks, JY - Junkyard soil, MA - combined sewer overflow site, MC - downstream from a wastewater treatment plant discharge, RR diesel fuel contaminated soil, and Rav - ravine soil.)

The study was conducted in two segments. The first segment was carried out with the marine sediment enrichment cultures. Experimental flasks were filled with 100 mL artificial seawater (Spotte et al. 1984) and 0.5 g of NS521 oil. The flasks were then inoculated with one mL of culture. All flasks were incubated on a rotary shaker at 200 rpm at 20°C. At each sample time quadruplicate flasks were sacrificed for microbiological and hydrocarbon analysis. Sample times were 0, 1, 2, 4, 8, 12, 20, and 28 days of incubation. The second segment was carried out with freshwater or soil consortia using Bushnell Haas. Triplicate flasks were set up, and sample times were 0, 3, 7, 11, 21 and 28 days of incubation.

Prior to oil analysis, a 5mL subsample from each flask was withdrawn for microbiological analysis. Oil analysis was carried out according to the procedures published in the National Environmental Technology Application Center protocol. Briefly, methylene chloride and recovery standards were added to each flask to extract the oil hydrocarbons. The extracts were dried with sodium sulfate and exchanged into hexane. Internal standards were then added and the samples were analyzed by GC-MS. The concentrations of individual hydrocarbons were calculated and normalized to hopane.

The removal rates were determined using the first order rate equation:

$$C = C_0 e^{-kt}$$

where C = analyte concentration, C_0 = initial concentration, t = time in days, and k = the degradation rate constant, day^{-1}. Rates are reported for 48 compounds even though 69 oil components were measured, because 20 of the analytes were not present in sufficient quantity in the source oil to determine their rates. Hopane was used as a conserved biomarker. Rates are also reported for summed n-alkanes, branched alkanes (pristane and phytane), and summed aromatic compounds.

Differential estimation of alkane and aromatic hydrocarbon degraders was performed using the MPN method (Wrenn and Venosa, 1996) on the 5 mL subsamples taken from the flasks prior to chemical analysis. The procedure uses 96-well microtiter plates to provide an 8 tube MPN with 11 dilutions. The sample is transferred into the first row of wells and serially diluted across the plate. After incubation at 20°C for 2 or 3 weeks (alkane or aromatic plates respectively), positive wells were scored and MPNs calculated (Klee, 1996).

RESULTS AND DISCUSSION

Biodegradation removal rates were calculated both individually and by summation of compound categories. All cultures were determined to be competent alkane degraders. Rates for summed n-alkanes ranged from 0.14 to 0.61 day $^{-1}$ with r^2 values greater than 0.83. The rates for pristane and phytane were lower, ranging from 0.085 to 0.33 day $^{-1}$ with r^2 values greater than 0.80 for all but one consortia. Figure 1 shows the summed rates for each of the culture consortia. Two of the freshwater cultures had significantly higher n-alkane degradation rates than the other cultures. However the degradation rate constants of all the freshwater cultures for pristane - phytane were 2 to 6 times lower than the n-alkane degradation rates. The marine consortia showed a different trend as the degradation rates for the branched chain alkanes were only slightly lower than the respective n-alkane degradation rate. There are also differences in PAH

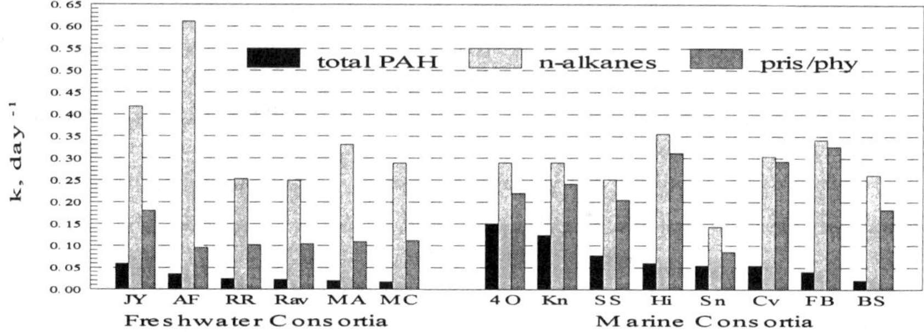

Figure 1. Removal rates for summed oil components, sorted by culture type and PAH k value.

removal rates. The degradation rates ranged form 0.02 to 0.14 day^{-1} with r^2 greater than 0.76 for 9 of the consortia. The consortia with the lower r^2 values were the least competent degraders. The degradation rates for total PAH removal for 6 of 8 marine consortia were greater than for 5 of 6 freshwater consortia.

Figure 2 depicts the removal rates of dibenzothiophene (DBT) and its alkylated homolog series by each consortium. A similar pattern is seen for all aromatic compounds. In all cases the rates decreased with increasing substitution. The addition of each alkyl-substituent group seems to have had a greater effect on the degradative ability of the freshwater consortia than on that of the marine consortia. However with the exception of two of the Alaskan cultures (4Org and Kn) which proved to be superior PAH degraders overall, the freshwater cultures demonstrated a greater potential to degrade DBT and C1 DBT. For the C2 and C3 DBT, the marine cultures yielded greater degradation rate constants than did the freshwater cultures.

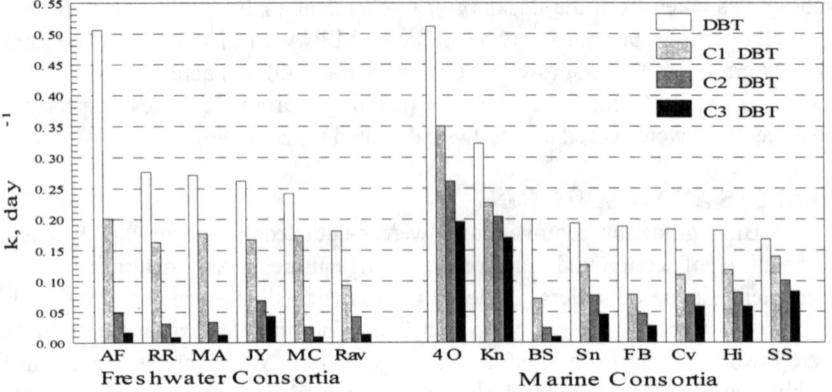

Figure 2: Degradation rates for dibenzothiophene and its homologs.

The degradation rates for 48 measured components for the best freshwater and best marine consortia are given in Figure 3. While the rates are greater for AF than 4Org for the low to mid-range alkanes, they decrease dramatically above C32 for AF. A similar decrease was observed for all the freshwater consortia. Removal rates of the various aromatic substituted series (naphthalenes, phenanthrenes, fluorenes, dibenzothiophenes, and naphthobenzothiophenes) showed a strong disparity between freshwater and marine cultures.

The numbers for the differential MPN plates for the two cultures are given in Figure 4. AF grew to a peak population by Day 3 for alkane degraders and by Day 7 for aromatic degraders and then declined. 4Org, the marine consortium, did not demonstrate as dramatic an increase in numbers nor the subsequent decay. The alkane degrading population numbers were comparable (between 10^{-6} and 10^{-8} /mL) for both consortia shown as well as for all consortia. For the aromatic degrading bacteria, the marine consortia started lower and the peak numbers tended to remain lower. 4Org had a much lower aromatic degrading population than AF but had higher removal rates.

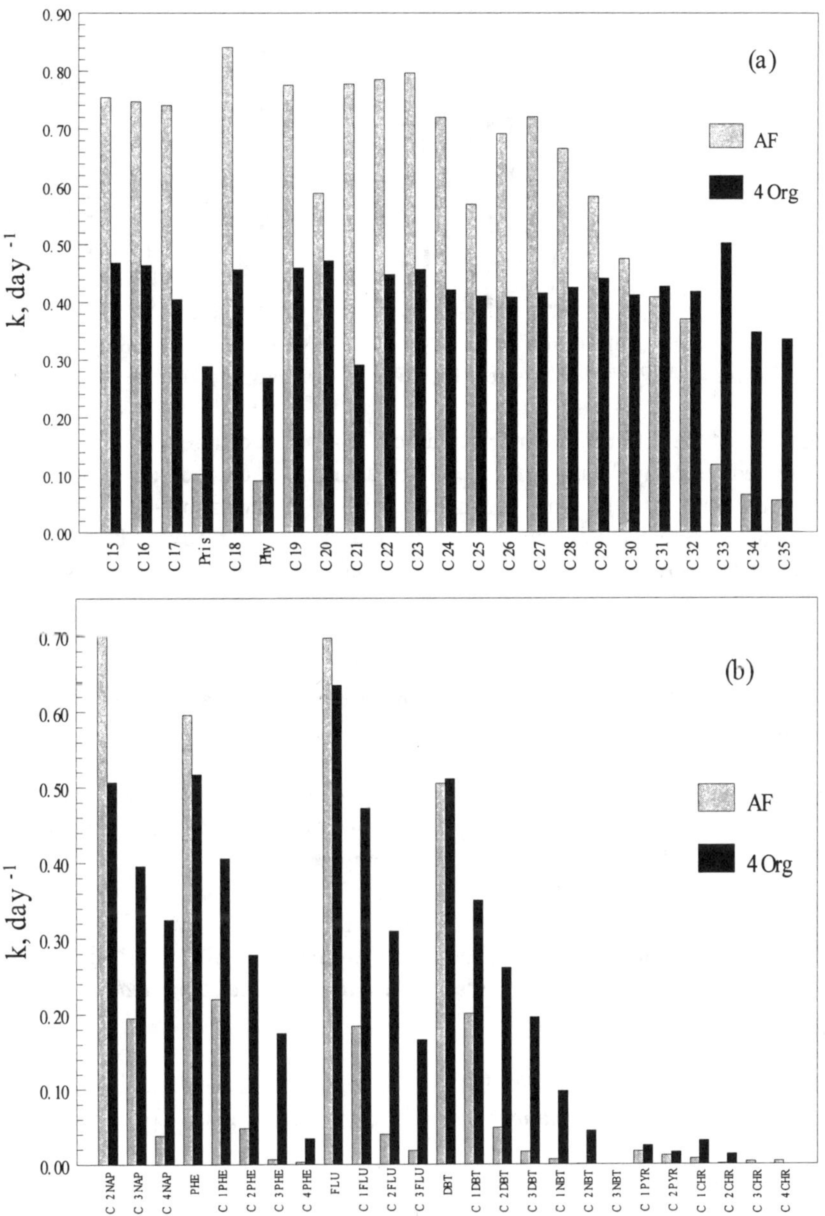

Figure 3: K values for 42 analytes a) 23 alkanes and b) 25 PAHs for the best freshwater and best marine consortia

The consortia showed large variance in activity, but there were some overall trends. The differences between the marine and the freshwater consortia

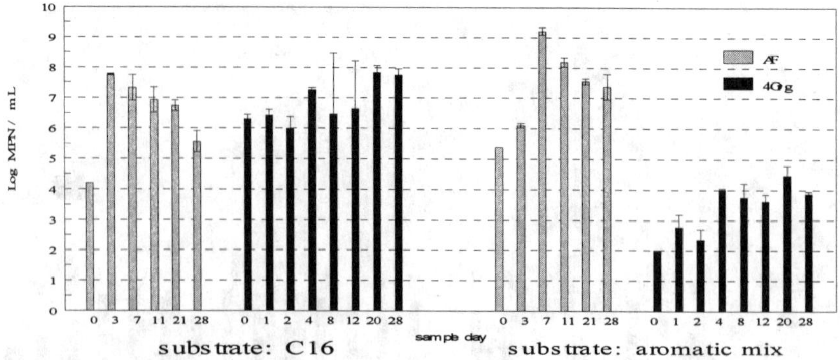

Figure 4: MPN's for the best freshwater and marine culture consortia

may reflect the differing inputs of hydrocarbons to the different environments. In bioremediation applications the ability of a culture to degrade large quantities of substrate with little biomass production, such as is shown by 4Org, is a benefit and could be relevant in the ability of a consortia to effectively bioremediate a site. The pollutants can be degraded without retaining excess carbon in the environment.

REFERENCES

Klee, A. J. 1993. "A Computer Program for the Determination of Most Probable Number and its Confidence Limits." *J. Microbiol. Methods* 18:91-98.

National Environmental Technology Applications Center 1991. *Oil Spill Bioremediation Products Testing Protocol Methods Manual.* University of Pittsburgh Applied Research Center, Pittsburgh, PA

Rogers, J. D., K. F. Reardon, and N. M. DuTeau. 1997. "Microbial Population Dynamics and Biodegradation Kinetics of Aromatic Hydrocarbon Mixtures." In B. C. Alleman and A. Leeson (Eds.) *In Situ and On-Site Bioremediation:* Vol. 4, p. 81 Battelle, Columbus, OH.

Spotte, S., G. Adams, and P.M. Bubucis. 1984. "GP2 Medium is an Artificial Seawater for Culture or Maintenance of Marine Organisms." *Zoo Biology* 3:229-240.

Venosa, A. D., G. A. Davis, J. R. Stephen, S. J. MacNaughton, Y. J. Chang, and D. C. White. 1998. "Microbial population changes during bioremediation of an experimental oil spill." *Eighth International Symposium on Microbial Ecology.* September 1998. Halifax, Nova Scotia.

Wrenn, B. A. and A. D. Venosa. 1996. "Selective Enumeration of Aromatic and Aliphatic Degrading Bacteria by a Most-Probable-Number Procedure." *Canadian J. Microbiology* 42:252-258.

FORMATION OF CARBOXYLIC ACIDS DURING BIODEGRADATION OF CRUDE OIL

J.S. Watson and D.M. Jones (University of Newcastle-upon-Tyne, UK)
R.P.J. Swannell (AEA Technology, Oxford, UK)

INTRODUCTION

This study investigates, using samples from laboratory and field biodegradation experiments, the hypothesis that carboxylic acids can form a significant part of degraded crude oil residues.

During the biodegradation of crude oil by micro-organisms, the concentration of hydrocarbons decreases but there may be the incorporation of partially metabolised hydrocarbons and possibly biologically-derived material into the resin and asphaltene fractions. Carboxylic acids are known to be produced as intermediates during the biodegradation of petroleum hydrocarbons, although it is unclear whether these organic acids contribute to the degraded crude oil residue that is resistant to further biodegradation. The detailed analysis of carboxylic acid distributions in degraded oil residues may make it possible to differentiate between metabolites and biologically-derived acids and those originally present in the undegraded oil.

The objectives of the experiments reported here were to analyse hydrocarbon and carboxylic acid distributions in order to examine their relationships during the microbial decomposition of a crude oil in the laboratory and in field samples. Bulk analysis of the oil or solvent extractable organic matter (EOM) using Iatroscan TLC-FID was used to detect changes in concentrations of the saturated hydrocarbon, aromatic hydrocarbon, resins and asphaltene fractions. Fourier Transform Infra-Red spectroscopy (FT-IR) of the EOM was used to monitor the development of functionalised (e.g. carbonyl) groups in the oil during biodegradation.

MATERIALS AND METHODS

The biodegradation of a light Arabian crude oil (~350mg) was carried out in Erlenmeyer flasks using two different naturally-occurring hydrocarbon-degrading microbial populations in seawater (~150ml), with and without nutrient supplement over an 80 day time period. Killed (using mercuric chloride as biocide) controls and duplicates were also included. Sample abbreviations were related to the medium used as follows; seawater-only (S), seawater + nutrients (SN) and seawater + sediment + nutrients (SSN). The nutrients used were $NaNO_3$ (200ppm), KH_2PO_4 (20ppm) and other trace nutrients. Degradation of a light Arabian crude oil was carried out in the field on a fine grained sand, initial oil concentration was 1.8l/m^2. Samples were analysed at weeks 0, 13, 31 and 39 after oil application.

Prior to extraction and analysis, the samples were spiked with hydrocarbon and carboxylic acids surrogate recovery standards. The total solvent extract was split into aliquots for different analysis. The hydrocarbons were fractionated and analysed in a way similar to that described in U.S. EPA method #8270. The carboxylic acid fractions were separated from the total extract using a new, rapid method using non-aqueous ion-exchange column chromatography, developed in this laboratory. The separated carboxylic acid fractions were methylated using BF_3-methanol (Fisher Scientific) and analysed by GC and further characterised by GC-mass spectrometry. FT-IR analyses of extract aliquots were carried out using thin films between KBr discs.

RESULTS AND DISCUSSION

Hydrocarbons. The laboratory oil degradation samples inoculated with seawater-only (S) displayed negligible hydrocarbon degradation over the 80 day period of the experiment and, using comparison with the killed control samples, most hydrocarbon losses could be accounted for by evaporation. Both the seawater + nutrients (SN) and the seawater + sediment + nutrients (SSN) samples displayed rapid losses of *n*-alkanes and two ring aromatic compounds, concurrently. The SSN samples displayed more rapid and more extensive degradation than the SN samples. Depletion for the total GC detectable hydrocarbons were 24%, 64% and 88% over the 80 day period for the S, SN and SSN samples respectively. The depletions for the total extract (EOM) measured gravimetrically were 0%, 30% and 45% for the S, SN and SSN respectively. The S samples are thought to nutrient limited; the more extensive and rapid degradation for the SSN samples compared to the SN samples, is thought to be due to the larger initial number of microorganisms present and a larger consortium introduced with the sediment.

For the SSN samples, by Day 10 of degradation all naphthalene, methylnaphthalenes and dimethylnaphthalenes had been removed; by day 20 all the *n*-alkanes, trimethylnaphthalenes, phenanthrene and dibenzothiophene had been removed; by Day 80 the only detectable 3 ringed aromatic compounds were the dimethylphenanthrenes and dimethyldibenzothiophenes. This order of degradation was consistent with many previous reports (e.g. Rowland *et al.*, 1986). In the most degraded samples, sterane hydrocarbons were found to have been degraded relative to the hopane hydrocarbons, with the order of degradation for the steranes being $C_{27}>C_{28}>C_{29}$. From simple ratios, it was possible to detect that some of the hopane hydrocarbons had been degraded, with the C_{33} components depleted relative to C_{29} and C_{30} compounds. No degradation of the triaromatic steroid hydrocarbons was detected.

Bulk analysis. Iatroscan TLC-FID data showed that there was a reduction in the proportion of hydrocarbons (saturated and aromatic) with biodegradation but the amounts of the asphaltene fractions remained fairly constant. The relative amounts of resin fractions in the SSN samples displayed a decrease followed by an increase with degradation time. FT-IR spectra clearly showed the development of a carbonyl stretching band with biodegradation. The SSN samples displayed a rapid increase in the C=O/C-H absorbance ratio up to Day 5, followed by a slight decrease and then a gradual increase, which appeared to still be increasing at Day 80 (Figure 1). Trends for O-H stretching were difficult to measure because the O-H peak was very broad and poorly resolved. Other information gained from FT-IR analysis suggested that there was an overall decrease in aliphatic chain length and an increase in overall aromaticity with degree of biodegradation.

Carboxylic acids. For the SSN samples at Day 0, hexadecanoic acid was the major peak in the gas chromatogram and the carboxylic acids were at very low concentrations (Figure 2). By Day 5 there were a significant number of chromatographically resolvable carboxylic acids (<C_{20}) at relatively high concentrations present but these acids were significantly reduced in concentration by Day 10 and greatly depleted by Day 20. By Day 40, an unresolved complex mixture (UCM) or 'hump' (>C_{20}) of branched and cyclic acids had developed in the chromatograms, with hopanoic acids (identified from GC-MS analysis) present on the UCM, which were also prominent in the Day 80 samples. The seawater-only samples displayed no increase in carboxylic acids concentrations.

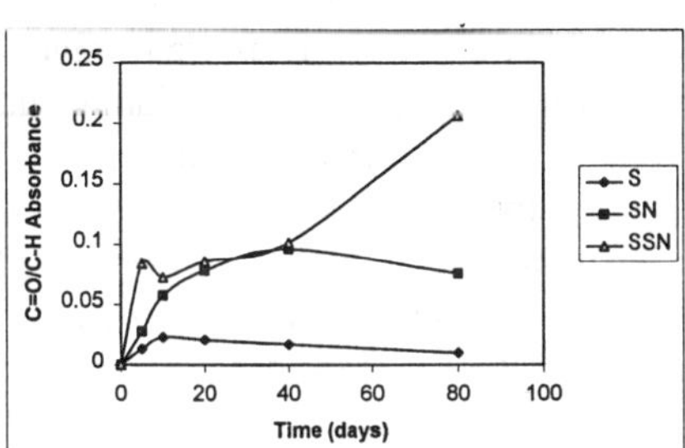

FIGURE 1. Plot of C=0/C-H (1706 cm^{-1}/2924cm^{-1}) absorbance against degradation time.

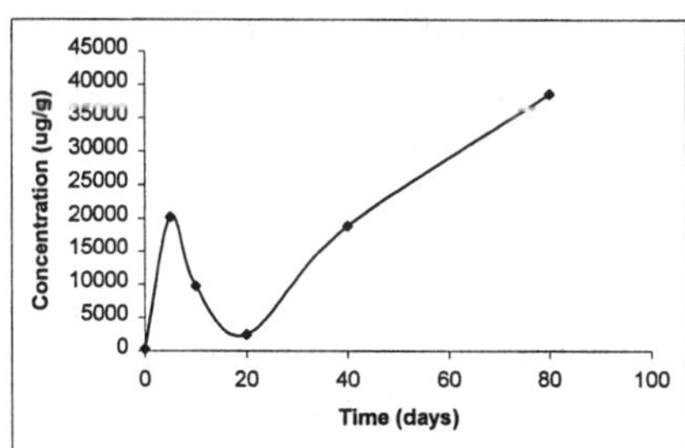

FIGURE 2. Concentration of total GC detectable carboxylic acids (as methyl esters) for the SSN samples with degradation time.

The identification of the hopanoic acids in the samples was based on a comparison of their relative retention times in published reports (e.g. Jaffé and Gallardo, 1993). The most abundant of these hopanoic acids in the laboratory degraded oils were C_{29}, C_{31} 22R+S and C_{32} 22R+S compounds which had the 17α(H),21β(H) configuration, suggesting that they originated from the microbial oxidation of the hopane hydrocarbons in the oil. However, no peaks were detected for either of the C_{30} αβ epimers suggesting that either the C_{30} hopane is resistant to biodegradation or that the carboxylic acids produced are rapidly further biodegraded or that the degradation pathway does not involve attack of the side chain.

The concentrations of the hopanoic acids represent only a very small fraction of the equivalent hopane hydrocarbon originally in the sample. For the SSN sample at Day 80 which has the greatest overall concentration of these acids, if one assumes that these acids are derived by microbial oxidation of the equivalent hopane, the percent transformation of hydrocarbon to hopanoic acid are approximately C_{29} αβ, 1.2%; C_{31} αβ 20S, 3.2%; C_{31} αβ 20R, 3.5%; C_{32} αβ 20S, 2.9% and C_{32} αβ 20R, 1.0%.

Field samples. The oil in the samples from a field degradation experiment was significantly less degraded (approx. 10% *n*-alkanes still present after 9 months) than the Day 10 SSN laboratory degradation experiment but they also displayed significant increases in their carboxylic acid concentrations. However, interestingly unlike the laboratory biodegraded samples the field sample distributions were dominated by C_{10}-C_{30} *n*-acids. The only hopanoic acids detected were C_{31} and C_{32} ββ 20R isomers, which are the "biological" configurations and are unlikely to have been derived from the oxidation of oil-derived hopane hydrocarbons.

CONCLUSIONS

During the biodegradation of crude oil, significant amounts of carboxylic acids were produced. Medium chain length ($<C_{20}$) carboxylic acids were rapidly produced, which coincided with removal of the *n*-alkanes but these medium chain acids were then themselves rapidly biodegraded. After extensive biodegradation of the hydrocarbons (removal of all *n*-alkanes and the majority of the three ringed aromatics) there was an increase in the concentration of larger ($>C_{20}$) branched and cyclic carboxylic acids. These latter acids appeared recalcitrant to further biodegradation during the experiment. The presence of hopanoic acids with the 17α(H), 21β(H) configuration in the most degraded laboratory samples, indicated that biodegradation had occurred of hopane hydrocarbons. During degradation, the carboxylic acid concentrations increased from approximately 300µg/g oil to over 38,000µg/g oil and appeared to be still increasing at the time the degradation was stopped. This suggests that these acids may be useful for monitoring biodegradation after the more commonly used hydrocarbon compounds have been completely removed.

Further analyses of field samples from different environments and containing oils of varying degrees of degradation are required to confirm the generality of the findings from these laboratory and field experiments. However, the increased carboxylic acid concentrations detected in the actively biodegrading field samples studied, indicates that our new method of analysing the intermediates of crude oil biodegradation may be a valuable new tool for monitoring the progress of bioremediation in the field, and for confirming natural attenuation processes in environmental samples.

REFERENCES

Jaffé, R. and M. T. Gallardo. 1993. "Applications of carboxylic acid biomarkers as indicators of biodegradation and migration of crude oils from the Maracaibo Basin, Western Venezuela". *Organic Geochemistry* 20: 973-984.

Rowland S. J., R. Alexander, R I. Kagi, D. M. Jones and A. G. Douglas. 1986. "Microbial degradation of aromatic components of crude oils: A comparison of laboratory and field observations." *Organic Geochemistry 9*: 153-161.

POTENTIAL OF BIOREMEDIATION OF MARINE FUEL OIL ON SINGAPORE SHORELINES

Lee-Ching Ng, Wei-Fang Ng, Sok-Kiang Lau, Yoke-Cheng Tan, ***Alex Chang***
DSO National Laboratories, 20 Science Park Drive, Singapore 118230.

ABSTRACT: The collision of a very large crude carrier and an oil tanker in the Singapore Strait on October 15, 1997, resulted in a spillage of 28,000 tonnes of marine fuel oil, the worst spill in Singapore's history. The bioremediation of petroleum and petroleum products has not received attention commensurate with its capability as a clean-up response to oil spill or leakage, even though the warm and humid tropical climate of Singapore and the surrounding region, promises a diverse microflora that allows the application of such technology.

In this laboratory study, chemical and microbial analyses using batch slurries were performed on beach samples contaminated by the marine fuel oil spilled into the Singapore Straits. It was demonstrated that, with supplement of nitrogen and phosphorus sources, 90% of the alkanes monitored were biodegraded within 4 days. This result is in contrast to the negligible biodegradation in the absence of nutrient supplement even after 7 days. Through the laboratory investigations, we determined that:

1. Microorganisms indigenous to Singapore water and shores have the capability to degrade the hydrocarbons in the spilled heavy marine fuel oil.
2. The nitrogen and phosphorus content is indeed limiting, both in the oil and in the environment.
3. The addition of nitrogen and phosphorus sources is able to significantly accelerate the biodegradation process.

The results from a field test performed on one of the contaminated beaches also are discussed.

INTRODUCTION

The collision of a very large crude carrier and an oil tanker in the Singapore Strait on October 15, 1997, resulted in a spillage of 28,000 tonnes of marine fuel oil, the worst spill in Singapore's history. Major cleanup of the spillage was achieved by spraying chemical dispersants, collecting oil off the surface of the sea with skimmers, and manual removal of contaminated beach material. Bioremediation technology was not employed as a treatment technology in response to the oil spill, even though the technology has been employed by some local companies for the remediation of grounds contaminated by oil leakage from tanks. In the local scene, the biodegradation of petroleum and petroleum products has not received attention commensurate with its potential as an oil spill clean-up response.

The application of mineral nutrients that contain nitrogen and phosphorus sources has been shown to enhance the degrading capabilities of microorganisms and accelerate the biodegradation process. Biodegradation of the heavy marine fuel oil, which is less susceptible to microbial attack, has not been reported in this region.

Singapore, with a warm and humid tropical climate, promises to allow the application of such technology. In this study, chemical and microbial analyses were performed on beach samples, contaminated by the marine fuel oil spilled into the Singapore Straits.

Objectives. Through the laboratory investigations, we sought to determine if:
1. Microorganisms indigenous to Singapore water and shores have the capability to degrade the hydrocarbon in the spilled heavy marine fuel oil.
2. The nitrogen and phosphorus content is indeed limiting, both in the oil and in the environment.
3. The addition of nitrogen and phosphorus sources is able to accelerate the biodegradation process.
4. The acceleration of bioremediation could be readily extrapolated to the field.

MATERIALS AND METHODS

Sample Collection and Preparation. Beach samples contaminated with marine fuel oil was collected from a beach of an off-shore island. Uncontaminated seawater was collected from a seaside of the mainland, which was not affected by the oil spill.

Batch culture experiments were performed to determine changes in the content of the oil and to monitor the number of marine heterotrophs, over 7 days. Prior to inoculation, the sand samples were thoroughly homogenised by vigorous shaking in a bottle for an hour. Each batch of culture contained 4 g of sand in 20 ml of autoclaved uncontaminated seawater. To study the effect of the addition of nitrogen and phosphorus sources, mineral salts NH_4Cl and KH_2PO_4 were added to a final concentration of 1 g/L and 3 g/L respectively. Equal amounts of autoclaved sand samples were used as controls to determine the extent of non-microbial degradation. Cultures were incubated with shaking at 30°C.

To determine the initial concentration of hydrocarbon contamination, 4g each of the homogenised sand samples were used for chemical analysis. At four days and seven days after the start of the batch cultures, a set of duplicates consisting of 2 batches with autoclaved sand sample; 2 batches with sand sample; and 2 batches with sand sample and added nutrients, were sacrificed for chemical analysis.

Chemical Analysis. Hydrocarbons from sand samples were extracted with dichloromethane. Liquid components of batch cultures were separated from the solid sand before extraction was performed on each component. For the solid sand, a sample of about 10 g (dry weight) was extracted by shaking it multiple times for 10 min with 10- ml aliquots of dichloromethane. The extraction was continued until a clear extract was obtained. The extracts were combined and concentrated if necessary under a stream of nitrogen. Tributylphosphate was added to the concentrated extract as the internal standard.

The amount of C17 and C18 in the sample was quantified with a standard solution of C17 and C18 containing the internal standard. Another calibration plot was made with this standard solution. The detector response was expressed as the response for the amount of carbon present. The calibration plot was used in the calculation of the total amount of alkanes present in the sand sample.

To ensure that all alkane hydrocarbons were accounted for in our batch culture experiments, the aqueous layer of each sample also was extracted by performing liquid-liquid extraction with dichloromethane. Twice the volume of the aqueous layer of dichloromethane was added to a separation funnel containing the sample, which then was manually agitated for 2-3 minutes before the organic layer was recovered. This procedure was repeated and the organic extracts were combined, concentrated, and analysed on a gas chromatograph fitted with a flame ionisation detector (GC-FID).

Recovery studies were performed with clean sand from an uncontaminated site of the island. Six 10 g sand samples were spiked with C17 and C18 at a concentration of 10 ppm. The samples were shaken for 30 min and allowed to stand for several hours before extraction was carried out. The recovery was better than 90% with the extraction procedure.

The conditions of the instrument used for the analysis of the extracts were as follows. The injection port temperature was held at 285°C, and the extracts were injected splitless by closing the split vent for 0.5 min when the sample was injected. The oven temperature was held at 40°C for 1 min and ramped at a rate of 10°C/min to 280 °C and held at this temperature for 20 min. The detector temperature was 300°C. The column used was DB5 (30 m x 0.25 mm inside diameter, 0.25 µm film thickness) from J & W. The carrier gas was helium at a flow rate of 35 cm/s.

Marine Heterotrophs Count. The number of marine heterotrophs was monitored by plating serial dilutions of the cultures on modified Luria Bertani agar plates which contain 20g/L of NaCl, 5g/L of yeast extract, 10g/L of tryptone and 15g/L of Grade A agar. Number of bacteria on the plates was scored after 3 days of incubation at 30°C.

RESULTS AND DISCUSSION

Negligible Effect of Autoclaving. Analysis of hydrocarbon content in sand samples before and after autoclaving was performed to determine if heat and pressure have any effect on the hydrocarbons. Figure 1 shows that, although the content of lighter hydrocarbons like C13 to C15 were reduced by autoclaving, the other hydrocarbons were measured after autoclaving at concentrations close to 100% of non-autoclaved samples and therefore were negligibly affected by the autoclaving process. Because autoclaving eliminates microorganisms, autoclaved samples were used in subsequent studies as controls to account for the hydrocarbon losses resulting from other means other than microbial degradation.

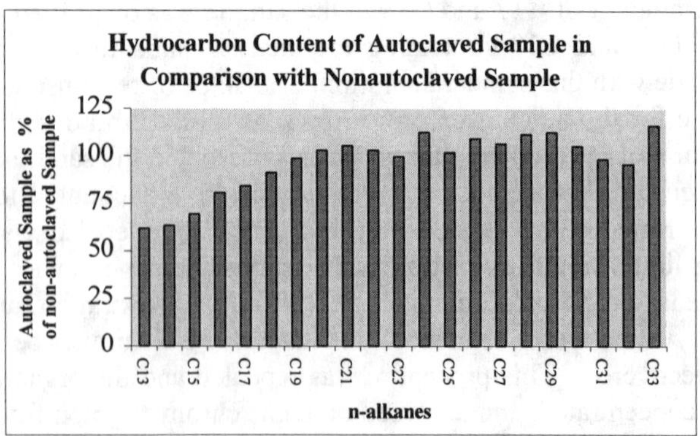

Figure 1. Hydrocarbon content of autoclaved samples as percentage of that of non-autoclaved samples.

Effect of Nutrients on Composition of Oil. The visible oil contamination went as deep as 15 cm below the beach surface, with the hydrocarbon content in deeper regions being 10% of that of the surface. Soil samples at the two levels of contamination were collected.

Batch cultures containing slurry made up of autoclaved seawater and sand from the beach were prepared. The hydrocarbon profile and content of the batch cultures of both samples Surface and Deep were determined by GC at day 0, 4, and 7. For each day, hydrocarbon composition of batches of culture with 1) autoclaved sand, without nutrients, 2) sand without nutrients, and 3) sand with nutrients were analysed in duplicates. Figures 2a and b show the enhanced biodegradation of the hydrocarbons in the surface and deep samples on Day 4 effected by the addition of phosphorus and nitrogen sources.

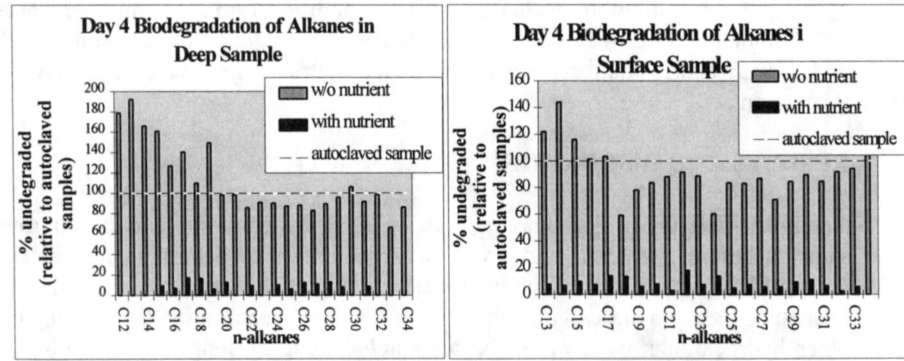

Figures 2a and 2b. Enhanced biodegradation of n-alkanes in deep and surface samples, affected by additon of phosporous and nitrogen sources, seen on Day 4.

In each sample, more than 90% of the hydrocarbons monitored were degraded within 4 days after addition of nutrients, as compared to 0 to 40%

reduction of oil content in the absence of added nutrients. The biodegradation capabilities of the indigenous microorganisms spans through the range of alkanes detected, with no apparent preference for any of the compounds. The apparent higher content of the lower molecular weight alkanes in the sample without nutrient as compared to the negative control (autoclaved sample) is most likely the result of vaporisation during autoclaving.

Degradation was monitored by measuring amounts of C17 and C18 over 7 days. The result is especially conspicuous in the deep sample, where a drastic decrease in C17 and C18 content was seen only when nutrients were supplemented (Figures 3a and b). Incubation from day 4 to 7 saw no further significant reduction in the alkane content in the samples. As the nitrogen and phosphate sources added were in excess, it is unlikely that they were limiting the biodegradation of the hydrocarbons. Limited availability of other mineral salts or bio-inavailability of the hydrocarbons could explain the observation.

Enhanced biodegradation also was visually evident by the disappearance of oil clumps after 4 days in flasks where nutrients were added. This result is in contrast to the presence of oil clumps in the autoclaved batches and the batches without nutrients, which persisted until the 7th day (Figure 4).

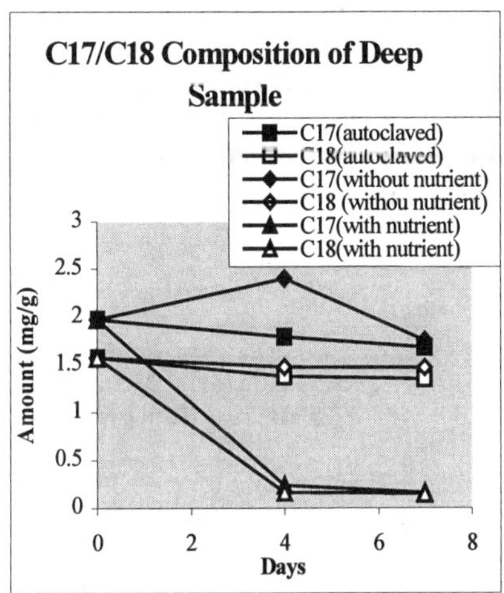

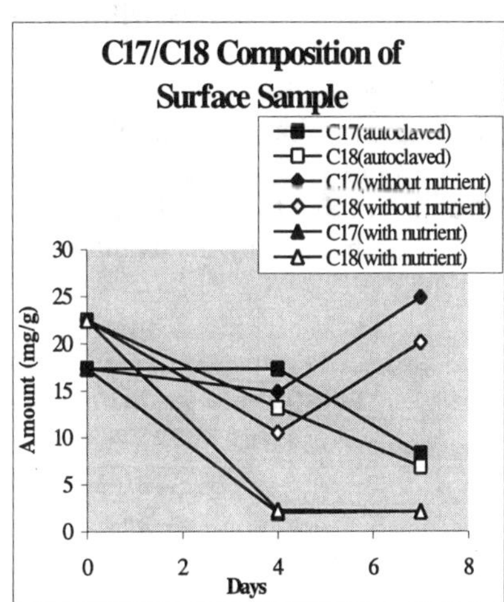

Figures 3a and b. C17 and C18 composition of deep sample and surface sample over 7 days. Enhanced biodegradation seen in samples with nutrients.

Effect of Nutrients on Heterotroph Counts. The role of microorganisms in the degradation is further substantiated by total heterotroph counts. Results for both samples showed that the addition of phosphorus and nitrogen caused a 10-fold increment in the number of bacteria (Figures 5a and b). It led us to ascertain that, although the carbon source provided by the spilled oil is in excess, other elements like phosphorus and nitrogen are limiting in the environment. Supplementing slurries with these elements encouraged proliferation of microorganisms necessary for biodegradation, thus enhancing the usage and breakdown of the oil.

Figure 4. Visual evidence of enhanced biodegradation by addition of phosphorus and nitrogen sources. From left: slurry with nutrients, slurry without nutrient, autoclaved slurry.

While the number of bacteria in deep sample peaked at Day 2 of the experiment, the number in surface sample was still increasing at Day 7. Considering that the amount of hydrocarbons were 10-fold more in the surface in the surface sample, it is apparent that the initial higher population and the increasing number sample are being maintained by the available hydrocarbons.

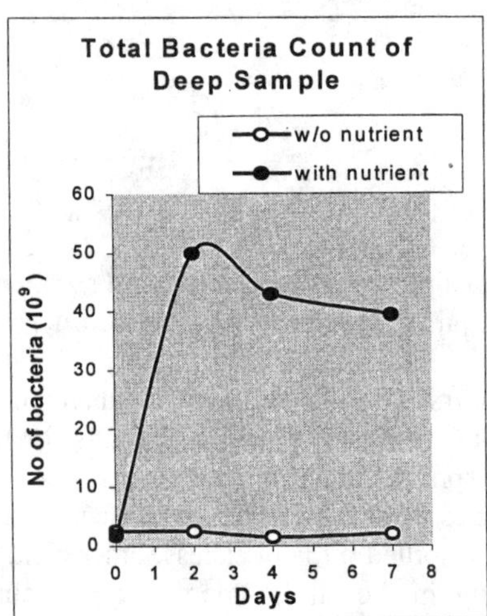

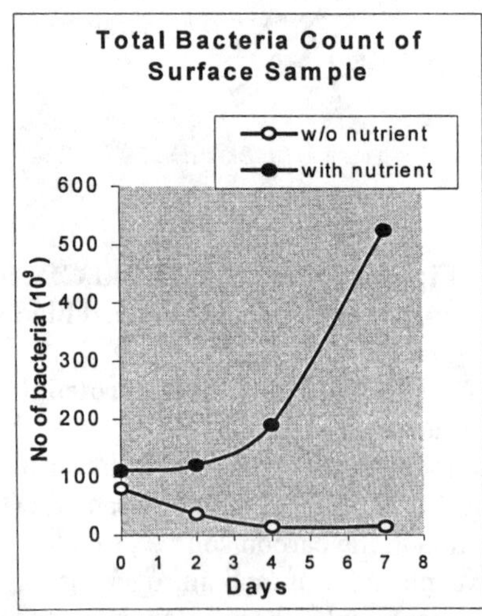

Figure 5a and b. The total heterotroph counts for deep and surface samples, in the presence and absence of nutrients.

Potential of Indigenous Microorganisms. This study demonstrates the potential of indigenous microorganisms of Singapore to break down heavy marine fuel oil. The acceleration of the breakdown effected by the addition of nitrogen and phosphorus shows that the concentration of these elements offered by the local environment are limiting the biodegradation process. It is thus concluded that enhancement of bioremedation by the addition of nutrients has great potential as a treatment technology in response to the oil spill in the local scene. Several isolates of bacteria that are capable of degrading marine fuel oil are in the process identification and analyses.

Effect of Commercial Fertiliser on Degradation of Oil on Beach. To extrapolate the above laboratory study to the field, a commercial fertiliser was used. A plot of the contaminated beach of 10m by 6m, within the intertidal zone was used as a test site. Another plot which is about 50 m away from the test site with a similar level of contamination was used as a control site. The commercial fertiliser was sprayed to a concentration of 0.4 kg/m^2, on Day 0 and Day 14. Sand samples were collected first before and then weeks after the test site was sprayed with the commercial fertiliser,

Figure 6 shows the rate of disappearance of the oil from the beach during the weeks following the introduction of nutrients. The monitoring of C17 and C18 levels demonstrates that the enhanced degradation effected by the nutrients is evident during the first few weeks following the spraying of the nutrients. The effect wore off in the subsequent weeks and natural degradation eventually brought the oil content down to similarly low levels. Results obtained from total oil extraction correlate with that from C17 and C18.

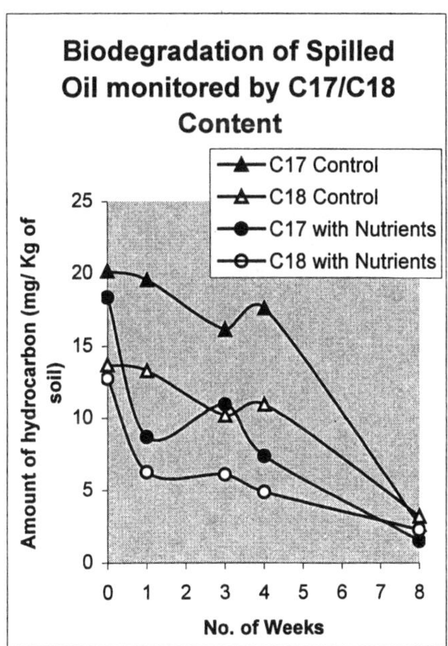

Figure 6. Decrease of oil content in beach material with and without the application of nutrients.

The results from the field trial show that, in the long term, natural degradation plays a major role in the cleaning up of spilled oil. However, in the shorter term, enhancement by manipulation can be achieved.

We have demonstrated the great potential of bioremediation of an oil spill, rendered possible by the natural microflora in local environment. The laboratory trial established that bioremediation, with local indigenous microorganisms, could be achieved within days in controlled conditions. However, the enhancement seen in the field trial was not as conspicuous as those seen in the laboratory tests. It is evident from our study that commercially-available fertiliser may not be the ideal nutrient to be used for local beaches contaminated with heavy marine fuel oil. To fully exploit the potential offered by local microoganisms, more research needs to be done to develop strategies and nutrients appropriate for local conditions.

IN-SITU BIOREMEDIATION OF A HIGHLY ORGANIC CONTAMINATED SOIL

Jami B. Walsh
Worcester Polytechnic Institute
Worcester, MA USA
James C. O'Shaughnessy Ph.D.
Worcester Polytechnic Institute
Worcester, MA USA

Abstract: This study is focused on the use of bioremediation as the primary method of decontamination for a soil contaminated with industrial waste oils. The area from which the samples were taken was used as a disposal site for oily wastewater for a period of more than 20 years. The study was performed in two phases. The first phase utilized a control condition and three experimental conditions: anoxic condition; aerobic condition with very little mixing; and aerobic condition with a high mixing rate. The samples were tested using Gas Chromatography (GC). After 60 days of testing results indicated a significant reduction in the amount of contamination present. The second phase of the study focused on quantifying the reduction of contamination due to biological activity. An apparatus was designed that pumped air through the soil and a series of traps. The traps and the soil were then analyzed for TPH and BTEX and a comparison of the contamination level aided in the quantification of biological activity.

INTRODUCTION

This study focused on the use of bioremediation, as the primary method of decontamination for a soil contaminated with industrial waste oils. The area from which the samples were taken was used as disposal site for oily wastewater for a period of more than 20 years. During this time the soil became severely contaminated. For the past 30 years, additional wastes have entered this location.

The area contaminated by the oily wastes is approximately 1-acre. Prior to its use as a disposal site, the area was a natural freshwater wetland. Currently the area is solid to the touch and has a 2 to 3 foot thick peat top layer. The peat is underlain by a relatively non-porous solid soil strata. The soil is unique in that the organic content is 94%. The high organic content complicates the use of other remediation options such as thermal desorption, landfill disposal, or other stabilization methods. In-situ bioremediation was evaluated as an acceptable treatment technology. The site is surrounded on three sides by natural wetlands. The ground water table is located 1 foot below the ground surface. Free product oil has been observed in the soil and on the groundwater. When the surface of the area was permeated, the escape of the gases trapped below the peat layer was noticeable. The area does have some vegetation beginning to grow which may indicate that natural attenuation is occurring on the site. The goal of this study was to determine the feasibility of bioremediation as an in-situ process under these conditions, rather than ex-situ technologies.

METHODOLOGY

The experiments conducted in this study were conducted in the laboratory rather than in the field. Samples were gathered using the grab method from various sampling locations around the site. A composite sample was then made utilizing the soil taken from each sampling location. (Figure 1)

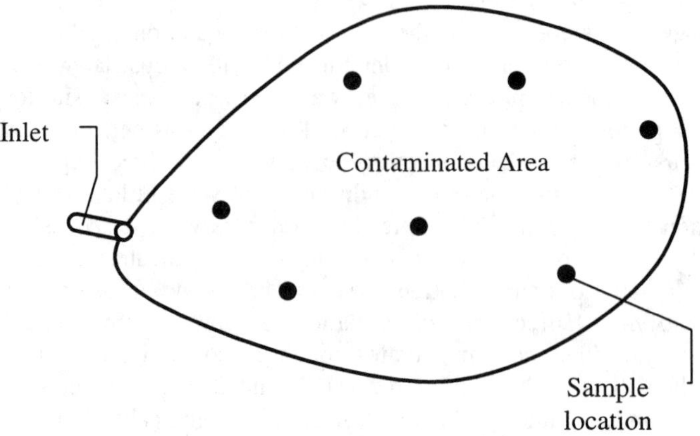

Figure 1 Sample Locations for Phase I

The composite sample was then divided into three 500 ml reactors and one anoxic reactor. The first reactor acted as the control for the experiment, no nutrient addition and a minimum of agitation took place. The moisture content of the soil was maintained at the in-situ level of 25%. The second reactor, containing the contaminated soil, was the first of three experimental situations. This reactor had nutrient addition and a high mixing frequency. A commercial fertilizer was used to inoculate the sample. The amount of fertilizer needed was based on a 100:10:5 Carbon:Nitrogen:Phosphate ratio (C:N:P) by weight. The water content in the high mix reactor was also maintained at the in-situ level of 25%. The mixture in this reactor was agitated a minimum of three times a week for 1-2 minutes using a stirrer. The third reactor of contaminated soil represents the second experimental condition, nutrient addition and low mixing. Again the nutrient addition was performed based on the C:N:P and the moisture content was maintained at the in-situ level. This reactor, was also agitated but only once or twice a week for 30 seconds to 1 minute with a stirrer. The final experimental condition in phase 1 was an anoxic reactor. Again the moisture content was maintained at the in-situ level and nutrients were added according to the C:N:P. This reactor was also agitated on a regular basis. All four reactors were placed side by side and maintained at ambient air temperature for the two month duration. Samples were taken at various time intervals and the results of these samples will be discussed in the following section.

The samples were analyzed for Total Petroleum Hydrocarbons (TPH) and volatile organic compounds (VOCs) such as benzene, toluene, xylene, and ethylene (BTEX). The samples were analyzed through the use of gas chromatography (GC). An 8260 test was performed to identify the BTEX present, while an 8100 test was utilized to identify the TPH.

Phase 2 of the study focused on the quantification of biological activity. An apparatus was designed to determine the amount of removal that is chemically based versus the amount of removal that is biologically based. The system consisted of 4 series with: 1 - 1000 ml flask followed by 4 - 250 ml flasks as indicated in Figure 2. Air was pumped through the system and the off gases were collected in the water and methanol traps. The laboratory air supply was filtered through a Cole-Parmer air filter prior to flow into the soil reactors. Flow was regulated using a Cole-Parmer four tube flowmeter with stainless steel floats.

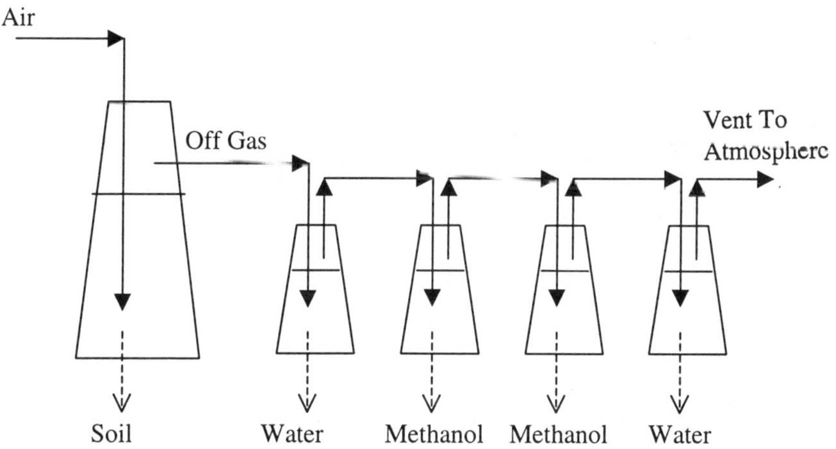

FIGURE 2 Apparatus For Off Gas Collection

Samples were collected from the soil as well as the methanol and water traps and tested for TPH and BTEX by GC method.

RESULTS AND DISCUSSION

Phase 1 of the study investigated the affect of nutrient addition, oxygen, and agitation on the biological remediation of the soil. An initial sample was taken to determine the starting TPH level. Samples were also taken on day 1, 3, 5, 10, 20, 30 and 60. Table 1 provides the levels of TPH found in each situation.

TABLE 1 Results of Phase 1 TPH Study

DAY	CONTROL	HIGH	LOW	ANOXIC
0	79600	79600	79600	79600
1	47800	81800	51100	36800
3	85500	65700	70800	53400
5	71000	44200	53200	50700
10	76900	77500	66200	44300
21	15800	10700	11300	13700
30	11300	8240	8440	10100
60	16300	9050	7930	12200

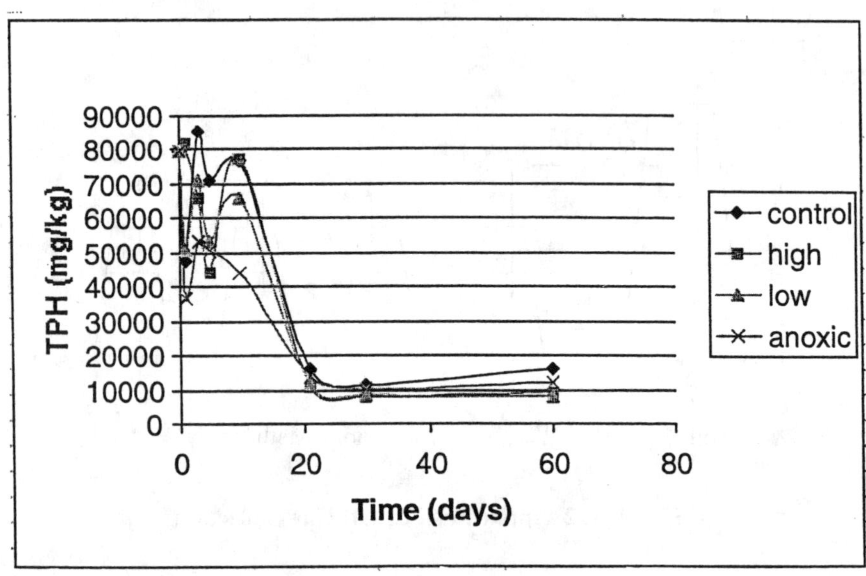

FIGURE 3 TPH vs. Time for Phase 1

In interpreting the data it is important to remember that the sample was taken from an actual site, not prepared in the laboratory, and therefore the samples are neither homogenous nor isotropic. Therefore, variations in TPH levels are expected to some degree. The samples did show an overall downward trend that followed a typical removal curve. No volatile organic compounds were detected during the first five days of the trial, therefore, VOC tests were not conducted after day 5.

The focus for phase 2 of the study was to quantify the portion of removal that was biological in nature. Air was bubbled through the soil and the off gases produced were collected in water traps and methanol traps. This test was also run for two months. Based on the TPH and BTEX contained in the soil and the traps under the various conditions, a determination could be made as to the percent of the removal that can be attributed to biological activity. At the time of publication phase 2 was still being completed.

CONCLUSIONS

The data from phase 1 indicates a strong possibility that bioremediation is a feasible method of clean-up for the site in question. Significant removals of TPH were achieved, however, they were well above the acceptable levels set forth by the U.S. Environmental Protection Agency (EPA) and the Massachusetts Department of Environmental Protection (MADEP). Phase 2 will quantify the amount of biological activity present. Through this quantification a prediction may be made as to the extent to which bioremediation will be a factor in the removal of contamination. As previously mentioned, the site does appear to be undergoing natural attenuation. Combining the natural attenuation with biological treatment, may provide an acceptable level of treatment.

ENHANCED AEROBIC CLEANUP OF HOME HEATING OIL

Gil Oudijk (Triassic Technology, Inc., Princeton, New Jersey)
Baxter E. Duffy (Inland Pollution Services, Inc., Elizabeth, New Jersey)
L. Donald Ochs (Regenesis, San Juan Capastrano, California)

ABSTRACT: Releases of no. 2 heating oil from residential storage tanks have become an increasing and costly problem in New Jersey. In many cases, common remediation technologies will cause severe disruption to homeowners and the household. Often, the source of the problem is in hard to reach locations because petroleum migration extends to beneath the residence and its foundation. The introduction of Oxygen Release Compound (ORC®) to soil and ground water is a cost effective method of remediating petroleum releases with minimal disruption to the household.

The injection of ORC to the subsurface will passively remove residual concentrations of petroleum trapped within the soil and dissolved in the ground water. Over the past two years, the ORC remediation method has been applied on four residential petroleum cases where both soil and ground water have been impacted to concentrations above the regulatory standards. After the injection of ORC into the subsurface, concentrations of volatile organic compounds (VOCs) and base/neutral extractable compounds (B/Ns) were found to decrease to within the regualtory standards. Two of the four cases have been closed by the regulatory agency and the remaining two cases are still within the review phase.

INTRODUCTION

Oxygen Release Compound (ORC®) is a mixture of magnesium oxide, magnesium hydroxide and magnesium peroxide. The introduction of these compounds into ground water cause an increase of the oxygen content to occur. The increased oxygen levels allow the indigenous petroleum-degrading bacteria to multiply. The bacteria then use the petroleum as a food source causing the dissolved petroleum concentrations to decrease.

Petroleum releases at residential sites occur from leaks in underground, aboveground and basement storage tanks. Residential tanks are commonly used for the storage of no. 2 heating oil, but may also contain gasoline and kerosene. In general, releases occur because of tank or line failures, improper filling, improper tank installation or poor maintenance.

Because the tanks are located within a residential setting, remediation efforts may often be difficult and cumbersome. In many cases, the petroleum migrates to hard to reach locations, such as beneath foundations, within drainage systems, underneath living quarters or around expensive structures such as pools. The use of excavating methods is often impossible or highly disruptive to the household. In addition, the installation of pump-and-treat systems, to remove the petroleum and the impacted ground water, may not be physically possible or not a cost effective remedial alternative.

The introduction of ORCs into the subsurface is a cost effective and relatively non-intrusive method of remediating difficult residential petroleum releases. The ORC is injected into the subsurface in a slurry through borings. The borings may be completed by hand or with a small drilling rig, such as a direct-push Geoprobe rig. The ORC slurry is then injected with a high-pressure pump or pressure grouter. In addition, ORC may be obtained in a sock form which is hung within monitoring wells, if wells happen to be present at the site.

REGULATORY CRITERIA

The state of New Jersey is heavily dependent on ground water as a source of potable water. Many communities outside of the metroplitan areas of Newark, Paterson, Camden or Trenton do not have municipal water systems. Many municipal water systems rely on ground water for at least a portion of their supply. Private residences outside of these areas normally rely on individual potable wells for their water needs. In these same areas, private residences commonly use underground or aboveground tanks for the storage of petroleum products. Releases of petroleum from residential tanks may cause severe ground-water contamination sufficiently serious to render the wells unfit for use as a water supply. Accordingly, the New Jersey Department of Environmental Protection (NJDEP) oversees all residential cleanups to insure that ground-water supplies in these areas remain fit for consumption.

In 1994, the State of New Jersey passed regulatory criteria for contaminants in soil and ground water. Three types of cleanup standards were provided for soil: (1) a Residential Direct-Contact Soil Cleanup Criterion; (2) a Non-Residential Direct-Contact Soil Cleanup Criterion, and (3) an Impact to Ground Water Soil Cleanup Criterion. The direct-contact criteria apply to surficial soils, while the impact to ground water criteria applies to soils at deeper depths. On residential cases, the impact to ground water criteria are normally applied and they are generally the most stringent. A responsible party may petition the NJDEP for a less stringent cleanup standard. In these cases, an alternative criterion is used and a deed restriction may be placed on the property. Table 1 presents the impact to ground water soil cleanup criteria for selected petroleum compounds.

TABLE 1. Impact to Ground Water Soil Cleanup Criteria for Selected Petroleum Compounds

Compound	Criterion (mg/kg)	Compound	Criterion (mg/kg)
Benzene	1	Total Xylenes	10
Toluene	500	Total Organics	10,000
Ethyl benzene	100	Total Volatile Organics	1,000

In new Jersey, soil samples to be collected in response to a release of no. 2 heating oil must be analyzed for total petroleum hydrocarbons (TPHCs). If the TPHC concentration exceeds 1,000 milligrams per kilogram (mg/kg), 25% of the samples must also be analyzed for volatile organic compounds (VOCs). The results are then compared to the criteria for the individual VOCs. If the TPHC concentration exceeds 10,000 mg/kg, additional analyses are not required because the criterion for total organic contaminants has been exceeded.

In 1994, the state of New Jersey also passed ground-water quality standards. The state was essentially divided into three classes: Class I, the Pine Barren where no degradation of the ground water is allowed; Class II, the majority of the state where the promulgated ground-water quality criteria are in effect, and Class III, heavily industrialized areas or portions of the state where low-permeability formations are present, where degradation of ground-water quality may be allowed with restrictions. Table 2 presents the ground-water quality criteria for the most common petroleum compounds.

TABLE 2. Ground-Water Quality Criertia for Selected Petroleum Compounds

Compound	Criterion
Benzene	1 microgram per liter (μg/l)
Ethylbenzene	700
Toluene	1,000
Total xylenes	1,000

In Class II areas, which encompasses the majority of the state, responsible parties may petition the NJDEP for a less stringent ground-water quality criterion. In these cases, a Classification Exception Area (CEA) may be delineated. The installation of wells would be restricted within the CEA and its extent would be attached to the deed for the property.

CASE HISTORY NO. 1

In 1997, a 280-gallon underground storage tank (UST), previously containing leaded gasoline, was excavated and removed from an abandoned farm in Burlington County, New Jersey. The UST has been out of service since the mid-1950s and was found during an inspection of the property prior to its sale. Because corrosion holes were found in the base of the UST, a ground-water sample was also collected from the on-site potable well for laboratory analysis of volatile organic compounds (VOCs). The potable well was hand dug and is located within the basement of the farmhouse. The sample exhibited a slight odor of gasoline and it was subsequently submitted to a laboratory for analysis of VOCs. The laboratory analytical results revealed that targeted VOCs detected in the sample included chloroform, benzene, toluene and *o,m,p*-xylenes at concentrations of 2.5 micrograms per liter (μg/l), 1.5 μg/l, 3.1 μg/l and 3.3 μg/l, respectively. Non-targeted VOCs were detected at a total concentration of 376 μg/l.

In December 1997, five monitoring wells were installed at the farm. Gasoline odors were detected in two of the boreholes: MW-3 and MW-4. Evidence of petroleum contamination was not present in the remaining boreholes and a product layer was not present in any of the wells. Subsequent sampling of the wells confirmed a small plume originating from the former UST location and extending to the farmhouse potable well (Table 3).

TABLE 3. Concentrations of BTEX in ground-water samples, Burlington County site

Well	Pre-ORC (7/97)	Post-ORC (7/98)
MW-1	ND	ND
MW-2	7	ND
MW-3	164	ND
MW-4	1.6	ND
MW-5	ND	ND
Potable Well	386	0.79

Remedial activities were conducted at the farm in January 1998 to insure that the shallow aquifer remains potable. A series of thirteen borings were completed with a Geoprobe rig. An ORC slurry containing 10 pounds of ORC and 3.5 gallons of water was injected into each boring at a depth interval of 10 to 20 feet bgs. A carbon treatment unit was then installed onto the existing potable well and the treated effluent from the carbon units was discharged into a wooded area behind the farmhouse. A replacement potable well was also installed at a location east of the farmhouse and within an area known to be free of gasoline contamination.

In July 1998, ground-water samples were collected from the potable well and the monitoring wells. The data reveal that a significant reduction in the concentrations of gasoline constituents in the ground water had occurred after the introduction of ORC. Accordingly, since the installation of the carbon treatment unit and the treatment of the ground water with ORC, VOC concentrations detected in all the wells have decreased to concentrations well below NJDEP criteria. The ground-water sampling data were subsequently submitted to the NJDEP with a request for case closure. The NJDEP is presently reviewing the data and case closure is expected shortly.

CASE HISTORY NO. 2

In 1996, a 550-gallon UST was excavated and removed from a residential site in Middlesex County, New Jersey. The UST was located immediately adjacent to the foundation of the residence and additional excavating beyond the removal of the UST could not be performed. The excavation was subsequently backfilled to prevent undermining of the residence's foundation. Because the water table was within 4 feet of the ground surface, impacted soil was not of concern. However, petroleum concentrations in excess of 6,000 μg/l were detected in the ground water. Hand auger borings were completed at locations surrounding the residence and it was determined that the plume was localized to the vicinity of the former UST excavation.

In late February 1997, three hand auger borings were completed at locations surrounding the excavation and adjacent to the foundation. A two-inch diameter PVC screen with a well point was manually driven into each borehole to a depth of approximately 8 feet. A slurry of ORC was then injected into each well with a pressure grouter. The slurry for each well contained 20 pounds of ORC with 10 gallons of water. After the injection of the ORC, each PVC screen was removed.

In July 1997, a temporary monitoring well was installed and a ground-water samples collected for analysis of VOCs and B/Ns. The sample analytical results revealed that neither targeted nor non-targeted VOCs and B/Ns were present in the sample. A second confirmation sample was collected with the same results. The ground-water sampling data were subsequently forwarded to the NJDEP and the case was closed.

TABLE 4. Concentrations of VOCs and B/Ns in ground-water samples, Middlesex County site

Well	Pre-ORC VOCs (2/97)	Pre-ORC B/Ns (2/97)	Post-ORC VOCs (7/97)	Post-ORC B/Ns (7/97)
TW-1	191	6,605	ND	ND

CASE HISTORY NO. 3

In January 1997, a 275-gallon UST was abandoned in place within a condominium complex in Monmouth County, New Jersey. The UST was abandoned and not removed because it was located between a foundation and a retaining wall. Its removal would have undermined both structures and would have prevented entrance through the only door to the condominium for an extended period of time. The UST was taken out of service because petroleum product was found to be seeping into the basement of the condominium.

During the abandonment process, numerous holes were detected in the base of the UST and it was apparent that a significant discharge of petroleum had occurred. Three borings were completed through the base of the UST and an ORC slurry was injected into the underlying soils. In addition, ORC injections were conducted at 7 additional locations surrounding the abandoned UST.

In February 1997, six monitoring wells were installed at locations surrounding the abandoned UST. Two wells were installed with a Geoprobe rig within the garage, while two additional wells were placed within the driveway. A fifth well was installed immediately adjacent to the abandoned UST, while the sixth well is hydraulically downgradient of the site. Since 1997, ORC socks have been placed in the wells to enhance the removal of dissolved petroleum.

Since February 1997, ground-water samples have been collected from the six monitoring wells on four occasions and analyzed for TPHCs (Table 5). In general, TPHC concentrations have decreased significantly. However, beads of petroleum product continue to recharge wells MW-3 and MW-4 and it is apparent that soils underlying the abandoned UST have not been adequately remediated. It is likely that a residual saturation of petroleum is present in the soil beneath the abandoned UST. Additional injections of ORC are proposed for this location and additional monitoring will be conducted.

TABLE 5. Concentrations of Total Petroleum Hydrocarbons (mg/l) in Ground-Water Samples, Monmouth County site

Well	2/2/97	8/15/97	11/20/97	6/16/98
MP-1	106	5	12	1.93
MP-2	122	3	3	7.79
MP-3	442	915	170	5.87*
MP-4	304	2	41	4.09*
MP-5	NA	NA	NA	1.18
MP-6	NA	5	ND	1.82

CONCLUSIONS

The injection of ORC into the subsurface is a cost effective and relatively non-intrusive method of reducing residual petroleum concentrations in ground water. It is often a preferred remedial alternative for residential cases because it alleviates the need for excavating or the use of expensive pump-and-treat systems. However, it should be noted that the use of ORC may have little or no effect when large quantities of petroleum product are present. Therefore, for the optimum use of this technology, the source of the contamination must be removed. After the source has been removed, ORC may be applied to the ground water through injections or as socks in monitoring wells.

COMBINING HVME AND ENHANCED BIOREMEDIATION FOR GASOLINE CONTAMINATION IN GROUNDWATER

Patrick M. Hicks (Regenesis, Kingwood, Texas)
Gerald Eshbaugh (The Environment Company, Plano, Texas)
Craig Sandefur (Regenesis, San Juan Capistrano, California)

ABSTRACT: A leaking underground storage tank at a commercial petroleum retail site located in Haskell, Texas resulted in soil and groundwater petroleum hydrocarbon contamination. This contamination, including benzene, toluene, ethylbenzene, and total xylenes (BTEX) and total petroleum hydrocarbons (TPH) migrated down-gradient below a road and neighboring property. Access below the adjacent roadway was limited, and consultants at The Environment Company sought a solution that would allow for simultaneous treatment of on-site and off-site sections of the plume. As a result of the presence of separate phase hydrocarbons (SPH) a high vacuum multi-phase extraction (HVME) was used to extract product vapor and contaminated groundwater in a limited area of the site. Oxygen Release Compound (ORC®) was applied in the dissolved section of the petroleum plume for treatment on-site and off-site. The HVME treatment performed in October, 1998 successfully removed approximately 360 liters of gasoline equivalent as vapor, and over 19,000 liters of contaminated groundwater. ORC was applied via injection in March, 1998. The most recent sampling event results indicate a substantial reduction in the concentration of benzene (92%), total BTEX (76%) and TPH (81%).

INTRODUCTION

Groundwater at a commercial petroleum retail center in Haskell, Texas had separate phase hydrocarbons (SPH) and dissolved hydrocarbons as a result of a leaking underground storage tank system (LUST). The dissolved hydrocarbon plume had apparently migrated below a roadway and adjacent properties. Roadway access was limited, and a remedial solution was sought that would allow for simultaneous treatment of on-site and off-site sections of the plume. Due to off-site access restrictions, a remedial system that could be applied on-site and yet have a positive remedial effect off-site was deemed advantageous.

This project involves cleanup of gasoline released into a major aquifer, the Seymour Formation, which is unique in west central Texas due to relatively high groundwater velocity and shallow depth. This groundwater resource has been used as the domestic water supply for the city of Haskell and is currently used for domestic and agricultural irrigation.

In August of 1997, The Environment Company evaluated numerous remediation options, and recommended that Oxygen Release Compound (ORC®) be used to enhance the natural biological activity at the site. ORC is manufactured by Regenesis Bioremediation Products in San Juan Capistrano, California and is a patented formulation of magnesium peroxide that slowly releases molecular oxygen when hydrated, thereby facilitating aerobic bioremediation of organic contaminants (Koenigsberg et al., 1997). The ORC releases oxygen slowly over a period of approximately 6 to 9 months. Oxygen delivery is the most important aspect of in situ bioremediation (Norris, 1995). The indigenous aerobic microbes utilize this oxygen during respiration and can metabolize petroleum contaminants, resulting in reduction of petroleum hydrocarbon mass in groundwater. ORC enhances bioremediation of contaminants in-situ, without exchanging one environmental problem for another. One of the most important attributes of bioremediation is that harmful contaminants are metabolized by microbes, producing harmless byproducts (CO_2 and water) and no further treatment is necessary. Other remediation methods may transfer contaminants to another medium which requires removal, transportation, and possibly additional cleanup efforts. While other treatment options may release volatile petroleum components to the atmosphere, ORC enhances bioremediation *in-situ*, thereby eliminating the potential for respiratory or dermal exposure to these contaminants.

The appearance of phase separate hydrocarbon (PSH) in some of the on-site wells was noted in July, 1998. Over a period of 3-6 months the PSH column continued to increase as groundwater levels declined due to local drought conditions. Additional wells were installed to facilitate additional high vacuum multi-phase extraction (HVME) treatment operations in October, 1998.

MATERIALS AND METHODS

A detailed ORC application design was performed using data from monitoring wells installed within the boundary of the plume. Additionally, biological and chemical oxygen demand analyses were performed to more accurately estimate total oxygen demand. The project approach called for the ORC installation in two separate injection phases, with the majority of the ORC installed during the first phase. This type of application approach optimizes the operational flexibility that can be realized with the ORC technology.

ORC was injected into the on-site aquifer in March, 1998 per the design plan. At that time 3,232 pounds of ORC were injected via 70 direct push points in the area downgradient of the UST's and existing dispenser islands. The amount of ORC delivered to the subsurface within the dissolved phase plume was specific to the BTEX mass measured. To enhance degradation of the off-site plume, ORC was applied along the down-gradient property boundary. Natural groundwater velocity and flow direction was used to deliver oxygenated water to the off-site hydrocarbon mass.

An aggressive monitoring schedule was performed immediately prior and following ORC injection in order to measure the efficacy of the bioremediation process. The monitoring program included analysis of BTEX, TPH as well as field parameters such as dissolved oxygen (DO).

A second application of ORC has been delayed due to the unexpected presence of SPH. Upon completion of product recovery operations, ORC will be reapplied to polish off the residual hydrocarbon levels prior to closure request.

RESULTS AND DISCUSSION

The following table illustrates the average reduction in dissolved petroleum concentrations measured in all on-site and off-site monitoring wells eight months after ORC application.

TABLE 1. Average decrease in dissolved petroleum constituent concentrations following ORC treatment.

Dissolved Contaminant	Average Initial Concentration (mg/L)	Concentration After 8 Months (mg/L)	Percent Reduction
Benzene	4.11	0.33	92%
BTEX	13.86	3.33	76%
TPH	11.15	2.12	81%

Figures 1 through 3 illustrate the reduction in dissolved petroleum concentrations in selected individual monitoring wells over the 8 month treatment period. Well designate MW-16 and MW-17 are located at the downgradient property boundary, wells MW-6, MW-12 and MW-13 are located in the central section of the on-site plume.

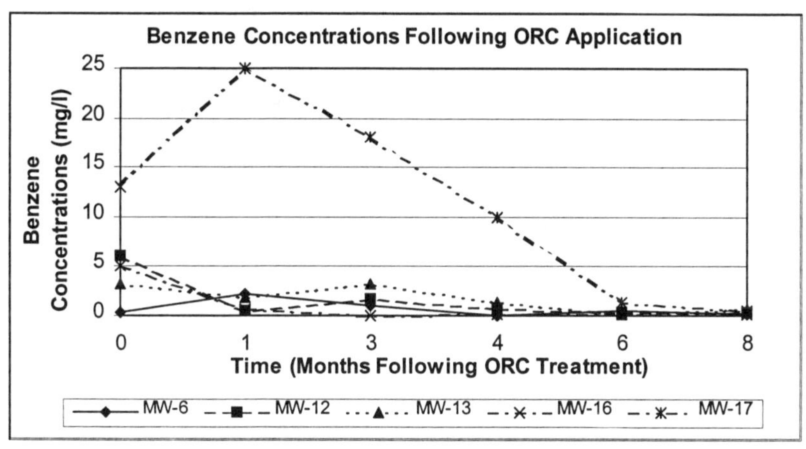

FIGURE 1. Benzene concentrations following ORC application.

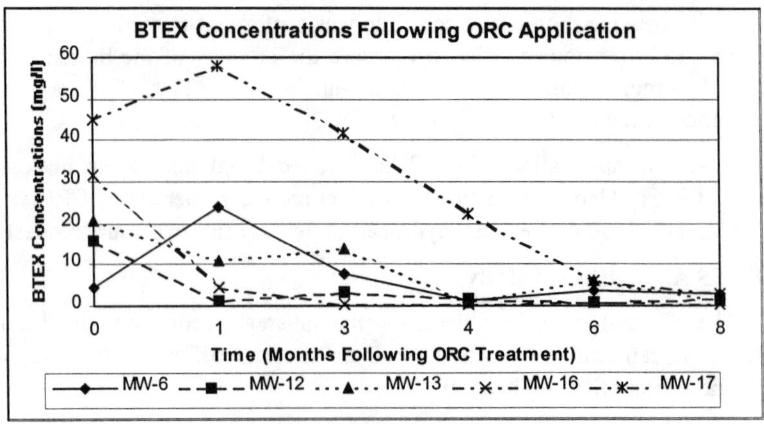

FIGURE 2. BTEX concentrations following ORC application.

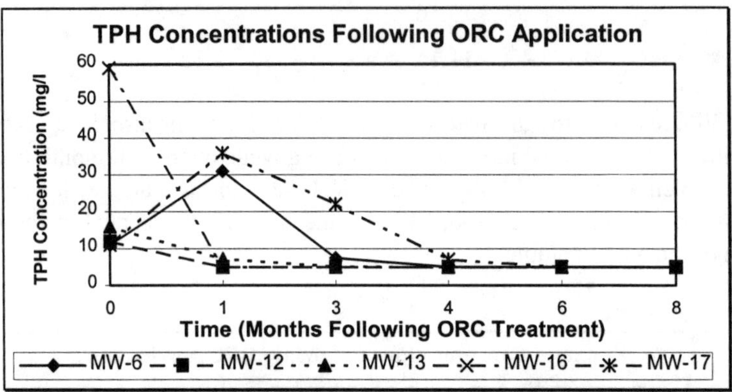

FIGURE 3. TPH concentrations following ORC application.

Figure 4 illustrates the increase in dissolved oxygen concentrations in selected individual monitoring wells over the 8 month treatment period.

Cost Analysis. Implementation of a traditional remediation technology air sparging with soil vapor extraction (AS/SVE) at this site would have cost substantially more, as well as produced much greater site and operational disruption. In addition, an AS/SVE system would likely not have been as effective within the same time frame. For example, the estimated cost savings in the application ORC vs. installation of an AS/SVE system is approximately $75,000. As a result of no operation and maintenance costs using the ORC technology, an additional $25,000 savings was realized. Finally, the reduced

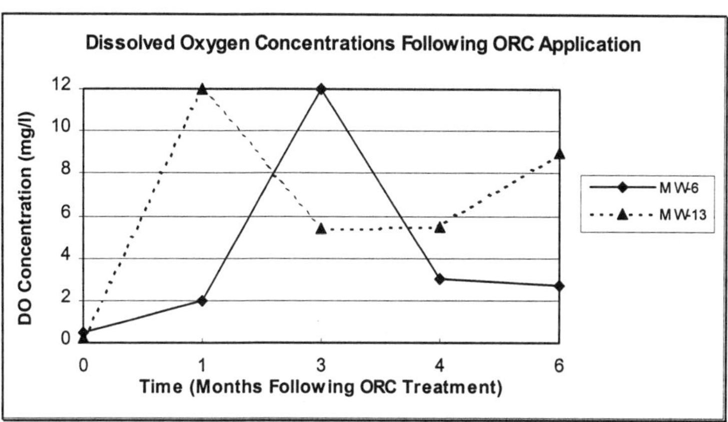

FIGURE 4. Dissolved oxygen concentrations following ORC application.

frequency and extent of groundwater monitoring at the site provided an estimated savings of $20,000. The total estimated cost savings for this project are approximately $120,000.

The unexpected appearance of free product in several monitoring wells at the site resulted in plans for aggressive product removal via high vacuum multi-phase extraction events. Additional wells were installed in the product area to facilitate the process. The extraction events were scheduled to occur prior to additional ORC applications. The HVME did increase the remedial costs, but will ultimately reduce total project costs by decreasing the contaminant mass treated via bioremediation. To date the HVME has removed 360 liters of gasoline and 19,000 liters of contaminated groundwater.

If required, additional ORC applications will be designed to polish residual dissolved petroleum concentrations to closure limits following product recovery. To date, DO concentrations are sufficient in most of the treated area to support continued bioremediation.

CONCLUSIONS

This project involved application of two technologies. Treatment with HVME removed SPH primarily in soil and groundwater, as well as some dissolved mass in groundwater. This process was enhanced by a non-invasive and cost-effective means to enhance natural bioremediation of contaminants in groundwater. The two technologies have worked synergistically, thereby increasing the efficacy of the remedial approach.

The ORC offers a flexible and environmentally friendly method of enhancing nature's own cleanup process: bioremediation (Morin, 1997). Since remediation with ORC is an in-situ approach, operations and maintenance requirements were minimal. The application of HVME and ORC injection in tandem was more cost-effective than conventional technologies such as air sparging and vapor extraction. ORC is an in-situ (non-invasive) and passive remedial option; this minimizes disruption of business activities at the site. The project approach did not require trenching, piping, and on-site equipment. Faster site cleanup was achieved compared to conventional technologies, saving both time and money.

REFERENCES

Koenigsberg, S., C. Sandefur, and W. Cox. 1997. The Use of Oxygen Release Compound (ORC) in Bioremediation. *In Situ and On-Site Bioremediation*. Vol. 4(4): 247.

Morin, T. C. 1997. "Enhanced Intrinsic Bioremediation Speeds Site Cleanup". Pollution Engineering. February, 1997.

Norris, R.D. 1995. Selection of Electron Acceptors and Strategies for In Situ Bioremediation. 1995. *In Situ and On-Site Bioremediation*. Vol. 3(6): 483.

REMEDIATION OF BTEX IN GROUNDWATER WITH LNAPL USING OXYGEN RELEASING MATERIALS (ORM)

Eric S. Mysona (Parsons Engineering Science, Inc., Cincinnati, Ohio)
William D. Hughes (Parsons Engineering Science, Inc., Cincinnati, Ohio)

Abstract: A field scale study was conducted at a former gasoline service station to evaluate the remediation of groundwater contaminated with dissolved benzene, toluene, ethylbenzene, and xylenes (BTEX) using oxygen-releasing materials (ORM). The injection of oxygen through ORM slurry was the remedial alternative chosen to enhance the in-situ microbial degradation of the BTEX in groundwater. Direct-push drilling methods were used to inject the ORM into the source area of impacted groundwater in a grid pattern with five-foot (1.5 m) centers. The remediation of residual pockets of light, non-aqueous phase liquids (LNAPL) was enhanced by the injection of the ORM slurry, which forced the LNAPL to flow into existing monitoring wells where it could be collected with a bailer. The results of groundwater sampling show increased post-injection dissolved oxygen (DO) concentrations and a 16% to 87% decrease in BTEX concentrations in the wells located within the injection grid. The disappearance of dissolved iron (Fe^{2+}) in groundwater within the injection grid indicates that there is sufficient oxygen to oxidize iron.

INTRODUCTION

A field study was conducted at a former gasoline service station in Cincinnati, Ohio to remediate groundwater contaminated with dissolved benzene, toluene, ethylbenzene, and xylenes (BTEX). The site was operated as a gasoline service station until operations ceased in the late 1970's. The underground storage tanks (USTs) were removed in 1981. The USTs were located near the center of the site, and a gasoline dispenser was located northeast of the USTs (Figure 1). A total of seven groundwater monitoring wells, MW-1 to MW-7, were installed at the site between May and October 1996 for site assessment activities. Soil samples were collected during the installation of the wells, and groundwater samples have been collected from each well.

The geology of the site consists of silty clay from surface grade to depths ranging from 8 to 12 feet (2.4 to 3.7 m) below ground surface. The soil grades into silt with variable amounts of sand below the silty clay from 12 to 29 feet (3.7 to 8.8 m), then the soil grades back into silty clay from 29 feet to 31 feet (8.8 to 9.5 m), the greatest depth sampled at the site. The former UST pit consists of silty clay fill material with some gravel and large concrete fragments. The monitoring wells are screened through the bottom of the silty clay layer into the saturated silt and sand layers. Groundwater generally flows to the southwest, parallel to the road, with a hydraulic gradient of 0.05 feet per foot (Figure 1).

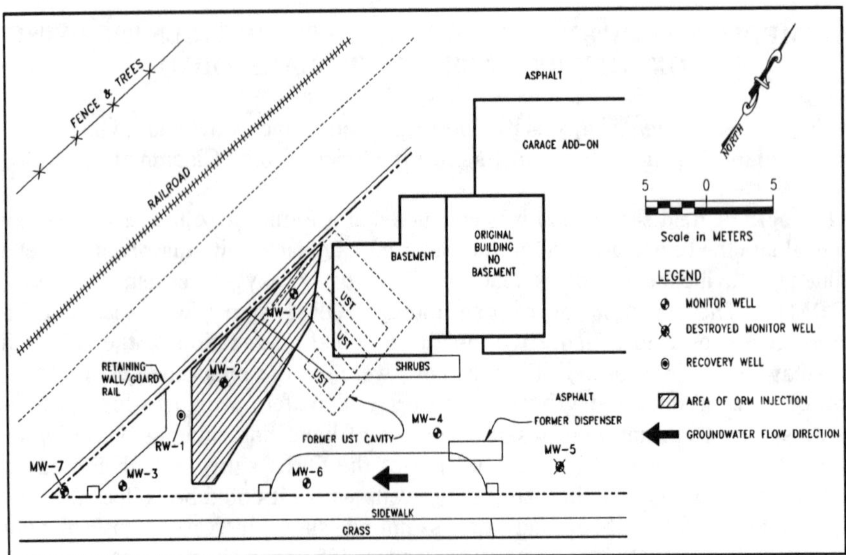

FIGURE 1. Site Plan showing site features and monitoring well locations.

Elevated concentrations of petroleum hydrocarbons (up to 0.750 mg/kg benzene and 103.4 mg/kg total BTEX) were encountered in the soil samples collected from the former UST pit. Total BTEX concentrations were 0.652 mg/kg, 0.025 mg/kg, and 4.15 mg/kg, respectively, in the soil samples collected from MW-2, MW-3, and MW-4. The source areas for the petroleum hydrocarbon contamination appear to be the former UST pit and the former dispenser island.

Dissolved benzene concentrations as high as 6,300 µg/l and total BTEX concentrations were detected as high as 79,800 µg/l in groundwater from MW-2, which is located down-gradient of the former UST pit. The presence of light, non-aqueous phase liquids (LNAPL) in MW-1, MW-2, and MW-3, and elevated BTEX concentrations in the groundwater samples collected during site assessment activities warranted remedial action. A vacuum enhanced groundwater recovery pilot test was conducted using MW-1, MW-2, and MW-3 as test wells. The positive displacement blower used in this pilot test attained a vacuum of 5 inches (2 cm) of mercury with a maximum flow rate of less than five cubic feet per minute in all three test wells. The vacuum enhanced groundwater recovery was determined to not be an effective remedial alternative at this site.

The remedial alternative chosen for this site consisted of in-situ enhanced bioremediation using oxygen-releasing materials (ORM). This in-situ remedial alternative was chosen due to the minimal interference with current site activities. The oxygen added to the groundwater was intended to enhance aerobic microbial degradation of petroleum hydrocarbons. The objective of this study was to evaluate the effectiveness of the remediation of groundwater contaminated with BTEX using ORM.

METHODOLOGY

Background data was collected from all existing monitoring wells on 19 February 1998. Field data included depth to LNAPL, depth to groundwater, dissolved oxygen (DO), oxidation-reduction potential (ORP), dissolved iron (Fe^{2+}), temperature, pH, and specific conductance. Groundwater samples were sent to the laboratory for BTEX analysis by EPA Method 602 (Table 1).

TABLE 1. Groundwater elevations, BTEX, dissolved oxygen, and dissolved iron concentrations in MW-1, MW-2, and MW-3.

Sample ID	Date	Groundwater Elevation (meters)	Benzene (ug/l)	Total BTEX (ug/l)	Dissolved Oxygen (mg/l)	Dissolved Iron (mg/l)
MW-1	24-Sep-97	27.18	3500	21200		
	19-Feb-98	27.33	1300	14900	3.6	4
	20-Jul-98	27.56	410	8810	15	0
	26-Oct-98	27.15	1100	8700	4.9	
	5-Jan-99	27.28	1100	10030	5.7	0
MW-2	24-Sep-97	27.08	6300	79800		
	2-Mar-98	27.24	2700	34400	3.2	10
	20-Jul-98	27.58	2000	17300	32	0
	26-Oct-98	27.13	370	10070	9.8	
	5-Jan-99	27.24	340	11240	15.4	0
MW-3	24-Sep-97	26.98	22	4322		
	19-Feb-98	27.27	130	5130	3.3	9.4
	20-Jul-98	27.47	52	1822	0.1	9
	26-Oct-98	27.06	44	2202	3	
	5-Jan-99	27.23	37	2087	1.1	6.6

The injection of ORM at the site occurred between 27 and 29 April 1998. Direct push drilling methods were used to inject the ORM into the source area on a grid pattern with 5-foot (1.5 m) centers (Figure 1). There were 32 boreholes advanced to approximately 20 feet (6.1 m) below grade. The ORM solution was injected from the bottom of each borehole to approximately 10 feet (3 m) below grade while the sampling rods were being slowly withdrawn. A grout pump was used to inject the solution.

A proprietary form of slow release magnesium peroxide, called oxygen-releasing compounds (ORC®), was used as the oxygen source. Each batch mix consisted of 60 pounds (27.3 kg) of ORC® powder mixed with 30 gallons (114 liters) of water, creating a 25% weight/weight solution. This less viscous solution was used to increase the radius of influence in the fine-grained soils of the aquifer. The consistency of this solution was similar to latex paint. Approximately 55 pounds (25 kg) of ORC® powder was injected in each borehole, and the total mass of ORC® injected was 1,830 pounds (832 kg). The boreholes were filled with bentonite pellets immediately after the ORM solution was injected.

The existing monitoring wells were used to monitor the effectiveness of the ORM injection. The monitoring wells near the injection points were visually observed for bubbling and water level fluctuations during ORM injection. The existing monitoring wells have been monitored and sampled three times since the ORM injection for depth to water, depth to LNAPL, DO, ORP, dissolved iron (Fe^{2+}), temperature, pH, conductivity, and BTEX. The monitoring wells were purged prior to sampling using a peristaltic pump with a flow rate of less than 0.5 gallons (1.9 liters) per minute. The LNAPL in MW-2 and MW-3 was removed with a disposable bailer prior to purging with the peristaltic pump. The field measurements were made in a flow-through cell after each volume of water purged to ensure that representative formation water was being analyzed.

RESULTS AND DISCUSSION

The injection of ORM solution into the boreholes within the grid pattern encompassing the source area affected the monitoring wells within the grid (MW-1 and MW-2). Groundwater in MW-1 was observed to be bubbling while injecting ORM solution within 5 feet (1.5 m) of the well. The groundwater in MW-2 was bubbling while injecting ORM within 5 feet (1.5 m) of the well, and LNAPL along with ORM solution was forced from MW-2 while injecting within 3 feet (0.9 m) of MW-2. The ORM solution was also forced from several other boreholes within 5 feet (1.5m) of each other while injecting ORM solution in the southern portion of the grid. An in-situ DO measurement greater than 20 mg/l was made in MW-1 on 28 April 1998, which shows that the groundwater was supersaturated with respect to dissolved oxygen within 24 hours of ORM injection.

The DO concentrations had increased to supersaturated levels in MW-1 and MW-2 three months after the injection. Nine months after the injection, the DO concentration remained supersaturated in MW-2 and above pre-injection levels in MW-1. Figure 2 shows the DO and benzene concentrations in MW-1, MW-2, and MW-3 before and after ORM injection. The benzene concentration in MW-1 decreased from 1300 μg/l before treatment to 410 μg/l after the injection, and the benzene concentration in MW-2 dropped from 2700 μg/l to 2000 μg/l over the same time interval (Figure 2). The total BTEX concentration in MW-1 dropped from 14900 μg/l prior to ORM injection to 8810 μg/l afterwards, and the total BTEX concentration dropped from 34400 to 17300 over the same time interval in MW-2. Nine months after ORM injection, the benzene and total BTEX concentrations remained below pre-injection levels in MW-1 and MW-2, and benzene concentrations continue to decrease in MW-2.

Monitoring wells MW-1, MW-2, and MW-3 had periodically contained LNAPL prior to ORM injection. There has never been more than 0.02 feet of LNAPL in MW-1 and MW-3, but there has been up to 0.17 feet of LNAPL in MW-2. The last time LNAPL was encountered in MW-1 was in October 1996. Approximately 0.2 feet of LNAPL was observed in MW-2 in October 1998, and

LNAPL globules were encountered in MW-3 during all three post-injection sampling events. The LNAPL was removed with a disposable bailer each time it was encountered.

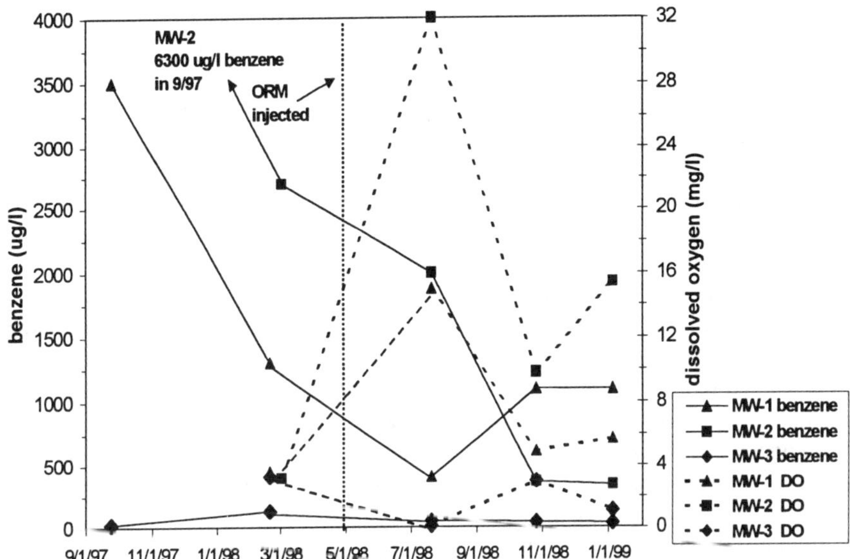

FIGURE 2. Plot of benzene concentration and dissolved oxygen (DO) concentration in MW-1, MW-2, and MW-3 before and after the ORM injection.

The benzene concentration in MW-3 dropped from 130 μg/l before injection to 52 μg/l after injection, and is steadily decreasing. The total BTEX concentration in MW-3 dropped from 5130 μg/l to 1822 μg/l over the same time interval, and the total BTEX concentration was 2087 μg/l nine months later. Dissolved oxygen concentrations in MW-3 dropped from 3.3 mg/l prior to ORM injection to 0.1 mg/l three months afterwards. The DO concentration rose to 3 mg/l six months after injection, and fell to 1.1 mg/l nine months after the ORM injection. Groundwater flows to the south and southwest with a hydraulic gradient of 0.05 feet per foot. The average groundwater velocity is 15 feet (4.6 m) per year, which means that groundwater from the closest injection points should reach MW-3 in approximately 16 months (Freeze, Cherry, 1979). The groundwater elevations increased by at least 0.5 feet (0.2 m) in all monitoring wells in July 1998, but fell to below pre-injection levels on 26 October 1998.

Dissolved iron (Fe^{2+}) was present in the groundwater in MW-1 and MW-2 above 10 mg/l prior to ORM injection. Dissolved iron (Fe^{2+}) has not been detected in the groundwater from MW-1 and MW-2 since the ORM injection. The dissolved iron concentration in MW-3 has dropped from 9.4 mg/l prior to injection to 6.6 mg/l nine months after the injection. The presence of dissolved iron indicates reducing conditions. The insoluble, oxidized form of iron (Fe^{3+}) acts as an electron acceptor when there is no dissolved oxygen present, and the iron is reduced to dissolved iron (Fe^{2+}) (Drever, 1988). The disappearance of dissolved iron (Fe^{2+}) in the groundwater from MW-1 and MW-2 after the ORM injection indicates that dissolved oxygen is present.

CONCLUSIONS

The BTEX concentrations in groundwater within the source area injected with ORM have decreased by 26% to 69% within three months after treatment. BTEX concentrations in groundwater samples collected from MW-1 and MW-2 nine months after ORM injection are 16% to 87% lower than pre-injection levels. The decline in BTEX concentrations in MW-1 and MW-2 can be attributed to the addition of ORM to the groundwater. The BTEX concentrations have also decreased by 60% to 67% in MW-3. The declining BTEX concentrations in MW-3 may be attributed to natural attenuation due to the distance from the injection grid. There is insufficient evidence to show that the ORM solution reached MW-3, which is located 20 feet (6.1 m) down-gradient of the edge of the injection grid.

Groundwater within the injection grid became supersaturated with respect to dissolved oxygen. The DO concentrations remained above pre-injection levels for up to nine months in MW-1 and MW-2. There was no apparent increase in DO concentrations in MW-3, located 20 feet (6.1 m) down-gradient of the treated area. Advective transport of ORM in groundwater would have taken 16 months to reach MW-3, and the oxygen would most likely be used up by aerobic microbial degradation before it reaches MW-3. The disappearance of dissolved iron in MW-1 and MW-2 shows that the increased levels of DO in those wells are oxidizing iron.

Residual LNAPL still appears to persist in isolated pockets. The injection of ORM under pressure forced these pockets toward monitoring wells located within the grid. The ORM injection expedited the recovery of LNAPL, but there may still be isolated pockets of LNAPL. The LNAPL and adsorbed petroleum hydrocarbons in the smear zone may continue to desorb and dissolve into the groundwater. Continued quarterly groundwater sampling events will monitor whether the dissolved BTEX concentrations rebound to pre-injection levels.

REFERENCES

Chapman, S.W., Byerley, B.T., Smyth, D.J., and Douglas Mackay. 1997. "A Pilot Test of Passive Oxygen Release for Enhancement of In-Situ Bioremediation of BTEX-Contaminated Ground-Water." *Ground Water Monitoring and Remediation*. VII(2): 93-105.

Drever, J.I. 1988. *The Geochemistry of Natural Waters: Second Edition.* Prentice Hall.

Freeze, R.A., and John A. Cherry. 1979. *Groundwater.* Prentice Hall.

ENHANCED *IN-SITU* BIOREMEDIATION OF GROUNDWATER AT MACDILL AIR FORCE BASE, FL

Don Boyle, Marshall Goers, and Shanthi Mandava (Earth Tech, Greenville, SC)
Patrick Hicks (Regenesis, Inc., Kingwood, TX)
Ken Lemons (Black & Veatch, Inc., Charlottesville, VA)
Doug Pendrell (U.S. Army Corps of Engineers, Omaha, NE)
Mark Canfield (MacDill Air Force Base, FL)

ABSTRACT: Soils and shallow groundwater underlying a former gasoline service station at MacDill Air Force Base (AFB), FL were contaminated by petroleum hydrocarbons. Prior to remediation, total benzene, toluene, ethylbenzene, and xylenes (BTEX) concentrations in the sandy surficial aquifer exceeded 15,000 µg/l at the center of the dissolved groundwater plume. In 1997, the Base excavated contaminated soils for off-site thermal treatment, and selected enhanced *in-situ* bioremediation using Oxygen Releasing Compound (ORC®) to treat groundwater. ORC is a proprietary time-release formulation of magnesium peroxide. In May 1998, 662 kg of ORC were injected at 87 locations throughout the plume. Initial monitoring results (through six months) indicated that the periphery of the plume was fully remediated; however, recalcitrant hydrocarbon concentrations at the center of the plume remained well above State action levels. After nine months, groundwater BTEX concentrations began to decrease significantly near the center of the plume. Due to slow BTEX removal rates, a second application using 671 kg of ORC was conducted in January 1999. During this event, the Base conducted more extensive soils and groundwater analyses to further evaluate the subsurface conditions that impact the effectiveness of ORC. Soils analyses conducted immediately prior to the second application indicated an increase in biological activity within the center of the plume. Subsequent groundwater monitoring (using discrete sampling probes located around the injection points) also indicate decreased BTEX concentrations, one week following the ORC re-application. Further analyses of groundwater at the site will be used to evaluate the effectiveness, and limitations, of ORC-enhanced bioremediation for potential use at other sites.

INTRODUCTION

A site investigation in 1994 indicated the presence of hydrocarbon contamination in soils and groundwater at Site 56, Former AAFES Service Station, MacDill AFB. The dissolved groundwater plume covered an area of approximately 1,200 square meters (see Figure 1). The site is underlain by sands and silty sands. Groundwater occurs at a depth of 1.5 meters and extends to an underlying clay and limestone aquitard at approximately 5 meters below ground surface (bgs). The hydraulic gradient is relatively low, resulting in an approximate discharge velocity of 8 to 12 meters per year. Groundwater flow is generally to the northwest. An intrinsic bioremediation study conducted for the

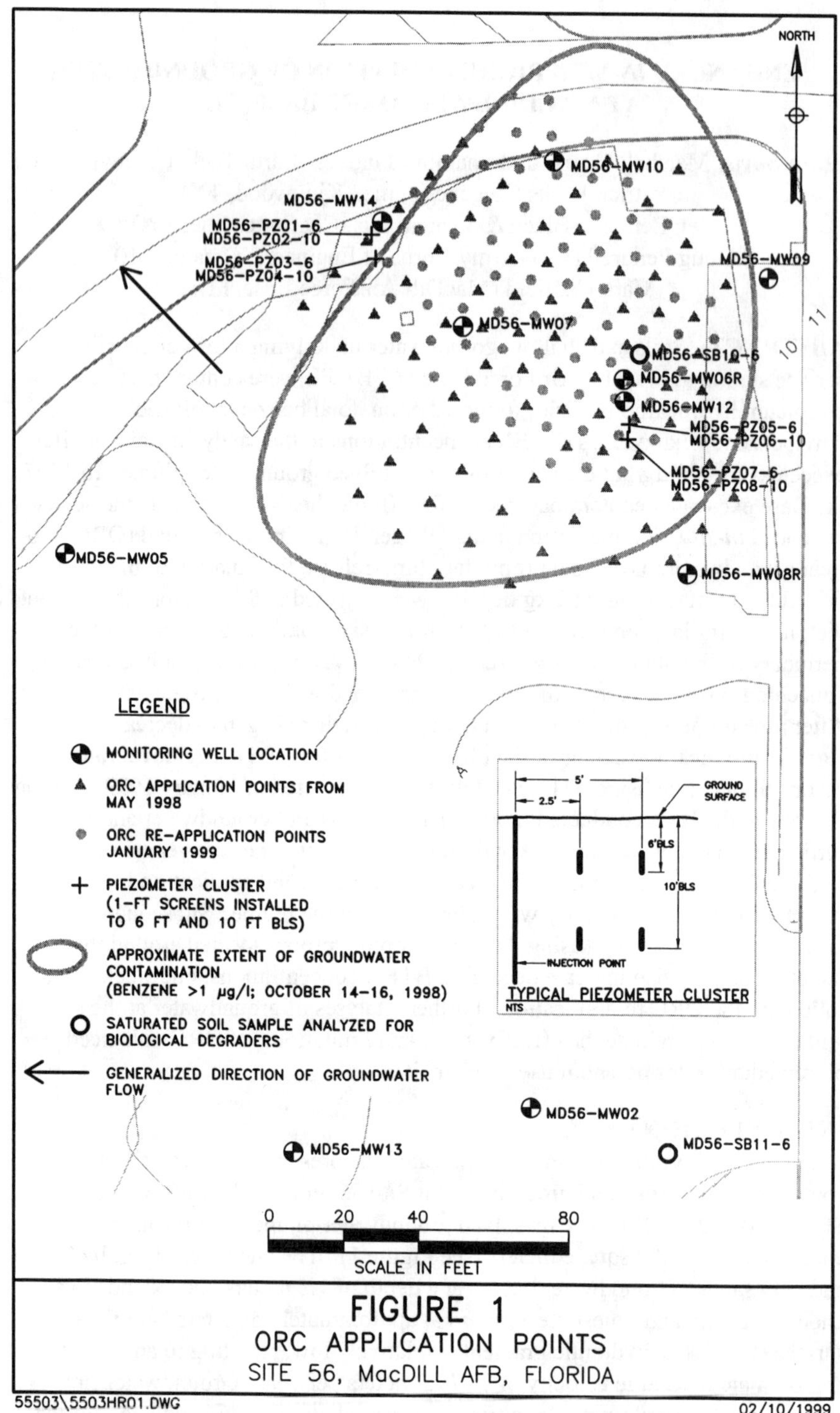

FIGURE 1
ORC APPLICATION POINTS
SITE 56, MacDILL AFB, FLORIDA

55503\5503HR01.DWG 02/10/1999

Base in 1996 indicated that natural attenuation was a feasible remedy for the site, but may require up to 50 years to reach State regulatory levels.

To expedite cleanup, the Base elected to excavate contaminated soils for off-site treatment, to be followed by enhanced bioremediation of groundwater using ORC. In October 1997, 1,800 cubic meters of petroleum-contaminated soils were excavated for off-site treatment by thermal desorption. Soil was excavated to a depth of 1.5 meters, corresponding to the approximate elevation of the low seasonal groundwater table. Laterally, the excavation included all soils exceeding 50 ppm Volatile Organic Compounds (VOC) in headspace screening.

ORC APPLICATION METHODS

ORC Design. A conservative mass-balance approach was used to determine ORC requirements based on oxygen demand. First, the groundwater plume was divided into three areas: the source area, the peripheral area adjacent to the source, and the surrounding edge of the plume. Theoretical oxygen demand was calculated for each zone based on historical BTEX, total petroleum hydrocarbons (TPH), sulfide, and ferrous iron concentrations in monitoring wells. Additional oxygen demand was added to account for potential oily-phase TPH contamination in the smear zone (assuming residual concentrations of 1 mg/kg in the source area and 0.1 mg/kg in the peripheral areas). The total theoretical oxygen demand, in kilograms, was then used to estimate ORC requirements, using a safety factor of 5.0. The total design ORC requirement using this method was 581 kg.

Initial ORC Application and Monitoring. Based on the design analysis, a total of 617 kgs ORC grout was injected at 87 locations in May 1998. The grout was injected through direct push rods, starting at 4 meters in depth and slowly raising to 1 meter in depth (approximately corresponding to the seasonal high water table). At the center of the plume, 20.4 kg (45 lb) of ORC were injected at each point, spaced 3 meters apart, while at the outer edges of the plume, 4.5 kg (10 lb) were injected at a 4.5-meter spacing (see Figure 1 for initial locations).

To evaluate the initial results, groundwater samples were collected from six site monitoring wells for a period of nine months following ORC application. Groundwater samples were analyzed for VOCs (including BTEX compounds and methyl tert butyl ether, or MTBE), TPH, total organic carbon (TOC), chemical oxygen demand (COD), magnesium (Mg), ferrous iron, alkalinity, hardness, oxidation-reduction potential (ORP), and dissolved oxygen (DO).

Follow-up ORC Application and Monitoring. Based on the results of the first six months of groundwater analyses (described below), and in accordance with ORC manufacturer's recommendations, it was determined that re-application of ORC may be required to adequately treat groundwater at the site. In January 1999, 671 kg ORC grout was injected at 74 locations (9 kg (20 lb) of ORC at each point). The reinjection points were concentrated near the center of the groundwater plume, off-set from the original, May 1998, locations (see Figure 1).

In conjunction with the follow-up application, more extensive analyses of soils and groundwater was conducted to further evaluate the effectiveness of

ORC. Prior to sampling, two soil samples were collected and analyzed for total heterotrophs, and for specific degraders of BTEX and TPH. One soil sample (SB10-6) was collected from the center of the contaminated area, and the second sample (SB11-6) was collected from a background area. Both samples were collected from a depth of 6 feet, below the ambient water table. Also, four soil samples were collected from the center of the groundwater plume for TPH analysis prior to re-injection. These samples were collected at one-foot intervals starting from 5 feet bgs to 8 feet in depth, and analyzed for TPH to estimate mass of hydrocarbon in the soil matrix within a potential smear zone.

To evaluate the effects of ORC on groundwater near the injection points, discrete sampling points (piezometers), each with 0.3-meter screens, were installed 0.75 and 1.5 meters away from two injection points (as shown in Figure 1). The temporary piezometers were sampled prior to re-injection, then one week after re-injection, and analyzed for BTEX compounds, TPH, DO, total alkalinity, COD, hardness, and TOC. Also, six monitoring wells were sampled prior to re-injection (corresponding to the ninth-month sampling event described above) and analyzed for VOCs, sulfides, sulfites, total alkalinity, COD, ammonia, phosphorus, hardness and TOC. Sulfide and sulfite analyses were added to this event to evaluate potential inorganic sources of oxygen demand.

Prior to ORC re-application in January, four oxygen uptake tests were conducted to evaluate the relative rate of oxygen utilization throughout the plume. The *ex-situ* tests were conducted using groundwater extracted from monitoring wells MW06R, MW07, MW10, and MW14 (see Figure 1). For each test, 300 ml of groundwater were thoroughly aerated, then decanted to a BOD bottle. DO measurements were then recorded for 15 minutes. As a qualitative analysis, the slopes of the DO degradation curves were used to estimate the instantaneous oxygen utilization rate for each location (ref. Standard Method 2710 B). The tests were repeated following ORC injection, using the same wells, but with the addition of a 1-tablespoon aliquot of soils collected from a soil boring adjacent to each well.

RESULTS

Residual BTEX concentrations in peripheral monitoring wells were reduced to below detection limits within six months of initial ORC application. However, recalcitrant BTEX concentrations near the center of the plume were not significantly affected until the ninth month following initial application (Table 1).

TABLE 1. BTEX in Source-Area Monitoring Wells (μg/l)

Date	MW06R	MW07	MW10
April 27-28, 1998 (baseline)	523	886	2361
June 10, 1998	879	483	29
July 23, 1998	1,373	2,120	87
October 16, 1998	2,060	460	930
January 6, 1999	140	84	130

The apparent increase in BTEX concentrations during the months following the initial application (in May 1998) may be due to variations in groundwater levels and resultant flushing of TPH within the smear zone. The data also indicate successful bioremediation of MTBE (Table 2).

TABLE 2. MTBE in Source-Area Monitoring Wells (μg/l)

Date	MW06R	MW07	MW10
April 27-28, 1998 (baseline)	39	46	110
June 10, 1998	32	25	9.7
July 23, 1998	21	33	4.5
October 16, 1998	<50	20	<1
January 6, 1999	<20	15	<20

Continued analyses of monitoring well samples are scheduled for February and April, 1999.

With respect to other groundwater parameters, DO measurements have been consistently below 1 mg/l, both before and after ORC application. COD concentrations have ranged from <50 to 140 mg/l over the 9-month duration. TOC concentrations have ranged from 1.2 to 73 mg/l over the same period. Ferrous iron ranged from 0.1 to 1.0 mg/l; alkalinity ranged from 100 to 400 mg/l; hardness ranged from 100 to 800 mg/l; and ORP ranged from –100 to –200 mV. These values were typical throughout the site, and did not exhibit a strong correlation with respect to locations within the plume. There was no observable increase in magnesium concentration in groundwater following ORC injection. Sulfide concentrations (during January 1999) ranged from 3.7 to 12 mg/l, and sulfite concentrations ranged from <2 to 12 mg/l. The highest sulfide and sulfite concentrations occurred outside of the plume (away from ORC injection points).

Results of the *ex-situ* oxygen uptake tests ranged from 1 to 5 mg/l/hour. These semi-quantitative tests confirmed that instantaneous oxygen utilization was occurring within the plume. More extensive analyses, such as biological oxygen demand (BOD) tests or *in-situ* oxygen utilization tests are needed to provide more quantitative and definitive analysis of ORC effects on oxygen uptake. Also, for other sites, pre-application and concurrent background oxygen utilization tests are recommended to fully evaluate the effectiveness of ORC on oxygen utilization.

Results of soils analyses for heterotrophs and hydrocarbon degraders, in colony forming units per gram (CFU/g), are summarized in Table 3:

TABLE 3. Bacterial Count Results for Soil Samples (CFU/g)

Parameter	SB10-6 (source)	SB11-6 (background)
Total Heterotrophs	1.5×10^5	$<1.0 \times 10^4$
BTEX Degraders	1.1×10^5	$<1.0 \times 10^3$
Gasoline Degraders	7.4×10^4	$<1.0 \times 10^3$

These results indicate increased bacterial growth within the source area, as compared to background.

TPH analyses of soils samples from within the source area indicated 32 and 30 mg/kg of TPH were present at depths of 1.8 and 2.1 meters (6 and 7 feet), respectively. These are not extremely high TPH concentrations, and could be associated with dissolved hydrocarbons within the soil matrix beneath the water table. Nonetheless, TPH concentrations in this horizon may also indicate the presence of adsorbed oily-phase petroleum: a potential continuing source of BTEX contamination – and oxygen demand – in groundwater. Samples from 1.5 and 2.4 meters (5 and 8 feet) in depth were non-detect for TPH.

Analyses of groundwater samples from piezometers near the center of the plume indicated slight decreases in BTEX and TPH concentrations during the first week following ORC re-application. None of the piezometers exhibited any significant increase in DO during the first week. The piezometers will be sampled again in February 1999, one month after ORC re-application.

CONCLUSIONS

Based on diminishing BTEX and MTBE concentrations, the use of ORC appears to have enhanced *in-situ* bioremediation at Site 56. This is supported by increased bacterial counts and measurable oxygen uptake rates within the plume. However, these parameters were not measured prior to ORC application. Thus, while increased biological activity and BTEX degradation rates are suspected, these effects due to ORC application alone cannot be confirmed.

The lack of measurable DO increases indicates that the rate of oxygen release from ORC is equal to or less than the total (organic and inorganic) oxygen uptake rate. This can be expected based on how ORC is designed to release oxygen. ORC is designed to quickly release approximately 10 percent of the oxygen content, immediately following application, to overcome reducing conditions and stimulate biological growth. Then the slow release of oxygen from the remaining ORC allows biological degradation to occur in anoxic conditions, such that sufficient oxygen is provided to meet the demands of increased biological activity, without necessarily increasing DO concentrations. Nonetheless, consistently low ORP measurements at Site 56 indicate reducing, anaerobic conditions persist. Thus, aerobic biodegradation remains oxygen limited. As a result, the time to reach cleanup goals will be longer than expected.

DO measurements alone were not an adequate indicator of ORC effectiveness because of the slow-release formulation. For future sites, pre- and post-application oxygen uptake tests and bacterial counts will be recommended to confirm that ORC has effectively improved *in-situ* bioremediation.

Ultimately, the effectiveness of ORC at Site 56 has been limited by: strong reducing conditions, high initial BTEX concentrations in groundwater, and the potential presence of a oily-phase smear zone. Due to reducing conditions at MacDill AFB, a more aggressive oxygen delivery system will be recommended for other highly contaminated sites. Still, based on initial results at Site 56, ORC could prove to be useful for sites with lower levels of groundwater contamination.

STUDY OF NATURAL BIOREMEDIATION PROJECTS USING TIME-RELEASE OXYGEN COMPOUNDS

Susan L. Race, E.I.T. and Patrick M. Goeke, P.E. (BE&K/Terranext, Lenexa, KS)

ABSTRACT: Enhanced natural bioremediation studies are being conducted on a series of sites throughout the midwest United States. Each site exhibits a unique geologic condition, contaminant concentration, enhancement application, and remediation goal. Based on a review of 14 sites, the usage and effectiveness of enhanced bioremediation has been categorized to provide general guidelines for applications and monitoring. The enhancement application reduced petroleum hydrocarbons in 65% of the sites; however a rebound in petroleum hydrocarbons was observed in 41% of the sites. The most significant factor affecting the bioremediation effectiveness is the presence of free product and source material.

INTRODUCTION

In the early 1990s, BE&K/Terranext ground water BTEX data from gasoline underground storage tank (UST) sites indicated natural bioremediation was occurring downgradient of source areas. At finite distances downgradient of the source, an equilibrium was reached between the BTEX mass and the ability of natural organisms to bioremediate the mass. Ground water data indicated the dissolved oxygen levels in source areas were generally zero, whereas the dissolved oxygen levels downgradient were higher. BE&K/Terranext approached state regulators about conducting site trials to test a time-release oxygen compound to enhance this bioremediation process at sites which could not be aggressively remediated.

In 1995 and 1996, BE&K/Terranext used the Regenesis oxygen release compound (ORC®) at five sites to evaluate the effectiveness of enhanced natural bioremediation. Varying ORC application methods were utilized at sites with different geology and contamination levels. During that time, it became apparent that several factors affected the outcome of these projects. In an effort to improve the performance of future bioremediation projects, BE&K/Terranext collected data from 14 sites to identify factors which lead to successful, as well unsuccessful projects.

Site Characteristics. In addition to five BE&K/Terranext sites, data were collected from nine project files at state regulatory agencies. Each site is located within a 450 kilometer radius of Kansas City, Missouri. The site names and locations have been omitted at the request of the state agencies. Details regarding the site status, maximum BTEX concentrations, geology, ground water depths, and plume descriptions are summarized in Table 1.

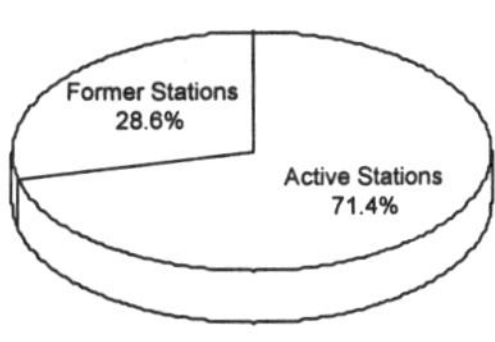

Figure 1 - Site Types

The geology of 82% of the sites generally consists of fine grained soils overlying coarse grained sands. At the remaining 18% of the sites, clay overlies a silty clay, a shale, or fractured limestone. Ground water depths ranged from 1 to 6 meters, but was less than 4 meters at 78% of the sites. All of the sites are believed to fall within a ground water velocity category of less than 0.08 meters per day. At six sites, the maximum BTEX concentration was less than 5,000 micrograms per liter (µg/l) and at eight sites the maximum BTEX concentration was greater than 5,000 µg/l. Plume sizes range from 0.04 to 4.5 hectares, with an average of 1 hectare. Free product was identified at three sites. The first treatment occurred in 1995, but the majority occurred in 1996 and 1997. Regenesis' ORC product was used at each site.

TABLE 1 - SITE INFORMATION TABLE

Site Number and Current Status	Maximum Site VOCs Levels Before ORC Treatments, µg/l	General Geology and Depth to Ground Water (DTW)	Contamination and Plume Description	Plume Area (m^2)
Site 1 Active Station	Benzene: 15,500 Total BTEX: 42,400	1.2m of clay over coarse sand DTW: 2.5m	Significant soil source material. Ground water impacted 90m downgradient. Possible upgradient source.	350
Site 2 Active Station	Benzene: 6,300 Total BTEX: 9,000	2m of clay over coarse sand DTW: 2.7	Significant soil source material. Ground water impacted 90m downgradient. Possible upgradient source.	580
Site 3 Former Station	Benzene: 2,600 Total BTEX: 9,300	1 to 2m of clay over coarse sand DTW: 2.7m	Minor soil impacts limited to source area. Ground water impacted 60m downgradient. Upgradient sources.	1,400
Site 4 Active Station	Benzene: 2,400 Total BTEX: 51,000	.6 to 1.3m of clay over coarse sand and gravels DTW: 2.4m	Minor soil impacts limited to source area. Ground water impacted 75m downgradient. Plume defined by 4 monitoring wells.	2,300
Site 5 Active Coop and 2 Active Stations	Benzene: 19,800 Total BTEX: 65,000	2.4 to 3m of clay over coarse sand and gravels DTW: 3.3 to 4m	Extent of soil impacts unknown. Ground water impacted 300m down-gradient. Known additional sources cross-gradient.	46,500
Site 6 Active Station	Benzene: 10,000 Total BTEX: 44,000	1.5 to 2.m of clay over sandy, silty clay DTW: 6 to 9m	Minor soil impacts limited to source area. Ground water impacts limited to on-site areas, 30 x 60m.	1,860
Site 7 Former Station	Benzene: 5,500 Total BTEX: 7,000	0.9m of clay over coarse sand and gravels DTW: 1.8 to 2.1m	Minor soil impacts, portion of impacted soil removed. Ground water impacts approximately 90 x 90m. Known upgradient ground water source.	8,400
Site 8 Active Station	Benzene: Free Total BTEX: Product	1.5 to 2.4m of silts and clays over sand DTW: 4.6m	Significant soil impacts limited to source areas. Ground water impacts approximately 90 x 90m.	8,300
Site 9 2 Active and 2 Former Stations	Benzene: Free Total BTEX: Product	9m feet of clay over sands and gravels DTW: 5.5 to 6.1m	Extent of soil impacts unknown. Ground water impacts extend 900m downgradient. Free product exists over entire site.	FP plume: 1,860
Site 10 2 Active Stations	Benzene: 15,000 Total BTEX: 45,000	1 to 1.5m of clay over shale DTW: 2 to 2.7m	Extent of soil impacts unknown, but extend into shale. Ground water impacts approximately 120 x 120m.	15,000
Site 11 Active Station	Benzene: 1,500 Total BTEX: 7,200	2.4 to 3m of clay over sand (approximately 30m thick) DTW: 4 to 4.5m	Significant soil impacts limited to source areas. Ground water impacts extend approximately 30m downgradient.	1,600
Site 12 2 Active Stations	Benzene: Free Total BTEX: Product	6.1 to 6.7m of clay over sand with shale at 11.3m. DTW: 6.1m	Limited soil impacts at two source areas. Ground water plume extends 45 and 37m downgradient. Free product present in both source areas.	1,500 1,500
Site 13 Former UST site	Benzene: 3,800 Total BTEX: 28,500	1.2 to 2.7m of clay over sand DTW: 2.1m	Excavate soil impacts during UST removal. Ground water impacts 45 x 15m	675
Site 14 3 Former Stations	Benzene: 800 Total BTEX: 13,000	2.5 to 3.7m of clay over fractured limestone DTW: 1 to 4.3m	Soil impacts extend into limestone and are limited to source areas. Ground water impacted by migration through limestone fractures.	11,000

Note: Thickness of geologic unit unknown if not indicated

MATERIALS AND METHODS

ORC consists of magnesium peroxide, which slowly releases oxygen as water passes through the compound. The ORC provides oxygen to naturally occurring bacteria that use hydrocarbons as a food source. The objective in using ORC is to increase the amount of oxygen in an oxygen-deficient environment, thereby increasing the aerobic bioremediation process. The oxygen time-release and flow through the subsurface is generally achieved by advection and dispersion. Site specific conditions, such as geology, ground water velocity and gradient, BTEX concentrations, and background oxygen and bacteria levels are significant factors in the application and effectiveness of ORC enhanced bioremediation.

Three ORC application methods were used: filter socks, slurry injection using a Geoprobe®, or slurry injection prior to plugging monitoring wells. Table 2 summarizes the application dates, methods, quantities, and locations for the sites.

TABLE 2 - REMEDIATION DATA

Site Number and Current Status	Remedial Activity Other Than ORC	ORC Installation Data		
		Date	Method and Quantity	Location
Site 1 Active Station	None	July 1996 Oct 1997	Injected as a slurry using a Geoprobe twice, 440 kg then 560 kg	66 injection points in the source area, followed by 12 injections points
Site 2 Active Station	None	July 1996	Injected 393 kg as a slurry using a Geoprobe	59 injection points in source area
Site 3 Former Station	None	1995, 1996 1998	Installed 8 feet of filter socks in existing monitoring wells	2 monitoring wells in the source area
Site 4 Active Station	None	April 1998	Installed unreported length of filter socks in existing monitoring wells	2 monitoring wells in the source area, 1 monitoring well downgradient
Site 5 Active Coop and 2 Active Stations	SVE at Coop None at Gas Stations	July 1997	Injected 9.3 kg of slurry in each well prior to plugging monitoring well	1 well plugged downgradient of site 3 wells plugged in the source areas
Site 6 Active Station	SVE System	March 1997	Injected 11.2 kg of slurry in each well prior to plugging monitoring well	3 wells on upgradient edge of plume
Site 7 Former Station	UST removed & limited soil excavation	June 1997	Injected 571 kg as a slurry using a Geoprobe	69 injection points in the source area and slightly downgradient
Site 8 Active Station	SVE System operating	Oct 1997	Injected 185 kg as a slurry using a Geoprobe	17 injection points in downgradient plume to protect domestic and public wells
Site 9 2 Active and 2 Former Stations	None	May 1997	Injected 1530 kg as a slurry using a Geoprobe	66 injection points forming 2 barrier lines in downgradient portion of plume.
Site 10 2 Active Stations	None	July 1997	Injected an unreported weight as a slurry using a Geoprobe	24 injection points along downgradient portion of plume
Site 11 Active Station	None	Aug 1996 Jan, Aug 1997 Apr 1998	Installed 2.4 m of filter socks in existing monitoring wells	2 monitoring wells in the source area and 2 downgradient monitoring wells
Site 12 2 Active Stations	Limited soil removal at both sites	June 1998	Injected 350 kg as a slurry using a Geoprobe at each site	30 injection points in each source area (60 total)
Site 13 Former UST Site	Soil source removed	June 1998	Injected 54 kg as a slurry using a Geoprobe	15 injections points in source area
Site 14 3 Former Stations	None	Aug 1996 Oct 1996	Injected 235 kg as a slurry using a Geoprobe into limestone fractures	5 injection points in the source area

Retrievable filter socks were placed in selected monitoring wells from approximately 25 cm above the water table to the total depth of the well. Filter socks were generally placed in monitoring wells that contained the highest BTEX concentrations. Based on quarterly measurements of dissolved oxygen and BTEX concentrations, the filter socks were replaced or discarded, as appropriate. Generally, the filter socks required replacement every six months.

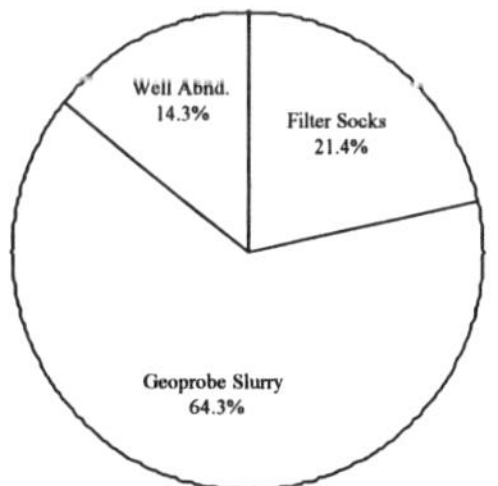

Figure 2 -Treatment Methods

For ORC slurry injections, ORC was mixed with water and pressure injected into the subsurface using a Geoprobe. Injection locations were generally within the source areas. At three sites, injection locations were in the downgradient portion of the plume to form barrier lines for potential receptors. The general slurry mixture at each site was an approximate 60% dry weight mixture. The amount of ORC utilized at each injection point ranged from 5.6 to 47 kilograms (kg). The slurry was generally injected uniformly into the saturated zone from 3 meters below the water surface to the top of the water. A follow-up injection activity was conducted at two sites, to continue remediation in the source areas.

For well abandonment applications, an ORC slurry mix was injected into existing monitoring wells immediately prior to plugging. The abandoned wells were located within source areas, upgradient, or downgradient of the source. In general, 11 to 14 kg of ORC were utilized for each well abandonment.

RESULTS AND DISCUSSION

Evaluation of Outcomes. The data collected for this study was obtained from on-going remediation projects. Therefore, site conditions, other remediation activities, installation methods, treatment amounts, available monitoring data, and other factors complicate the evaluation of the data. Table 3 summarizes the monitoring data used to evaluate the effectiveness of ORC in enhancing natural bioremediation.

TABLE 3 - DATA, EFFECTIVENESS, AND FACTORS AFFECTING OUTCOME

Site Number and Status	Well Loc	Hydrocarbon Concentration, µg/l			% Reduction	Effectiveness	Factors Affecting Outcome
			Before	After			
Site 1 Active Station	S	B	5000	713 / 5300 I 1200 / 6900	85/ -10/80/ -30%	Reduction+Rebound	Soil source material present
	S	TB	13200	1684/7480 I 3900/17400	90/ -40/50/-130%		Possible upgradient ground water source
	D	B	650	305 / 2000 I 1100	50/ -200/45%	Reduction+Rebound	
	D	TB	1100	965 / 3400 I 2300	10/ -200/30%		
Site 2 Active Station	S	B	950	250 / 2200	75 / -130%	Reduction+Rebound	Soil source material present
	S	TB	1900	450 / 4300	75 / -125%		Possible new release
	D	B	1150	170 / 1000	85 / 15%	Reduction+Rebound	Frequency of treatment
	D	TB	1800	250 / 1500	85 / 15%		
Site 3 Former Station	S	B	1700	720	60%	Significant Reduction	Upgradient ground water source
	S	TB	2100	940	55%		Limited source material present
	D	B	25	9	65%		Limited radius of influence of filter socks
	D	TB	75	9	85%		
Site 4 Active Station	I	B	100	5	95%	Inconclusive -	No monitoring wells available other than
	I	TB	1500	55	95%	Reduction measured	treatment wells
						in treatment wells	
Site 5 3 Active Stations	U	B	2000	210 / 1900	90 / 5%	Reduction+Rebound	Multiple ground water sources
	U	TB	7400	660 / 5500	90 / 25%		Number of ORC locations
	D	B	100	1000	-900%	Minimal Reduction	Limited amount of ORC used
	D	TB	9000	6000	33%		Frequency of ORC treatment
Site 6 Active Station	U	B	100	<1 / 2090	100 / -2000%	Reduction+Rebound	Location relative to source
	U	TB	800	<6 / 13000	100 / -1500%		Limited amount of ORC used
							Frequency of treatment
Site 7 Former Station	S	B	5500	2300 / 6100	60 / -10%	Reduction+Rebound	Soil source material present
	S	TB	7000	2800 / 7500	60 / -7%		Possible upgradient source
	S	B	4200	384	90%	Significant Reduction	Well location not in soil source area and
	S	TB	5200	470	90%		on D/G edge of ORC treatment zone..
Site 8 Active Station	S	B	41000	2000 / 18700	95% / 55%	Reduction+Rebound	Soil source material present
	S	TB	81000	6000 / 62000	90% / 20%		Presence of free product
							Insufficient downgradient wells to evaluate
							cutoff reduction
Site 9 2 Active	D	B	5200	790	85%	Significant Reduction	Downgradient treatment wall
	D	TB	10000	1200	95%		On-going source remediation
Site 10 2 Active Stations	D	B	2270	1430	40%	Significant Reduction	Soil source material present
	D	TB	3400	3000	10%		Ground water in shale; uncertain ORC
	D	B	55	17	70%	Significant Reduction	contact with ground water flow paths
	D	TB	70	17	75%		
Site 11 Active Station	S	B	976	1700	-75%	Reduction+Rebound	Soil source material present
	S	TB	4000	5000	-25%		Number of ORC locations
	D	B	250	280	-10%	Minimal Reduction	Limited amount of ORC
	D	TB	630	485	25%		
Site 12 2 Active Stations	S	B	3170	1590	50%	Ineffective	Insufficient source removal
	S	TB	33000	70000	-110%		Presence of free product
	S	B	3000	4850	-60%	Ineffective	Limited time duration since treatment
	S	TB	14000	35000	-150%		
Site 13 Former UST Site	S	B	600	ND	100%	Significant Reduction	Source removed
	S	TB	1900	ND	100%		Limited impact area
							Low initial concentrations
Site 14 3 former Stations	S	B	420	1400	-230%	Ineffective	Soil source material present
	S	TB	2800	10000	-260%		Ground water flow in fractured bedrock,
	D	B	266	180 / 345	32 / -30%	Reduction+Rebound	uncertain ORC contact with flow paths
	D	TB	3500	2700	23%		Limited amount and locations

Notes

Well Loc: Location of Monitoring Well for data shown (S) = Source Area; (D) = Downgradient; (U) = Upgradient; (I) = Injection Well

Hydrocarbon Concentration -

Before: B=benzene TB=total BTEX

After: 500 / 1000 I 400 / 800 = Lowest Conc after ORC / Rebound Conc then Lowest Conc following 2nd ORC injection/ Rebound Conc

% Reduction: Percent reduction and rebound computed from change for BEFORE Conc.

At the five BE&K/Terranext sites, including two sites which were extensively studied in conjunction with Regenesis, installation of ORC resulted in a definitive increase in the ground water dissolved oxygen content, even in the source areas. Since published data document the ability of ORC to oxygenate the ground water, and since dissolved oxygen data were not available for some of the sites, categories were established to rate the effectiveness of the ORC in reducing the ground water plume, not increasing the dissolved oxygen level in the ground water.

The monitoring data were used to group each site into one of four outcome categories: 1) significant reduction, 2) reduction followed by rebound, 3) minimal to no reduction, and 4) inconclusive. As indicated in Table 3, the enhancement resulted in a reduction of the ground water BTEX concentrations at 11 sites (65%). Four sites (24%) showed a permanent decrease in the BTEX plume and seven sites (41%) showed a decrease followed by a rebound in the BTEX plume. Minimal to no reduction occurred at five sites (29%), and the data at one site was inconclusive. ORC treatment was applied at 12 source areas on 11 sites, four downgradient locations, and one upgradient location. Figure 3 illustrates the project outcome compared with treatment location.

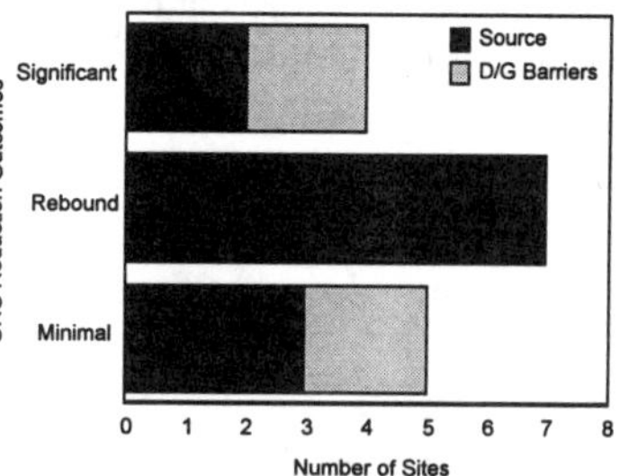

Figure 3 - Outcome for Treatment Location

Factors Affecting the Outcome. Factors which potentially affect the enhanced natural bioremediation process were compared to project outcomes to determine if a correlation could be identified. These factors include: the presence of free product, source material present on site, ORC application method, ORC usage volume, BTEX concentration, ground water flow characteristics, and injection location.

The data indicate the most significant factors affecting remediation outcomes are the presence of free product and/or source material. Free product was encountered at three sites. At two sites, a BTEX increase was observed following ORC treatment. At the third site, a BTEX reduction was followed by an increase to levels greater than initial concentrations.

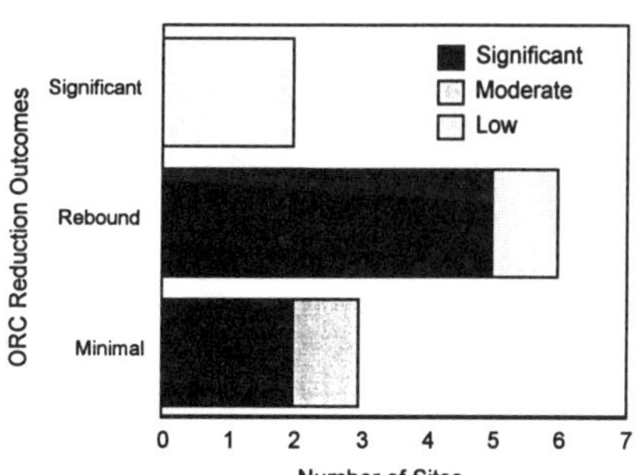

Figure 4 - Outcome Verses Source Material

Figure 4 illustrates the outcome versus the source impacts on site. Significant reductions occurred where source impacts were low. Conversely, when significant source impacts were present, either minimal reduction, or a reduction followed by a rebound was observed. The rebound was often greater than the reduction.

Other factors affecting the outcome include upgradient sources, method of ORC application, geologic conditions, ORC volume, and BTEX concentrations. Bar charts in Figures 5 through 7 compare the outcome versus application method, volume, and site BTEX concentration. Geoprobes were the most effective application method, due to the greater ORC dispersion through the subsurface. Filter socks were effective on smaller plumes with minimal source material. Application volume has little impact on outcome, however, it is likely that the application volume cannot overcome the presence of free product or source material. The outcomes were less successful at sites with higher BTEX concentrations, likely associated with on site source material.

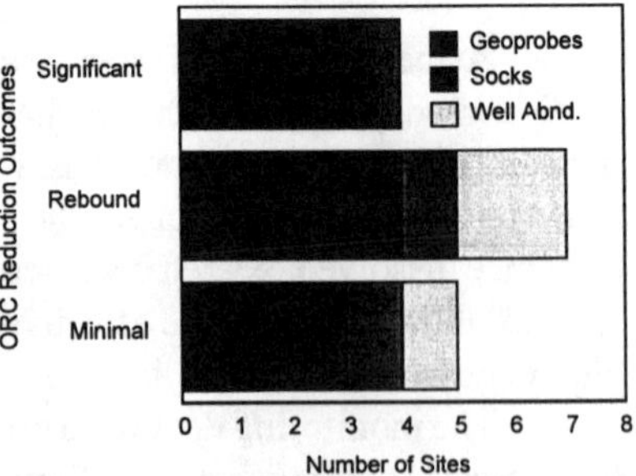

Figure 5 - Outcome Versus Application Method

Enhanced bioremediation was effective when used as a downgradient barrier to prevent further migration of BTEX plumes. ORC barriers effectively reduced the BTEX plume downgradient of the treatment zone at two sites.

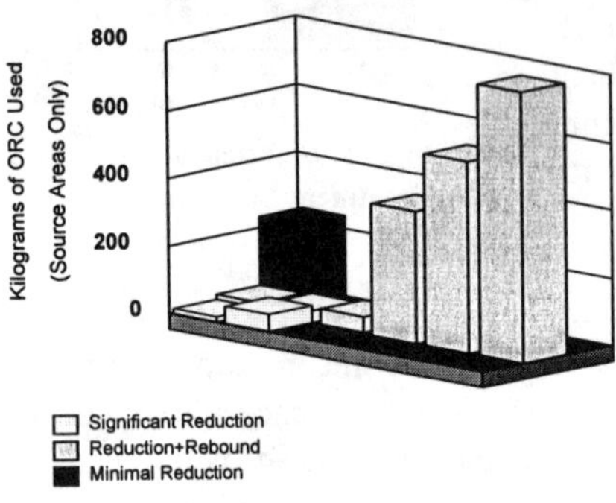

Figure 6 - Outcome Versus ORC Volume

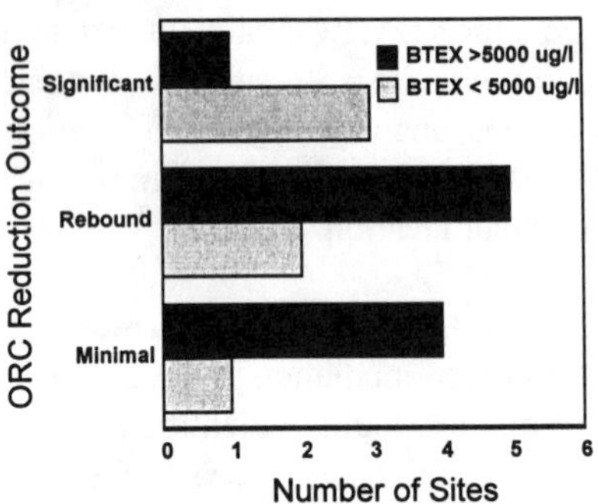

Figure 7 - Outcome Versus BTEX Concentration

CONCLUSIONS

Time-release oxygen compounds effectively increase the dissolved oxygen in ground water and enhance natural bioremediation. Site studies show that ORC can reduce the BTEX concentrations in ground water; however, the reductions may be transient when free product or significant source material, either in the soil or a "smear" zone, are present. Applications utilizing ORC filter socks can be effective in limited, low concentration areas; however, ORC slurry injection using Geoprobes are generally more efficient in distributing the ORC through the treatment area. ORC application volumes are important to successful reduction, however, large volumes of ORC cannot overcome the presence of free product or significant source material.

ACKNOWLEDGMENTS

Due to the confidentiality of the projects, references cannot be provided. The authors acknowledge the assistance of project managers at the state agencies, for providing access to the project files, and Regenesis for technical support.

FATE AND TRANSPORT OF OXYGEN FROM OXYGEN RELEASE COMPOUND

David Kelly, P. E., IT Corporation, Martinez, California USA
Roger Henderson, P.E., U.S. Army Corps of Engineers, Sacramento, California USA

ABSTRACT: A field study was conducted at a petroleum hydrocarbon release site to evaluate the fate and transport of oxygen in groundwater from Oxygen Release Compound® (ORC). The field study included the direct injection of ORC into the shallow groundwater zone and the monitoring of dissolved oxygen (DO) concentrations in temporary wells installed in a diamond pattern at distances of 1.5 to 5.0 feet (ft) downgradient of the injection point. The results of the study indicate that the introduction of the ORC was effective at increasing the DO concentration at all monitoring points. Biodegradation rates were estimated from the data obtained and correlated well with the DO concentrations. As the DO concentrations increased initially, the biodegradation rates increased. When the ORC was depleted, both the DO concentrations and the biodegradation rates decreased.

INTRODUCTION

When bioremediation of organic contaminants in groundwater is completed in situ, biological processes are enhanced by engineered systems. The engineered systems supply chemicals that are limited by site conditions, but required by microorganisms to efficiently degrade the contaminants. At most petroleum hydrocarbon contaminated sites the limiting chemical is oxygen.

One method for supplying oxygen to groundwater in situ is the use of Oxygen Release Compounds® (ORC). The ORC is a proprietary formulation of magnesium peroxide that slowly releases oxygen when hydrated. The ORC can be directly injected into the subsurface as a slurry to react and increase the DO concentrations in groundwater.

A field study was conducted to evaluate the effectiveness of ORC at remediating petroleum hydrocarbons at the Building 1065 Area at the Presidio of San Francisco, California. The study was intended to provide basic parameters for development of a conceptual design for the full-scale remediation of the site.

Site Description. Building 1064, a small building previously located southwest of Building 1065, served as a post exchange gasoline service station at the Presidio. In the summer of 1996, one 375-gallon and two 550-gallon steel underground storage tanks were removed from the building location. Evidence discovered during removal activities indicates that a petroleum hydrocarbon release had occurred from the tanks.

A site investigation was conducted to evaluate site geology, hydrogeology, and concentrations and extent of chemicals of concern (COCs). Site stratigraphy consisted of fill at land surface underlain by alternating layers of sand and clayey silt

(Bay Mud). Depth to the shallow water table averages approximately 5 to 6 ft below ground surface (bgs). The shallow groundwater zone is approximately 1 to 3 ft thick and occurs in a laterally discontinuous sand underlying the vadose zone. Below the shallow groundwater zone lies a Bay Mud layer. Groundwater flow direction is to the north-northeast. Seepage velocities estimated during the investigation averaged 0.32 ft per day.

The COCs detected in groundwater were predominantly petroleum hydrocarbons as gasoline. The area where the ORC study was conducted had the highest concentrations of gasoline. The concentrations of detected chemicals from a groundwater sample collected from the shallow groundwater zone near the ORC study are presented in Table 1. Samples collected during the ORC study had a sheen of petroleum hydrocarbons on the surface of the water.

TABLE 1. Petroleum hydrocarbon concentrations in groundwater.

Compound	Concentration (μg/L[a])
Gasoline	98,000
Benzene	1,200
Ethylbenzene	1,400
Toluene	150
Total Xylenes	490

[a] μg/L - micrograms per liter

MATERIALS AND METHODS

The field study included the injection of ORC and monitoring of the DO concentrations in temporary wells installed downgradient. The DO concentrations were measured twice weekly for the first 6 weeks, then periodic monitoring was conducted until sufficient data were generated. The DO measurements were taken using a hand-held DO meter on groundwater samples collected by a peristaltic pump.

Figure 1 shows the layout of the ORC injection point and temporary monitoring wells. The ORC and wells were both installed using direct-push methods. Approximately 40 pounds (dry weight) of ORC was injected as a slurry under pressure into a 2-inch boring from a depth of approximately 12 ft bgs to the ground surface. The four temporary monitoring wells were constructed immediately downgradient of the ORC and were screened from 5 to 10 ft bgs with 0.010-inch slotted screen prepacked with sand.

RESULTS AND DISCUSSION

The results of the DO monitoring are presented in Figure 2. The initial DO concentrations in all wells averaged 0.2 mg/L. The data indicate a rapid increase in DO concentrations, initially in the well closest to the injection point (1065TMW1) and then later down gradient (1065TMW3). This was followed by a period where the DO concentrations in all wells level off at maximum concentrations ranging from 1.60 to 1.86 mg/L. As the ORC was depleted, a gradual decrease in DO to

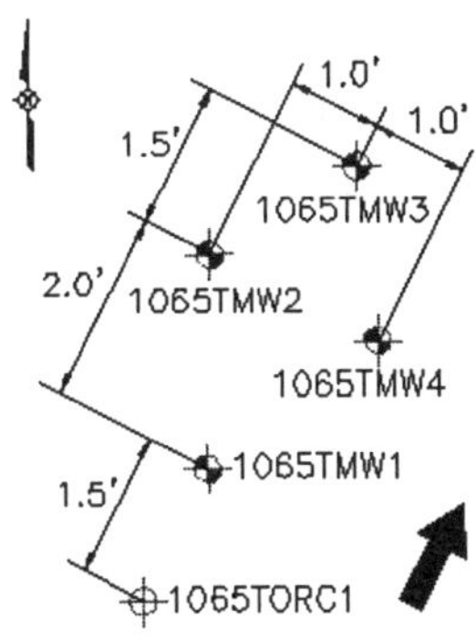

FIGURE 1. Layout of ORC and temporary monitoring wells with groundwater flow direction shown.

concentrations at or below the initial concentrations was observed in all wells. The elevated DO concentrations were observed in the well closest to the ORC for approximately 100 days.

The DO concentrations that were measured downgradient of the injection point and 1 ft crossgradient of the centerline of flow were similar to the concentrations at the centerline measured 1.5 ft further downgradient. This is consistent with an elliptical plume that would be generated by a point source of material (Freeze and Cherry, 1979).

Estimation of Extent of Enhancement. The furthest well from the ORC injection point (1065TMW3) was a distance of 5 ft away. The DO concentrations at this well were at levels high enough to indicate that oxygen had likely transferred farther than 5 ft from the ORC injection point. Assuming that a DO concentration of 1.0 mg/L is sufficient to maintain the biodegradation activity and that the DO distribution is linear from the injection point, the estimated extent of enhancement of biodegradation ranged from 11.4 to 19.2 ft downgradient from the injection point. Also, assuming that the DO plume extending from the ORC injection point would have a width that was half of the length, then the plume ranges from 5.7 to 9.6 ft wide.

These dimensions are somewhat larger than those predicted by fate and transport models that assume similar conditions (Regenesis, 1997). The difference in values predicted by the model and those observed may be due to the assumptions used in the model and actual field conditions. In addition, the increased biological activity due to the introduction of oxygen is expected to increase the rate of oxygen usage and reduce the extent that the oxygen will travel from the injection source (oxygen use up faster than it can be transported).

Estimation of Biodegradation Rates. A biodegradation rate can be estimated from the change in oxygen concentrations over time assuming mineralization of a single compound that represents the contaminant. Using hexane to represent gasoline, and

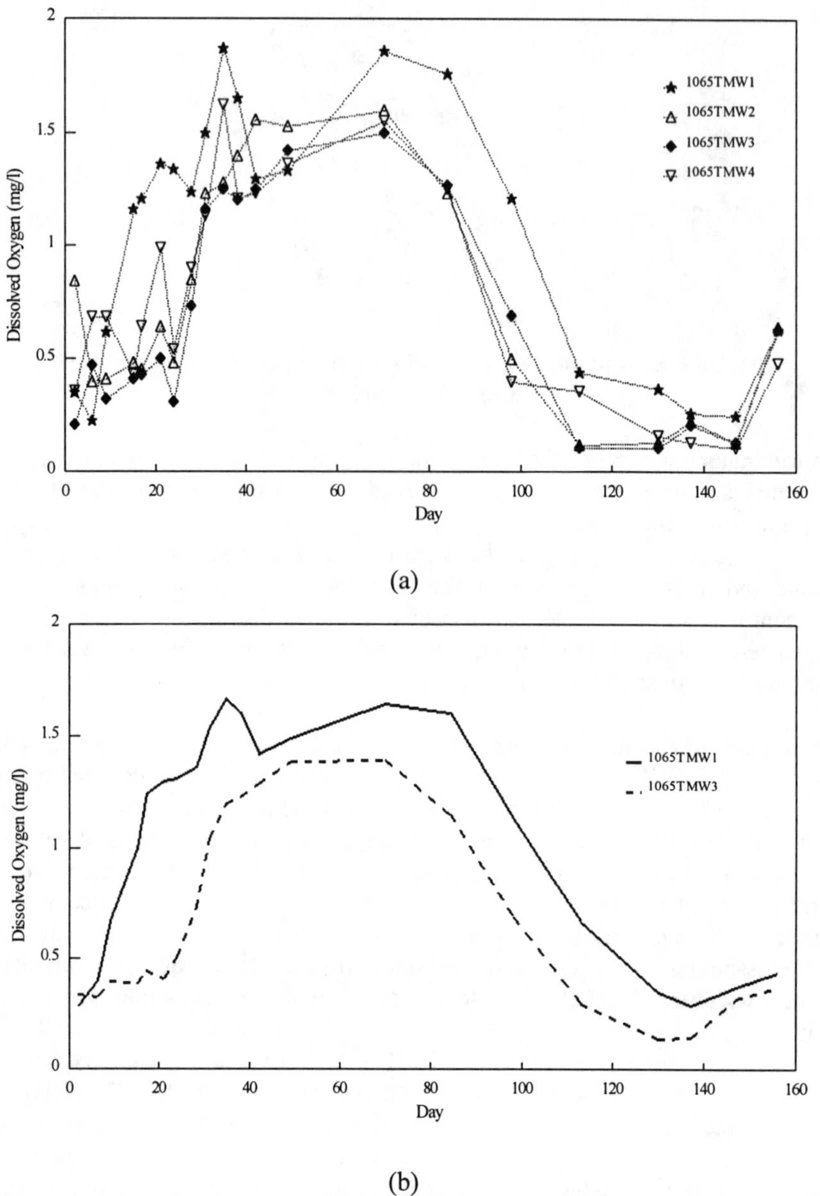

FIGURE 2. Dissolved oxygen concentrations over time. (a) All data. (b) Three point moving average for data from monitoring wells closest to and farthest from the ORC injection point.

calculating a change in oxygen over time, a biodegradation rate can be calculated for site-specific conditions. Based on the stoichiometry of the mineralization of hexane, 3.5 grams of oxygen are required to degrade 1 gram of hexane.

Once the DO concentration in each well reaches its maximum (approximately steady state), as a volume of groundwater flows through the soil matrix, oxygen is lost from the volume through adsorption, dispersion, diffusion and degradation. Adsorption, dispersion, and diffusion losses are assumed to be minimal. The majority of the DO is used by the bacteria to degrade organics. Assuming that the ORC releases oxygen at a constant rate, then at any given time after steady state is reached the difference in DO concentrations of the nearest and farthest well downgradient of the ORC is due largely to biodegradation.

By comparing the time required for the oxygen concentrations to increase in the temporary well closest to the ORC to the time required for the oxygen concentrations to increase in the farthest well downgradient (as can be seen in Figure 2b), an oxygen advective transport rate can be estimated. This assumes that the biodegradation is minimal initially (before bacteria can acclimate to the elevated DO concentrations). The time for the DO concentration in 1065TMW3 to rise to the concentrations in well 1065TMW1 averaged 16 to 18 days. With a distance of 5 ft between the sample points, the advective transport rate ranged between 0.31 and 0.28 ft per day. These values are consistent with the groundwater seepage velocity estimated for the site (0.32 ft per day).

Using the DO data from the study, the biodegradation rates were estimated to range from 1.7×10^{-3} mg/L·day^{-1} initially to 8.3×10^{-3} mg/L·day^{-1} when the dissolved oxygen concentrations were at the maximum levels.

As the study progressed, the biodegradation rates increased when the DO concentrations were at the maximum, presumably as biological activity increased. The biodegradation rates then decreased as the ORC became depleted and the DO concentrations decreased. A sight time lag was observed between changes in DO concentrations and changes in biodegradation rates, presumably due to the time necessary for the bacteria to acclimate to the new conditions.

CONCLUSIONS

This study demonstrates a straight forward method for providing information on the effectiveness of the use of ORC under site-specific conditions. The method is similar to those used for bioventing (Hinchee et al., 1992). The estimates of extent of enhancement and biodegradation rates were used to prepare a conceptual design for full-scale implementation. The data were also used to develop a petroleum hydrocarbon fate and transport model for site remediation to estimate the duration and costs of remediation.

For this study it was assumed that, due to the high concentrations of petroleum hydrocarbons where the study was conducted, the reduction in concentrations from a single injection of ORC could not be effectively measured. If these concentration changes could be measured, the changes over time could be used to compare to the biodegradation rate that was calculated.

The study could be further refined by adding a tracer material to the ORC to better estimate the losses due to adsorption, dispersion, and diffusion. In addition, as a control, a second ORC injection point could be placed outside the contaminated plume with monitoring wells to estimate the use of oxygen in the degradation of non-petroleum organics that might be present in the soil matrix. This information could be used to refine the biodegradation rates of the petroleum hydrocarbons alone, as well as the estimated quantities of ORC required and the time required to remediate the site.

ACKNOWLEDGMENT

The installation of the ORC was completed as part of a workshop sponsored by Regenesis, Geoprobe® Systems, Fast-Tek Engineering Support Services, IT Corporation, and the U.S. Army Corps of Engineers, Sacramento District. Also, special thanks to Dr. Robert Ellgas and Rachel Hess for their review and comments on the manuscript.

REFERENCES

Freeze, R. Allan and John Cherry. 1979. *Groundwater*. Prentice-Hall. New Jersey.

Hinchee, R. E., S. K. Ong, R. N. Miller, D. C. Downley, and R. Frandt. 1992. *Test Plan and Technical Protocol for a Field Treatability Test for Bioventing*. Prepared for the U.S. Air Force, Center for Environmental Excellence.

Regenesis. 1997. *Oxygen Release Compound, ORC® Computer Modeling Results for an Iterative Cut-Off Design*. Technical Bulletin #4-1.3.

ACTUAL GROUNDWATER AND SOILS REMEDIATION AT A PETROLEUM STORAGE TERMINAL

Craig Dockter, ENTRIX, Inc., Houston, TX, USA
Daniel Ransbottom, ENTRIX, Inc., Houston, TX, USA

Abstract: The site is a 180-acre products pipeline terminal containing four major areas of petroleum-affected soils and groundwater. Three of these areas currently contain a measurable amount of free-phase petroleum (product) on the shallow groundwater. This paper focuses on the results of two air sparging/soil vapor extraction (AS/SVE) pilot studies, four full-scale AS/SVE systems, and groundwater recovery from the three areas containing product. The AS/SVE technology injects clean air into the groundwater, stripping volatile organics as the bubbles rise through the groundwater and into the unsaturated (vadose) zone. The vapor extraction system then removes the petroleum-affected vapors from the vadose zone. The four AS/SVE act as barriers by treating dissolved petroleum in situ prior to migrating off site, while the groundwater recovery wells are removing product from the source areas. The contaminants of concern (COCs) identified during a Remedial Investigation (RI) conducted at the site are benzene, toluene, ethylbenzene and xylenes (BTEX) and methyl-*tert*-butyl-ether (MTBE).

INTRODUCTION

Historical operations at the products pipeline terminal have resulted in the release of petroleum products into subsurface soils and shallow groundwater. Several groundwater investigations have been conducted identifying four major areas on the terminal containing petroleum-affected soils and groundwater. Three of these areas currently contain a measurable amount of free-phase petroleum (product) on the surface of the first water-bearing zone. These three areas include the Western Terminal Boundary, the Manifold Area, and the Northern Terminal Boundary. The fourth area of significant petroleum-affected soils and groundwater is located along the Southern Terminal Boundary.

Initial investigations identified the primary COCs as BTEX. MTBE was later identified after pilot studies had been conducted.

Local groundwater is generally encountered from 4 to 18 feet below the ground surface in a fine to medium sand, under unconfined conditions. The underlying confining layer consists of silt at depths ranging from 13 to 40 feet.

PILOT STUDY DISCUSSIONS AND RESULTS

AS/SVE pilot studies were conducted to determine if a full scale AS/SVE system would be effective and feasible for the remediation of petroleum-affected groundwater at the site. This was accomplished through the collection and interpretation of chemical, biological, geologic water level, vacuum, and pressure data from an on-site pilot study. The pilot study was conducted in December 1993, and the results of the pilot study are presented in the following paragraphs.

Based on information collected during previous site investigations, it was necessary to conduct pilot studies in two areas in order to evaluate the effects of the system performance on stratigraphy unique to each area. These two areas are referred to as the Western Terminal Boundary and the Manifold Area.

Each pilot study location consisted of three AS wells (SP-1, SP-2, and WP-1), two SVE wells (SVE-1 and SVE-2) and three monitor wells (MW-1, MW-2, and MW-3) arranged as shown in Figure 1. Independent AS and SVE pilot tests were performed to predict the most effective flow rates for the 96-hour combined pilot study. BTEX samples were collected as part of the combined AS/SVE pilot study to establish the reduction in dissolved contaminants.

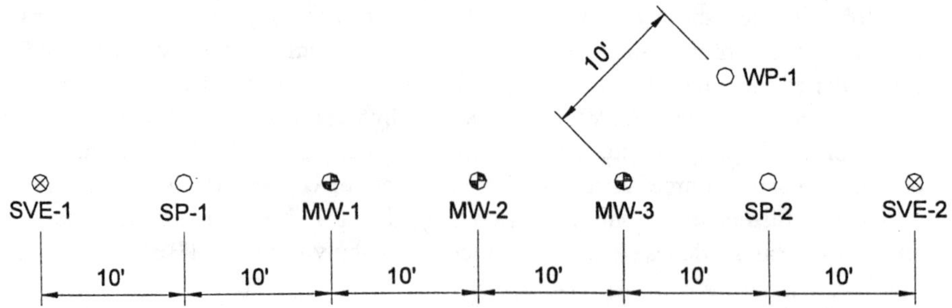

LEGEND:

⊗ SOIL VAPOR EXTRACTION WELLS

○ SPARGE WELLS

MONITOR WELLS

FIGURE 1. Pilot Study Layout

Independent AS / SVE Initial Pilot Tests Results. AS points SP-1 and SP-2 consisted of 3/4-inch and 2-inch slotted PVC, respectively, and were installed with a hollow-stem auger rig. The third air sparge point (WP-1) was installed by driving a 1.25-inch well point to the appropriate depth. These tests were developed to evaluate the well point performance against installing conventional PVC wells with the hope of eliminating soil cuttings from dozens of AS locations during the full-scale system installation. The air sparging radius of influence was determined upon detecting a positive pressure in the monitor well. In addition, direct observations of air bubbles in the three monitor wells confirmed that the air was transported through the groundwater, and not simply creating back pressure at the monitor wells by short-circuiting through the vadose zone.

Both SVE wells (SVE-1 and SVE-2) were constructed of 2-inch PVC slotted screen installed in the vadose zone with a hollow-stem auger rig. A vacuum was applied to the well by a portable vacuum blower and the radius of influence was estimated by monitoring the magnehelic gauges placed on each monitor well and both SVE wells. These observations determined the maximum distance from the SVE well that a negative pressure could be achieved, thereby capturing the vapors produced by the air sparging points.

Sieve analyses performed on samples collected from the Western Terminal Boundary area confirmed the soils consist of fine to medium grained sands from the ground surface to a depth of approximately 35 feet.

The AS performance was monitored at varying flow rates during the 1-hour test. An optimum flow rate of 2.5 actual cubic feet per minute (acfm) established an estimated radius of influence of 20 feet during the independent test and was selected for the combined pilot study. The SVE performance was monitored at various flow rates during the 1-hour test. An optimum flow rate of 10 acfm established an estimated radius of influence of 60 feet from the SVE well during the independent test and was selected for the combined pilot study.

Sieve analyses performed on samples collected from the Manifold Area confirmed the soil consists of silty and clayey fill material from the ground surface to a depth of approximately 15 feet, with fine to medium grained sand material from approximately 15 to 32 feet.

The AS performance was monitored at varying flow rates during the 1-hour test. An optimum flow rate of 1.25 acfm established an estimated radius of influence of 20 feet during the independent test and was selected for the combined pilot study. The SVE performance was monitored at various flow rates during the 1-hour test. An optimum flow rate of 20 acfm established an estimated radius of influence of 10 feet during the independent test and was selected for the combined pilot study.

The results of the independent pilot tests also indicated that it would be feasible to use the 1.25-inch well point for both combined pilot study locations and the full-scale systems.

Combined AS / SVE Pilot Study Results. A 96-hour pilot study was conducted to evaluate the effects of combining the AS and SVE systems at the optimum flow rates at the Western Terminal Boundary and Manifold Area. Air flows were held constant while the pressures, vacuums and water levels were adjusted and monitored over the 96-hour pilot study. BTEX samples were collected from the monitor wells to measure the reduction in dissolved contaminants. MTBE had not yet been identified as a COC and was therefore not tested as part of this pilot study.

The Western Terminal Boundary combined pilot study was conducted by injecting 2.5 acfm into the 1.25-inch AS well point (WP-1) and removing 10 acfm from the SVE well (SVE-1). The analytical results of groundwater samples collected from monitor well MW-1, at a distance of approximately 30 feet from WP-1, are presented in Figure 2. The significant reduction in BTEX concentrations at MW-1 confirmed an AS radius of influence of 30 feet. The negative pressure achieved at SVE-2 (60 feet from SVE-1) confirmed an SVE radius of influence of 60 feet.

The Manifold Area extended pilot study was conducted by injecting 1.25 acfm into the 1.25-inch well point (WP-1) and removing 20 acfm from the SVE well (SVE-1). The BTEX concentrations presented in Figure 3 are the analytical results of groundwater samples collected from monitor well MW-1, at a distance of 30 feet from WP-1. The lack of reduction in BTEX concentrations at MW-1 is apparently due to the presence of free-phase product in the Manifold Area which continuously dissolves into the groundwater. The static pressures achieved at MW-2 (20 feet from SVE-1) suggests an SVE radius of influence of 20 feet.

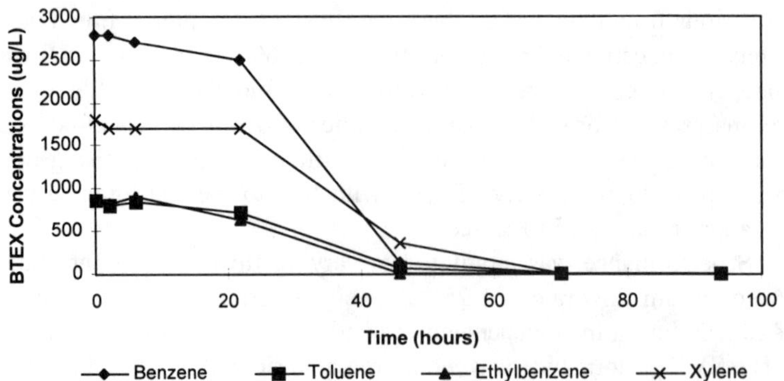

FIGURE 2. BTEX Analytical Results - Western Boundary

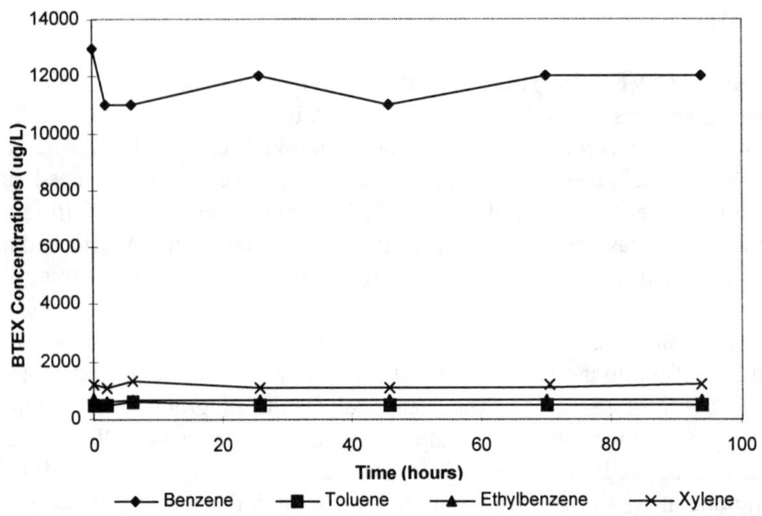

FIGURE 3. BTEX Analytical Results - Manifold Area

INTERIM AS/SVE REMEDIAL SYSTEMS RESULTS

Based on the results of the pilot studies, interim AS/SVE systems were designed, installed, and are effectively operating along the Western, Northern and Southern Terminal Boundaries. At the request of the state agency, the SVE system adjacent to the Manifold Area is operating without the AS system until the free-phase product has been removed from the groundwater.

The Western Terminal Boundary interim AS system consists of 48 wells equally spaced 40 feet apart, and was started in October, 1996. Based on the results of several sampling events, the average AS flow rate has been reduced from 2.5 to 1.5 acfm, at injection pressures ranging from 1 to 25 pounds per square inch (psi). AS locations which correspond to monitor well locations exhibiting more persistent dissolved-phase petroleum have been adjusted to 2.0 acfm. The interim SVE system consists of 15 wells equally spaced 120 feet apart, with three AS wells between each SVE well. The SVE wells have been operating at an average flow rate of 10 to 20 acfm since October 1996.

The dissolved BTEX/MTBE concentrations have been significantly reduced along the Western Terminal Boundary as a result of operating the interim AS/SVE system. The average BTEX and MTBE concentrations ranged from 1 to 10 micrograms per liter (mg/L) prior to startup. These concentrations currently range from 0.00 to 0.020 mg/L for BTEX, and 0.010 to 0.100 mg/L for MTBE, well within acceptable limits.

The Manifold Area interim AS system consists of 18 wells equally spaced 40 feet apart. The AS wells have been installed, but are not currently operating due to the presence of product on the surface of the first water-bearing zone. The interim SVE system consists of 19 wells equally spaced 40 feet apart, with one AS well between each SVE well. The SVE wells have been operating at an average flow rate of 10 to 15 acfm since March 1998.

The dissolved BTEX/MTBE concentrations have not been significantly reduced in the Manifold Area as a result of operating the SVE system. The average BTEX and MTBE concentrations have remained at 10 to 20 mg/L due to the presence of free-phase product on the groundwater surface.

The Northern Terminal Boundary interim AS system consists of 34 wells equally spaced 40 feet apart, and was started in October, 1977. Based on the results of several sampling events, the average AS flow rate has been reduced from 2.0 to 1.0 acfm, at injection pressures ranging from 1 to 25 psi. AS locations which correspond to monitor well locations exhibiting more persistent dissolved-phase petroleum have been adjusted to 1.5 acfm. The interim SVE system consists of 12 wells equally spaced 120 feet apart, with three AS wells between each SVE well. The SVE wells have been operating at 10 to 20 acfm since October, 1997.

The dissolved BTEX/MTBE concentrations have been significantly reduced along the Northern Boundary as a result of operating the interim AS/SVE system. The average BTEX and MTBE concentrations ranged from 1 to 20 mg/L prior to startup. These concentrations currently range from 0.00 to 0.020 mg/L for BTEX, and 0.010 to 0.100 mg/L for MTBE, well within acceptable limits.

The Southern Terminal Boundary interim AS system consists of 28 wells equally spaced 25 feet apart. The average flow established since the startup in March 1998 ranges from 2.0 to 3.0 acfm at pressures ranging from 1 to 25 psi. The closer well spacing and higher air flow rate is required to reduce the less-strippable MTBE concentrations along the Southern Terminal Boundary. The interim SVE system consists of 15 wells equally spaced 50 feet apart, with two AS wells between each SVE well. The SVE wells have been operating since March 1998 and have been effectively removing BTEX, and to a lesser extent, the MTBE. The average flow at each well averages from 10 to 15 acfm.

The plume along the Southern Terminal Boundary consists primarily of the less-strippable MTBE. The average BTEX and MTBE concentrations ranged from 0 to 20 mg/L prior to startup. These concentrations currently range from 0.00 to 0.010 mg/L for BTEX, and 0.010 to 1 mg/L for MTBE. While the reduction in dissolved BTEX/MTBE concentrations is less significant than in other areas of the terminal, they are well within acceptable limits.

INTERIM GROUNDWATER RECOVERY SYSTEMS

Free-phase product has been detected in three of the four areas of the terminal in which interim AS/SVE systems are installed. Twenty-one pneumatic pumps were installed in 14 recovery wells from August 1997 through December 1998 in the following locations: 1) Western Terminal Boundary (3 wells); 2) the Manifold Area (8 wells); and 3) the Northern Terminal Boundary (3 wells). Free-phase product has not been detected along the Southern Terminal Boundary.

The recovered groundwater is pumped from each area to a central oil/water separator, and treated by an air stripper before being discharged through a National Pollutant Discharge Elimination System (NPDES)-permitted outfall. The recovered free-phase product is recycled.

The groundwater system has treated over 13,000,000 gallons of groundwater and has recovered 13,000 gallons of free-phase product. The system is currently recovering 35 gallons per minute (gpm) of groundwater and producing approximately 50 gallons per day of free-phase product.

CONCLUSIONS

The successful operation of the four full-scale AS/SVE systems has justified the decision to conduct quality pilot studies to ensure in situ performance. To account for in situ heterogeneity, the most conservative assumptions were used in designing the full-scale systems. It has also been demonstrated that AS/SVE alone does not work for all conditions, such as in the presence of free-phase product. However, when properly applied, in situ AS/SVE systems are extremely effective in removing dissolved BTEX, and with additional air flow and closer well spacing, they can be effective in reducing MTBE. The in situ full-scale systems have proven to reduce the BTEX/MTBE concentrations to acceptable limits along the terminal perimeter prior to migration of the groundwater off site.

IN SITU BIOREMEDIATION OF A SOIL CONTAMINATED BY MINERAL OIL: A CASE STUDY

D. Pinelli, ***M. Nocentini,*** and F. Fava (University of Bologna, Bologna, Italy)

ABSTRACT: The present paper shows how a limited set of well-focused laboratory tests can help in selecting a correct remediation technology, which can result in better resource utilisation. The main contaminant of the considered soil was mineral oil; the contamination involved only the superficial soil. Our department was commissioned to carry out laboratory tests to study the feasibility of a bioremediation treatment. Our job included process design and the evaluation of possible, additional enhancing treatments (such as supply of nutrients, addition of an inoculum, soil pH correction, etc.). The results of the tests showed very low contaminant depletion rates. All the attempts to increase the bioremediation rate failed. This fact suggested that pollutant desorption and transfer phenomena were probably involved as rate-limiting steps. On the basis of qualitative considerations of the risk associated with the contamination, an intrinsic bioremediation approach was suggested to the local environmental agency.

INTRODUCTION

It is well-known that bioremediation can be effectively applied to reclaim soils contaminated by complex mixtures of hydrocarbons. Nevertheless, it is sometimes reported that a residual fraction of contaminants remains undegraded in the soil even when optimal biodegradation conditions are provided (Huesmann, 1997). This does not mean that the "residual" part of the pollutants cannot be biodegraded further, but only that the kinetics of the process have changed so that biodegradation proceeds in a much slower way. One of the most widely accepted hypotheses to explain this phenomenon is that pollutant transfer phenomena are probably involved as a rate-limiting step (Williamson et al., 1997). Under these circumstances, the residual concentration depends mainly on the soil characteristics (Allen et al., 1997) and contamination aging (Williamson et al., 1997).

The present paper reports on experimental laboratory work done to support the design of a bioremediation project in which the "residual concentration" is higher than the regulatory concentration limit. The sediments of a channel were excavated and were placed on the ground beside the channel. The presence of pollutants in the excavated sediments caused the contamination of the first 20-30 cm of the soil for a width of about 3 m and a length of about 2 km. The sediments contained a complex mixture of organic substances due to the repeated, illegal disposal of craft and domestic wastes. According to a first characterisation of the pollutant mixture carried out by the local environmental agency, no particularly toxic substances were present. The main constituent of the mixture was found to be mineral oil in concentrations from 1,000 to 5,000 mg/kg – the regulatory concentration limit being 100 mg/kg.

Among the applicable reclamation technologies, bioremediation was considered an economical alternative to soil excavation and disposal in a landfill. Our department was commissioned to carry out laboratory tests and to study the feasibility of the treatment. The experimental work was conducted in three steps. The first phase of the work was exploratory in nature and comprised slurry condition microcosm tests. The second phase consisted of solid-phase microcosm tests meant to evaluate the effectiveness of the treatment and to estimate the time needed to reach the target level of 100 mg/kg fixed by the local environmental agency. The third phase consisted of helping the remediation company to select the appropriate technology, of choosing the strategy for its application in the field and of monitoring the site during treatment.

Due to the limited space available, the present paper summarises the results obtained and focuses on how they were used to make final decisions about the remediation strategy. A more complete presentation of the results of the project will be published in the future.

MATERIALS AND METHODS

Contaminated Soil. The tests were conducted using three different soils sampled after about one year from the initial site investigation: i) a soil named S2 with an average contaminant concentration (420 mg/kg) which was obtained by mixing five different samples taken in the contaminated area at different positions along the channel side in the zone where the maximum contamination had been observed, ii) a soil named S1 coming from a single position exhibiting the lowest oil concentration (250 mg/kg), and iii) a soil named S3 coming from a single position exhibiting the highest mineral oil concentration (920 mg/kg). The soil used in this study was of a silt-loam texture (clay $13_{wt}\%$, silt $72_{wt}\%$, sand $15_{wt}\%$). The contaminated soil was characterised as follows: total organic carbon (TOC) = $3.0_{wt}\%$, total N = 1.5 g/kg, total P = 2.4 g/kg, pH = 7.6, field capacity = $37_{wt}\%$. The total heterotrophic bacterial viable biomass was about 10^{-6} colony forming units (CFU)/g dry soil. Before setting up the microcosms, the soil samples were preliminarily air dried and sieved on a 2-cm sieve.

Microcosms and Their Set Up. Two parallel sets of experiments in microcosms were performed, namely: i) 13 slurry condition microcosms consisting of 3-L Erlenmayer flasks with baffles loaded with about 300 g of soil suspended in 1.5L of water, incubated at 30°C and constantly agitated at 120 rpm, and ii) 28 solid-phase microcosms consisting of 1-L bottles loaded with about 300 g of contaminated soil, incubated at 30°C, and periodically agitated (hand shaken). The microcosms were different from each other as per nutrients concentration (NH_4NO_3 as N, K_2HPO_4 and KH_2PO_4 as P), temperature, addition of exogenous micro-organisms specialised in degrading heavy molecular-weight hydrocarbons (four commercial mixtures), soil moisture (only in solid-phase microcosms). In addition, some microcosms were modified after a set period of treatment by introducing additional hydrocarbon mixtures (kerosene, motor oil and a concentrated mixture of the pollutants solvent-extracted from the contaminated soil). Finally, the addition of surfactants (Triton X100) was also tested – in concentrations corresponding to half the CMC (critical micellar concentration) and double the CMC – in order to increase the bioavailability of the heavier, less water-soluble contaminant fraction. The microcosms were monitored by periodically determining contaminant concentration, O_2 and CO_2 concentrations, and total viable bacterial heterotrophic biomass concentration. The collected data were used to calculate degradation yields, contaminant consumption rates, O_2 consumption and CO_2 formation rates, and to estimate mineralisation yields.

RESULTS

Results for the Slurry Condition Microcosms. The values of average depletion rates and depletion yields for the slurry condition microcosms are reported in Table 1. The depletion rates were calculated by linear interpolation of the concentration data. The data reported in table 1 show very low consumption rates in the microcosms in which exogenous specialised micro-organisms, N and P as nutrients were added (ID. No. 2,4,5,6,7). Nevertheless, some degradation was obtained in 8 months of treatment – irrespective of which additional treatment was applied.

Table 1 - Slurry condition microcosms: average mineral oil depletion rate and yield.

ID. No.	Soil Type and Additions	Averaging Period *(d)*	Average Rate *(mg/kg/d)*	Average Yield *(%)*
2,4,5,6,7	S2 - average values - (initial conc.: 420 mg/kg)	0-235	0.8	22
1	S1 (initial conc.: 250 mg/kg)	0-39	< 0.5	<10
3	S3 (initial conc.: 920 mg/kg)	0-235	1.1	30
1,8	S1 & S2 + motor oil (ca. 4200mg/kg)	39-123	19	36
	(1,8 average values)	123-260	2.8	15
10	S2 + contaminant mixture (2830 mg/kg)	0-100	9.4	34
		100-207	7.7	46

Since the nature of the contaminants was only partly known and a detailed characterisation of the mixture was beyond the limits imposed on our experimentation, some specific tests were conducted to verify the absence of inhibitory phenomena due to the possible action of toxic substances. Kerosene addition (slurry No. 5 after 25 days of treatment, 2900 mg/kg) was used to stimulate the hydrocarbon degrader microorganisms present in the soil, as this is a mixture of hydrocarbons which is well known to be readily biodegradable in the soil (Nocentini et al., 1997). After only 11 days, the kerosene concentration had declined to 110 mg/kg, corresponding to a depletion yield of about 96%. After kerosene addition, the O_2 consumption rate increased considerably (Figure1); but after 20 days it returned to the initial value. In the same period, no appreciable increase in the mineral oil depletion rate was observed. Furthermore, unused 10W-30 motor oil (4200 mg/kg) and the contaminant mixture extracted from the S2 soil (2830 mg/kg) were added to some of the microcosms (slurry Nos. 1, 8 and 10) in order to reveal the actual remedial capabilities of the microorganisms present in the soil. The results in terms of consumption rate and yield are reported in Table 1. The addition of the motor oil yielded a very high initial depletion rate, though after 100 days the rate decreased considerably (at this point, 36% of the added oil was degraded). On the other hand, the addition of the contaminant mixture yielded a high initial depletion rate that remained roughly constant up to the end of the test after 200 days (final concentration about 1000 mg/kg). As regards respirometric measurements, the motor oil addition yielded a strong increase in O_2 consumption and CO_2 formation rates (the former passing from less than 1 mmol/kg/d before the addition to about 15 six days after the addition); however, these higher values rapidly diminished in less than another 50 days. The total biomass concentration increased considerably after the addition, reaching values two orders of magnitude higher than the initial ones (from 1E+06 to 1E+08 CFU/g dry soil) in about 150 days. Even higher rates were obtained with the addition of the contaminants extracted from the S2 soil – especially in terms of O_2 consumption rate, which reached values of about 60 mmolO_2/kg/d.

The rapid and vigorous response obtained after the addition of different mixtures of hydrocarbons demonstrated the absence of inhibitory effects by unknown compounds possibly present in the soil; it also showed that the microorganisms present in the soil were well equipped and adapted to degrade hydrocarbon mixtures.

Table 2 - Solid-phase microcosms: average mineral oil depletion rates and yields.

ID. No.	Soil Type and Additions	Averaging Period *(days)*	Average Rate *(mg/kg/d)*	Average Yields *(%)*
11	S1 (22% soil moisture)	0-293	<0.5	<10
1	S2 (19% soil moisture)	0-263	0.8	51
2	S2 (22% soil moisture)	0-263	0.7	41
3	S2 (24% soil moisture)	0-263	0.3	20
4	S2 (27% soil moisture)	0-263	0.4	24
12	S3 (22% soil moisture)	0-293	1.2	48
7	S2 + motor oil (2750mg/kg) (22% soil moisture)	0-142	8.5	24
8	S2 + contam. mix. (2860mg/kg) (22%soil moist.)	0-142	12.0	33

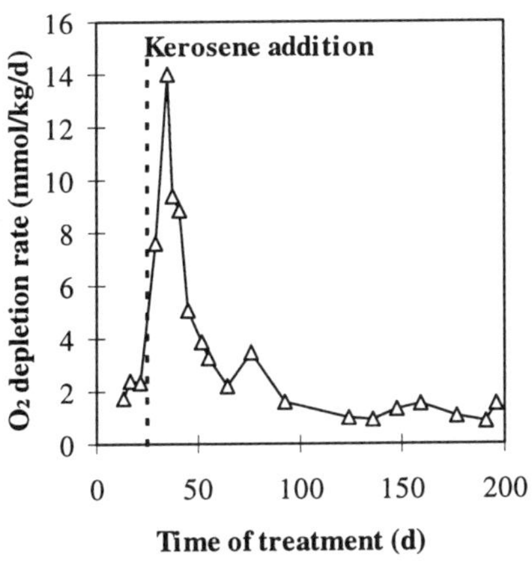

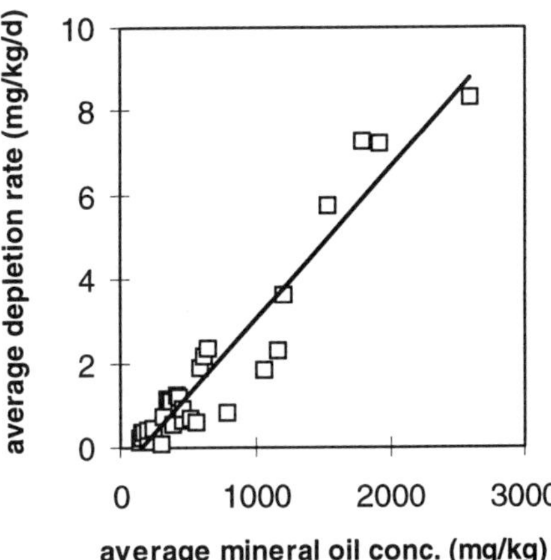

Figure 1 - Time evolution of the O_2 depletion rate in the slurry No. 5. Addition of kerosene (2850 mg/kg) after 25 days.

Figure 2 – Depletion rate dependence on mineral oil concentration; in-field monitoring data.

Results for the Solid-Phase Microcosms. Several tests were carried out in order to study the same parameters that were investigated in the slurry conditions. The results obtained were comparable with those obtained in the slurry tests. The most important results are summarised in Table 2. As for the tests with the addition of mineral oil and the contaminants extracted from S2 soil, the same effects described for the slurry condition tests were also observed – though less pronounced.

Furthermore, in the solid-phase microcosms the soil moisture was varied in the range between 19 and $29_{wt}\%$. A remarkable influence of this parameter was observed. It should be noted that the depletion rate decreases with increasing the soil moisture – contrary to what would occur if the biological process were the rate limiting step.

Site Monitoring Results. The third part of our project comprised site monitoring. In particular, two sampling campaigns were conducted over the course of two years. The results are reported in Figure 2 in terms of depletion rate dependence on average oil concentration. It should be noted that the observed trend is compatible with a first order kinetic model. Nevertheless, Figure 2 reveals that the intercept of the interpolation line with the x-axis is different from 0. This would mean that – for a certain concentration (about 200 mg/kg) – a nearly zero depletion rate is expected. This observation agrees with what was found in the microcosm tests and has been reported above.

DISCUSSION

All attempts to increase the natural bioremediation rate failed. The addition of exogenous specialised microorganisms yielded no appreciable increase in performance, even when a second massive inoculation was done (3-5 times larger that the first). The experimental data collected in the microcosm tests showed that the depletion rate becomes much slower as a residual concentration of pollutants of about 500 mg/kg is reached. A possible explanation for this observation is that this residual contamination part is hardly bioavailable to the microorganisms that, in turn, proved to be very active in utilising hydrocarbons as a carbon source. It should be noted that the fast degradation of the contaminant mixture extracted from the S2 soil seems to exclude the intrinsic recalcitrance of the pollutants as an alternative explanation (Huesmann, 1997). The degradation that still occurs in the soil seems not to respond to any typical biological parameter. This fact suggests that pollutant desorption and other transfer phenomena are probably involved as a rate-limiting step.

In view of this, a different approach to the problem of site decontamination was adopted. Attention was focused on the estimation of the risk associated with the contamination. Some experiments were conducted to estimate the potential diffusion of the pollutants into the environment. In particular, some additional analyses of soils sampled in different positions at a depth of 50 cm (about 20 cm under the contaminated layer) were carried out. The results showed that no contamination of the subsoil had occurred, although about two years had passed from the original contamination event.

On the basis of these results and considerations, an intrinsic bioremediation approach was suggested to the local environmental agency. A time extent longer than two years was estimated as necessary to reach acceptable concentration limits. Monitoring of the site is still in progress.

REFERENCES

Allen, K.A., B.E. Herbert, P.J. Morris, and T. J. McDonald. 1997. "Biodegradation of Petroleum in two Soils: Implications for Hydrocarbon Bioavailability". In *"In-situ and On-Site Bioremediation: Volume 5"*; pp. 629-634, Battelle Press, Columbus, OH.

Huesmann, M.H. 1997. "Incomplete Hydrocarbon Biodegradation in Contaminated Soils: Limitations in Bioavailability or Inherent Recalcitrance?" *Biorem. J. 1*(1): 27-39.

Nocentini, M., D. Pinelli, G. Pasquali, F. Fava, and A. Prandi. 1997. "Biotreatability and Feasibility Studies for a Bioremediation Process of a Kerosene Contaminated Soil". *Proceedings of International Symposium Environmental Biotechnology*, Part II: 307-310, Technologisch Instituut, Antwerpen, Belgium.

Williamson, D.G., R. C. Loehr, and Y. Kimura. 1997. "Measuring Release and Biodegradation Kinetics of Aged Hydrocarbons from Soils". In *"In-situ and On-Site Bioremediation: Volume 5"*, pp. 605-610, Battelle Press, Columbus, OH.

ENHANCED AEROBIC BIOREMEDIATION OF PETROLEUM UST RELEASES IN PUERTO RICO

Baxter E. Duffy, Inland Pollution Services, Inc., Elizabeth New Jersey
Gil Oudijk, Triassic Technology, Inc., Princeton, New Jersey
John H. Guy, Inland Pollution Services PR, Inc., Bayamón, Puerto Rico

ABSTRACT: Releases of petroleum compounds from underground storage tanks (USTs) containing diesel fuel and gasoline are a serious and increasingly prevalent problem in Puerto Rico. Remediation of groundwater impacted by these releases utilizing common methodologies can be quite costly and require long periods of time to implement. Oxygen is often the limiting factor retarding optimal biodegradation of petroleum compounds in groundwater. The delivery of dissolved oxygen to impacted groundwater zones will enhance the biodegradation of dissolved phase benzene, toluene, ethyl benzene and xylene (BTEX) and total petroleum hydrocarbon (TPH) compounds by microorganisms which occur naturally in the subsurface. Oxygen Release Compound (ORC®) is an efficient and effective method of supplying dissolved oxygen directly to the impacted groundwater. In a pilot scale study, ORC was delivered to subsurface groundwater bearing zones of a gasoline service station in Puerto Rico by slurry injection with direct push equipment and by ORC socks. Concentrations of BTEX and TPH were found to decrease inversely to the increase in dissolved oxygen. Concentrations of BTEX were reduced by 89 to 90% and TPH concentrations were reduced by 75 to 96% within eight weeks of installing ORC into the subsurface water bearing zones.

INTRODUCTION

BTEX and TPH contamination in groundwater caused by leaking USTs is a significant problem in Puerto Rico. The objective of this pilot study is to demonstrate the effectiveness of the use of ORC in increasing dissolved oxygen levels in groundwater thereby enhancing the natural biodegradation of BTEX compounds and petroleum hydrocarbon contaminants. The goal is to accomplish the introduction of ORC and effect the enhancement of natural biodegradation in a passive manner without altering groundwater flow or the installation of mechanical equipment.

Site Description. The site selected for the implementation of the pilot study is a Texaco Puerto Rico, Inc., Texaco Service Station, located in Toa Baja, Puerto Rico. There are three active underground storage tanks (USTs) at the operating service station. These USTs include two twelve thousand gallon capacity gasoline and one ten thousand gallon capacity diesel fuel tanks. USTs previously in service at this location were confirmed to have discharged gasoline and diesel fuel at the site.

The site is primarily flat and level lying at approximately 33 feet above Mean Sea Level. There are no drainage features within the boundary of the site. The site lies on alluvial deposits of Pleistocene age consisting primarily of silt and clay with some sand and gravel within the Rio Plata River's alluvial valley. During site activities, groundwater was observed to fluctuate from as shallow as 9.05' to 16.32'. The average groundwater level was 12.77'.

MATERIALS AND METHODS

Two pilot scale test plots were established at the site. ORC was installed in an existing monitoring well utilizing ORC socks and in an "Oxygen Barrier" array at a second existing well location. At the location of existing well MW-1, a 2" diameter well 20' in total depth with 15' of screened interval, was fitted with seven, 2" diameter ORC Socks. A 1" diameter well point was installed down gradient from MW-1. At the location of existing well MW-4 a passive barrier of ORC was installed as a slurry by direct push techniques. A 1" diameter well point was installed between MW-4 and the oxygen barrier.

ORC is a patented formulation of phosphate intercalated magnesium peroxide, MgO_2, produced by Regenesis Bioremediation Products, which when hydrated releases oxygen slowly. The hydrated product is magnesium hydroxide, $Mg(OH)_2$. The basic chemistry of ORC is as follows: ORC is magnesium peroxide, "oxygenated magnesia", which gives off oxygen upon contact with water (at 3% moisture content). The magnesium peroxide is converted into magnesium hydroxide, ($Mg(OH)_2$) as oxygen is released. This is also the fate of the magnesium oxide, which hydrates to form the hydroxide. The reactions are:

$$MgO_2 + H_2O \rightarrow 1/2\ O_2 + Mg(OH)_2\text{; and}$$
$$MgO + H_2O \rightarrow Mg(OH)_2$$

Therefore, the uniform end point of ORC, from both compounds, is magnesium hydroxide.

Existing Well ORC Sock Application. At MW-1, the existing 2" PVC monitor well was utilized and fitted with ORC socks. A monitoring point was installed approximately 3 feet from the well downgradient and designated as GP-1 (Figure 1). This monitoring point was constructed of 1" diameter schedule 40 PVC with .010" slotted screen installed in the aquifer to a depth of 16.3'.

Seven 2" socks were installed into the existing well MW-1. Each 2" sock is 12" in length and contains approximately 0.75 pounds of ORC. The total weight of ORC installed is approximately 5.25 pounds. Only those socks, which are suspended in water and hydrated, will provide dissolved oxygen to the subsurface.

Oxygen Barrier Installation. During the implementation of the ORC installation at MW-4, a series of five soil borings were installed to a depth of 19'. These borings lie along an arc aligned at an orientation as to surround the subject well from the source area at a distance of ten feet on center in relation to the well (Figure 1). These borings were designated as the oxygen barrier. ORC was injected into the borings as a slurry, composed of 150 pounds of ORC powder and approximately 42 gallons of potable water divided evenly across the five boring locations. The slurry was pressure grouted into the borings from 14 to 19' below ground surface into groundwater utilizing direct push drilling equipment. A monitoring point was installed approximately 5 feet from the well between the well and the oxygen barrier and designated as GP-3 (Figure 1). This monitoring point was constructed of 1" diameter schedule 40 PVC with .010" slotted screen installed in the aquifer to a depth of 19'.

FIGURE 1. SITE PLAN

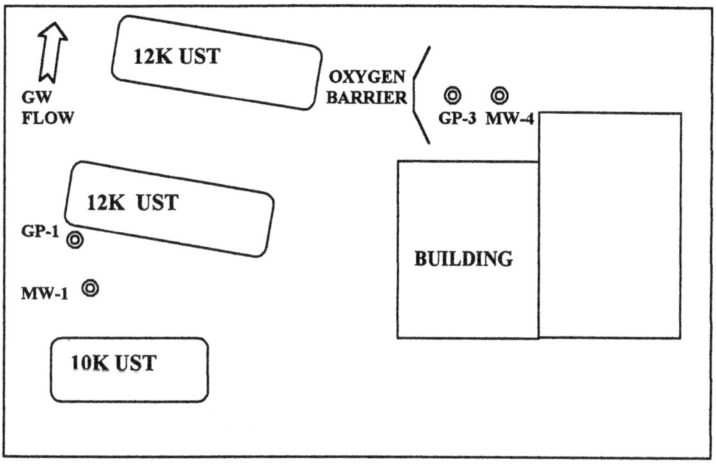

GROUNDWATER SAMPLING AND ANALYSIS

In October of 1997, the study was initiated by collecting base line samples. Groundwater samples were collected from wells MW-1 and MW-4 along with GP-1 and GP-3. The reaction of BTEX, DO and CO_2 in groundwater at the monitoring point locations was monitored every two weeks and at the conclusion of the study (twelve weeks). TPH and pH were analyzed for at the base line sampling and final sampling events.

Groundwater grab samples were collected from the existing 2" monitor wells and 1" Geoprobe well points. Groundwater samples collected were analyzed for Total Petroleum Hydrocarbons (method 8015M), total BTEX (method 8020), DO (method 360.1), Alkalinity (method 310.2) (for CO_2), and pH (method 150.1).

The base line sampling was conducted in October of 1997. The six subsequent monitoring events were performed from November of 1997 to February of 1998 collecting samples every two weeks. The results for BTEX and DO are shown on Figures 2 and 3.

RESULTS AND OBSERVATIONS

At MW-1, base line sampling indicated concentrations of BTEX and TPH of 10.37 and 47.06 ppm respectively. No free phase product was observed in MW-1. At the location of existing well MW-4, base line sampling indicated BTEX and TPH concentrations of 31.54 and 1,774.39 ppm respectively. No free phase product was observed in MW-4. On November 5, 1997, IPSI re-mobilized to the site to install the ORC in the test areas.

FIGURE 2
DISSOLVED OXYGEN CONCENTRATIONS
(IN PPM)

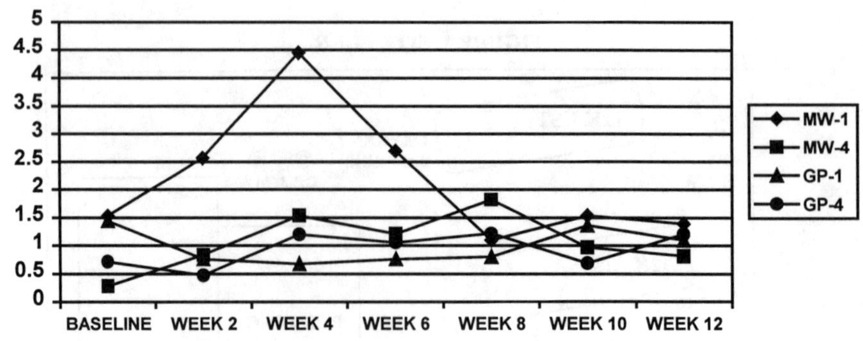

FIGURE 3
BTEX CONCETRATIONS (IN PPM)

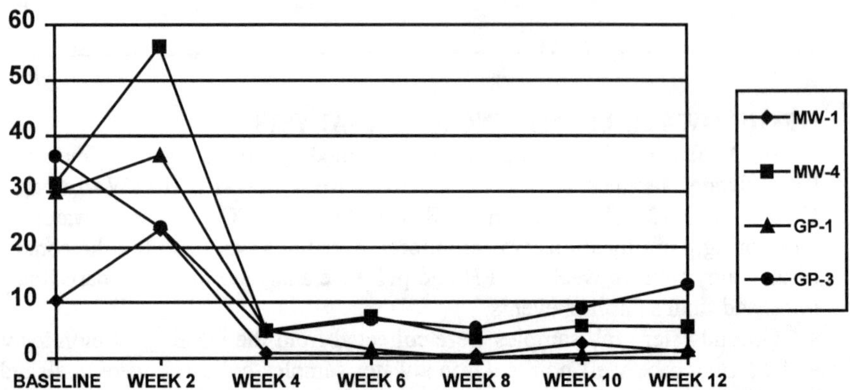

Benzene, Toluene, Ethyl Benzene and Total Xylenes. BTEX were tracked as the prime indicator at all monitoring points as this is commonly a wide spread contaminant on gas station sites and therefore easily observed. The results indicate a sharp rise in BTEX concentrations in the first monitoring event conducted two weeks following the installation of ORC. This trait has been observed by Regenesis on many other sites and may be attributed to mobilization of contaminants during the injection process and biosurfactant production by the microbes caused by the sudden increase in the release of enzymes brought on by the proliferation of the native bacteria. The biosurfactant enzymes will act to desorb organics from the saturated soil matrix. If desorption rates are greater than the rate of biodegradation, an increase in dissolved BTEX concentrations will be observed. BTEX concentrations then decreased significantly in the following sampling event as the bioactivity increases. This response is exaggerated in the tropical environment by higher mean groundwater temperatures.

The BTEX concentrations continued to fall to the week eight sampling event. The dissolved oxygen levels also drop with the increased bacterial activity. At week ten, the levels begin to show a small rise in BTEX and dissolved oxygen indicating the beginning of the die off of the bacterial population stimulated by the addition of the ORC.

The overall reduction of BTEX compounds at MW-1 from base line sampling to week eight is 9.87 ppm from 10.37 to 0.5 ppm a reduction of 95%. The overall reduction of BTEX compounds at MW-4 from base line sampling to week eight is 27.93 ppm from 31.54 to 3.61 ppm a reduction of 89%.

Dissolved Oxygen. Dissolved oxygen is an important indicator of the effectiveness of the ORC itself as a source of dissolved oxygen, as well as the method of installation in terms of the dispersion of the ORC material. Dissolved oxygen was analyzed for and tracked throughout the study at all test locations and monitoring points. In the tropical environment, background DO levels are normally low as the solubility of oxygen in water decreases with higher temperatures. The results show the DO rising through weeks two and four from 1.53 to 4.45 ppm in MW-1 and 0.28 to1.54 ppm in MW-4. The DO is then observed declining with the BTEX concentrations through week eight as the DO is consumed with the available BTEX (1.54 ppm at MW-1 and 0.97 ppm at MW-4). By week ten the DO begins to rise back near background concentrations with the die off of the bacteria population.

TPH, pH and CO_2. TPH compounds were analyzed as a base line indicator prior to the installation of ORC and at the close of the study to observe the effectiveness of ORC treatment on these compounds. The TPH concentrations responded to the increase in dissolved oxygen and resultant increase in microbial activity with decreasing concentrations. The concentrations of TPH dropped from 47.06 to 11.59 ppm at MW-1 from base line to week ten. a reduction of 75%. The concentrations of TPH dropped from 187.85 to 37.87 ppm at MW-4 from base line to week ten, a reduction of 80%. The concentrations of TPH dropped from 1,774.39 to 63.6 ppm at GP-3 from base line to week ten, a reduction of 96%.

pH was analyzed for as a base line indicator prior to the installation of ORC, at week four and at the close of the study to observe the impact of ORC treatment on pH values in the environment. The change in pH levels ranged from approximately 0.5 to 1.5 across the site during the study.

Carbon dioxide (CO_2) is a by-product of microbial activity and is an indicator of the increase in such activity. As the microbial degradation increases, so should the concentrations of CO_2. To determine CO_2 , Alkalinity in groundwater was analyzed for and tracked throughout the study at various test locations and monitoring points. These results were compared with concurrent pH results to calculate the CO_2 concentrations. Samples collected to determine CO_2 concentrations prior to installation of any ORC as a base line value was zero. All subsequent samples collected at two week intervals during the study and at the close of the study for CO_2 were zero. This is likely due to geochemical reactions, consuming CO_2 as it is generated. This indicates that the use of CO_2 as an indicator of biodegradation in tropical environments may not be reliable and, although costly, biological plate counts should be utilized in the future.

CONCLUSIONS

The sampling results show that as DO is supplied to the impacted zone by ORC, the BTEX compounds are first mobilized and then consumed by the proliferating naturally occurring bacteria. The levels of BTEX dropped by as much as 95% and TPH by 96% at test locations during the study.

As the ORC is hydrated, the DO rises above background concentrations and then fluctuates with the bacterial population. As the DO is exhausted, the bacterial population dies off and the BTEX levels rise again. The pH levels in groundwater rise as the ORC is hydrated. The overall increase in the pH level is between 0.5 and 1.5 across the study areas as expected. As the ORC is depleted, the pH levels begin to return to background levels immediately. The drop in pH levels occurs concurrently with the return of DO to near background levels and the rise in BTEX levels indicating that the amount of ORC provided in the pilot study has been exhausted. It is assumed that the lack of detection of CO_2 at any point of the study indicates a geochemical reaction may be masking any production of CO_2 at this site.

REFERENCES

Driscoll, F.G.. 1986. *Groundwater and Wells.* 2nd ed. Johnson Division, St. Paul, MN.

Howard, P.H., R.S. Boethling, W.F. Jarvis, W.M. Meylan and E.M. Michalenko. 1991. Hand Book of Environmental Degradation Rates. H.T. Printup (Ed.). Lewis Publishers, Chelsea, MI.

U.S. Geologic Survey. *Atlas of Ground-Water Resources in Puerto Rico and the U.S. Virgin Islands.* Water-Resources Investigations Report 94-4198.

Wilson, J.T. and M.J. Noonan. 1984. "Microbial Activity in Model Aquifer Systems." In G. Bitton and C.D. Gerba (Eds.), *Groundwater Pollution Microbiology.* John Wiley and Sons, NY, NY.

HYDROCARBON DEGRADATION IN FINE-TEXTURED MARSH SOILS UNDER AEROBIC AND DENITRIFYING CONDITIONS

M. Brecht (Institute of soil science, Kiel, Germany)
S. Hüttmann (Groth company, Itzehoe, Germany)
L. Beyer (Institute of soil science, Kiel, Germany)

ABSTRACT

The goal of the laboratory studies was to prove hydrocarbon degradation in fine-textured marsh soils under denitrifying conditions and compare it with the degradation under aerobic conditions. In order to accelerate the bioremediation process in soils with redox potentials below -200 mV (pH 7) the use of different additives and oxidising agents was tested. The testing programm lasted four weeks and included treatments with KNO_3, fertilizer and a mixture of KNO_3 and a fertilizer for the hydrocarbon degradation under anoxic denitrifying conditons or the addition of hydrogen peroxid or agents to improve the soil structure such as oil binder for the aerobic hydrocarbon degradation. During the incubation period the respiration (CO_2-production) was measured nearly every day, the content of hydrocarbons at the beginning of the test and after two and four weeks, as well as the number of hydrocarbondegrading microbes (aerob, anaerob), NH_4^+-N, NO_3^--N and N at all. In the soil treated with KNO_3 / KNO_3+fertilizer / fertilizer the redox potential was also measured.

There were several indicators for the hydrocarbon degradation under denitrifying conditions: high denitrification rates, increase in the number of denitrifying bacteria, higher respiratory quotients and higher degradation rates than the reference. At the end of the test period the oil binder treated soil had the highest degradation rate, but there were severe variations in the degradation rate between the first and second fortnite and the treatments. The investigation suggest that the application of KNO_3+fertilizer is the most promising treatment to accelerate biological hydrocarbon degradation in fine-textured marsh soils.

INTRODUCTION

Marshland soils form various places along the northern German coast have been contaminated with mineral oil due to leakages of underground storage tanks containing heating oil and had to be excavated and cleaned using a bioremediation system. The degradation rate of hydrocarbons in fine-textured marsh-soils with redox potentials below -200 mV (pH 7) is very low because of a deficit of electron acceptors. There are two possibilities to accelerate the bioremediation process:

1. Using additives to provide alternative electron-acceptors (others than oxygen) such as nitrate (Norris, 1993) to raise the redox potential.

2. Trying to bring oxygen into the soil with hydrogen peroxid (Franzius/Bachmann, 1996) or with components to improve the soil structure (Schulte, 1995) such as oil binder.

In laboratory the effects of the following additives on the oil degradation rate were tested and compared in a four week incubation: H_2O_2 (3% solution), oil binder (10%), KNO_3 (0,15% N), KNO_3 + fertilizer (0,15% N), fertilizer (0,075% N). To find the optimal amount of every additive several respiration tests were necessary before the incubation could be started.

MATERIALS AND METHODS

Respiration. Microbial respiration was measured using the Isermeyer-method by trapping the ewolved CO_2 with NaOH and basal respiration ujsing a sapromate by measuring the O_2 consumption and the CO_2 evolution (Alef, 1994).

Content of hydrocarbons. The content of hydrocarbons was measured in an soil extract with 1,1,2-trichlorotriflouroethane by FTIR.

Microbial biomass. The microbial biomass was calculated using the viable cell counts determined with the microdrop method (Drews, 1974) in a mineral oil hydrocarbon-containing broth (aerobic hydrocarbon degradating microbes). For the measuring of anaerobic hydrocarbon degradating microbes the broth contains additional KNO_3 and the broth were incubated under N_2-atmosphere.

NH_4^+-N / NO_3^--N-content. NH_4^+-N and NO_3^--N were measured according to Schinner, 1995.

N-content. The N-content was measured in a CHN-analyser (Leco, CHN 600).

Redox potential. The redox potential was measured with platine electrodes according to Schlichting, 1995.

RESULTS AND DISCUSSION

After the addition of KNO_3 / KNO_3+fertilizer the redox potential could be raised up to 200 - 250 mV and denitrifying conditions could be maintained depending on the amount of nitrate added (diagram 1.).

Surprisingly there was little difference between the low and the high KNO3-application rate, but the 1,5% treatment had a more longlasting effect.

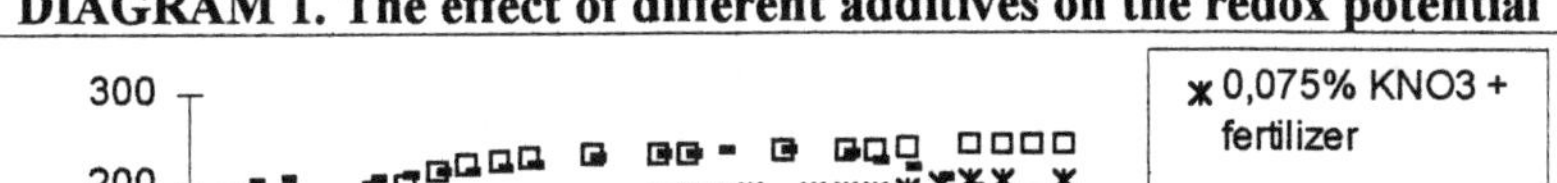

DIAGRAM 1. The effect of different additives on the redox potential

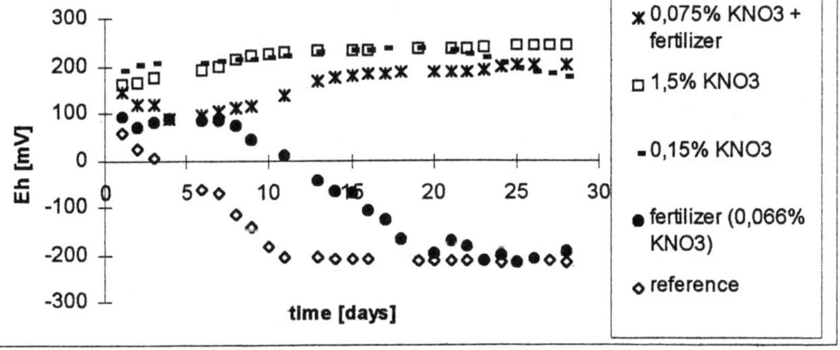

However, the respiration test in which the effect of different KNO_3-concentrations was demonstrated showed that it is not necessary to apply extremly high amounts of nitrate to promote microbial respirations (table 1.).

TABLE 1. Microbial CO_2-production depending on the added amount of KNO_3

treatment [mg KNO_3 $100g^{-1}$soil]	CO_2-production [mg kg^{-1}dm 72 h^{-1}]	CO_2-production [% of reference]
reference	233	100
50	342	147
100	432	185
150	456	196
200	321	138
250	317	136
500	206	88
1000	170	73
1500	166	71

Microbes were most stimulated by an addition of 0,15% KNO_3. To maintain the redox potential at the level of 200 mV a new application of KNO_3 will be necessary every three weeks. The data of the NO_3-N investigations provide the same result: If 0,15% KNO_3 was added all nitrate was consumed by micobes at the end of the four week incubation (diagram 2.).

DIAGRAM 2. Composition of the N-pool

(a) treatment with KNO_3+fertilizer (b) treatment with oil binder

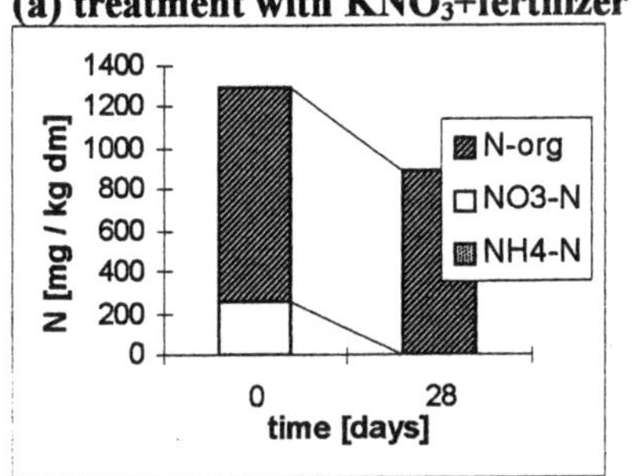

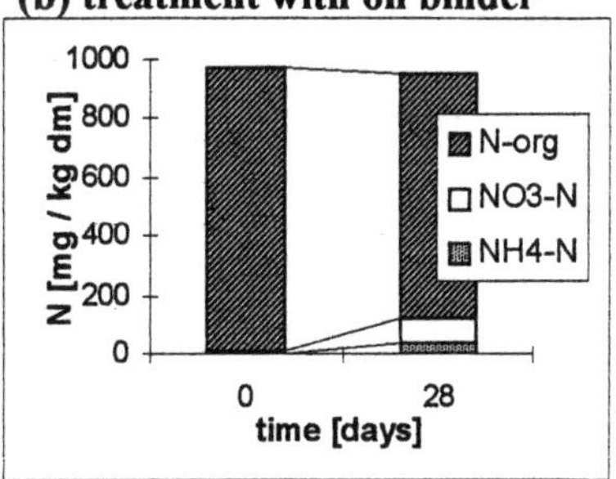

The different development of the two possibilities of treatment (1. stimulation of anaerob micobes by an addition of KNO_3, 2. stimulation of aerob microbes by improving the soil structure) shows that denitrification must have been started after the soil was treated with KNO_3 or KNO_3+fertilizer because the amount of N that left the soil system was as much as the regression of the NO_3-N. Another hint on denitrification was NO_2 that could be qualitativly established after two weeks. If the soil was treated with oil binder or H_2O_2 (oxic conditions) the NO_3-N-/NH_4-N-content was higher at the end of the incubation period because of the mineralization of organic N-compounds due to the activity of heterotrophic microorganisms.

DIAGRAM 3. Development of the oil degrading microbial biomass
(a) denitrifying microbes b) aerobic microbes

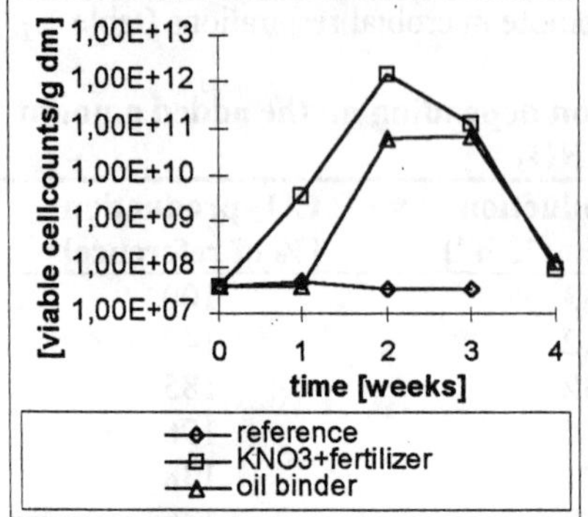

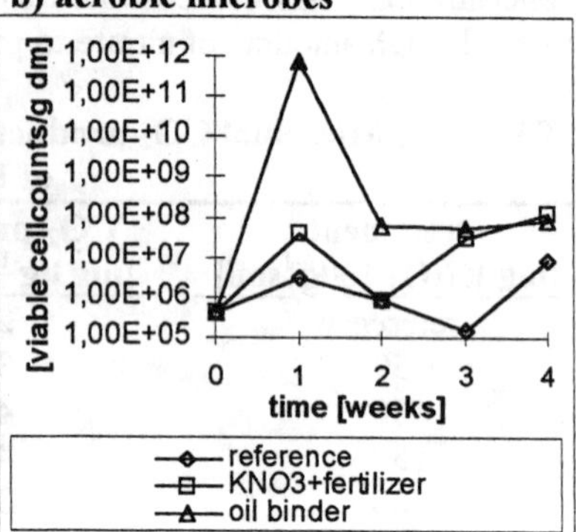

An increase of the number of oil degrading denitrifying microorganisms was the effect of an application of KNO_3+fertilizer. The decrease of the denitrifying microorganisms in the third and fourth week can be attribute to the deacreasing NO_3^- content. Oil binder had the most increasing effect on the aerobic oil degrading microorganisms. The decrease of aerobic oil degrading microorganisms in the second week went with an increase of the denitrifying microorganisms due to the consumption of O_2.

Degradation rate: The addition of nitrate, fertilizer, nitrate+fertilizer accelerated the bioremediation process as well as the addition of H_2O_2, (62% $\pm$ 2% hydrocarbon left after 4 weeks). Only 50% of the hydrocarbon were found after the test period if the soil was treated with oil binder. However, there were severe variations in the degradation rate between the treatments and the first and second fortnite (Table 2.).

TABLE 2. Development of the degradation rate [mg*kg^{-1}dm*d^{-1}]

Treatment	1st - 14th day	15th - 28th day
-	70	29
KNO_3	78	38
KNO3 + fertilizer	60	51
fertilizer	76	34
H_2O_2	84	31
oil binder	108	40

The procentual regression of the degradation rate from the first fortnite to the second fortnite of the test period was used to calculate a further development of the hydrocarbon degradation rate. Diagram 4 gives an idea about the time schedule for the bioremediation process.

DIAGRAM 4. Hypothetic development of hydrocarbon degradation

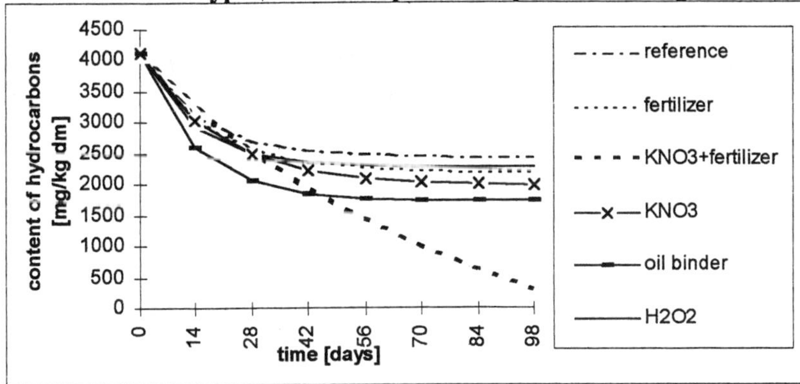

The projected assessment suggests that with a KNO_3+fertilizer application the hydrocarbon degradation is more efficient and mathematically linear compared to all other treatments.

CONCLUSIONS

According to these lab investigation we propose the application of nitrate+fertilizer to promote biological hydrocarbon degradation in fine-textured marsh soils. However, an additional field study is necessary to confirm the results in biopile treatments.

REFERENCES

Alef. 1994. *Bioremediation methods (German: Biologische Bodensanierung)*. Methodenhandbuch. VCH Verlagsgesellschaft mbH. Weinheim.

Drews. 1974. *Microbiological studies (German: Mikrobiologisches Praktikum)*. 2nd edition. Springer-Verlag. Berlin. Heidelberg. New York.

Franzius/Bachmann. 1996. *Remediation of contaminated sites and soil protection (German: Sanierung kontaminierter Standorte und Bodenschutz)*. Erich Schmidt Verlag GmbH & Co. Berlin.

Norris/Brown/Melasty. 1993. *Handbook of bioremediation*. Boca Raton. Florida.

Schlichting/Blume/Stahr. *Soil science studies (German: Bodenkundliches Praktikum)*. 2nd edition. Blackwell Wissenschafts-Verlag. Berlin.

Schulte. 1995. „Biological clean-up of extremly silty soils." (German: „Biologische Reinigung von stark schluffigem Boden."). *TerraTech 3/1995: 52-54.*

Shinner/Öhlinger/Kandeler/Margesin. 1995. *Methods in Soil Biology*. 2nd edition. Springer-Verlag. Berlin. Heidelberg.

INFLUENCES OF HYDROGEN PEROXIDE AND *Azospirillum* sp. HYDROCARBONS BIODEGRADATION IN SOILS

Claudio O. Belloso (Facultad Católica de Química e Ingeniería. Rosario. Argentina)

ABSTRACT: The desperate career against the time and the environmental legislations make seem too slow any type of procedure of restoration of contaminated soils with hydrocarbons by means of bioremediation.

In order increase the speed of the degradation of the pollutant, diverse technologies have been introduced, some of which consist on the biostimulation and/or bioaugmentation of the microbial strains which produce the biodegradation processes. To achieve this made, different types of nutrients, helping substances, modification of the experimental physical, chemical and biological conditions are being tested.

The influence of the hydrogen peroxide and of the *Azospirillum* sp. on hydrocarbons biodegradation was evaluated. It could be shown that hydrogen peroxide favors the degradation activity of an isolated strain from a contaminated site, on the other hand, the presence of *Azospirillum* sp. doesn't seem to contribute positively to the process, on the contrary, it seems to diminish the degradation activity of indigenous strain.

INTRODUCTION

Bioremediation, or use of microbial transformation of chemical compounds, has proven to be a cost-effective solution to the removal of toxic contaminants from wastes sites.

One of the most limiting factors in the success of bioremediation procedures has been a poor understanding of the ecology of the microorganisms being used in these processes.

For in situ bioremediation to be successful, the microorganisms must survive and proliferate in the areas targeted for cleanup. One of the characteristics influencing bacterial survival and distribution in contaminated environments is their ability to move to (motility), detect (chemotaxis), and use these contaminants. It has been demonstrated that species of *Azospirillum* besides fixing nitrogen can use aromatic compound as sole carbon source. (Chen et al., 1993).

The most common bioremediation approach is based on aerobic processes. As currently practiced, conventional in-situ biorestoration of petroleum-contaminated soils, aquifer solids, and ground water relies on the supply of oxygen to the subsurface to enhance natural aerobic processes to remediate the contaminants. It has been recognized that the rate at which oxygen can be introduced by sparging air in a ground-water injection well limits the effectiveness of the technology.

Since the amount of oxygen which can be added to water from air is limited (8-10 mg/L), other sources of oxygen have been used. These include pure oxygen and hydrogen peroxide.

Hydrogen peroxide is commonly used as a method of introducing oxygen (Brown et al., 1984). Because of this using hydrogen peroxide can, theoretically, provide oxygen 5-50 times faster than could sparging air or pure oxygen into injection wells and should result in shorter remediation times.

However, the efficiency of delivering oxygen by this method has been quite variable even when favorable results were obtained from laboratory screening test (Huling et al., 1990).

In this research we investigate the influence of the *Azospirillum* sp. and hydrogen peroxide on hydrocarbons (HC) biodegradation in soils with fertilizers.

Objective. To determine the influence of hydrogen peroxide and *Azospirillum* sp. on the degradation activity of an isolated bacterial strain from a hydrocarbons contaminated soil.

MATERIALS AND METHODS

Six metallic containers of 50 L of capacity were used. Thirty Kg of soil per container was added.

Soil was obtained out of landfarming belonging to Petroleum Refinery San Lorenzo S. A. located in San Lorenzo City, Santa Fe Province, Argentina.

To avoid contact of the soil with the metal of the container, proceeded to recover the containers inner walls with a thick layer of non-biodegradable resin under the conditions of the test.

Initial Total Organic Carbon of the landfarming soil was 1.02% w/w.

TABLE 1. Percentual composition of hydrocarbon residual contained in landfarming soil.

Hydrocarbon distribution	Percentage
Saturates	41
Aromatics	19
Resins	15
Asphaltenes	25

Containers inoculation. Containers N° 3, 4, 5 and 6 were inoculated with a hydrocarbons degrading bacteria denominated W2 isolated of the same landfarming (Belloso et al., 1997).

Just one application of fertilizer Z2, type NPK, in the containers N° 3, 4 and 6 was made, according to the following relationship: C/N=100; NPK= 20:20:1; source of nitrogen: $NH4^+$, except in container N° 4 where according to the relationship: C/N=200 was applied.

In container N° 5 a culture in exponential grow phase of *Azospirillum* sp. (Az) (Malik and Schlegel, 1980) was added.

Azospirillum sp. in liquid medium DSMZ - Medium 221 *Azospirillum* Medium were cultivated. Its formulation is detailed next:

DSMZ – Medium 221 *Azospirillum* Medium.

Yeast extract	0.05	g
K_2HPO_4	0.25	g
$FeSO_4.7H_2O$	0.01	g
$Na_2MoO_4.2H_2O$	1.0	mg
$MnSO_4.H_2O$	2.0	mg
$MgSO_4.7H_2O$	0.2	g
NaCl	0.1	g
$CaCl_2.2H_2O$	0.02	g
$(NH_4)_2SO_4$	1.0	g
Biotin	0.1	mg
Bromothymol blue	25.0	mg
Distilled water	950.0	ml

Container N°1 contains only soil as control test. Container N°2 contains soil and $HgCl_2$ (2g/100g of dry soil) in order to evaluate abiotics losses of hydrocarbons (Ferrari et al., 1994).

In containers N° 3, 4 and 5, hydrogen peroxide 3% twice per week, 2 L per container, was added.

In table 2 we can see that inoculated material in each container.

TABLE 2. Inoculated material.

Container N°	Inoculated bacterial strain	Inoculated *Azospirillum* sp.	Added additive	Applied fertilizer
1	-	-	-	-
2	-	-	$HgCl_2$	-
3	W2	-	H_2O_2	Z2
4	W2	-	H_2O_2	2 x Z2
5	W2	Az	H_2O_2	-
6	W2	-	-	Z2

All containers were maintained outdoors for 150 days, from December 1996 to May 1997 once inoculated. Soils were aired weekly by blend the whole content.

At the beginning of the trial and every 30 days the following parameters in each soil were determined: pH, moisture, total hydrocarbons, total aerobic heterotrophic bacteria (TAHB) and hydrocarbons degrading microorganisms (HDM).

The pH in suspensions of soil (40% w/v) in $CaCl_2$ 0.01M solution was measured.

Moisture in soils samples was determined by drying to 105°C until constant weight.

The hydrocarbons content was determined by extraction in Soxhlet with Ethylic Ether for 6 hs.

TAHB in soils were enumerated by the pour plate method (Clesceri et al., 1992) using Merck plate count agar. All plates were incubated at 30°C for 24 hours.

HDM in soils were enumerated by the more probable number method (Clesceri et al., 1992) using Bushnell-Hass mineral broth (Bushnell and Hass, 1941) with *n*-hexadecane as the sole carbon source. A five-tube MPN technique was used. Three sets of five screw-cap tubes, each containing 5 ml of Bushnell-Hass mineral broth were inoculated with 1, 0.1, and 0.01 ml of a sample dilution.

After inoculation, 50 µL of UV-sterilized *n*-hexadecane (Merck) and 100 µL of autoclaved resazurin solution (Merck, 50 mg/L) were added to each tube.

After incubation at 30ºC for 7 days, tubes giving positive reaction were counted. The color of positive tubes changed from blue to pink.

RESULTS AND DISCUSSION

In 150 days of studies the average temperature was 22 ºC. Moisture varied between 10 and 20%. The pH varied between 4.5 and 7.7.

In table 3 and 4 we can see results of counts in the different containers in 5 months of studies.

TABLE 3. TAHB count on each container during test

Container Nº	CFU/g	CFU/g	CFU/g	CFU/g	CFU/g
1	8.5×10^5	1.3×10^6	4.8×10^6	7.7×10^7	1.4×10^7
2	0	0	0	0	0
3	2.5×10^6	4.1×10^6	6.6×10^7	2.0×10^8	2.4×10^8
4	6.9×10^6	4.5×10^6	6.4×10^6	2.4×10^8	2.4×10^8
5	4.5×10^6	4.8×10^6	2.5×10^7	7.4×10^7	1.4×10^8
6	3.9×10^6	4.2×10^6	1.8×10^7	4.4×10^7	4.9×10^7

TABLE 4. HDM count on each container during test

Container Nº	MPN/g	MPN/g	MPN/g	MPN/g	MPN/g
1	1.2×10^7	1.4×10^7	9.7×10^6	1.4×10^8	1.4×10^8
2	0	0	0	0	0
3	1.3×10^7	5.7×10^6	9.5×10^6	$>1.4 \times 10^8$	$>1.4 \times 10^8$
4	1.2×10^7	$>1.3 \times 10^7$	9.6×10^6	$>1.4 \times 10^8$	$>1.4 \times 10^8$
5	1.3×10^7	1.3×10^7	9.6×10^6	5.9×10^7	5.7×10^7
6	8.9×10^6	1.5×10^7	1.2×10^7	2.5×10^7	3.3×100^8

In Table 5 we can see percentages of hydrocarbons obtained each month and the final reduction percentages after 150 days of studies.

TABLE 5. Percentages of hydrocarbons reduction on each container

	HC (% w/w) 0 days	HC (% w/w) 30 days	HC (% w/w) 60 days	HC (%w/w) 90 days	HC (%w/w) 120 days	HC (%w/w) 150 days	HC (%w/w) reduction obtained by linear regression
1	3.82	3.44	3.43	3.12	3.18	3.01	21.5
2	5.04	4.45	4.00	4.29	4.45	4.70	6.7
3	3.93	3.29	2.65	3.27	2.11	1.60	59.3
4	3.85	3.49	2.88	3.19	2.83	2.60	32.5
5	4.39	3.77	3.74	3.22	2.93	3.10	29.4
6	3.86	3.77	3.65	3.12	2.56	2.10	45.6

Significant differences were not observed in the counts among the different containers, in spite of the treatment with hydrogen peroxide and to the presence of *Azospirillum* sp..

In container N° 3 the biggest reduction, 59.3%, was obtained.

CONCLUSIONS

Hydrogen peroxide seems to increase the degrading activity of W2 strain according to the reduction percentages obtained in the containers N°3 and 6.

More quantity of fertilizer Z2 (container N°4) it doesn't improve the biodegradation.

Presence of *Azospirillum* sp. (container N°5) doesn't seem to contribute positively to the process, on the contrary, it seems to diminish the degrading activity of W2 strain.

ACKNOWLEDGEMENTS

The author thanks to the authorities of the Faculty and Refinery San Lorenzo that believed in the potentials of this project and supported it.

REFERENCES

Belloso, C. O., J. Carrario, and D. Viduzzi. 1997. "Hydrocarbons Biodegradation in Soils of Landfarming Contained in Containers" (in Spanish). In *Proceeding III Jornadas Rioplatenses de Microbiología*. Asociación Argentina de Microbiología. pp. 106, K3. Buenos Aires, Argentina.

Brown, R. A., R. D. Norris, and R. L. Raymond. 1984. "Oxygen transport in contaminated aquifers". In *Proceeding Petroleum Hydrocarbon and Organic Chemical in Groundwater: Prevention Detection and Restoration*. National Water Well Association. Houston, Texas, U.S.

Bushnell, L. D., and H. F. Hass. 1941. "The utilization of certain hydrocarbons by microorganisms". *J. Bacteriol. 41*: 653-673.

Chen, Y. P., G. Lopez de Victoria and C. R. Lovell. 1993. "Utilization of aromatic compounds as carbon and energy sources during growth and N_2-fixation by free-living nitrogen fixing bacteria". *Archives of Microbiology 159*: 207-212.

Clesceri, L. S., A. E. Greenberg, and R. Trussel Rhodes (Eds.). 1992. "Standards Methods for the Examination of Water and Wastewater". 18th. ed., American Public Health Association, N.Y.

Ferrari, M. D., C. Albornoz, and E. Neirotti. 1994. "Biodegradability on soil of residual hydrocarbons from petroleum tank bottom sludges" (in Spanish). *Revista Argentina de Microbiología, 26*: 157-170.

Huling, S. G., B. E. Bledsoe, and M. V. White. 1990. "Enhanced Bioremediation Utilizing Hydrogen Peroxide as a Supplemental Source of Oxygen: A Laboratory and Field Study". NTIS PB90-183435/XAB. EPA/600/2-90/006. 48 p.

Malik, K. A. and H. G. Schlegel. 1980. "Enrichment and isolation of new nitrogen-fixing hydrogen bacteria". *FEMS Microbiol. 8*: 101-104.

BIOREMEDIATION OF A MANUFACTURED GAS PLANT GROUNDWATER - A FIELD PILOT STUDY

M. Stieber (DVGW-Technologiezentrum Wasser, Karlsruhe, Germany)
P. Werner (Technische Universität Dresden, Germany)
A. Bächle and P. Melzer (Mannheimer Versorgungs- und Verkehrsgesellschaft mbH, Mannheim, Germany)
K. Piroth and E. Robold (Arcadis Trischler & Partner, Karlsruhe, Germany)

ABSTRACT: At the former manufactured gas plant (MGP) site in Mannheim, Germany, where soil and groundwater are heavily contaminated with monocyclic (BTEX) and polycyclic aromatic hydrocarbons (PAH), a field pilot study was realized to demonstrate the feasibility of bioremediation by pump and treat technology. Therefore contaminated groundwater was recovered downstream the pilot field, decontaminated in a multistage on-site treatment system, and re-injected into the contaminated aquifer upstream of the pilot field. To stimulate aerobic in-situ degradation processes, H_2O_2 was added into the injection well together with PO_4^{3-} as an important growth-factor for the indigenous microorganisms. Furthermore, NO_3^- was injected to enable the denitrifying degradation of the pollutants beside the aerobic process. Within a period of 9 months all added electron acceptors were completely transformed in the contaminated aquifer. Approximately 590 kg of hydrocarbons had been biodegraded in-situ, and 150 kg of hydrocarbons as well as 3,100 kg of NH_4^+ were removed through the recovery well and eliminated on site. In total, the pilot field experiment showed the possibility of an efficient elimination of the ground-water contaminants by combined biological on-site/in situ treatment, but technical problems concerning the continuous injection of the required electron acceptors were predominant limiting factors.

INTRODUCTION

The manufactured gas plant (MGP) of Mannheim was the largest gas plant of the federal state of Baden-Württemberg and produced town gas from coal between 1900 and 1968. The former production location enclosed an area of approx. 15 hectares. As is typical of most MGPs, during the production period tar oil and other production residues leaked into the aquifer and contaminated soil and ground-water with monocyclic (BTEX) and polycyclic aromatic hydrocarbons (PAH) as well as ammonium.

On account of the present risk potential, there is a need to remediate the MGP-groundwater within the next few years. Therefore, one of the relevant technologies is bioremediation combining on site and in situ treatment (Norris et al., 1994). Several works have shown that pollutants like BTEX and PAHs can be eliminated by indigenous microorganisms (Cerniglia & Heitkamp, 1989; Stieber 1997). Preliminary microbiological investigations of the contaminated groundwater confirmed that the indigenous microorganisms living in the aquifer of the MGP of Mannheim were able to metabolize the organic pollutants aerobically.

Though aromatics also are degradable anaerobically by denitrification and by iron- or sulfate-reduction, biodegradation of aromatic pollutants occurs most efficient under aerobic conditions. Therefore, it is necessary to transfer molecular oxygen into the contaminated aquifer to achieve a time and cost effective application of in situ bioremediation. To evaluate whether groundwater cleanup by bioremediation is technically feasible and cost effective in the case of the MGP of Mannheim, it was necessary to conduct a field pilot study directly at the contaminated site.

FIELD EXPERIMENT

Design of on-site treatment[(1)]. For the on-site cleanup contaminated groundwater was recovered downstream the pilot field with a flow rate of 6 m^3/h. of that amount, 3 m^3/h were drained off into the local sewage plant, and the remaining 3 m^3/h were treated in a multistage on-site system. In the first step, the contaminated groundwater passed through multi-layer sand filters to eliminate ferrous iron and manganese by oxidation with technical oxygen. Furthermore, the availability of molecular oxygen should stimulate indigenous microorganisms to mineralize the organic contaminants. Subsequently, two fluidized bed bio-reactors were installed to remove ammonium nitrogen by nitrification and denitrification processes. In conclusion, remaining pollutants and particles should be eliminated by filtration through sand and activated carbon filters.

Design of in-situ treatment. An area with an expanse of 20 m by 15 m and a depth extension of 32 m inside the contamination source has been selected for the field experiment. The aquifer characteristics were gravel and sandy layers with a permeability between $K_f = 2x10^{-4}$ m/s and $9x10^{-4}$ m/s. Figure 1 shows a cross-sectional view of the test field. The position of all different wells at the pilot field, used for ground-water recovery, injection, and monitoring, as well as the stream lines of groundwater flow indicating the hydraulic control of the field experiment, are presented in Figure 2. During the in situ treatment, the groundwater flowrate was about 1 m/d.

To stimulate aerobic in situ biodegradation of the organics, on-site purified groundwater was re-injected into the contaminated aquifer upstream the pilot field (maximum 3m^3/h) after an addition of H_2O_2 (at a maximum of 1,000 mg/L). Furthermore, the effluent concentration of the on-site denitrification bioreactor was adjusted to approx. 50 mg/L of NO_3^-, to stimulate denitrifying degradation of the pollutants beside the aerobic process. To supply the microorganisms with phosphorus as an important growth factor, 5 mg/L PO_4^{3-} were added to the re-injected groundwater.

[(1)] The on-site treatment system was designed and built by Preussag Wassertechnik GmbH, Zwingenberg, Platanenallee 55, 64673 Zwingenberg, Germany.

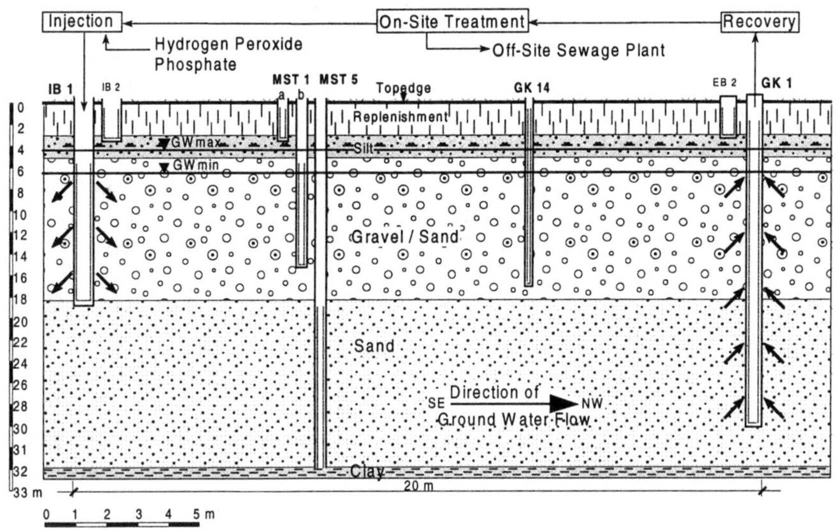

FIGURE 1. Cross-section of the test field, showing injection, monitoring, and recovery wells.

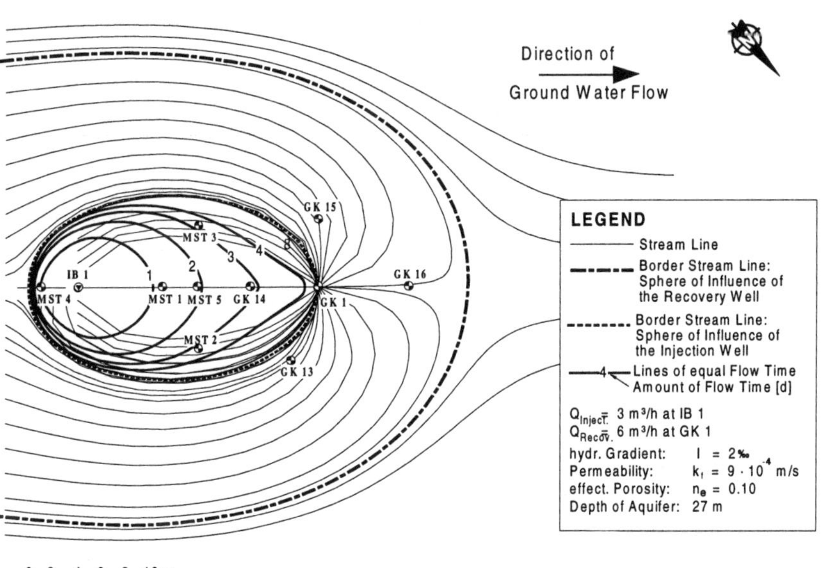

FIGURE 2: Test field with recovery, injection, and monitoring wells as well as the stream lines of the groundwater flow.

System Operation and Monitoring. The total operation time for the in situ field study was 9 months. During this period, 32,503 m^3 of contaminated groundwater were pumped off, of which 15,050 m^3 were cleaned up on-site and re-injected into the aquifer. Together with the treated groundwater, 2,920 kg H_2O_2 (corresponds to 1,372 kg O_2), 452 kg NO_3^- , and 75 kg PO_4^{3-} were injected.

A few months before the in situ field experiment was started, the on-site treatment system was tested and optimized, whereby the microorganisms in the multistage system adapted to the different environmental conditions and pollutant concentrations. With beginning of in situ treatment, every stage of the on-site system operated 100 %.

The chemical and biochemical processes during on-site and in situ treatment were monitored by measuring the following parameters at different points of the on-site system as well as at the recovery, monitoring and injection wells: aromatics (PAH, BTEX, phenol), dissolved organic carbon (DOC), dissolved O_2, H_2O_2, NH_4^+, NO_2^-, NO_3^-, PO_4^{3-}, pH, total cell number, number of heterotrophic bacteria, PAH- and BTEX-degrading microorganisms, denitrifying microorganisms, and bioluminescence inhibition testing.

RESULTS AND DISCUSSION

Pollutant Concentrations. At the beginning of the combined on-site/in situ treatment, the initial groundwater concentrations and loads of the contaminants at the recovery well were 2,000-2,500 µg/L and 12-15 g/h of BTEX(2), 2,000-2,300 µg/L and 12-14 g/h of PAHs(3) as well as 100,000 µg/L and 600 g/h of NH_4^+. At the monitoring wells pollutant concentrations from 60 µg/L up to 6,800 µg/L of BTEX, 30 µg/L up to 5,500 µg/L of PAHs, and 10,000µg/L up to 60,000 µg/L of NH_4^+ have been detected in the ground-water.

On-Site Treatment. Right from the beginning of the on-site treatment, 99% of the recovered aromatics were eliminated in the first stage by indigenous microorganisms settled on the sand grains of the multi-layer sand filters. The rest was removed in the following stages. More than 99 % of NH_4^+ was oxidized to NO_3^- in the nitrifying fluidized bed reactor and subsequently reduced to molecular nitrogen in the denitrifying stage with ethanol as an organic substrate. Merely approx. 50 mg/L NO_3^- remained in the effluent to stimulate denitrifying in-situ degradation processes.

In-Situ Treatment. Due to the addition of H_2O_2 to the re-infiltrated groundwater, concentrations up to 40 mg/L of dissolved O_2 were detected at the monitoring wells located close to the injection well (MST 4, MST 1b, see Figure 3). The O_2-supply of the previously anaerobic ground-water increased the numbers of BTEX and PAH degrading microorganisms of about 2-3 orders of magnitude.

---

(2) The total amount of BTEX includes methyl-, diethyl-, and trimethyl-benzenes as well as styrole, indene, indane, 1-, 2-Methyl-naphthalene and 1,1' Biphenyle.

(3) Naphthalene, acenaphthene, fluorene and phenanthrene were the dominant substances.

At the same time a decrease of BTEX and PAH concentrations as well as a decrease of the toxicity value occurred. As the DOC values showed, that no significant enrichment of dissolved metabolites appeared, these results indicate, that the aromatics were predominantly eliminated by means of aerobic microbial mineralization and biomass building.

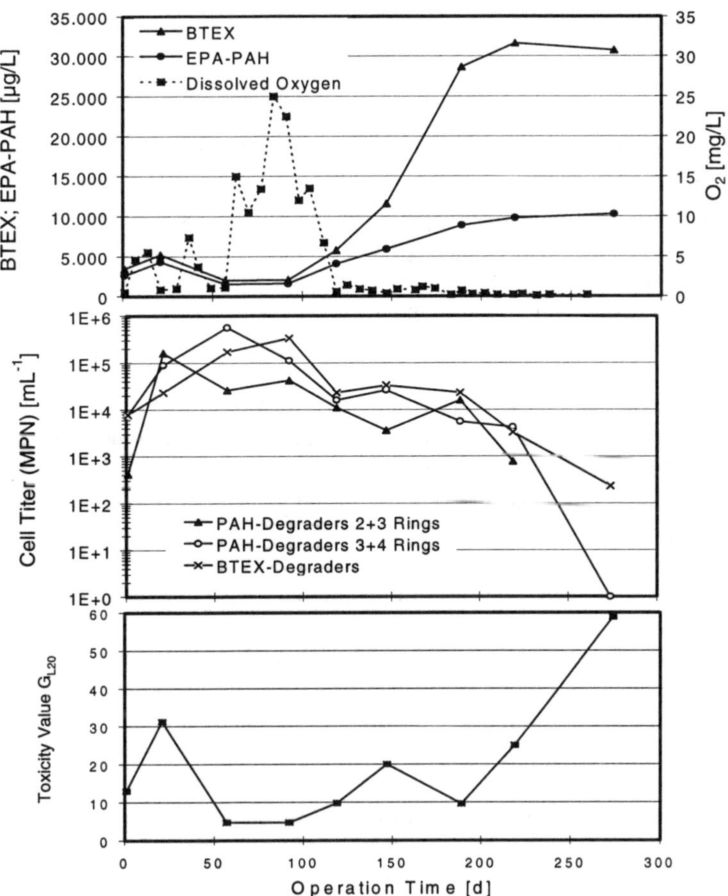

FIGURE 3: Concentrations of pollutants, dissolved oxygen, hydrocarbon degraders and toxicity value at the monitoring well MST 1.

After an operation time of 4 months, the infiltration capacity at the injection well declined. This fact required a considerably lower H_2O_2 dosage. Not being successful with the regeneration of the injection well, two additional monitoring wells (MST 2, MST 3) were used to inject H_2O_2. Nevertheless, only small amounts of H_2O_2 could be infiltrated into the contaminated aquifer during the further period of the pilot study. As a result, the in situ concentration of dissolved O_2 and the numbers of BTEX- and PAH-degraders declined, whereby the concentration of aromatics as well as the toxicity value increased again (Figure 3).

During the complete operation time of the pilot field study neither H_2O_2, O_2 nor NO_3^- were detected in the ground-water at the recovery well. These results indicate that the injected H_2O_2 was totally disintegrated to H_2O and O_2 or consumed for oxidation processes. It is obvious that molecular O_2 was used up completely by microbial degradation and chemical oxidation processes, and that denitrifying microorganisms must have used NO_3^- to oxidize organic pollutants in the aquifer.

Calculation of mass balances, considering aerobic and denitrifying metabolization of the aromatics, lead to approx. 590 kg of aromatic hydrocarbons to be biodegraded in-situ during the field pilot study. Microbial nitrification of NH_4^+ leads to NO_3^-, which was utilized by indigenous microorganisms to degrade organics. In this case it is not necessary to distinguish between the O_2 consumption for the biooxidation of the organic groundwater contaminants and NH_4^+. On the other hand, only approx. 150 kg of aromatic hydrocarbons as well as 3,100 kg of NH_4^+ were removed hydraulic through the recovery well.

CONCLUSIONS

The results of the field pilot study have demonstrated the possibility of an efficient elimination of the groundwater contaminants by bioremediation technology at the MGP of Mannheim. It was obvious that aerobic (together with denitrifying) in-situ degradation by indigenous microorganisms was much more effective than the hydraulic removal of the contaminants. The pilot study showed that the in situ bioremediation was mainly limited by technical problems concerning the continuous injection and transport of the required electron acceptors into the aquifer.

The high concentration levels of BTEX and PAHs measured at different monitoring wells (see Figure 3) indicate that there are nonaqueous-phase liquids (NAPLs) inside the aquifer. For the technical remediation of the MGP site it has to be recognized, that NAPL pools can be sources for ground-water contamination over a long period of time. NAPLs are not biodegradable, though only the water dissolved organic contaminants are available for the indigenous microorganisms. So it has to be considered, whether bioremediation is time and cost effective compared to other remediation technologies.

ACKNOWLEDGEMENTS

This work was funded by the federal state of Baden-Württemberg, Germany, and the Energie- und Wasserwerke Rhein-Neckar (RHE). We express our thanks to all participants for their professional and friendly cooperation.

REFERENCES

Cerniglia C.E. and Heitkamp M.A. 1989. "Microbial degradation of PAH in the aquatic environment". In Varanasi U. (ed.) *Metabolism of polycyclic aromatic hydrocarbons in the aquatic environment*, CRC Press Inc., Boca Raton: 41-68

Norris, Hinchee, Brown, McCarty, Semprini, Wilson, Kampbell, Reinhard, Bouwer, Borden, Vogel, Thomas, and Ward. 1994. *Handbook of Bioremediation.* Lewis Publishers, CRC Press, USA.

Stieber M. 1997. "Erfahrungen zum biologischen in-situ-Abbau von Kohlenwasserstoffen in Grundwässern". In *Möglichkeiten und Grenzen der Reinigung kontaminierter Grundwässer*, 12. DECHEMA-Fachgespräch Umweltschutz, ISBN 926959-80-0: 387-399.

MOLECULAR CHARACTERIZATION OF MICROBIAL COMMUNITIES IN A JP-4 FUEL CONTAMINATED SOIL

David C. White[†,‡], John R. Stephen[†], Yun-Juan Chang[†], Ying Dong Gan[†], Aaron Peacock[†], Susan M. Pfiffner[†], Michael J. Barcelona[§], Sarah J. Macnaughton[†]

[†] Center for Environmental Biotechnology, The University of Tennessee, Knoxville, TN 37932
[‡]Environmental Sciences Division, Oak Ridge National Laboratory, Oak Ridge, TN 37831[1]
[§]National Center for Integrated Bioremediation Research and Development, Dept. of Civil and Environmental Engineering, University of Michigan, MI 48109.

ABSTRACT: In this study, lipid biomarker characterization of the bacterial and eukaryotic communities was combined with PCR-DGGE analysis of the eubacterial community to evaluate correlation between JP-4 fuel concentration and community structure shifts. Vadose, capillary fringe and saturated soils were taken from cores within, up- and down-gradient of the contaminant plume. Significant differences in biomass and proportion of Gram negative bacteria were found inside and outside the plume. Sequence analysis of DGGE bands from within the spill site suggested dominance by a limited number of phylogenetically diverse bacteria. Together with pollutant quantification, these molecular techniques should facilitate significant improvements over current assessment procedures for determination of remediation end points.

INTRODUCTION

Shifts in microbial community structure provide a sensitive target for assay of the progress of bioremediation. The dominant organisms of contaminated sites are likely to be active in remediation of the contaminant. By combining PLFA analysis with PCR-DGGE analysis of the bacterial community we document herein shifts in a field population structure resulting from contamination with JP-4 fuel.

METHODS AND MATERIALS

Field site and sampling. The contaminated area was located at the KC-135 crash site at Wurtsmith Airforce Base (WAFB), Oscoda, Michigan. Soil samples were obtained using a Geoprobe piston corer. Cores were taken from 3 bore-holes, up-gradient, within, and down-gradient of the initial crash site. Each core was sectioned into vadose (7.5 ft; ~2.13 m), capillary fringe (8.5 ft; ~2.43 m) and saturated zones (~9.5 ft; 2.74 m). Samples were split for, a) lipid biomarker

---

[1] *Oak Ridge National Laboratory, managed by Lockheed Martin Energy Research Corporation, for the U.S. Department of Energy under contract number DE-AC05-96OR22464.

analysis (75 g cone line sections), b) DNA extraction and subsequent PCR (~1.5 g; in sterile whirlpacks), and c) volatile organic compound analysis (VOCs; 7.0 g; EPA vials). Samples for VOC analysis were preserved with 5 mL 40 % $NaHSO_4$ aqueous solution. Lipid and DNA samples were preserved on dry ice and shipped overnight to the University of Tennessee, Knoxville.

Volatile Organic Compounds. Samples were analyzed for VOCs on a HP-5890 series II Gas Chromatograph (GC) with an HP 5972 mass selective (MS) detector as described by Fang *et al.,* (1997). Separation was accomplished using an HP-624 GC column: 60 m x 0.25 mm i.d. (film thickness d_f = 1.8 µm; Hewlett-Packard). For all 43 compounds detected, calibration curves were linear between 1.0 µg/L to 200 µg/L. Compounds were identified based on relative retention time and verified by mass spectra. Concentrations of VOCs were calculated using the internal standard method, and are reported as µg/kg (Fang *et al.*, 1997).

Lipid analysis. Duplicate sub-samples of each zone from each soil core (six sub-samples per core) were extracted using the modified Bligh/Dyer as described previously by White *et al.,* (1979). The total lipids obtained were then fractionated into glyco-, neutral- and polar-lipids (Guckert *et al.,* 1985) with polar lipid then subjected to a sequential saponification/acid hydrolysis/esterification (Mayberry and Lane, 1993). The PLFA and dimethyl acetals (DMA) methyl esters were recovered. The PLFA and DMAs were separated, quantified and identified by gas chromatography-mass spectrometry (GC-MS; Ringelberg *et al.*, 1994). Fatty acids were identified by relative retention times, comparison with authentic standards (Matreya Inc., Pleasant Gap, Pa) and by the mass spectra (collected at an electron energy of 70 mV) Ringelberg *et al.,* (1989). Fatty acid is as described by Kates (1986).

DNA analysis. Nucleic acid was extracted directly from triplicate 0.5 g sub-samples from each zone from each soil core (9 sub-samples per core) using the method described in Stephen *et al.* (1999). PCR amplification and DGGE were carried out as described in Muyzer *et al*., (1993) using a D-Code 16/16 cm gel system.

Sequence analysis. PCR products from excised bands and cloned products were sequenced using the primer 516r (Lane *et al.*, 1985) and an ABI-Prism model 373 automatic sequencer with dye terminators. Sequences were compared to the GenBank database by use of the BLASTN facility of the National Center for Biotechnology Information (http://ncbi.nlm.nih.gov). Sequences were classified using the RDP release of 31-July-1998 (Maidak *et al.*, 1997).

Statistical analysis Analysis of variance (ANOVA) was used to determine whether there were significant differences between the lipid biomarker data obtained from the crash site (n=3) and that obtained from up- and down-gradient of the site (n=6). ANOVA was performed using Statistica Version 5.1 for Windows software.

RESULTS AND DISCUSSION

Geochemical and volatile organic compound analyses. Total VOCs from within and up-gradient of the crash site were detected in the saturated zone at 732, and 46 μg kg^{-1}, respectively. The VOCs were below detection limits in all vadose and capillary fringe zones and at all levels down-gradient of the site. These findings were contrary to those of Fang *et al.* (1997) in which the trace amounts of VOCs were detected down-gradient rather than up-gradient of the crash site.

Biomass content. The crash site biomass content was significantly higher than that of samples taken from either up- or down-gradient of the site ($P<0.05$). Bacterial cell numbers were calculated based from PLFA recovery data (Balkwill *et al.*, 1988). It is important to remember that with any conversion factor, the number of cells can vary by up to an order of magnitude (Findlay and Dobbs, 1993). Bacterial cell numbers for these samples per gram wet weight ranged from $5.2 \pm 0.2 \times 10^5$ in the saturated zone from the up-gradient sample, to $3.9 \pm 0.3 \times 10^7$ in the sample taken from the capillary fringe of the crash site sample.

Community structure. The microbial community structures of the samples differed dependent on both bore-hole location and depth (Figure 1). The microbial communities from the crash site samples contained significantly more monoenoic PLFA ($P<0.05$) indicative of Gram negative bacteria (Wilkinson, 1988), than did samples from up- and down-gradient of the site. Although containing significantly less biomass than samples taken from within the crash site($P<0.05$), samples from up-gradient and down-gradient contained significantly higher relative proportions of PLFA ($P<0.05$) indicative of sulfate-reducing bacteria (10me16:0, i17:1ω7c, Dowling *et al.*, 1986 and Edlund *et al.*, 1985, respectively). At all sites, the relative proportions of the biomarkers indicative of sulfate-reducing bacteria were significantly higher ($P<0.05$) in the capillary fringe and saturated zones. In all samples, the relative proportions of terminally-branched saturated PLFA, such as i15:0, i17:0 and cy17:0, indicative of anaerobic Gram negative bacteria (Wilkinson, 1988), increased with zone depth ($P<0.05$). Conversely, relative proportions of biomarkers typical of eukaryote PLFA (*e.g.* 18:2ω6 and 18:3) decreased with depth.

A hierarchical cluster analysis (HCA) of the bacterial PLFA profiles (arc sine transformed mol % data) showed the relatedness between samples (Figure 1). The bacterial PLFA comprised the total PLFA minus the polyenoic and normal saturate PLFA above 18 carbons in chain length, both of which are generally associated with eukaryote biomass. From the HCA it was apparent that the bacterial populations from within the crash site were dissimilar from one another and the up- and down-gradient samples, while the PLFA profiles from up- and down-gradient showed a higher level of relatedness, with the two vadose zones clustering together. The crash site vadose zone contained the most unique community, mainly due to the higher relative proportion of 2me13:0. A principle components analysis of the same data gave similar results, with two principal components derived, accounting for, sequentially, 76 and 15% of the variance inherent in the data set. Principle component 1 was most strongly influenced by

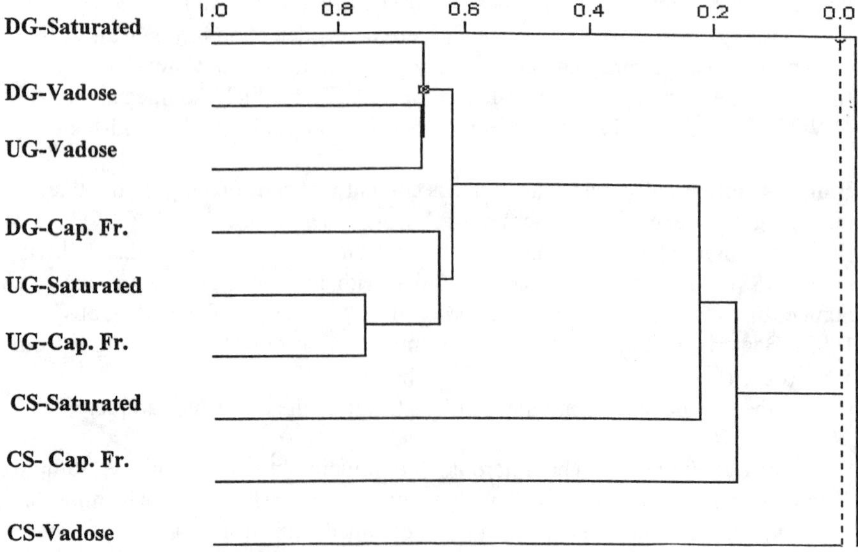

Figure 1: A dendogram representation of a hierarchical cluster analysis (single linkage based on euclidean distance) for the bacterial PLFA profiles. DG, down-gradient; CS, crash site; UG, up-gradient

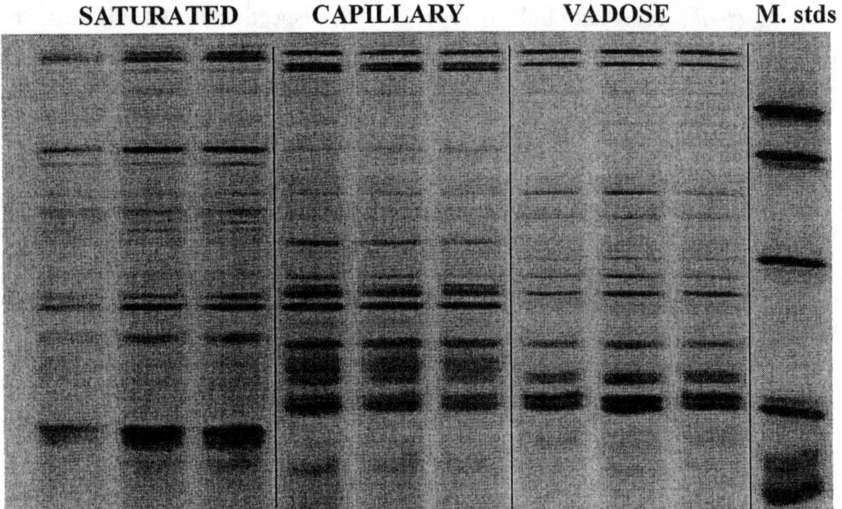

Figure 2: DGGE-eubacterial community profile of soil from within the crash-site. The portion of the gel shown represents the range 30-52 % denaturant, in which all visible bands were found.

2me13:0, cy17 and cy19 and accounted for the diffuse grouping of the crash-site samples. Of these PLFA, cy17:0 and cy19:0 are common components of Gram negative bacteria. Principle component 2 was most strongly influenced by 10me16:0, representative of the sulfate-reducing bacteria *Desulfobacter* (Dowling *et al.*, 1986), 16:0, 18:1w9c and a15:0. Representative of the presence of anaerobic bacteria, DMAs were present in significantly greater quantities in the samples taken from the crash site. In these samples the relative DMA concentrations (DMA/PLFA) followed the order, capillary fringe> saturated>vadose zone. The relative DMA concentration followed no discernable order in the remaining bore-holes.

DGGE analysis of microbial diversity. DGGE analysis of triplicate sub-samples taken within the heavily JP-4 impacted zone showed strong and reproducible banding and stratification at all depths (Figure 2). Two sequences were recovered from all three depths and represented unknown organisms from the β-subgroup proteobacteria. Four bands were absent from the saturated zone, representing an α-and 3 β-proteobacterial sequences. The capillary fringe displayed 3 unique bands, representing an uncultured bacterium associated with the Flexibacter-Cytophaga-Bacteroides phylum, an α and a β-proteobacterium. A member of the *Cytophaga*-subgroup was found in both capillary and saturated soil. Five sequences were recovered from only the saturated zone, and represented members of the α, β-and ε-subgroup proteobacteria, *Flexibacter-Cytophaga-Bacteroides*-phylum and the Gram-positive phylum. None of these bands were visible outside the plume, suggesting that these organisms were active in remediation of the JP-4.

CONCLUSION

Shifts in biomass content and community structure throughout the JP-4 contaminated soil samples were detected related to the increased VOC concentration. Measured as PLFA, the highest viable biomass levels were detected in the most highly contaminated site.

ACKNOWLEDGEMENTS

This work was supported by a grant from the Strategic Environmental Research and Development Program (SERDP contract number 11XSY887Y).

REFERENCES

Balkwill, D.L., Leach, F.R., Wilson, J.T., McNabb, J.F., and White, D.C. 1988. "Equivalence of microbial biomass measures based on membrane lipid and cell wall components, adenosine triphosphate, and direct counts in subsurface sediments." *Microbial Ecol 16*:73-84.

Dowling, N.J.E., Widdel, F., and White, D.C. 1986. "Phospholipid ester-linked fatty acid biomarkers of acetate-oxidizing sulfate-reducers and other sulfide forming bacteria." *J Gen Microbiol 132*:1815-1825.

Edlund, A., Nichols, P.D., Roffey, R., and White, D.C. 1985. "Extractable and lipopolysaccharide fatty acid and hydroxy acid profiles from *Desulfovobrio* species." *J Lipid Res 26*:982-988.

Fang, J., Barcelona, M.J., and West, C. 1997. "The use of aromatic acids and phospholipid ester linked fatty acids for delineation of processes affecting an aquifer contaminated with JP-4 fuel." In Egenhouse, R.E. (Ed), *Molecular markers in environmental geochemistry.* pp. 65-76. American Chemical Society Symposium 671, Washington DC. American Chemical Society.

Findlay, R.H. and Dobbs, F.C. 1993. "Quantitative description of microbial communities using lipid analysis." In Kemp, P.F., Sherr, B.F., Sherr, E.B., Cole, J.J. and Boca Raton, R.L. (Eds), *Handbook of methods in aquatic microbial ecology* pp. 271-284. Lewis Publishers.

Guckert, J.B., Antworth, C.P., Nichols, P.D., and White, D.C. 1985. "Phospholipid ester-linked fatty acid profiles as reproducible assays for changes in prokaryotic community structure of estuarine sediments." *FEMS Microbiol. Ecol. 31*:147-158.

Kates, M. 1986 (Ed). *Techniques in lipidology: isolation, analysis and identification of lipids*. Second edition, Amsterdam: Elsevier Press.

Maidak, B.L., Olsen, G.J., Larsen, N., Overbeek R., McCaughey, M.J., and. Woese, C.R. 1997. The RDP (Ribosomal Database Project). *Nucleic Acids Res. 25*:109-111.

Mayberry, W.R., and Lane, J.R. 1993. "Sequential alkaline saponification /acid hydrolysis/esterification: a one tube method with enhanced recovery of both cyclopropane and hydroxylated fatty acids." *J. Microbiol. Methods. 18*:21-32.

Muyzer, G., de Waal, E.C., and Uitterlinden, A.G. 1993. "Profiling of microbial populations by denaturing gradient gel electrophoresis analysis of polymerase chain reaction amplified genes codign for 16S rRNA." *Appl Environ Microbiol 59*:695-700.

Ringelberg, D.B., Davis, J.D., Smith, G.A., Pfiffner, S.M., Nichols, P.D., Nickels, J.S., Hensen, J.M., Wilson, J.T., Yates, M., Kampbell, D.H., Read, H.W., Stocksdale, T.T. and White, D.C. 1989. "Validation of signature polarlipid fatty acid biomarkers for alkane–utilizing bacteria in soils and subsurface aquifer materials." *FEMS Microbiol Ecol 62*:39-50.

Ringelberg, D.B., Townsend, G.T., DeWeerd, K.A., Sulita, J.M., and White, D.C. 1994. "Detection of the anaerobic dechlorinating microorganism *Desulfomonile tiedjei* in environmental matrices by its signature lipopolysaccharide branch-long-chain hydroxy fatty acids". *FEMS Microbiol Ecol 14*:9-18.

Stephen, J.R, Chang, Y-J, Macnaughton, S.J., Lowalchuk, G.A., Leung, K.T., Flemming, C.A and White, D.C. 1999. "Effect of toxic metal on indigenous soil β-subgroup proteobacterium ammonia oxidizer community structure and protection against toxicity by inoculated metal-resistant bacteria." *Appl Environ Microbiol 65*:95-101.

White, D.C., Davis, W.M., Nickels, J.S., King, J.D., and Bobbie, R.J. 1979. "Determination of the sedimentary microbial biomass by extractable lipid phosphate." *Oecologia 40*:51-62.

Wilkinson, S.G. 1988. Gram-negative bacteria. In Ratledge, C., and Wilkinson S.G. (Eds), *Microbial lipids*, pp. 299-488, London: Academic Press.

ASSESSING CRUDE OIL BIOREMEDIATION USING CATABOLIC AND METABOLIC MICROBIAL BIOSENSORS

Jake G. Bundy[1,2], Colin D. Campbell[2], Graeme I. Paton[1]

1: Aberdeen University, Aberdeen, UK
2: Macaulay Land Use Research Institute, Aberdeen, UK

Abstract: microbes can be used as whole-cell biosensors when transformed with a suitable reporter mechanism, such as bioluminescence. They can be used to measure either the toxicity of a system (metabolic sensors) or the presence of hydrocarbons (catabolic sensors). Six different bioluminescent biosensors – three metabolic and three catabolic – were used to assess the progress of bioremediation in a series of soil microcosms. The microcosms were spiked with five different oils: diesel oil; and four different crude oils spanning a range of physical properties from light to very heavy. The microcosms were nutrient-amended and incubated for 120 days, and the biosensors tested against extracts of the incubated soils. The catabolic biosensors responded positively and sensitively to hydrocarbons in all of the different oils; one was still recording a large positive response at the end of the incubation period. The responses were greatest for the light and medium crude oils. The metabolic biosensors were sensitive to all of the oils; there was a slight initial decrease in toxicity, followed by a sharp increase in toxicity after 30 days incubation, which was not ameliorated over the course of the experiment. The dualistic approach, utilizing both types of biosensor, offers advantages over traditional biosensor technology.

INTRODUCTION

The genes for bioluminescence can be cloned into bacteria to produce biosensors: the light output is a sensitive and quantifiable reporter mechanism (Meighen, 1988). These biosensors can be of two kinds: catabolic biosensors have the *lux* genes inserted behind genes for hydrocarbon catabolism. Exposing them to hydrocarbons activates the degradation genes, and stimulates bioluminescence. Conversely, metabolic biosensors express *lux* constitutively, and in this case bioluminescence is a measure of the metabolic activity of the cell: hence exposing them to toxic concentrations of hydrocarbons will cause a reduction in light output. Clearly catabolic biosensors are ideal for measuring oil pollution: they respond sensitively (mg/L level concentrations) and fairly specifically to different hydrocarbons or groups of similar hydrocarbons. In addition there is the potential of assessing bioavailable concentrations of hydrocarbons: this has the advantage over traditional chemical analysis of measuring the fraction of oil that is likely to be available for bioremediation. Metabolic biosensors do not respond to specific compounds, and thus may be less useful in quantifying general oil pollution; however their properties are particularly suitable for assessing ongoing bioremediation projects. They can be used to determine if microbial toxicity is high enough to constrain bioremediation (including toxicity caused by copollutants), and if bioremediation is successful in decreasing relative toxicity. Thus the two types of biosensor used together can provide complementary information on the progress of a bioremediation project.

Previous studies have shown that microbial biosensors can be successfully used in evaluating bioremediation sites (Burlage *et al.*, 1994). This study was performed to compare the behaviour of a wider range of biosensors and of oils than have previously been used: the relative performance of six different bioluminescent biosensors in assessing the bioremediation of a series of microcosms was evaluated. Four crude oils with a range of properties were used, to determine if biosensors are more suitable for use with a particular class of oil; one refined oil, diesel, was included for comparison.

MATERIALS AND METHODS

Microcosms. Soil: Insch series, Insch association; texture, sandy loam; dystrochrept. The soil was spiked at a level of 5% oil on a dry weight basis, and then amended with NH_4NO_3 and a mixture of K_2HPO_4/KH_2PO_4 at pH 7 to give a final ratio of oil:N:P of 100:6:1. Water was then added to bring the soil to a final moisture content of 80% of water-holding capacity. The soil was stored in glass jars sealed with stretched Parafilm: 15 replicate jars were used for each oil/soil combination, allowing destructive sampling of triplicate jars at 5 time points. The microcosms were incubated in the dark at 25°C. Uncontaminated soil was prepared in the same way and used as a control.

Oils: one refined product (diesel oil) and four crude oils were used. The crude oils were given the labels A, B, C, and D; they covered a range of physical properties. The lightest crude was A; they increased in density up to D.

Each soil/oil combination was extracted in triplicate for biosensor testing (giving 9 determinations for each biosensor/extraction combination: three soil replications, 3 biosensor replications). Water extraction: 20 ml of water added to 10 g soil (wet weight) and treated in an ultrasonic bath for 30 minutes. The supernatant was then cleared of soil particles by centrifuging at 2500*g* for 30 minutes. This was then used as the test solution. Solvent extraction: 10 ml of methanol or of dimethyl sulfoxide was added to 10 g of soil, and then extracted as for the water extraction. The solvent extract was then diluted with water (1:8 extract:water ratio) to give the final test solution used.

Metabolic Biosensors. 1. Microtox acute assay, *Vibrio fischeri*. A vial was rehydrated in 2 ml of 2.2 % NaCl solution held at 5°C and used immediately. Ten μl of cell suspension was then added to a luminometer cuvette containing: 100 μl of 22 % NaCl solution; and 890 μl of test solution. The response was then read on a Jade portable luminometer after 15 minutes.

2. *E. coli* HB101 (pUCD607). The *luxCDABE* genes, coding for all the biochemical requirements for light production, are present on the plasmid

pUCD607. A freeze-dried vial of cells was rehydrated in 10 ml of 0.1M KCl solution and shaken at 200 rpm, 25°C, for 60 minutes. Twenty-five µl of cell suspension was then added, using a multichannel pipette, to 225 µl of test solution in a microtitre plate well, and mixed by pipetting for five seconds. The plates were incubated at 25°C, and the response was measured after 30 minutes exposure using a Lucy Anthos 1 microtitre luminometer.

3. *P. putida* F1 (pUCD607). The experimental protocol was the same as for *E. coli* HB101 (pUCD607), except that: the freeze-dried cells were only rehydrated for 30 minutes; and 5 µl of 4.6M KCl solution was added to the microtitre wells, such that the final assay concentration of KCl was 0.1M. This was necessary to prevent osmotic shocking of the *P. putida*.

Catabolic Biosensors. 1. *P. fluorescens* HK44 (pUTK21). "Naphthalene biosensor"; the *luxCDABE* genes are inserted into the *nah* operon (King *et al.*, 1990). This was used as described in Burlage *et al.* (1994); except that the final cell suspension was concentrated by centrifuging 100 ml of medium, and resuspending the cell pellet in 10 ml of 0.1M KCl solution. The cell suspension was then used as for *E. coli* HB101 (pUCD607), and the plates read after 60 minutes exposure.

2. *E. coli* (pGEc74, pJAMA7). "Octane biosensor"; only contains *luxAB* genes, hence requires decanal to be added to enable the bioluminescence pathway (Sticher *et al.*, 1997). A freeze-dried vial of cells was rehydrated in 10 ml of 0.1M KCl solution and shaken at 200 rpm, 25°C, for 60 minutes. The cell suspension was then used as for *E. coli* HB101 (pUCD607). After 40 minutes exposure, 16 µl of decanal solution (freshly prepared using 140 µl decanal in 20 ml ethanol) was added to each well, and the plate read after a further 3 minutes.

3. *E. coli* (pOS25). "Isopropylbenzene biosensor"; contains *luxCDABE* genes (Selifonova and Eaton, 1996). A freeze-dried vial of cells was rehydrated in 10 ml of 0.1M KCl solution and shaken at 200 rpm, 25°C, for 240 minutes. The cell suspension was then used as for *E. coli* HB101 (pUCD607), and the plates read after a further 240 minutes exposure.

RESULTS AND DISCUSSION

The microcosms were sampled at 0, 14, 30, 70, and 120 days. It was found that only the methanol extracts elicited a response from all of the biosensors; hence only the methanol extract data will be reported here.

Metabolic Biosensors. These were slightly inhibited by all of the oils at the beginning of the incubation (day 0). There was a slight decrease in toxicity by the second time point, followed by a sharp jump in toxicity by the third time point (30 days). This increased toxicity remained approximately constant for the remainder of the experiment for *P. putida* F1 (pUCD607) (fig. 1). The *E. coli* HB101

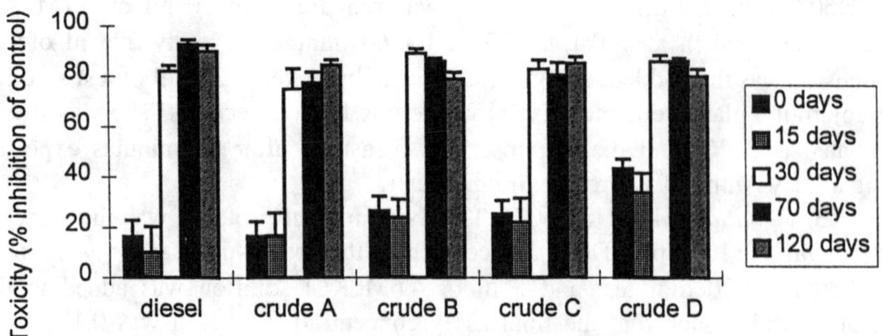

FIGURE 1. Changes in toxicity of methanol extractions to *P. putida* F1 (pUCD607)

(pUCD607) biosensor showed a slightly different pattern: overall the inhibition was less compared to the *P. putida*, and the toxicity tended to decrease with continued incubation (fig. 2). The refined product, diesel, retained higher toxicity compared to the unrefined oils.

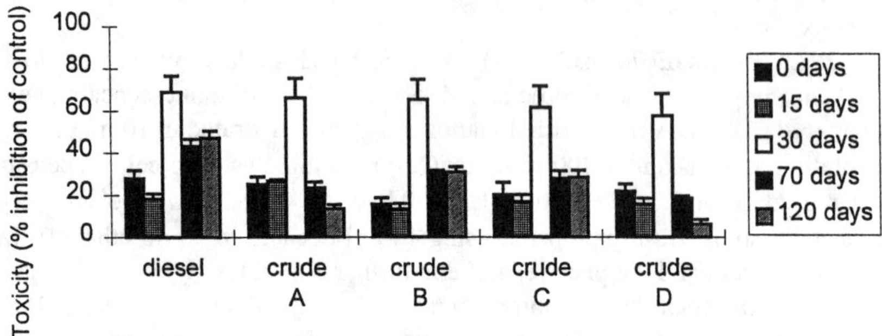

FIGURE 2. Changes in toxicity of methanol extractions to *E. coli* HB101 (pUCD607)

The data for the Microtox acute assay follow a similar trend to the *P. putida* F1 (pUCD607) biosensor results; hence they are not presented here.

This increase in microbial inhibition could be caused by an accumulation of toxic intermediates of hydrocarbon degradation. Shen and Bartha (1994) also observed that bioremediation was accompanied by an acute increase in microbial toxicity (as measured by Microtox), but that this toxicity then disappeared following lengthy bioremediation. These results have implications for practical bioremediation projects: in particular, the hydrocarbon degrader *P. putida* was inhibited. Further research would be necessary to determine what level of biosensor inhibition corresponds to a genuine toxic constraint on the indigenous microorganisms. The biosensors could then be used to determine if a slowly progressing bioremediation project was caused by toxicity of degradation intermediates, or if there was some other cause which could be addressed.

Catabolic biosensors. These sensors respond such that the light output is increased in the presence of hydrocarbons. These were very sensitive: the octane biosensor, in particular, had a response at day 0 of greater than 2500% of the control for some oils (fig. 3). The signal declined with time, indicating disappearance of the

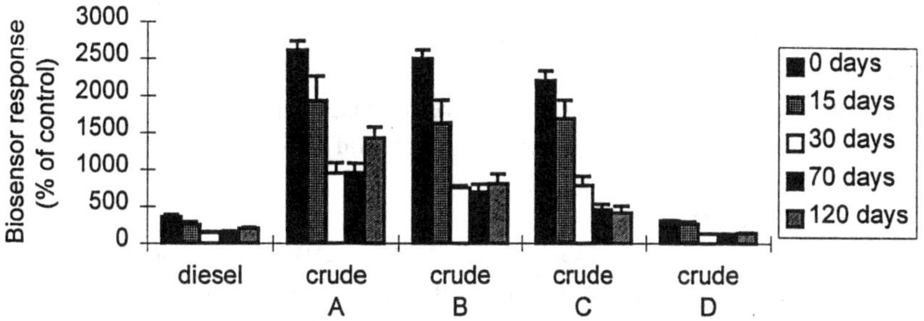

FIGURE 3. Response of E. coli DH5α (pGEc74, pJAMA7) (octane biosensor) to methanol extractions

readily biodegradable *n*-alkanes to which the biosensor responds. However, the response must be interpreted with care: the signal at the final time point increases significantly for crude A. This indicates that the response is a combination of biosensor induction and toxic inhibition.

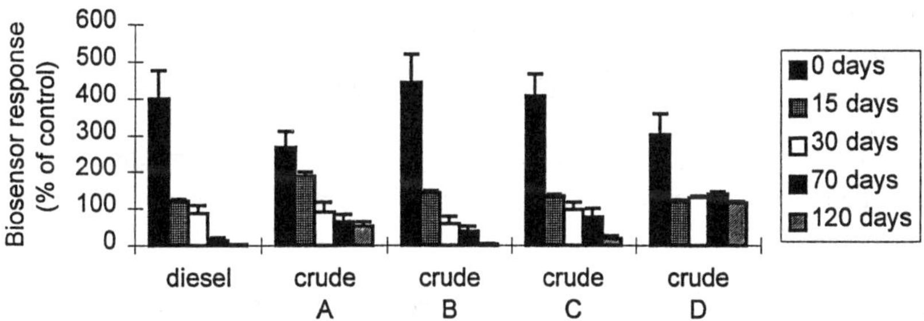

FIGURE 4. Response of P. fluorescens HK44 (pUTK21) (naphthalene biosensor) to methanol extractions

The naphthalene biosensor also showed initial induction by all of the oils, although to a lesser degree than by the octane biosensor (fig. 4). The signal declined rapidly with incubation time, and there was essentially no positive response by the day 30 time point or after.

Different oils elicited different responses. The octane biosensor had a poor response to the heaviest oil, crude D (fig. 3): this has a smaller fraction of short-chain alkanes. Conversely, the naphthalene biosensor responded slightly better to the medium-weight crudes B and C (fig. 4): the light crude A will contain few aromatic hydrocarbons, whereas the heavy crude D contains more three-ring and above aromatics (which do not cause induction of the biosensor). Similarly, the naphthalene biosensor responds well to the medium-range distillate diesel.

CONCLUSIONS

Microbial biosensors can be used to assess the progress of bioremediation on the microcosm scale; it should be possible to apply the same techniques to a field-scale project. Important complementary information is obtained by using a combination of metabolic (toxicity) and catabolic (hydrocarbon) biosensors. Integration of these methods with developments in real-time and on-line monitoring highlight the future of the environmental biosensor.

ACKNOWLEDGMENTS

Thanks are due to Professor GS Sayler, Professor RW Eaton, and Professor JR van der Meer, who supplied the naphthalene, isopropylbenzene, and octane biosensors. The crude oils were a gift from Shell International. The studentship for J Bundy was funded by Aberdeen Research Consortium.

REFERENCES

Burlage, R. S., A. V. Palumbo, A. Heitzer, and G. S. Sayler. 1994. Bioluminescent reporter bacteria detect contaminants in soil samples. *Appl. Biochem. Biotechnol.* 45/46: 731-740.

King, J.M.H., P. M. DiGrazia, B. Applegate, R. Burlage, J. Sanseverino, P. Dunbar, F. Larimer, and G. S. Sayler. 1990. Rapid, sensitive bioluminescence reporter technology for naphthalene exposure and biodegradation. *Science* 249: 778-781.

Meighen, E.A. 1988. Enzymes and genes from the *lux* operons of bioluminescent bacteria. *Ann. Rev. Microbiol.* 42: 151-176.

Selifonova, O. and R. W. Eaton. 1996. Use of an *ipb-lux* fusion to study regulation of the isopropylbenzene catabolism operon of *Pseudomonas putida* RE204 and to detect hydrophobic pollutants in the environment. *Appl. Environ. Microbiol.* 62: 778-783.

Shen, J. and R. Bartha. 1994. On-site bioremediation of soil contaminated by No. 2 fuel-oil. *International Biodeterioration & Biodegradation* 33: 61-72.

Sticher, P., M. C. M. Jaspers, K. Stemmler, H. Harms, A. J. B. Zehnder, and J. R. van der Meer. 1997. Development and characterization of a whole-cell bioluminescent sensor for bioavailable middle-chain alkanes in contaminated groundwater samples. *Appl. Environ. Microbiol.* 63: 4053-4060.

QUANTIFICATION OF CATECHOL 2,3-DIOXYGENASE GENES FOR MONITORING BIOREMEDIATION

Matthew B. Mesarch, Cindy H. Nakatsu, and Loring Nies
(Purdue University, West Lafayette, IN)

ABSTRACT: Current techniques for monitoring pollutant-degrading microorganisms are limited to standard culturing methods despite documented deficiencies. Fortunately, techniques based upon detection of cellular genetic elements can more sensitively monitor microbial populations without requiring culturing. Aerobic bioremediation of aromatic hydrocarbons generally requires the use of ring attacking and ring cleavage dioxygenase enzymes. Two types of ring cleavage dioxygenases exist, meta and ortho cleavage. Meta cleavage dioxygenases are more extensively characterized. Polymerase chain reaction (PCR) primers have been designed using these characterizations that can detect a subclass of dioxygenase genes encoding enzymes responsible for the degradation of toluene, phenol, xylenes, and the single ring metabolites of biphenyl and naphthalene. These primers were used to quantify meta cleavage dioxygenase genes. In pure cultures this technique has been used to detect and enumerate eight different bacteria with various substrate specificities. Through optimization of PCR conditions as few as 10^3 genes were detected, translating to a detection limit of at least 10^3 cells. Overall this technique proves to be more sensitive than standard culturing methods and can be used to more accurately monitor pollutant-biodegrading microorganisms. Thus, bioremediation in the field can be optimized to more efficiently use resources based upon the response of the appropriate microorganisms.

INTRODUCTION

Bioremediation has become an acceptable treatment method for many environmental contaminants including petroleum products such as diesel fuel, gasoline and other fuels. While intrinsic bioremediation has proven to be effective in cleaning up slow-moving contaminant plumes, more rapidly moving plumes require a more aggressive approach. In such instances engineered bioremediation is used. Often this entails amending the groundwater with nutrients or oxygen to increase the numbers and activity of bacteria with the capability to degrade petroleum contaminants. Unfortunately such amendments are often performed with little or no knowledge of whether the bacteria responsible for bioremediation actually benefit from such amendments. Typically only standard culture techniques are used to monitor microbial population sizes at bioremediation sites, but these techniques are known to be inaccurate. Less than ten percent of the total soil microbial community is believed to be culturable (Tate, 1995). Molecular genetic techniques such as the polymerase chain reaction (PCR) are more sensitive than culture techniques because they do not require the cultivation of bacteria. They can be applied to field samples directly, allowing faster access to site data that more completely characterizes the microbial community than previous techniques. These techniques can also be performed quickly since samples can be processed and analyzed in a day.

Our goal is to develop a technique to detect and quantify bacteria that can aerobically degrade aromatic hydrocarbons. Nearly all aerobic pathways for aromatic hydrocarbon degradation contain dioxygenase enzymes which are used to both activate and cleave the aromatic ring (Smith, 1990). Meta cleavage dioxygenase enzymes are involved in the degradation of many aromatics found in contaminated soils (Figure 1). Our long-term goal is to use sequences for meta cleavage dioxygenase genes to construct dioxygenase-specific PCR primers to quantify dioxygenase copy number in soil microbial DNA extracts. Using dioxygenase copy number we will be able to calculate numbers of aromatic hydrocarbon-degrading bacteria in soil samples.

MATERIALS AND METHODS

PCR primers were designed by comparing the sequences of eight different dioxygenase-containing bacteria that could degrade a wide spectrum of aromatic hydrocarbons (phenol, toluene, xylenes, biphenyl, and naphthalene). Regions of nucleotide conservation were chosen for use as primers. The primers chosen (DEG-F and DEG-R) should amplify a 238 bp dioxygenase fragment from all eight bacteria studied. The primers were tested on DNA extracted from the eight bacterial strains known to degrade different hydrocarbons (Table 1). Bacteria were grown in sterile selection media and DNA was extracted using BIO101 SpinKits (Bio101) after three consecutive freeze-thaw cycles to lyse the cells. DNA concentrations were measured by fluorometry.

TABLE 1. Isolates detected by the dioxygenase-specific primers.

Isolate	Substrate Specificity
Pseudomonas sp. strain CF600	phenol
Pseudomonas putida H	phenol
Pseudomonas putida P35X	phenol
Pseudomonas putida HS1	toluene
Pseudomonas putida MT-2	toluene
Pseudomonas sp. strain IC	biphenyl
Pseudomonas aeruginosa JI104	biphenyl
Pseudomonas sp. strain PpG7	naphthalene

PCR mixtures used contained (per reaction) 1x PCR buffer (Promega), 0.2 mM each dNTP (Pharmacia), 3.0 mM $MgCl_2$ (Promega), 0.250 μM each primer, 1 U *Taq* DNA polymerase, and between 1 and 10 ng template DNA. Amplification was performed in a PTC-100 Programmable Thermal Cycler (MJ Research, Inc.) beginning with an initial 95°C denaturation step for 5 minutes, 30 cycles of 94°C for 1 minute, 55°C for 1 minute, and 72°C for 2 minutes, and a final 72°C extension step for 10 minutes. All products were visualized on agarose gels stained with ethidium bromide and visualized under ultraviolet light.

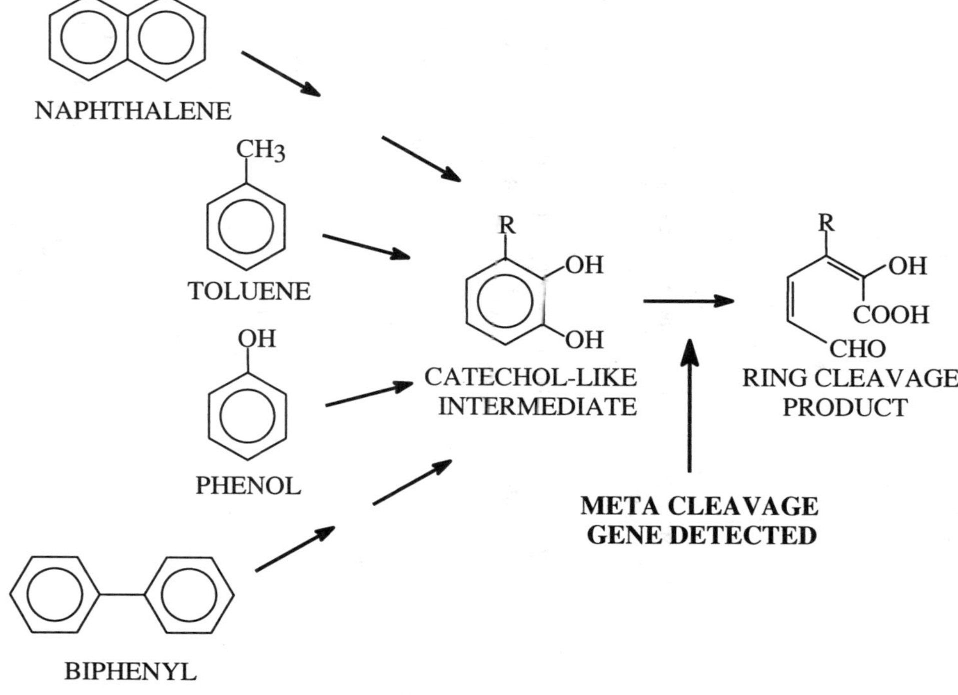

FIGURE 1: Meta cleavage attack on aromatic hydrocarbons

The competitor was constructed through the use of a composite primer (Jin et. al., 1994). This competitor acts as an internal standard. When known amounts of it are amplified with an unknown amount of target DNA, the target DNA can be quantified by comparing the relative amounts of target and competitor DNA before and after amplification. The competitor sequence was purified, ligated to a vector, and transformed into *E. coli* using standard methods (Sambrook et. al., 1989). Competitor plasmids were purified using standard methods and quantified fluorometrically. Known amounts of competitor were then spiked into PCR reactions using the same protocol described above. Competitive PCR experiments were run using several reactions containing a dilution series of competitor and a constant amount of the DNA to be quantified. These samples were run on agarose gels and relative signal intensities compared using image analysis.

RESULTS AND DISCUSSION

Using DNA isolated from the eight different bacteria, the expected 238 bp sequence was amplified indicating that the primers were capable of specifically amplifying dioxygenase sequences. These bacteria possessed a wide variety of substrate specificities ranging from single ring compounds such as phenol and toluene to double ring compounds like naphthalene and biphenyl (Table 1). Many of the isolates listed can also degrade or transform other aromatic contaminants. *Pseudomonas putida* MT-2, for example, is also capable of degrading *m*- and *p*-xylenes (Assinder and Williams, 1990). Because these primers can be used to detect organisms with broad substrate specificities, they can be used to check for the presence of aromatic-degrading bacteria at sites contaminated by different wastes such as fuels, wood preserving wastes, and PCBs. Additionally, many of these isolates originated from outside the United States implying that the primers have widespread geographic applicability.

A dilution series of both *P. putida* HS1 and *P. sp.* strain PpG7 was amplified using dioxygenase-specific primers to establish a detection limit. Approximately 10^3 dioxygenase gene copies per reaction could be detected on ethidium bromide-stained agarose gels. Assuming approximately one gene copy per cell this translates to a detection limit of 10^3 cells per reaction. Thus, in DNA extracted from contaminated soil, we would be capable of detecting as few as 10^3 cells per gram of soil.

Using a quantitative competitive PCR technique we were capable of quantifying dioxygenase gene sequences from pure cultures of the organisms listed in Table 1. Dioxygenase gene copy calculated with the quantitative competitive PCR protocol agreed with known values of dioxygenase gene copy number to within an order of magnitude. Quantifying tenfold dilutions of sample DNA resulted in a corresponding tenfold reduction in dioxygenase gene copy number. This suggests that this technique would be sensitive enough to detect changes in dioxygenase gene copy number (and therefore numbers of aromatic hydrocarbon-degrading bacteria) at sites where nutrient or oxygen amendment was used to stimulate aerobic bioremediation. Additionally, it could be used to demonstrate that populations of bacteria within a contaminant plume were greater than populations in uncontaminated zones. Such information would provide valuable evidence in support of natural attenuation where such an alternative was being considered.

CONCLUSIONS

Future work with this technique will involve quantifying dioxygenase gene copy number in DNA isolated from soil samples. This will determine whether or not different soil constituents (e.g., humics) affect competitive quantitative PCR. Once we are capable of quantifying dioxygenase copy number in soils, field conditions will be simulated in order to determine this technique's effectiveness in site characterization and monitoring.

REFERENCES

Assinder, S. J. and P. A. Williams. 1990. "The TOL Plasmids: Determinants of the Catabolism of Toluene and the Xylenes." In A. H. Rose and D. W. Tempest (Eds.), *Advances in Microbial Physiology*, pp. 2-64. Academic Press, New York, NY.

Jin, C., Mata, M., and D. J. Fink. 1994. "Rapid Construction of Deleted DNA Fragments for Use as Internal Standards in Competitive PCR." *PCR Methods and Applications*. *3*: 252-255.

Sambrook, J., Fritsch, E. F., and T. Maniatis. 1989. *Molecular Cloning, a Laboratory Manual*. Cold Spring Harbor Press, Cold Spring Harbor, NY.

Smith, M. R. 1990. "The biodegradation of aromatic hydrocarbons by bacteria." *Biodegradation*. *1*: 191-206.

Tate, R. L. 1995. *Soil Microbiology*. John Wiley & Sons, Inc., New York, NY.

RHEINGOLD PROJECT: ASSESSING REMEDIATION BY MOLECULAR AND ^{14}C-HYDROCARBON MINERALIZATION ASSAYS

Jeremy R. Mason (King's College London, UK)
Caroline Fenwick, Islam Haider, Christoph Henkler (King's College London, UK)
Eckart Bär (Environmental Expert Office, Rösrath, Germany)
L. Anne Glover, Ken Killham (University of Aberdeen, UK)

ABSTRACT: Samples taken from two boreholes on a BTEX contaminated site have been assessed for biodegradative potential and activity by two methods: *ex situ* (^{14}C-toluene mineralization) and *in situ* (DNA extraction and PCR amplification of genes encoding enzymes for aromatic hydrocarbon degradation). In samples taken from a borehole where the BTEX concentrations were high (F1), ^{14}C-toluene mineralization activity was higher than in samples taken from a borehole located downstream of the BTEX plume (F2). PCR amplification of DNA extracted from these sediment samples using primers specific for aromatic ring-hydroxylating Class IIB dioxygenases (*tod*C1) demonstrated that homologous DNA was present in all samples taken from F1. In contrast, samples from F2 did not always yield PCR products with these primers (although ^{14}C mineralization activity was detected at all depths). This suggests that genes other than those encoding Class IIB dioxygenases may be present in some samples, particularly where the BTEX concentration was low. Estimates of biodegradative activity in sediment samples were made, ranging from 1-14 kg toluene tonne sediment^{-1} year^{-1} and assisting in the design of a bioremediation strategy.

INTRODUCTION

In the past many industrial processes including paint manufacture, have used aromatic hydrocarbons such as benzene, toluene, ethylbenzene and xylene (BTEX) as solvents. Frequently these may be detected in the soil and groundwater of a manufacturing site and are therefore the subject of remediation programs. Remediation of contaminated sites may be achieved in a variety of ways, one of which is bioremediation using the indigenous microorganisms to detoxify or eliminate the pollutants *in situ*. Demonstration of the efficacy of bioremediation requires preliminary characterisation of this *in situ* biological activity and continued monitoring of this indigenous activity during the bioremediation process.

The majority of bioactivity assessment methods employed to date were based on *ex situ* measurements of microbial metabolism, which tend to be non-specific and conducted under conditions which are not representative of those in the environment. Molecular methods offer advantages by both their specificity and sensitivity and allow the detection of nucleic acids from both culturable and non-culturable microorganisms. Recently, methods have been developed for *in situ* monitoring of specific catabolic genes and their subsequent expression using direct isolation of DNA and mRNA from contaminated soils (Fleming *et al.*, 1993). Quantification of the former using the techniques of polymerase chain reaction (PCR) and reverse transcriptase-PCR (RT-PCR), enables the estimation of the *in situ* biodegradative capacity (presence of gene in DNA) and biodegradative activity (expression of gene as mRNA), respectively. Therefore, the progress of bioremediation may be monitored and the applicability of treatments such as sparging or nutrient addition may be assessed.

Site Description. The site under investigation has been used for paint manufacture for over 100 years. There are three main areas of contamination, which contain high levels of BTEX. The main studies and method development have concentrated on one plume, using sediment samples taken from two wells: F1 which is in the centre of the BTEX plume and F2, located 40 m downstream. These boreholes were drilled to a depth of 15 m and subsamples were taken at 1 m intervals.

Objectives.
1. To develop methods to assess the potential aerobic biodegradative activity towards BTEX compounds in samples from a contaminated paint manufacturing site.
2. A comparison of *ex situ* and *in situ* methods of bioactivity determination.

MATERIALS AND METHODS

Activity Determination by Hydrocarbon Mineralization. Sediment samples were taken under aseptic conditions and stored at 4°C prior to analysis. Mineralization assays were performed within 36 h of sampling following the method of Fleming *et al.* (1993) using ^{14}C-methyl labelled toluene (5.31 μCi mmol^{-1}). Sediment samples were sparged with moistened air prior to analysis to remove volatile hydrocarbons.

Nucleic Acid Extraction and Amplification. Sediment samples for nucleic acid extraction were taken under anaerobic conditions, snap frozen in liquid nitrogen and maintained at –70°C until analysed. The basic DNA extraction method used skimmed milk protein to block any potential DNA binding sites on the sediment prior to SDS lysis and ammonium acetate precipitation to remove proteins from the extract. The final extract was passed through a polyvinylpolypyrrolidone

(PVPP) spin column to remove any potentially inhibitory, co-purified humic acids (Berthelet *et al.*, 1996, Zhou *et al.*, 1996).

Estimation of Toluene Degrading Bacteria. Aliquots of serial dilution of sediment samples were plated on Hutner's minimal medium plates and incubated in the presence of toluene vapour for 72h at 25°C. Numbers were expressed as colony forming units (cfu) per kg of sediment (Cohen-Bazire *et al.*, 1957).

RESULTS

Design of PCR primers. BTEX hydrocarbons may be oxidised by a number of catabolic pathways. Pairs of PCR primers specific for genes encoding aromatic hydrocarbon-degrading enzymes (Mason *et al.*, 1998) have been designed. These have been screened against microorganisms isolated from the site under investigation. The *tod*C1 primers, specific for the α-subunit of Class IIB dioxygenases (Mason and Cammack, 1992), have been shown to have the broadest specificity, being able to amplify a 521bp band from over 60% of the strains examined (data not shown).

Extraction and amplification of DNA from sediment samples. The DNA extraction procedure was used to extract DNA from sediment samples with known amounts of bacteria that expressed Class IIB dioxygenases. It was determined that the technique was capable of PCR amplification and detection of the genes encoding these enzymes in samples containing down to 10^4 bacteria (Figure 1).

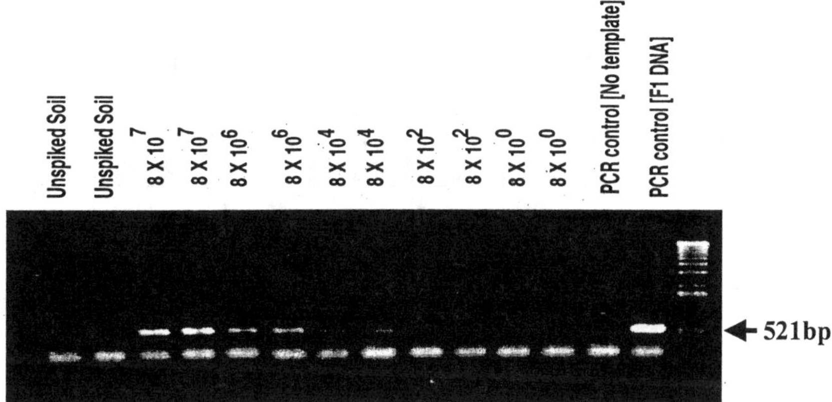

FIGURE 1. Agarose gel analysis of PCR products from DNA extracted from spiked sediments. DNA was extracted from sediment samples (0.25g) spiked with known numbers of the bacterium *Pseudomonas putida* F1 as indicated in lanes 3-12. PCR amplification was performed using *todC1* primers, specific to Class IIB aromatic ring hydroxylating dioxygenases (521bp product).

Comparison of *ex situ* (substrate mineralization) and *in situ* (PCR) estimation of biodegradative potential. *Ex situ* mineralization, measured as the rate of $^{14}CO_2$ evolution from ^{14}C-methyl-labelled toluene demonstrated activity in all samples taken from boreholes F1 and F2 down to 14 m (Figure 2).

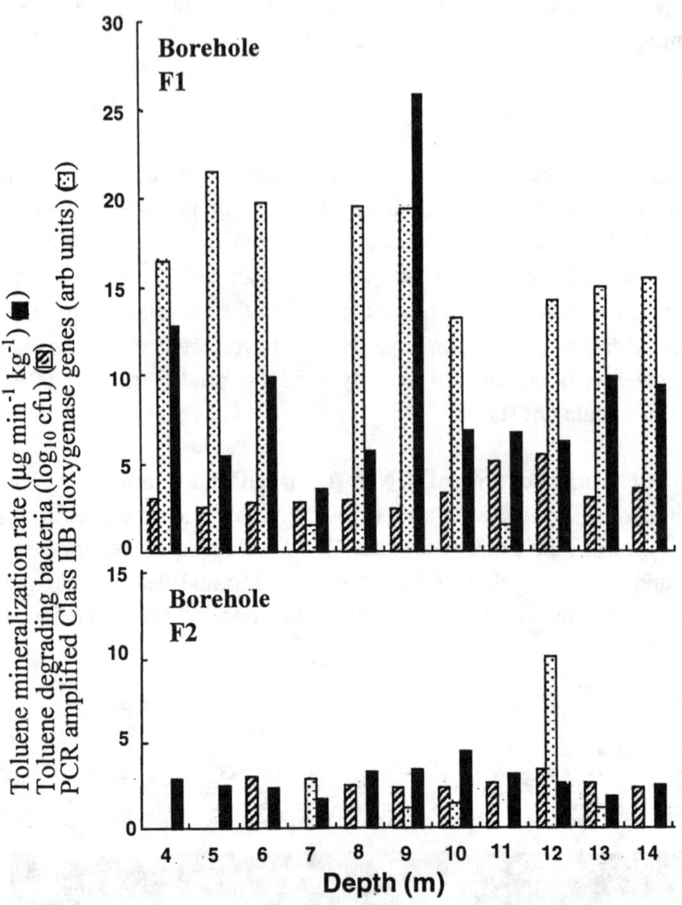

FIGURE 2. Analysis of biodegradative potential and activity in borehole sediment samples.

The activities ranged from 3.5 to 25.8 µg min^{-1} kg^{-1} in sediments from borehole F1 and from 1.7 to 4.5 µg min^{-1} kg^{-1} in sediment from borehole F2. This would correspond to a remediation rate of 1-14 kg tonne sediment^{-1} year^{-1}.

In borehole F1, maximum activity was measured at a depth of 9m although regions of high biodegradative potential were also detected at 4-6m and 14-15m. Areas where mineralization activity was highest: 9m and below 12m, has notably lower concentrations of BTEX than the rest of the sediment column (< 1000 μg kg^{-1}). The presence of free phase on the surface of the groundwater at 6-7 m is the most likely reason for low bacterial activity in this region. PCR amplification of DNA extracted from sediments of the F1 borehole demonstrated the presence of genes homologous to the *tod*C1 primers at all depths (Figure 2). The profile of distribution of these sequences bears a clear relationship to the profile of ^{14}C-toluene mineralization. In contrast the numbers of bacteria isolated by culture on Hutner's minimal medium in toluene vapour did not vary greatly with depth. Regions of high mineralization activity but low PCR amplification (for example 11 m) suggest the presence of microorganisms with alternative degradative pathways. Samples for which low ^{14}C-toluene mineralization activity but high potential activity (measured by PCR amplification) were demonstrated (for example 5 and 8m), indicate areas where biodegradation is being inhibited. Such information should be taken into consideration when designing a bioremediation strategy.

Although mineralization activity was detected in sediments of all depths taken from the F2 borehole, bacterial colonies were found only in those sediment samples taken from 6 m or below (Figure 2). BTEX concentrations in F2 sediments were much lower than in F1 (< 1000 μg kg^{-1}) at 8m or below. Above 8m, the BTEX concentrations were below the limit of detection (10 μg kg^{-1}). This would therefore leave the microbial community with only a low concentration of BTEX as a carbon source. PCR products were detected only at 3m and at 7m and below, although not at all depths. Clearly the microbial community present is capable of low mineralization activity but it appears that some of the pathways used are different to that encoded by the *tod*C1 gene.

CONCLUSIONS

1. A method for the extraction of nucleic acid from sediment samples has been successfully developed, which yields DNA of sufficient quality to be PCR amplified.
2. The presence of the *tod*C1 gene encoding the α-subunit of Class IIB dioxygenase enzymes may be successfully detected by PCR amplification of genomic DNA extracted from sediments contaminated with BTEX.
3. *In situ* measurements of the biodegradative capacity of sediment samples by PCR amplification of *tod*C1 genes were not related to the numbers of toluene-degrading microorganisms isolated from those samples.
4. With a few notable exceptions, *ex situ* measurements of biodegradative activity (measured by ^{14}C-methyl toluene mineralization) had a similar trend down the profile to that of the detection of *tod*C1 genes by PCR amplification.
5. Estimates of biodegradative activity in sediment samples were made, ranging from 1-14 kg toluene tonne $sediment^{-1}$ $year^{-1}$ and assisting in the design of a bioremediation strategy.

REFERENCES

Berthelet, M., L. G. Whyte, C.W. Greer, 1996. "Rapid, Direct Extraction Of DNA From Soils For PCR Analysis Using Polyvinylpolypyrrolidone Spin Columns." *FEMS Microbiology Letters* 138 (1): 17-22.

Cohen-Bazire G, W.R. Sistrom, R.Y. Stanier, 1957. "Kinetic Studies Of Pigment Synthesis By Non Sulphur Purple Bacteria." *Journal of Cellular and Comparative Physiology*, 49:25-68.

Fleming, J.T., J. Sanseverino, and G.S. Sayler, 1993. "Quantitative Relationship Between Naphthalene Catabolic Gene Frequency And Expression In Predicting PAH Degradation In Soils At Town Gas Manufacturing Sites." *Environmental Science and Technology*. 27: 1068-1074.

Mason, J.R. and R. Cammack, 1992. "The Electron-Transport Proteins Of Hydroxylating Bacterial Dioxygenases." *Annual Review of Microbiology*. 46: 277-305.

Mason, J.R. I. Haider, C. Henkler and C. Fenwick (1998). "Assessment Of Biodegradative Potential And Activity In Contaminated Sites: Molecular Approaches To Environmental Monitoring." *Biochemical Society Transactions*. 26: 694-697.

Zhou, J., M. A. Bruns and J. M. Tiedje. 1996. "DNA Recovery From Soils Of Diverse Composition." *Applied and Environmental Microbiology*. 62(2): 316-322.

ACKNOWLEDGEMENTS

This work was supported in part by a grant from ICI Paints Ltd.

PROJECT RHEINGOLD: RAPID BIOSENSOR ANALYSIS OF SOIL AND GROUNDWATER TOXICITY

L. Anne Glover, Hedda Weitz, H. Ling Kuan, Sofia Sousa and
Ken Killham (University of Aberdeen, UK)
E. Bär (Environmental Expert Office, Rösrath, Germany)
Rolf Henkler (ICI Paints, UK)

ABSTRACT: A toxicity survey can provide invaluable information prior to managed bioremediation of industrially contaminated sites. Furthermore, toxicity monitoring is often required during site remediation to monitor progress. Conventional toxicity assessment involves analytical techniques such as spectrometry, chromatography (HPLC and GC), GC-mass spectrometry and atomic absorption techniques for the determination of heavy metals. These techniques are powerful and sensitive to ppm and ppb levels. However, the ability to detect a compound does not provide information regarding the biological effects of that compound, its persistence in the environment, or information that can be employed in the design of a remediation strategy. *lux*-marked bacterial biosensors have been found to be rapid and sensitive indicators of toxicity of a wide range of organic and inorganic pollutants. This paper reports the use of *lux*- marked *Escherichia coli* (selected as a general indicator) and *lux*-marked pseudomonads (selected for site relevance) to assess the toxicity of soils and groundwater at a BTEX contaminated site. This approach, coupled with a regime of sample manipulation, allows the determination of the nature of the toxicity as well as providing information on which to base a remediation strategy.

INTRODUCTION

The toxicity assessment described here was carried out at a paint manufacturing site (approximately 250 acres), contaminated primarily with volatile organic pollutants such as benzene, toluene, ethylbenzene and xylene, collectively known as BTEX. These compounds are a concern at many locations worldwide due to their high toxicity and potential carcinogenic and teratogenic effects even at very low concentrations. The problems associated with any contaminated site can be approached in a variety of ways. Physical decontamination does not destroy contaminants but moves them to a designated area. In contrast, bioremediation has been promoted as a more environmentally and socially responsible approach to dealing with contaminated land and water which often provides a more cost-effective solution. Bioremediation processes depend upon the activity of the indigenous microbial population to metabolise the pollutants. Therefore bioremediation campaigns focus on promoting the maximum microbial activity at the site of contamination by alleviating any toxic constraints and removing any adverse environmental constraints including such factors as lack of oxygen, restrictive pH, absence of essential nutrients or lack of moisture.

Objective. The aim of our site toxicity assessment was as follows:

- to carry out a rapid determination of which areas of the site were contaminated
- to determine what the nature of the contamination was (e.g. volatile organic, non-volatile organic, inorganic)
- to compare biosensor analysis with traditional chemical analyses
- to determine whether there were any constraints to bioremediation and whether these constraints could be alleviated
- to design the best strategy for remediation of the site

MATERIALS AND METHODS

Samples were collected from boreholes to enable groundwater and sediment sampling at a variety of depths.

Soil and Groundwater Sampling. Sediment samples were mainly coarse sand with very low clay content (<1%). The sediment was obtained by pushing 500 g glass jars into the sediment core which was held under a nitrogen blanket. The jars were sealed with Teflon-faced metal lids and stored for no more than a week at 4°C before analysis. Water samples were collected from a pump, which had been running for 2 weeks, operating at 4 ml s^{-1}, into the bottom of 40 ml Wheaton® vials, filled to capacity and capped to avoid any loss of volatiles. Turbulence and bubble formation were avoided. Samples were stored at 4°C for no longer than 1 week before analysis.

Preparation of Biosensors. Biosensors (either *Pseudomonas fluorescens* 10586s pUCD607 [Amin-Hanjani *et al.*, 1993] or *Escherichia coli* HB101 pUCD607 [Rattray *et al.*, 1990]) were grown up to late log phase and freeze-dried in 1 ml aliquots and stored at –20°C until required (Sousa *et al.*, 1998). When required, freeze-dried biosensor was resuscitated for 1 h in 10 ml of 0.1 M KCl at 25°C (Paton *et al.*, 1995).

Biosensor Analysis. Triplicate 900 μl samples were added to 100 μl biosensor reagent in a 1 ml luminometer cuvette. The sample was mixed and exposed to a 5 or 15 minute bioassay at 20°C. Percentage maximum bioluminescence for each sample was calculated against a blank of double deionised water adjusted to pH 5.5 and the light output determined using a Bio-orbit 1251 luminometer with a Multiuse software package (Rattray *et al.*, 1990). Alternatively, uncontaminated groundwater was used as a blank. The organic content and the heavy metal content of the groundwater samples was carried out as described by Sousa *et al.* (1998). Sediment was prepared for analysis as described by Sousa *et al.* (1998). Statistical analysis of the data for both bioluminescence-based assays and heavy metal determinations were obtained using a Microsoft Excel 4 spreadsheet package.

Determination of Nature of Toxicity. The nature of the toxicity in each of the samples was dissected essentially by the method described in Sousa *et al.* (1998).

Figure 1 illustrates the procedures adopted to factor out toxicity contributed by volatile organics, non-volatile organics, heavy metals, pH constraints etc.

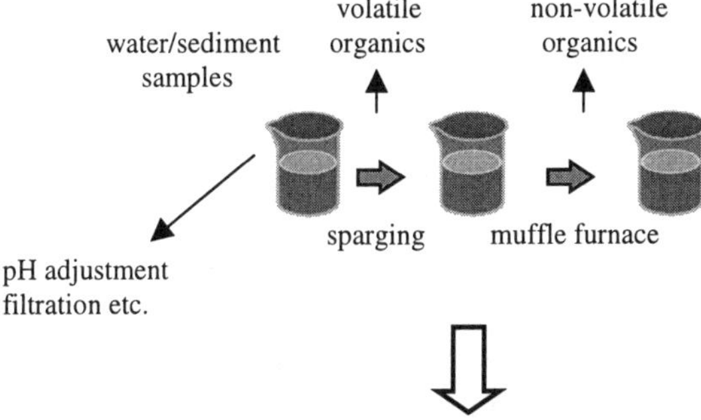

BIOSENSOR ANALYSIS BEFORE AND AFTER MANIPULATION

IDENTIFICATION OF CONSTRAINTS TO BIOREMEDIATION

MANAGEMENT OF CONTAMINATED SITES

FIGURE 1. Diagram of sample manipulation regime. Total toxicity of the sample can be derived from biosensor analysis of the un-manipulated sample. After sparging, biosensor analysis indicated what proportion of the total toxicity could be assigned to volatile hydrocarbon. After muffle furnace, biosensor analysis indicated what proportion of the toxicity was due to non-volatile organics. Any residual activity could be attributed to inorganic toxicity.

RESULTS AND DISCUSSION

Biosensor Analysis and Dissection of Groundwater and Sediment Toxicity. Samples of both groundwater and sediment were exposed to the sample manipulation regime illustrated in Figure 1 and the results are illustrated in Figure 2. Both the groundwater (W1 and W2) samples showed some toxicity with W1 being much more toxic. W1 had about 55% of toxicity due to volatile hydrocarbons that could be removed on sparging. Almost all the toxicity of W1 could be removed by muffle furnacing indicating that the remaining toxicity was due to non-volatile hydrocarbon with no underlying inorganic or environmental constraint. The toxicity of the sediment samples (S1 and S2) could be assessed in the same way. S1 had about 20% toxicity due to volatile hydrocarbon, with the remainder due to non-volatile hydrocarbon. S2 had some underlying inorganic constraints that became obvious with biosensor analysis after muffle furnace. S2 is also an example of a sample where pH posed a significant constraint to bioremediation.

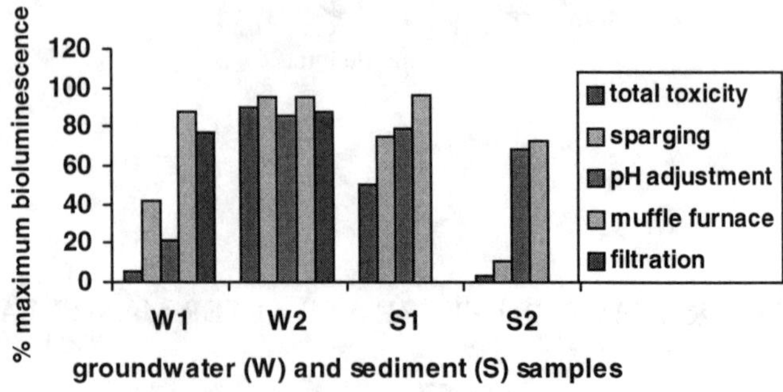

FIGURE 2. Biosensor analysis (*P. fluorescens* pUCD607, 5 min exposure) of groundwater and sediment samples. Samples were exposed to a variety of manipulations and then analysed by the biosensor. In this way, both the nature of the toxicity and whether the toxicity presented a constraint to bioremediation could be assessed.

This approach to toxicity assessment and manipulation indicates the unique way in which the nature of the toxicity can be assessed as well as providing rational information on which to base a bioremediation strategy design.

Comparison of Total BTEX Analysis and Biosensor Analysis. Groundwater samples from different depths were submitted to GC (FID) determination of BTEX and biosensor analysis. Samples were subjected to selective manipulation including sparging, pH adjustment and muffle furnacing to dissect the nature of the toxicity e.g. volatile organics, adverse pH and non-volatile organics (with residual toxicity due to inorganic pollutants). The results are shown in Figure 3.

The biosensors proved to be rapid and reliable indicators of BTEX contamination, providing EC50 values for toluene, for example, of 33 ppm for *lux*-marked *E. coli* cells. Greater sensitivity (EC50- 17 ppm) was achieved by cell washing prior to bioassay. Bioassay results correlated well with chemical analysis, confirming the value of the biosensors for rapid site screening.

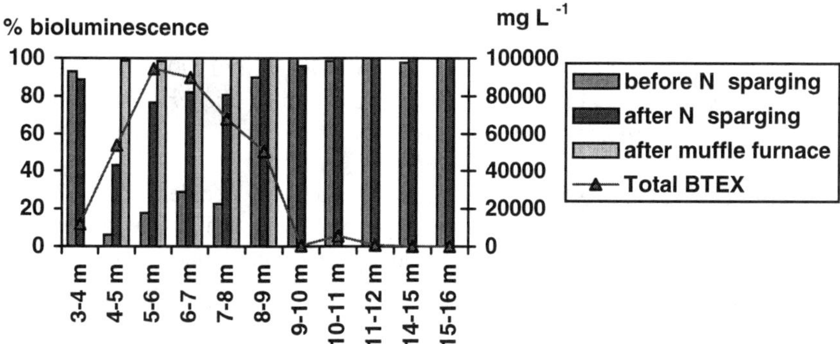

FIGURE 3. Comparison of total BTEX analysis and biosensor analysis. Biosensor analysis (*E. coli* pUCD607, 15 min exposure) was complemented with sample manipulation to reveal the nature of the toxicity.

In conclusion, application of *lux*-biosensors to an industrially contaminated site provided information on the total toxicity of the samples. In addition, biosensor analysis reported upon pollutant bioavailability which includes a combination of chemical, physical and biological characteristics (Paton *et al.*, 1995). This information is critical to allow a rational remediation strategy for contaminated land and cannot be obtained from chemical information alone. Similarly, constraints to bioremediation are difficult to assess by classic analytical methods whereas, the sample manipulations carried out in combination with the biosensor analysis, closely mimic the manipulations which would be imposed as part of a site remediation campaign. This allows a focussed approach to site remediation based on the knowledge of which intervention procedures are required and removes uncertainty from the site management plan.

As biosensor analysis is rapid and cost-effective it can be used not only as a decision support tool but also to monitor an on-going remediation campaign (through individual samples and on-line monitoring) and to provide information on end-points. As biosensor analysis is reporting on the biological impact of pollution, it is also invaluable in formulating risk assessment procedures.

REFERENCES

Amin-Hanjani, S., Meikle, A., Glover, L.A., Prosser, J.I. and Killham, K. 1993. "Plasmid and chromosomally encoded luminescence marker systems for detection of *Pseudomonas fluorescens* in soil." *Mol. Ecol.* 2: 47-54.

Paton, G.I., Campbell, C.D., Glover, L.A. and Killham, K. 1995. "Assessment of bioavailability of heavy metals using *lux*-modified constructs of *Pseudomonas fluorescens*." *Lett. Appl. Microbiol.* 20: 52-56.

Rattray, E.A.S., Prosser, J.I., Killham,K. and Glover, L.A. 1990. "Luminescence-based non-extractive techniques for *in situ* detection of *Escherichia coli* in soil." *Appl. Environ. Microbiol.* 56: 3368-3374.

Sousa, S., Duffy, C., Weitz, H., Glover, L.A., Baer, E., Henkler, R. and Killham, K. 1998. "Use of a *lux*-modified bacterial biosensor to identify constraints to bioremediation of BTEX-contaminated sites." *Environ. Toxicol. Chem.* 17: 1039-1045.

IN SITU ELECTROCHEMICAL MINERALIZATION OF TPH-RELATED COMPOUNDS

John J. Orolin, Vaughn A. Sucevich, Randy T. Drake, EverClear Environmental Technologies Corporation, Milwaukie, Oregon.

ABSTRACT: Biological treatment of water contaminated with total petroleum hydrocarbons (TPH) has become an important process in controlling the pollution of the aquatic environment from industrial sources. Electrochemical oxidation as a treatment of biologically refractory water has been accomplished using highly reactive oxidants such as ozone, hydrogen peroxide, and permanganate. In this report, however, we will compare the performance of a physical process involving a metallic alloy-coated in situ cell system to a chemical process utilizing Regenesis oxygen release compound (ORC) technology.

The application of electrocatalytic chemistry greatly increases the rate of oxidation of a contaminant. This is successfully accomplished by a greater involvement of highly reactive initiators (radicals such as the hydroxyl radical) created by the metallic alloys on the electrochemical cell. The hydroxyl radical typically reacts a million to a billion times faster than ozone and hydrogen peroxide, resulting in greatly reduced treatment costs and system size. The pilot study involved the in situ placement of an electrochemical cell and Regenesis ORC into two separate contaminated monitoring wells. This pilot test ran for a period of ninety (90) days and during this time intrinsic bioremediation parameters of the contaminated monitoring wells, along with the downstream and upstream wells, were analyzed.

INTRODUCTION

Oxidative electrochemical processes demonstrate promising versatility, environmental compatibility, and cost effectiveness. Electrochemical oxidation of organic compounds in aqueous solution is an anodic process. There is consensus among electrochemists that the first step in the oxygen evolution is the anodic discharge of water to produce absorbed hydroxyl radicals (OH), which are oxidizing agents for the majority of organic pollutants. Thus, the proper selection of metallic electrodes coated with active oxides can be utilized to oxidize many types of organic compounds. Also, these metallic electrodes when properly configured (residents and contact time), can be effective tools in oxidizing organic compounds. The electrodes also introduce copious amounts of oxygen into the aquifer necessary for in-situ bioremediation. The goals in field testing were to:

- Encourage destruction of organic compounds to carbon dioxide and water,
- Increase the availability of the electron acceptor, (oxygen), and
- Demonstrate the effectiveness of the process by monitoring contaminant concentrations in the downstream monitoring wells.

MATERIALS AND METHODS

Site Characteristics. The site is an operating gasoline station in Centralia, IL. The areas of environmental concern at the site include former gasoline and fuel oil underground storage tanks. Groundwater was contaminated with high concentrations of gasoline constituents, principally benzene, toluene, ethylbenzene, and xylenes (BTEX). The site geology is characterized by soft shales with interbedded harder sandstones and minor amounts of conglomerate, and the site soils were unsorted and unstratified material consisting of pebbles, cobbles, and boulders in a mixture of sand, silt, and clay. Hydraulic conductivity has been demonstrated between the monitoring wells that were tested. A site map is shown in Figure1.

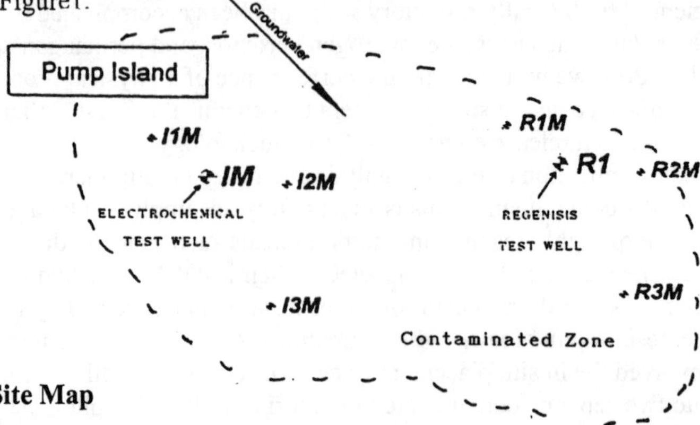

Figure 1. Site Map

Field Experimental Design. The first set of monitoring wells in the in-situ electrochemical cell zone are labeled IM, I1M, I2M, and I3M. IM is the primary monitoring well in which the electrochemical cell was placed into the groundwater just below the well screens at 12 ft (3.7 m) deep. Two other monitoring wells I1M (upstream) and I2M (downstream) are each located 4 ft (1.2 m) from the main well (IM). The last well, I3M, is located 15 ft (4.6 m) from the main well (IM). The monitoring wells in this set are all 12 ft (3.7 m) deep. The second set of monitoring wells in the in-situ Regenesis zone are labeled R1, R1M, R2M, and R3M. R1 is the primary monitoring well, in which the Regenesis ORC was placed into the groundwater just below the well screens. Two other monitoring wells R1M (upstream) and R2M (downstream) are each located 4 ft (1.2 m) from the main well (R1). The last well R3M is located 15 ft (4.6 m) from the main well (R1). The monitoring wells in this set are all 12 ft (3.7 m) deep.

RESULTS AND DISCUSSION

Field Study. Description: As described in the site map found in Figure 1, two (2) sets of monitoring wells, each 12 ft (3.7 m) deep, were used in the field study. Each set of monitoring wells, four in each set, was to compare a chemical process and a physical process for ninety (90) days.

The chemical process used a Regenesis ORC (magnesium peroxide) placed into MW-R1. The other monitoring wells associated with this process are labeled R1M, R2M, and R3M. After initial sampling of these monitoring wells, the Regenesis ORC was placed into MW-R1 and allowed to leach dissolved oxygen slowly into the aquifer. The analytical data on the monitoring well water R1, R1M, R2M, and R3M for a period of ninety (90) days is found in Tables 1 and 2.

Table 1. Regenesis Wells (Chemical Process) Day 0

Analyte mg/l unless noted	R1	R1M	R2M	R3M
Dissolved Oxygen	1.1	0.9	1.3	1.0
Conductance	913	896	903	900
Sulfate	10.6	10.5	12.9	11.0
Nitrate	0.3	0.4	0.4	0.5
Chemical Oxygen Demand	993	894	904	780
ORP	-83	-90	-88	-75
TPH	802	684	690	579
Benzene	0.352	0.261	0.276	0.093
Toluene	4.056	3.821	2.445	3.008
Ehtylbenzene	5.066	3.093	4.114	1.054
Total Xylene	11.953	12.046	3.952	2.054
Total Iron	5.9	4.2	2.5	1.5
Ferrous Iron	0.31	0.33	0.22	0.16
Chloride	14.2	12.2	13.1	10.4
Chlorine	0.0	0.0	0.0	0.0
Sulfide	12ug/l	13	11	12
pH Units	7.07	7.15	7.23	7.23
Manganese	0.6	0.6	0.5	0.3
Total Nitrogen	2.0	2.2	2.2	2.3
Temperature	56 F	55	54	55
Bacteria CFU	10^3		10^3	10^4

Table 2. Regenesis Wells (Chemical Process) Day 90

Analyte mg/l unless noted	R1	R1M	R2M	R3M
Dissolved Oxygen	6.8	2.4	2.3	0.8
Conductance	925	905	900	860
Sulfate	10.0	10.4	11.1	10.8
Nitrate	0.3	0.5	0.7	0.5
Chemical Oxygen Demand	750	489	710	895
ORP	-9	-17	-20	-79
TPH	580	563	477	658
Benzene	0.200	0.155	0.108	0.067
Toluene	1.748	1.632	1.700	1.736
Ethylbenzene	3.215	2.950	1.469	1.006
Total Xylene	9.089	9.120	2.673	1.883
Total Iron	1.4	3.0	1.1	3.4
Ferrous Iron	0.52	0.23	0.43	0.33
Chloride	12.1	12.9	12.0	10.0
Chlorine	0.0	0.0	0.0	0.0
Sulfide	12 ug/l	10	10	9
pH Units	8.40	7.73	7.70	7.33
Manganese	0.3	0.8	0.9	0.5
Total Nitrogen	2.1	1.2	0.8	2.0
Temperature	55 F	56	54	55
Bacteria CFU	10^4		10^5	10^4

The physical process used a electrochemical cell placed into MW-1M. The other monitoring wells associated with this process are labeled I1M, I2M, and I3M. The electrochemical cell used specific metallic cell coatings and a reversible polarity, constant-current D.C. power supply. The total time the cell and pump was activated each twenty-four (24) hour day amounted to one hundred-twenty (120) minutes. The constant current power setting was set at twenty (20) volts and five (5) amps or one hundred (100) watts per hour of energy use. During the twenty-four (24) hour period, the calculated oxygen gas production was 2,984 mg. The analytical data on the monitoring well water IM, I1M, I2M, and I3M for a period of ninety (90) days is found in Table 3 and 4.

Table 3. Electrochemical Wells (Physical Process) Day 0

Analyte mg/l unless noted	IM	I1M	I2M	I3M
Dissolved Oxygen	0.9	0.7	0.6	0.8
Conductance	877	850	817	835
Sulfate	11.2	12.0	12.9	12.2
Nitrate	0.4	0.9	0.4	0.9
Chemical Oxygen Demand	905	910	893	844
ORP	-121	-94	-98	-89
TPH	947	993	907	841
Benzene	0.721	0.483	0.746	0.127
Toluene	4.078	4.957	6.923	1.563
Ethylbenzene	8.951	8.056	9.012	2.845
Total Xylene	11.113	11.880	11.067	2.801
Total Iron	3.4	3.4	4.7	3.6
Ferrous Iron	0.25	0.44	0.22	0.30
Chloride	14.0	13.1	12.3	10.6
Chlorine	0.0	0.0	0.0	0.0
Sulfide	10 ug/l	11	14	12
pH Units	7.10	7.07	7.09	7.20
Manganese	0.5	0.9	0.4	0.8
Total Nitrogen	1.9	2.8	2.5	2.6
Temperature	55 F	54	55	56
Bacteria CFU	10^3		10^2	10^3

Table 4. Electrochemical Wells (Physical Process) Day 90

Analyte mg/l unless noted	IM	I1M	I2M	I3M
Dissolved Oxygen	17.2	10.3	11.1	4.0
Conductance	745	760	781	770
Sulfate	4.9	3.9	7.0	7.2
Nitrate	0.3	0.6	0.2	0.3
Chemical Oxygen Demand	321	308	278	496
ORP	+40	+20	+31	-11
TPH	195	507	372	590
Benzene	0.012	0.036	0.015	0.085
Toluene	0.946	1.563	0.673	0.990
Ethylbenzene	0.452	2.968	0.044	1.100
Total Xylene	0.733	2.040	0.200	1.113
Total Iron	0.9	1.3	0.3	1.5
Ferrous Iron	0.9	1.8	1.5	0.9
Chloride	11.4	10.9	15.2	11.7
Chlorine	0.0	0.0	0.0	0.0
Sulfide	2 ug/l	7	9	11
pH Units	7.14	7.19	7.08	7.03
Manganese	0.4	0.2	0.4	0.7
Total Nitrogen	1.2	1.4	0.4	1.3
Temperature	55 F	55	56	55
Bacteria CFU	10^2		10^7	10^6

Once generated, the hydroxyl radicals aggressively attack virtually all organic compounds. Depending upon the nature of the organic species, two (2) types of initial attacks are possible. As in Figure 2, the hydroxyl radical can abstract a hydrogen atom to form water, as with alkanes or alcohols. Or as in Figure 3, it can add to the contaminant, as in the case of olefins or aromatic compounds.

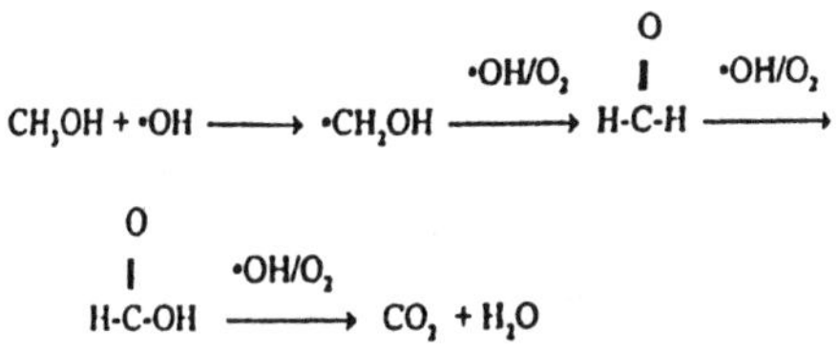

$$CH_3OH + \cdot OH \longrightarrow \cdot CH_2OH \xrightarrow{\cdot OH/O_2} H\text{-}\overset{O}{\overset{|}{C}}\text{-}H \xrightarrow{\cdot OH/O_2}$$

$$H\text{-}\overset{O}{\overset{|}{C}}\text{-}OH \xrightarrow{\cdot OH/O_2} CO_2 + H_2O$$

•OH
•OH
Hydrogen Abstraction

Figure 2. Hydrogen Abstraction 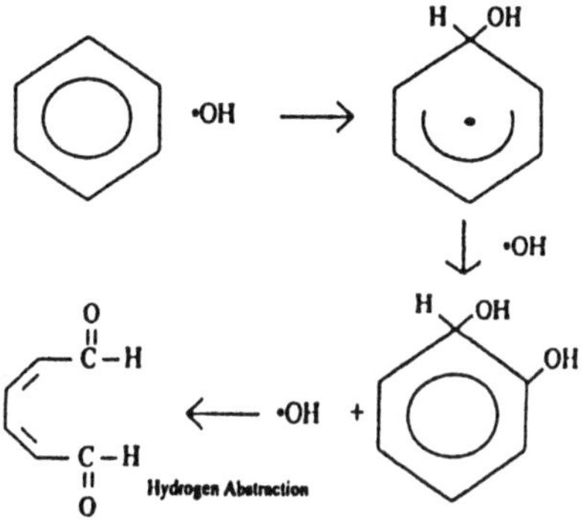 **Figure 3. Hydrogen Addition**

Dissolved oxygen plays an integral part in the aerobic biological remediation of a total petroleum hydrocarbon impacted site. Oxygen is a co-substrate for the only known enzyme (mono-and dioxygenase) that can initiate the metabolism of hydrocarbon and then can be used later as an electron acceptor for energy generation. In many cases, the major limitation on aerobic biodegradation in the subsurface is the low solubility of oxygen in water. For example, aerobic benzene biodegradation can be represented by the theoretical reaction shown in Figure 4.

$$7.5\ O_2 + C_6H_6 \xrightarrow{\text{BACTERIA}} 6\ CO_{2(g)} + 3\ H_2O$$

ΔG^{o}_{r} = - 3566 kJ/mole Benzene

Mass Ratio of O_2 to C_6H_6 = 3.1:1

0.32 mg/L C_6H_6 Degraded per mg/L O_2 Consumed

Figure 4. Benzene Oxidation

The dissolved oxygen produced by the electrocatalytic cell each day was calculated to be 2,984 mg. Apparently, most of this dissolved oxygen was able to leave MW-IM and begin to follow the groundwater flow. But most significantly was the increase in the heterotrophic plate counts for aerobic bacteria in the surrounding monitoring wells. The increase in bacterial counts and physical application of dissolved oxygen is a direct correlation to the decrease in total petroleum hydrocarbons previously found in the monitoring wells. As the total petroleum hydrocarbons decrease, in time so would the bacterial populations with a diminished organic carbon source.

CONCLUSION

Field Study. *The chemical process* used Regenesis ORC (magnesium peroxide) placed into MW-R1. Over the course of ninety (90) days the product was allowed to slowly leach oxygen into the aquifer. The analytical results are found in Tables 1 and 2. These results demonstrate that changes to the groundwater in MW-R1, R1M, and R2M occurred, especially the results observed in the ORP and dissolved oxygen measurements. Also, a decrease in the total petroleum hydrocarbon load was evident in these monitoring wells. This likely was occurring because of the increase in the heterotrophic bacterial numeration. The electron acceptor (oxygen), the carbon source (TPH), and carbon degrading microorganisms were in the process of mineralizing the organic compounds. Monitoring well R3M, 15 ft (4.6 m) downstream from R1, did not demonstrate any significant changes. The dissolved oxygen produced by the Regenesis ORC product was being utilized by the upstream aerobic microorganisms.

The physical process used a electrochemical cell placed in MW-1M, and over the course of ninety (90) days the cell was periodically pulsed on to produce dissolved oxygen and hydroxyl radicals (OH). The analytical results are found in Tables 3 and 4. The energy requirement of the electrochemical cell during the course of the field trial was twenty (20) volts and five (5) amps or one hundred (100) watts per hour. During a twenty-four (24) hour period the electrochemical cell was in the "on" mode for a total of one hundred and twenty (120) minutes or two (2) hours. The total energy used by the cell for ninety (90) days was eighteen (18) kilowatts.

General observations of the analytical results over a period of ninety (90) days indicated a reduction of TPH in all of the monitoring wells (IM, I1M, I2M, and I3M). This is due to the electrocatalytic cell production of the hydroxyl radicals (OH) in MW-IM. Hydroxyl radicals within this highly contaminated well were mineralizing a broad range of organic compounds. The groundwater leaving this impacted well then would contain less organic compounds and also would have reduced the chemical oxygen demand in the aquifer. In the application of electrocatalytic chemistry, the overall rate of oxidation of a contaminant is greatly increased over that obtained by the simple addition of oxidizing agents. The main reason is the much greater involvement of highly reactive initiator radicals such as the hydroxyl radical created by the metallic alloys on the electrochemical cell. Table 5 compares the reaction rates of various organic contaminants with ozone and the hydroxyl radical.

Table 5. Reaction Rate Constants

Compound	O_3	•OH
Chlorinated Alkenes	10^{-1} to 10^{3}	10^{9} to 10^{11}
Phenols	10^{3}	10^{9} to 10^{10}
N-containing Organics	10 to 10^{2}	10^{8} to 10^{10}
Aromatics	1 to 10^{2}	10^{8} to 10^{10}
Ketones	1	10^{9} to 10^{10}
Alcohols	10^{-2} to 1	10^{8} to 10^{9}
Alkanes	10^{-2}	10^{6} to 10^{9}

ACKNOWLEDGEMENTS
This work was supported in part by Carol Rowe P.E. of the CW3M Company Inc., Springfield, IL, and their client Shell Oil. Also, field support was contributed by Jerry Kellgren Ch.E and James Anderson of Land Chek Inc., Batavia, IL.

BIODEGRADATION OF JP-5 IN BARIUM-IMPACTED SOIL FROM THE NAVY'S FIRE-FIGHTING TRAINING FACILITY, NAVAL STATION SAN DIEGO

M. E. Losi, Jamshid Sadeghipour and Abram Eloskof
(Foster Wheeler Environmental Corporation, Costa Mesa, CA)
William T. Frankenberger, Jr. (Center for Environmental Microbiology, Riverside, CA)
Darren Belton (Naval Facilities Engineering Command, SWDIV, San Diego, CA)
Theresa Morley (Naval Station, San Diego, CA)

ABSTRACT: A laboratory study was conducted to assess the potential for *in-situ* bioventing of JP-5 jet fuel in barium (Ba)-impacted soil respirometry studies indicated that Ba did not inhibit JP-5 mineralization at concentrations up to 500 mg/kg. An initial column study suggested that although conditions appeared to be optimal, biodegradation of JP-5 in this soil was impeded. A subsequent column study indicated that the high concentration of JP-5 in the soil was potentially responsible for the initial lack of observed biodegradation. These studies demonstrated that biodegradation of JP-5 will occur in this soil, but could be very slow initially in areas with high JP-5 concentrations, with the rate probably increasing in these areas as JP-5 concentrations are lowered.

INTRODUCTION

Bioventing is an *in-situ* bioremediation technique in which highly volatile compounds of petroleum mixtures may be removed through a soil venting phase, while less volatile, higher molecular weight hydrocarbons are subject to microbial attack as O_2 is provided as the terminal electron acceptor. To help in determining whether this approach may be feasible at a specific site, basic information is needed such as the nature and the concentration of the residual hydrocarbons, the population density of the hydrocarbon-degrading microorganisms, various chemical and physical properties of the contaminated soil, the nature and concentrations of other contaminants with the potential to inhibit biological processes, and the biodegradation potential with (and without) the application of nutrients. Because these parameters are highly variable, the feasibility of any bioremedial approach must be studied on a case-by-case basis.

Soils contaminated with various jet fuels (JP-4, JP-5) are generally considered treatable using in-situ bioremediation techniques. However, reports of JP-5 biodegradation in the literature are rare relative to those of JP-4, and very little data is available regarding effects of barium (Ba), a common constituent in fire-fighting chemicals, on biodegradative processes. This study was conducted to investigate at the bench-scale the biodegradation of very high levels of JP-5 jet fuel in soil also impacted with barium (Ba) from fire-fighting training activities.

Objectives. Specific objectives of this study were as follows: (i) characterize relevant initial soil parameters, (ii) assess the response of the JP-5-degrading microbial population to various nutrient loads, (iii) study the effects of varying concentrations of Ba on mineralization of JP-5, and (iv) quantify removal of JP-5 from the contaminated soil under simulated field conditions.

MATERIALS AND METHODS

Soil Characterization. Soil samples were collected at the Fire-Fighting Training Facility (Site 8), Naval Station San Diego with a hand auger in areas previously identified as having high total petroleum hydrocarbon (TPH) concentrations. The samples were maintained at 4°C, and were sieved and homogenized (in the laboratory cold room) in the field moist condition to ensure uniformity. The following parameters were measured: pH (EPA 150.1), ammonium-nitrogen (NH_4-N) (EPA 350.1), nitrate-nitrogen (NO_3-N) (steam distillation method), inorganic phosphorous (P) (EPA 365.2), moisture content (gravimetric method), TPH (EPA modified 8015), benzene, toluene, ethylbenzene and total xylenes (BTEX) (EPA 8020), and CAM metals (EPA 6010/7000).

The biological analyses included an enumeration of the total native heterotrophic bacteria and the hydrocarbon-oxidizing bacteria capable of using actual free product (extracted from the plume at the site) as a sole carbon and energy source. Total heterotrophic plate counts were performed using the spread plate method, and JP-5-degrading microorganisms were enumerated using the most probable number method. Free product from the site was used as the sole carbon and energy source for the petroleum-degrading bacteria.

Effect of Nutrient Additions on JP-5 Degradation. The effect of nutrient additions on biodegradation of the JP-5 jet fuel was assessed in a respirometry study. This study consisted of monitoring O_2 uptake (as a measure of JP-5 degradation) by the contaminated soil subjected to different nutrient additions. Twenty grams of contaminated or background soil (moisture content was maintained at the previously measured field level) were added to sterile, 40-ml vials. The following treatments were applied to the contaminated soil: a sterile control (autoclaved for 12 hours at 121° C, 18 psi); application of water alone; treatment with nutrients consisting of N [100 mg N/kg as $(NH_4)_2SO_4$] or P (50 mg P/kg as K_2HPO_4); and the combination of N and P (50 mg N plus 25 mg P/kg; 100 mg N plus 50 mg P/kg; and 200 mg N plus 75 mg P/kg). The vials were capped with sterilized Mininert valves and were incubated statically at 25° C up to 5 days. Gas samples were periodically drawn with a syringe, and O_2 uptake was determined by gas chromatography using thermal conductivity detection (GC-TCD).

Inhibition of JP-5 Degradation by Barium. Inhibition of JP-5 degradation by Ba was assessed using respirometry as described in the preceding section, except that instead of monitoring O_2 uptake, CO_2 (as the final product of degradation) was measured. Barium chloride was added to the contaminated soil at the following concentrations (mg Ba/kg): 0, 100, 250, 500 and 1,000. Gas samples were periodically drawn with a syringe, and CO_2 concentrations were determined using GC-TCD. Immediately after sampling, the vials were opened to ambient air and the headspace was flushed with air for 5 minutes to insure that O_2 would not be limiting with long-term incubation. After aeration, water was added to the soils to maintain the moisture level at the field level and then the vials were recapped. The amount of CO_2-C released in each treatment was expressed on a cumulative basis over the specified incubation period.

Soil Column Study I. Plexiglas columns, 91.5 cm x 7.6 cm ID, were used for the soil column studies. In the first study, contaminated soil (4.5 kg) was added to each of the columns and subjected to three treatments as follows: (i) a static control, (ii) injection of air to a moist soil column and (iii) injection of air to a moist soil column supplemented with nitrogen [100 mg N/kg as $(NH_4)_2SO_4$]. Oil-less air was injected (via positive pressure) into the bottom of the columns through a porous diffuser and allowed to diffuse upward throughout the soil at a flow rate of 2 ml/min. Soil columns were incubated in the vertical position at ambient temperature for 10 weeks. During incubation, the loss of water in the soil columns was monitored with a moisture meter and added back as necessary. Activated carbon (C) traps were used to capture volatile organic components released from the soil and were extracted and analyzed for TPH (EPA modified 8015) and BTEX (EPA 8020) to control for non-biological losses. Composite soil samples (from eight on-column sampling ports) were analyzed for TPH (EPA modified 8015) and BTEX at 0, 2, 4, 6, 8 and 10 weeks. The hydrocarbon-degrading microbial population, as well as inorganic N (NH_4-N and NO_3-N) and orthophosphate-P were also monitored periodically.

Soil Column Study II. The following soil preparations were used for the second soil column study:

1. A 1:1 mixture of contaminated soil and clean, on-site soil (final TPH concentration: 4,400 mg/kg soil).
2. A 1:4 mixture of contaminated soil and clean, on-site soil (final TPH concentration: 2,600 mg/kg soil).
3. Free product from the site spiked into clean, on-site soil (final TPH concentration: 2,700 mg/kg soil).
4. Fresh JP-5 spiked into clean, on-site soil (final TPH concentration: 2,200 mg/kg soil).

Plexiglas columns were packed, treated and sampled as described above (Soil Column Study I). Subsamples from each soil column were analyzed for TPH, moisture content, and the JP-5-degrading microbial population at 0, 2, 4, 6, 8, 11 and 14 weeks (or until trends were evident).

RESULTS AND DISCUSSION

Physical, Chemical, and Biological Properties of the Contaminated Soil. Results of physical, chemical and biological analyses of the JP-5 contaminated soil are shown in Table 1.

TABLE 1. Initial physical, chemical and biological properties of the JP-5 contaminated soil.

Analyses	Results
pH	8.4
Total Inorganic Nitrogen, NH_4-N + NO_3-N (mg/kg)	3.6
Inorganic Phosphorus, Orthophosphate-P (mg/kg)	4.3
Moisture Content (%)	12.0
Total Petroleum Hydrocarbon (mg/kg)	17,600 (avg.)
Total BTEX (EPA 8020)	3.5
Total heterotrophic microorganisms (cfu/g soil)*	20×10^6
JP-5 degrading microorganisms (mpn/g soil)*	92×10^6

*Colony-forming units (cfu)/g soil or most probable number (mpn)/g soil.

Data in Table 1 suggest that JP-5 degrading microorganisms were selectively enriched in the impacted soil. Overall, the initial characterization of the contaminated soil suggested that chemical and biological parameters were within acceptable limits for promoting biodegradation.

Results of the CAM metal analysis of the contaminated soil showed that Ba was present in the soil at a concentration of 48 mg/kg, and, no other metals or metalloids appeared to be of concern (other metals detected included: chromium, 2.7 mg/kg; copper, 4.8 mg/kg; vanadium, 14 mg/kg; and zinc, 22 mg/kg). Because BTEX levels were very low relative to TPH, results of BTEX analyses are not discussed further.

Respirometry Studies. Results of the first respirometry study in which the effect of nutrient additions was assessed are presented in Figure 1. It is evident from the data in Figure 1 that O_2 uptake was stimulated in treatments receiving N, but not P. Combinations of N and P at various concentrations did not enhance O_2 uptake over the treatment receiving N at 100 mg/kg soil. Nitrogen was therefore determined to be a limiting nutrient in degradation of JP-5 in this soil, and was added at a level of 100 mg/kg soil in the column studies.

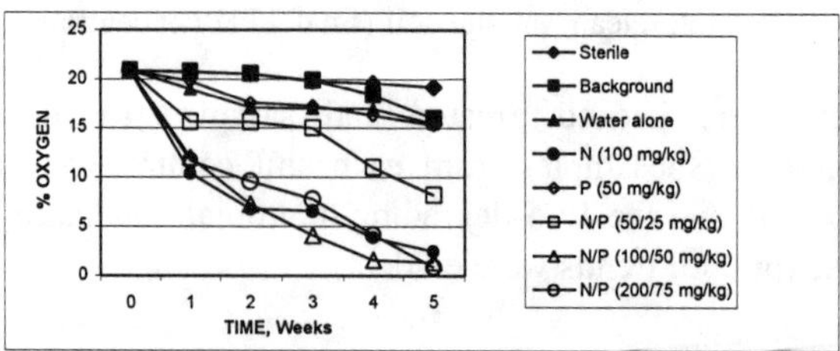

FIGURE 1. Effects of nitrogen and phosphorous additions on biodegradation of JP-5 in JP-5-contaminated soil as determined through uptake of O_2.

Results of the second respirometry study in which the effect of varying Ba concentrations on JP-5 mineralization rates were investigated are presented in Figure 2. The only treatment showing significant inhibition (46% of the control) in CO_2 evolution was 1,000 mg Ba/kg soil. Although the form of the Ba in the contaminated soil is not known, and therefore strict comparisons are not possible, it is evident that in this soil, only Ba levels in excess of 500 mg/kg have the potential to inhibit biodegradation processes. Because the Ba level in the contaminated soil was measured at 48 mg/kg, its effects on the hydrocarbon-degrading microbial population are probably minimal.

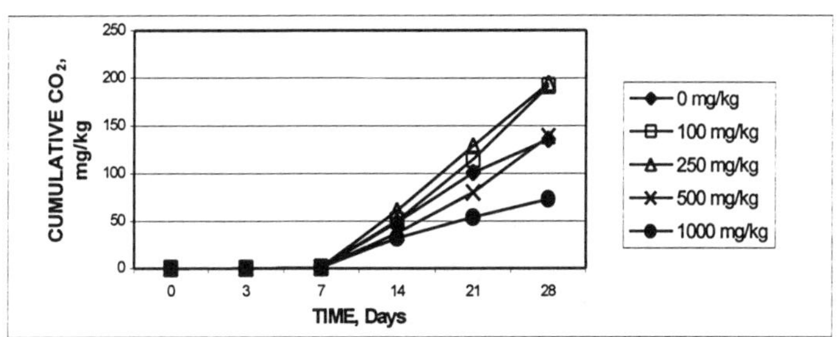

FIGURE 2. Effects of various concentrations of Ba on degradation of JP-5 in JP-5-contaminated soil as determined through evolution of CO_2.

Biodegradation of JP-5 Under Simulated Field Conditions: Column Studies. Results of the first column study are shown in Figure 3. The initial TPH content of the contaminated soil ranged from 15,000 to 21,000 (avg. 17,600) mg/kg soil. TPH varied considerably with each treatment throughout this study without showing notable degradation. Analysis of the carbon traps showed that volatilization from the system accounted for approximately 1-2% of the hydrocarbon losses. Periodic evaluation of physiochemical and biological parameters suggested favorable conditions for biodegradation of JP-5 jet fuel in this soil (the JP-5-degrading microbial population remained on the order of 10^6 to 10^7 cells/g soil throughout the experiment).

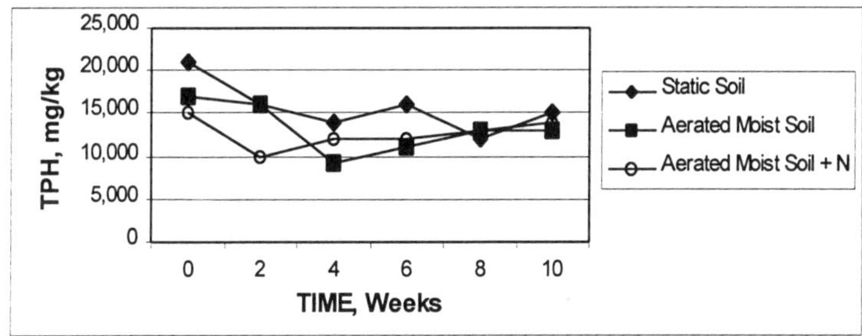

FIGURE 3. TPH concentrations over time in initial column study.

After 10 weeks of incubation, the experiment was terminated. It was hypothesized that the high concentration of JP-5 in the soil may have been responsible for the lack of observed biodegradation, possibly by causing the formation of large globules that minimized the amount of TPH surface area available for microbial attack, or restricted O_2 flow within the soil pore space.

The second column study was designed to investigate whether lowering the TPH concentration would stimulate biodegradation of the jet fuel. Results showing TPH losses during the second column study are presented in Figure 4. Data shown in Figure 4 reveal that fresh JP-5, and free product recovered from the site were substantially degraded (80% and 90%, respectively, over 8 weeks) when spiked into uncontaminated soil (from the same site) at lower concentrations (approximately 2,500 mg/kg).

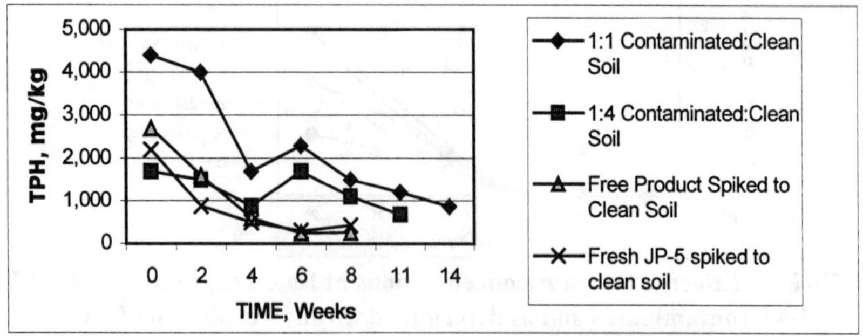

FIGURE 4. TPH concentrations over time in second column study.

Furthermore, when the contaminated soil was diluted with uncontaminated site soil to TPH concentrations of 4,400 and 1,700 mg/kg, 73% and 60% (respectively) of the JP-5 was degraded in 14 and 11 weeks. Volatilization was again negligible and physicochemical and biological parameters were very similar to those measured in the first column experiment. These results suggest that the initial high TPH concentrations limited hydrocarbon bioavailability and possibly the flow of O_2, therefore restricting biodegradative processes.

Other researchers have found that degradation of JP-5 in soil was inhibited at TPH concentrations of approximately 800 mg/kg and greater (Kittel et al., 1994). Our results confirmed JP-5 biodegradability and demonstrated JP-5 biodegradation in soil at relatively high concentrations. Results also suggest that Ba levels as high as 500 mg/kg will have little effect on biodegradative processes in this soil. This study indicated that biodegradation of JP-5 will occur in this soil, but could be very slow initially in areas with very high concentrations, with the rate increasing in these areas as concentrations are lowered.

REFERENCES

Kittel, J.A., R.E. Hoeppel, R.E. Hinchee, T.C. Zwick, and R.J. Watts. 1994. "In-Situ Remediation of Low Volatility Fuels Using Bioventing Technology." In E.J. Calabrese, P.T. Kostecki and M. (Eds.), *Hydrocarbon Contaminated Soils*, Volume IV, pp. 43-68. ASP, Amherst, MA.

EFFECT OF HEAVY METALS ON THE BIOREMEDIATION OF PETROLEUM HYDROCARBONS

Y-P Ting, C Fang and ***H-M Tan*** (National University of Singapore, Singapore)

ABSTRACT: A relatively clean soil was amended with dodecane, pristane and phenanthrene at a final concentration of 3% (w/w). The contaminated soil was then seeded with hydrocarbon-degrading microorganisms and the biodegradation of the hydrocarbons investigated in the presence of either one (Cd) or two (Cd and Pb) heavy metals. Under abiotic conditions, the loss of the hydrocarbons was in the order dodecane>phenanthrene>pristane. Under biotic conditions, however, the rate was in the order dodecane>pristane>phenanthrene. The presence of Cd at 130 ppm had no appreciable effect on the biodegradation of dodecane and pristane. However, higher concentrations of Cd singly or in combination with Pb at 800 and 1300 ppm, led to reduced rates of biodegradation of dodecane and pristane, being more pronounced for dodecane. No significant effect on phenanthrene degradation was observed. Taken together, the results indicate that Cd, compared to Pb, had a more deleterious effect on microbiota, and thus had a more important role in hydrocarbon bioremediation. However, the effects of heavy metals become less apparent with increasing recalcitrance of the hydrocarbon compound.

INTRODUCTION

Petroleum and petroleum products are used widely and large amounts of these compounds find their way into the environment either through industrial emissions or accidental spills and leaks. This has resulted in the contamination of soil and groundwater, posing a threat to human health and groundwater resources since many of the petroleum compounds are hazardous.

Various clean-up methods have been used to manage petroleum contamination. These include incineration, landfills and bioremediation. Compared to the other methods, bioremediation is environmentally attractive as it uses natural (indigenous) or inoculated microbial activities for the elimination or detoxification of toxic environmental pollutants into non-toxic and simpler end products such as water and carbon dioxide (Skinner et al., 1991; Atlas, 1993). Bioremediation is a complex process being governed by factors that include temperature, concentration and type of contaminants, concentration of oxygen and nutrients, activities and composition of microbial populations (Leahy and Colwell, 1990; Edgehill, 1992).

The bioremediation of petroleum-contaminated sites has been widely studied, but most investigations appear to focus on soil contaminated with petroleum hydrocarbons with little reported on the effect of heavy metals on the bioremediation process. The common heavy metals causing environmental concern include cadmium, lead, nickel, copper and chromium. Heavy metals potentially affect the rate and mechanisms of biodegradation since microorganisms are known to interact with heavy metals. Bioremediation does not remove these metals. Moreover, the metals are known to be toxic to the surrounding biota when present in excess concentration in the soil. However, microorganisms exposed to high levels of heavy metals can develop tolerance to them.

Objective. The objective of this study was to assess the extent of bioremediation of soil contaminated with petroleum hydrocarbons and heavy metals, and to investigate the influence of these metals (either singly or in combination) on the biodegradation of the petroleum hydrocarbons.

EXPERIMENTAL SET-UP

The experiment was carried out in 8 soil plots each of dimensions 42 cm by 24 cm by 11 cm with 6.1 kg soil (dry wt) from a plant nursery with no previous history of hydrocarbon or heavy metal contamination. Each plot was amended with three hydrocarbons: dodecane, pristane and phenanthrene, representing the different types of petroleum hydrocarbons, at a concentration each of 1% (w/w).

An abiotic control was set up by adding 5500 ppm (w/w) $HgCl_2$ to one of the plots (plot 1). To the other plots, Cd ($Cd(NO_3)_2.4H_2O$) was added at 130 ppm to 1300 ppm either singly or in a 1:1 ratio with Pb ($PbCl_2$). All the plots were seeded with inocula of hydrocarbon-degrading bacteria ($1x10^9$) previously isolated from a sludge farm (Ting et al., 1999). In addition, the plots were supplemented with NH_4NO_3 and K_2HPO_4 to give a C:N:P ratio of 100:10:1. The moisture content in each plot was in the range of 10-15%. The plots were maintained at 27-32°C and tilled twice weekly for a period of three months. A summary of the experimental plots is given in Table 1.

TABLE 1. Set-up of the experimental plots

Plot	1	2	3	4	5	6	7	8
Soil (dry, kg)	Each plot with 6.1 kg dry soil and supplemented with 1% each of dodecane, pristane and phenanthrene							
Hg (ppm, w/w dry soil)	5500	-	-	-	-	-	-	-
Pb (ppm, w/w dry soil)	-	-	-	-	-	66	400	650
Cd (ppm,w/w dry soil)	-	-	130	800	1300	65	400	650
Seeding with bacteria	Yes							
Addition of nurients	NH_4NO_3 & K_2HPO_4 added at 57 g and 11 g, respectively, for each plot at the start of experiment.							
Watering	Watered weekly with deionised water, moisture content in the range of 10-15%.							
Tilling	Twice weekly.							
pH control	No							
Temperature and Humidity	Ambient condition (Temperature: 27-32°C, Relatively humidity: 62-92%).							

RESULTS AND DISCUSSION

Plot 1, amended with $HgCl_2$, represented an abiotic control as no microorganism was detected throughout the experimental period from it. Under such abiotic conditions, the loss of the hydrocarbons was observed to be in the order dodecane>phenanthrene>pristane (Fig. 1). With the addition of hydrocarbon-degraders, both biotic and abiotic factors contributed towards the reduction of the hydrocarbons in plot 2 (Fig. 1). The results clearly indicate that a significant part of dodecane was biodegraded while phenanthrene was relatively resistant to micorbial attack. The biodegradability of the three hydrocarbons was in the order dodecane>pristane>phenanthrene, reflecting the influence of the molecular weight and polyaromatic nature of the hydrocarbons.

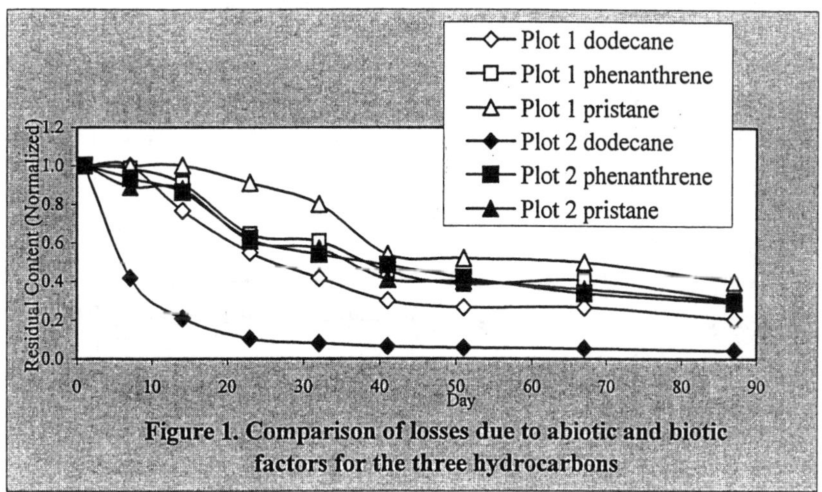

Figure 1. Comparison of losses due to abiotic and biotic factors for the three hydrocarbons

Effect of heavy metals on biodegradation of hydrocarbons. Dodecane ($C_{12}H_{26}$) is a saturated straight-chain alkane and a common component of petroleum. The effects of Cd, or Cd together with Pb, on the biodegradation of dodecane are shown in Fig. 2. When compared to the biotic control, it was observed that 130 ppm Cd, or Cd and Pb, did not affect the degradation of dodecane significantly. Plots 4, 5, 7 and 8, amended with higher concentrations (800 ppm and 1300 ppm) of the heavy metals showed a slower degradation rate, although for the same concentrations, Cd alone exerted a larger negative effect on the degradation of dodecane than when combined with Pb (plots 4 and 7; plots 5 and 8). This is consistent with the effects of the heavy metals on the number of hydrocarbon-degraders measured in the respective plots (Brown and Braddock, 1990), where plots with increasing Cd concentrations were found to have lower counts of hydrocarbon-degraders (results not shown). Interestingly, the presence of Cd, either singly or in combination with Pb, at 1300 ppm did not completely inhibit microbial degradation of dodecane (plots 5 and 8)

The effects of Cd alone on the biodegradation of pristane ($C_{19}H_{40}$), a branched-chain alkane, are only apparent at concentrations of 800 ppm and above, similar to those observed for dodecane (Fig. 3). With Cd and Pb together, no clearly discernable effect on pristane degradation was observed. This suggests that compared to Pb, Cd had a stronger influence on the biodegradation of Pb, since the concentration of Cd never exceeded 700 ppm (plots 6 to 8). Furthermore, it indicates that the deleterious effect of Cd on bioremediation is less pronounced on a hydrocarbon with a less biodegradability nature. Using TCLP (toxicity characteristic leaching procedure) it was found that the extractable Pb from the experimental plots were much lower when compared with Cd. This is due to the retention of Pb by the soil, especially in the presence of PO_4^{3-}, which resulted in a lower mobility of Pb. Consequently, the toxic effects of Pb are diminished.

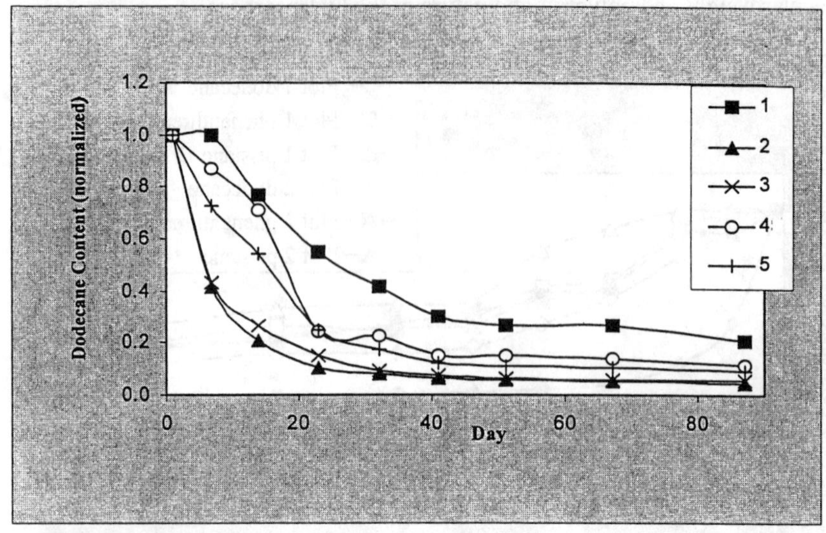

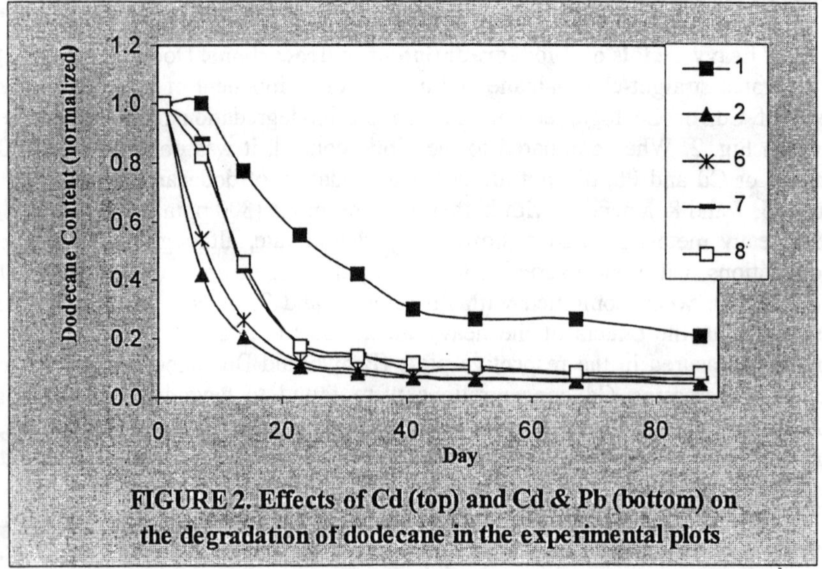

FIGURE 2. Effects of Cd (top) and Cd & Pb (bottom) on the degradation of dodecane in the experimental plots

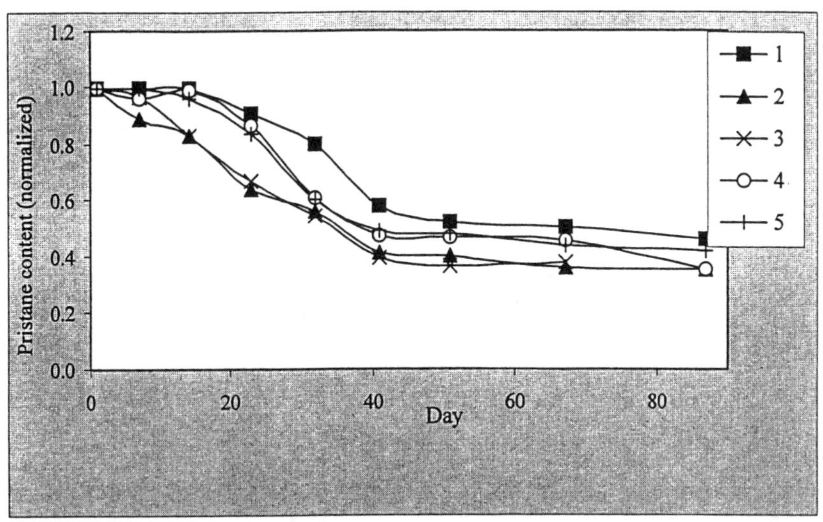

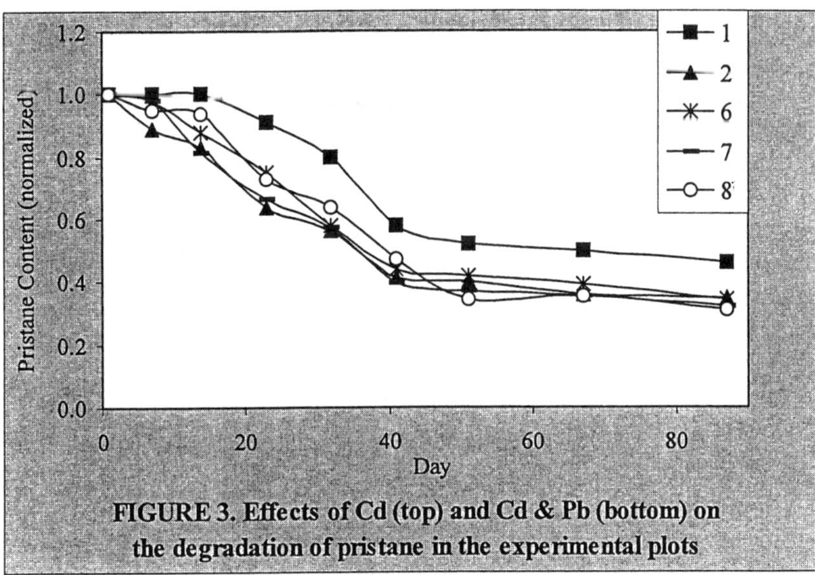

FIGURE 3. Effects of Cd (top) and Cd & Pb (bottom) on the degradation of pristane in the experimental plots

The presence of Cd over the range of 130-1300 ppm had no appreciable effects on the degradation of phenanthrene (Fig. 4). Similar results were also observed for Cd and Pb. This is consistent with the previous observation that the loss of phenanthrene was due mainly to abiotic factors (Fig. 1), and reflects the recalcitrance of the hydrocarbon to biodegradation. Thus, the effects of heavy metals on the biodegradation of this three-ring PAH (polyaromatic hydrocarbon) become largely insignificant.

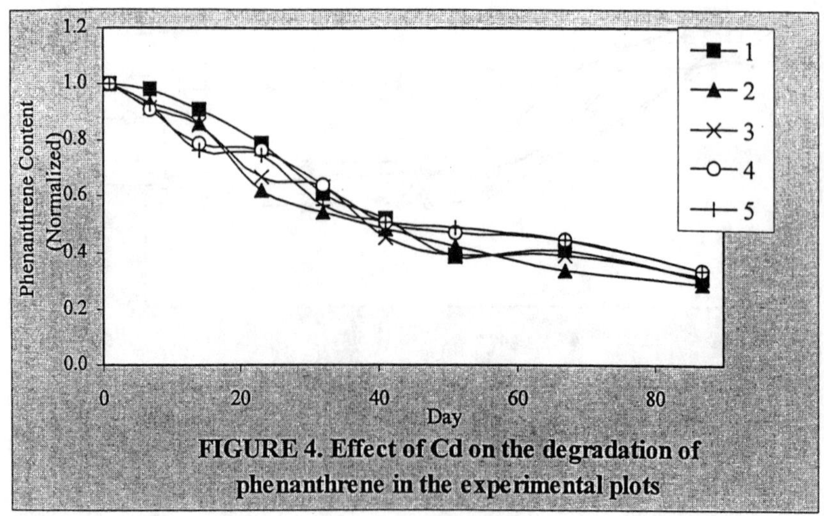

FIGURE 4. Effect of Cd on the degradation of phenanthrene in the experimental plots

REFERENCES

Atlas, R. M. 1993. Bioaugmentation to Enhance Microbial Bioremediation. In *Biotreatment of Industrial and Hazardous Waste*, ed. M. Gealt and M. Levin, McGraw-Hill, New York. pp. 19-37.

Brown, E. J., and J. F. Braddock. 1990. "Sheen Screen, a Miniaturized Most-Probable Number Method for Enumeration of Oil-Degrading Microorganisms." *Appl. Environ. Microbiol.* 56: 3895-3896.

Edgehill, R. 1992. "Factors Influencing the Success of Bioremediation." *Aust. Biotechnol.* 2; 297-305.

Leahy, J. G., and R. R. Colwell. 1990. "Microbial degradation of Hydrocarbons in the Environment." *Microbiol. Rev.* 54: 305-315.

Skinner, J. H., G. G. Ondich, and T. L. Baugh. 1991. US EPA Bioremediation Research programs. In *On-Site Bioreclamation Processes for Xenobiotic and Hydrocarbon Treatment*, ed. R.E. Hinchee and R. F. Olfenbuttel, Butterworth-Heinemann, Battelle, Columbus, OH., pp. 1-15.

Ting, Y. P., H. L. Hu, and H. M. Tan. 1999. "Bioremediation of Petroleum Hydrocarbons in Soil Microcosms." *Res. Environ. Biotechnol.* 2: 197-218.

SOIL TOXICITY ASSESSMENT USING ACUTE AND CHRONIC BIOASSAYS

Jaap van der Waarde[1], Anja Derksen[2], Eline van der Hoek[3], Martin Veul[4], Kees van Gestel[5], Sandra Bouwens[5] , Rene Kronenburg[6] en Gerard Stokman[7]
[1]Bioclear, Groningen; [2]AquaSense, Amsterdam; [3]KEMA, Arnhem; [4]Witteveen en Bos, Deventer; [5]Free University Amsterdam, Amsterdam; [6]Harbor Authority Amsterdam, Amsterdam; [7]SBNS, Utrecht, The Netherlands.

ABSTRACT: Soil ecotoxicity was determined using a range of bioassays. The aim of this study was to demonstrate the effectiveness of bioassays for assessment of the ecotoxicity of soil contaminated with mineral oil. The acute bioassay Microtox was successful in detecting soil toxicity. Chronic bioassays, in which earthworms, springtails or lettuce seeds were directly exposed to soil, showed a good correlation with contaminant levels. On average, the chronic bioassays resulted in a EC50 value of 2,000-3,000 mg mineral oil/kg dw, thus giving ecotoxicological support to the site specific remediation target level of 1,000-1,500 mg/kg dw. In conclusion a set of bioassays was developed that has been proven effective in assessing toxicity of mineral oil soil contamination.

INTRODUCTION

Criteria for soil remediation are based on chemical analyses of contaminant levels. Recent changes in Dutch and other environmental policies towards a more functional remediation approach, based on anticipated land use, have increased the demand of more site specific information on local effects of present contaminants. Contaminant levels are indicative of potential risks, and form the basis for remedial actions. Biological assays may give more direct evidence of acute effects of soil contaminants. As such, biological assays can be used to assess the urgency of a (bio)remediation action and can be used to determine the acute risks of residual contaminant levels after closing of the bioremediation operation. Low level residual contaminants may be poorly bioavailable which eliminates the need of continuation of the often fruitless bioremediation efforts of these residuals. In this study the use of bioassays to determine the toxic effects of mineral oil fractions in contaminated soil was evaluated. The aim of the study was to demonstrate the effectiveness of bioassays to determine biological effects of mineral oil soil contamination and to give empirical support for the anticipated bioremediation target levels for the contaminated site.

MATERIALS AND METHODS

The study was performed with soil samples from the Petroleum Harbor location in Amsterdam, The Netherlands. This location has been used for transport and storage of fuels for more than a century and has consequently a rich history of contamination events. The soil at the site consists of sandy clay, with a water table ranging from 0 to –2 m below surface.

Soil samples were analyzed for mineral oil (GC), organic content, dry weight, heavy metals (AAS), EOX (GC) and PAH (HPLC). Soil was extracted with demineralized water containing 1 mM $CaNO_3$ to produce extracts for acute bioassays.

The root elongation test with lettuce (*Luctuca sativa*) was performed according to Toussaint et al (1995). The ECHA test was performed according to the manufacturers guideline (ECHA, 1997).

The mobility test with nematodes was performed according to Kammenga (1995). The mobility test with the springtail *Folsomia candida* was performed according to AquaSense (1996). The growth inhibition test with the freshwater algae *Raphidocelis subcapitata* was performed according to ISO 8692 (1989). The Microtox test was performed according to NVN 6516 (1993). The mobility test with *Daphnia magna* was performed according to ISO 6341 (1996). The effects in the acute bioassays were evaluated against a blank consisting of the extraction liquid, for the test with algae and the Microtox assay standards prescribed in the test protocol were used.

The growth inhibition test with lettuce (*Luctuca sativa*) was performed according to OECD 208 (1984). The growth and reproduction inhibition test with earthworms (*Eisenia fetida*) was performed according to ISO guidelines (1991, 1997). The growth and reproduction inhibition test with the springtail *Folsomia candida* was performed according to ISO/DIS 11267 (1998). The bait lamina test was performed according to Kratz (1996). The effects in the chronic bioassays were evaluated against the results in non-contaminated OECD standard soil.
Bacterial numbers were determined with the MPN assay using diesel fuel or R2A medium as the carbon and energy source. Dehydrogenase levels were determined according to Alef (1991).

RESULTS AND DISCUSSION

Acute bioassays. The acute bioassays showed marked differences in response to contaminant levels. The root elongation test with lettuce showed no inhibition in water extracts from soil contaminated with mineral oil. A minor effect was seen in sample 7, possibly due to the presence of heavy metals (Cu, Zn) (data not shown). The mobility tests with nematodes and springtails were also unable to detect negative effects of contaminants present in the extracts. Reproduction of nematodes within the assay did show differences between the different samples, but these results were inconclusive due to the limited number of organisms. In sharp contrast, the ECHA test and the algae growth tests showed inhibition in almost every sample of soil extract as compared to the reference. These assays apparently respond to compounds extracted from soil, other than the contaminants that were present. Daphnia mobility was effected in some samples, all containing increased contaminant (mineral oil, heavy metals) levels, but not all contaminated samples were toxic. Only the Microtox assay showed a discriminative response to contaminant levels and showed toxicity at mineral oil levels above 2,000 mg/kg dw (Figure 1).

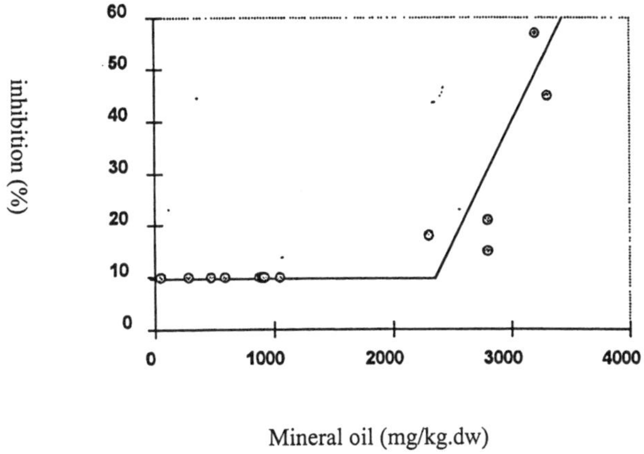

Figure 1. Toxicity of soil extracts in the Microtox bioassay

Chronic bioassays. The chronic bioassays with soil showed different responses in comparison with the acute bioassays with soil extracts. Germination of lettuce seeds was effected by the presence of mineral oil in the soil. The four samples with the highest concentration of mineral oil showed the lowest germination percentages (Figure 2). Both the standard reference soil and the soil with the lowest concentration of mineral oil (sample 10) showed poor germination rates, which can not be explained. Biomass production was also lowest on the soils with the highest contaminant levels.

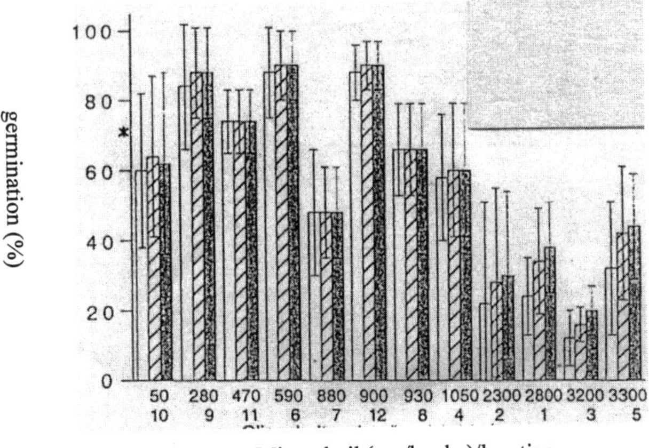

*: germination in non-contaminated OECD test soil

Figure 2. Germination of lettuce seeds in contaminated soil

Table 1. Toxicity of soil contaminated with mineral oil as determined with bioassays

Acute bioassays			Chronic bioassays					
Microtox	Daphnia	algae biomass	Lettuce germination	Worm reproduction	springtail survival or reproduction	Bait lamina	Mineral oil (mg/kg dw)	Toxicity evaluation
-	-	+	-	-	-	-	< 50	-
-	-	+	-	-	-	-	280	-
-	-	-	-	-	-	-	470	- (reference)
-	-	+	-	-	-	-	590	-
-	+	+	+/-	+	-	-	880	+ (metals)
-	-	+	-	-	-	-	900	-
-	-	+	-	+/-	-	-	920	-
-	-	+	-	-	-	-	1050	-
+/-	+	+	+	+/-	-	+	2300	+
+/-	-	+	+	+	-	+	2800	+
+	+	+	+	+	+	+	3200	++
+	-	+	+/-	+	+	+	3300	++

- : non toxic; +/- : moderately toxic (<50% inhibition); +: toxic (>50% inhibition)

Earthworms seemed to mature better in soils with low concentrations of mineral oil than in more contaminated soils. Biomass production was not affected by contaminant levels, but this can part be explained by the use of energy by the worms for reproduction in some lower contaminated soils. Reproduction as monitored with cocoon production was clearly effected by mineral oil levels and the lowest reproduction levels were found in the three soils with the highest concentration of mineral oil (Figure 3).

Survival and biomass reproduction of springtails was not effected by mineral oil at levels up to 2800 mg/kg dw. A dramatic decrease in viable organisms and juvenile production was found in the two soils with the highest level of mineral oil (Table 1). The bait lamina test was consistent with these results in that the four soil samples with the highest levels of mineral oil supported the lowest consumption rates. At mineral oil levels from 2,300 mg/kg dw and higher a significant negative effect was observed (Table 1).

The results from the chronic bioassays can be used to derive a test specific EC50 (mg/kg dw). It appears that for most bioassays, the corresponding EC50 for mineral oil varies between 2,000 and 3,000 mg/kg dw. Only the reproduction test with earthworm seems to indicate lower EC50 values, ranging between 700 and 1,500 mg/kg dw. It must be stressed that these EC50 values do not take into account any other negative effects of co-contaminants in the soil like heavy metals.

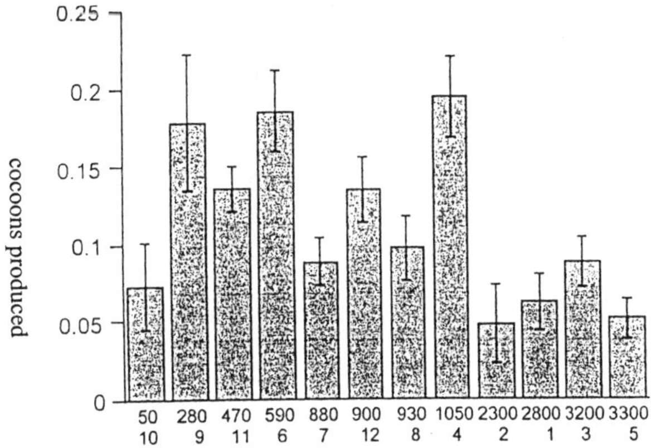

Mineral oil (mg/kg.dw)/location

*: cocoon production in non-contaminated OECD test soil

Figure 3. Reproduction of the earthworm *Eisenia fetida* in contaminated soil

Bacterial counts. The bacterial community in the soil showed total MPN values around 10^8-10^{10} cells/g dw. Only a small minority of these cells (<1%) was able to grow on mineral oil, indicating that the bacterial population in the soil was not adapted to mineral oil biodegradation. This was confirmed by the low dehydrogenase levels in the samples, on average 10 μg TPF/g dw. There was no clear correlation between contaminant levels, MPN_{total}, $MPN_{mineral\ oil}$ and dehydrogenase activities. This effect is most likely caused by the fact that at the time of sampling bioremediation had not commenced yet and the soil was water saturated and mostly anoxic.

CONCLUSIONS

In this study the ecological effects of a mineral oil soil contamination were evaluated using a series of biological tests. Both primary producers (lettuce, algae), destructors (bacteria), and consumers (nematodes, springtails, earthworms) were assayed for their sensitivity towards the contaminated soil in laboratory bioassays. Acute toxicity in soil extracts was only found with the Microtox bioassay.

The other acute bioassays with soil extracts were not conclusive in detecting toxic effects of the mineral oil contamination on the organisms tested. In the chronic bioassays that measured more long term effects on growth and reproduction, negative effects of contaminated soil were found at values above 2,000 mg/kg dw mineral oil. The earthworm test with *Eisenia fetida* showed the highest sensitivity towards the contamination. On average, the chronic bioassays resulted in a EC50 value of 2,000-3,000 mg mineral oil/kg dw. These values are higher than the location specific bioremediation target level of 1,000-1,500 mg/kg dw, thus giving ecotoxicological support to this value. Present research is focussed on the effectiveness of bioremediation on removing toxicity of the mineral oil contamination on the site and determining the toxicity of a potential residual contaminant fraction after bioremediation.

ACKNOWLEDGEMENTS

This work was supported with grants from NOBIS, the Amsterdam Harbor Authority and SBNS.

REFERENCES

Alef. 1991. Methodenhandbuch Bodenmikrobiologie. Ecomed. ISDN 3-609-65960-2.

AquaSense. 1996. Acute bioassay met de terrestrische springstaartensoort *Folsomia candida*. Verbetering van de methode en onderzoek naar de toepasbaarheid. Reportnumber 96.0859a. AquaSense, Amsterdam, The Netherlands.

ECHA. 1997. Biocide monitor II. ECHA Microbiology LTD. Microbiological Testing and Advisory Services Suppliers of ECHA products, Cardiff U.K.

ISO 6341. 1996. Water quality - Determination of the inhibition of the mobility of *Daphnia magna* Straus (*Cladocera, Crustacea*) - Acute toxicity test.

ISO 8692. 1989. Water quality - Freshwater algal growth inhibition test with *Scenedesmus subspicatus* and *Selenastrum capricornutum*.

ISO NNI. 1991. Bodem - Bepaling van het effect van chemische stoffen op de reproductie bij regenwormen. (Ontwerp NEN voorschrift / draft NEN guideline Ontwerp NEN 5797).NNI, Delft, The Netherlands.

ISO NNI. 1997. International standard ISO FDIS 11268-2. Soil quality - Effect of pollutants on earthworms (*Eisenia fetida*). NNI, Delft, The Netherlands.

ISO NNI. 1998. International standard ISO DIS 11267. Soil quality - Inhibition of reproduction of Collembola (*Folsomia candida*) by soil pollutants. NNI, Delft, The Netherlands.

Kammenga, J.E. 1995. Phenotypic plasticity en fitness consequences on nematodes exposed to toxicants.PhD thesis, Agricultural University Wageningen, Wageningen, The Netherlands.

NEN. 1993. Water - bepaling van de acute toxiciteit met behulp van Photobacterium phispohoreaum.NNI, Delft, The Netherlands.

OECD. 1984. OECD guideline for testing of chemicals 208: "Terrestrial Plants, Growth test". 4 April OECD Guideline 208.

Toussaint, M.W., T.R. Shedd, W.H. van der Schalie & G.R. Leather. 1995. A comparison of standard acute toxicity tests with rapid-screening toxicity tests. *Environmental toxicology and chemistry* 14(5): 907-915.

LAB-SCALE RESISTIVE HEATING EFFECT ON SOIL MICROBIAL COMMUNITIES

Lisa Carmichael, Rhea Powell, Vinod Jayaraman, ***Keith Schimmel***
North Carolina A&T State University, Greensboro, NC 27411

ABSTRACT: Soil heating technologies are increasingly being used for the in situ remediation of contaminated subsurface sites that have large fractions of dense non-aqueous phase liquids (DNAPL). In the resistive heating method electrodes are placed in the contaminated site and the voltage increases the subsurface temperature to solubilize and oxidize the target contaminants. Typically resistive heating technologies maintain high subsurface temperature profiles even after the pollutant removal process is completed. The microbial ecology associated with extensive subsurface heating that occurs from such heating technologies along with repopulation of the soils after a heating event is not well understood.

This study determined the lab-scale effect of soil heating technologies on the activity of microorganisms in uncontaminated vadose zone soils from the Dover Demonstration Laboratory (Dover, DE) and the piedmont area of North Carolina (Greensboro, NC). After Preliminary pan studies, box experiments were undertaken in a 2-dimensional, insulated lexan box (0.5 m H x 0.5 m W x 0.1 m D) fitted with numerous thermocouples and soil sampling ports and filled with soil that was heated up to 100 °C with resistive heating. Microbial activity in the soils before and after heating was determined by plate counts and Acridine orange direct counts (AODC).

INTRODUCTION

Soil heating technologies are increasingly being used for the in situ remediation of contaminated subsurface sites with difficult contamination profiles or site characteristics (Soils with large fractions of clay or soils beneath paved surfaces). In these technologies, electrodes are placed in the soil and the water is used to conduct energy through the soil (Heron et al., 1998). The applied voltage can increase subsurface temperatures to 200 °C to volatilize the target contaminants, which are then collected through soil vapor stripping. Soil heating technologies appear to be successful in removing contaminants from the subsurface, but the heating will have a to this point undetermined impact on subsurface microbial communities during and after a heating regime (Bergsman and Peurring, 1997).

Objective. The objective of this project was to determine the effect of soil heating technologies on the activity of microorganisms in vadose zone soils. The microbial activity is closely linked to the temperature profiles existing inside the soil. It is therefore necessary to study the temperature and microbial effects concomitantly and compile a relationship establishing the effect of heat on microbial behavior in the heating and cooling regimes involved in resistive heating.

MATERIALS AND METHODS

Box Construction. The experimental setup has been designed to accommodate the temperature and microbial analysis (Figure 1). Three rectangular boxes (0.5 m H*0.5 m W*O.1 m D) made of Polycarbonate Lexan[R] were fabricated. An additional layer of polystyrene foam was wrapped around the boxes to ensure a high degree of insulation.

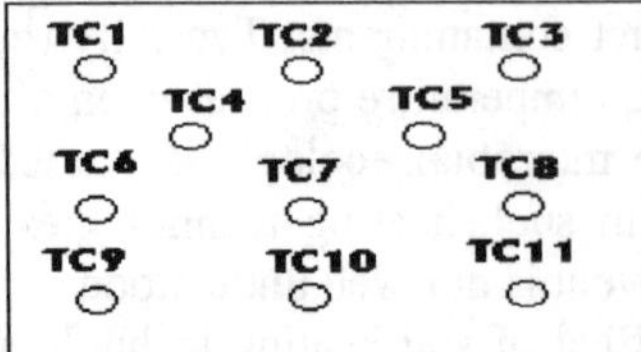

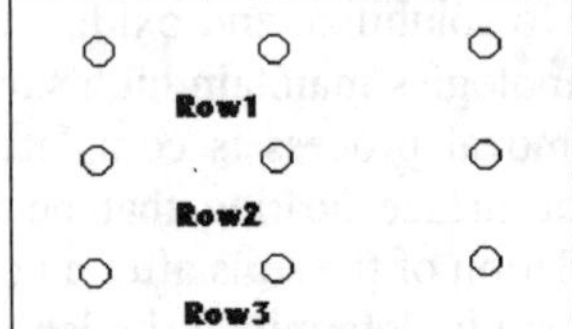

Figure 1. A schematic of the front and back view of the Lexan Box. The geometry of the temperature and soil sampling ports are shown in the figure.

The front sides of the boxes (Figure 1, left) are fabricated to accommodate numerous thermocouples (Cole Parmer, Type T) that have been arranged to monitor the temperatures in the various regions of the box. The thermocouples 4 &5 in the second row serve as reference temperature ports for soil sampling. A 12 port Cole Parmer Digisense temperature scanner reads the temperatures and interfaces with a computer to give real time data. Drainpipes at the bottom of the boxes are used in the nutrient addition and retrieval explained later in this section. The rear sides of the boxes are equipped with 3/8" soil sampling ports for microbial characterization. The soil sampling ports are arranged in a (3X3) matrix and named as $port_{ij}$, where i and j represent the row and column from top to bottom.

Power Supply and Electrodes. A variable power supply capable of supplying 240 V at up to 10 A was used to inject electric current at 60 Hz between the two electrodes. The tops of the boxes were sealed to prevent moisture loss from the box. The electrodes were custom fabricated out of stainless steel and placed at the ends of the boxes (Figure 2).

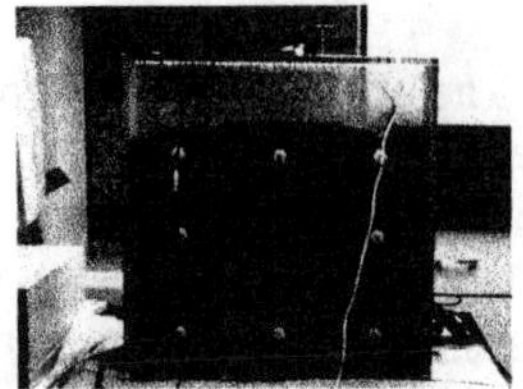

Figure 2. The resistive heating experimental setup.

Temperature Monitoring. The heating period in which the temperature increases from room temperature to around 100 °C was followed by a stagnation phase where the temperature was maintained at 100 °C for a period of approximately 8 hours. The duration of this phase depended on the moisture distribution in the soil. The Stagnation phase exposed the microbes to the harsh thermal environment for a prolonged period of time. The final step was the cooling phase in which the temperature fell to room temperature. A typical experiment lasted for 50-70 hours distributed between the three different phases (Figure 3).

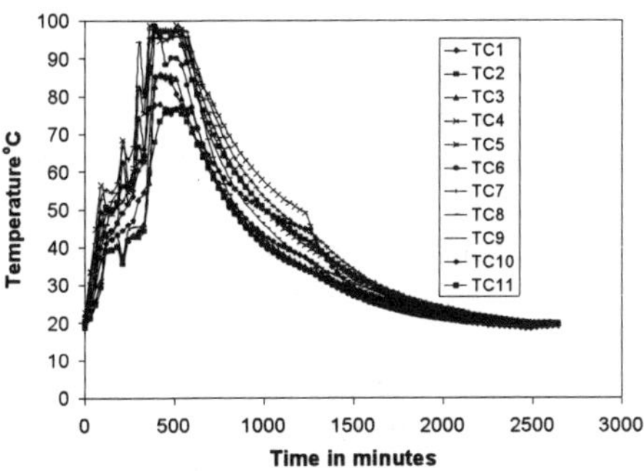

Figure 3. A typical resistive heating curve.

Soil Sampling and Analysis. The soil for the box experiments was collected from the piedmont region of North Carolina and preserved in cold storage. Soil was homogenized (Table 1) using a size 16 U.S.A standard testing sieve (A.S.T.M.E.-11 specification). The soil was sampled at different temperatures (25 °C, 50 °C, 75 °C and 100 °C). During the cooling period samples were taken at 100 °C, 75 °C, 50 °C and at room temperature. The sampling port ($Port_{ij}$) was changed at every sampling iteration. This ensured that the soil samples were representative of the events occurring in the box. CFU analysis and Acridine Orange Direct Counts (AODC) were used to analyze the microbial counts. R2A agar and Nutrient agar were used for the CFU analysis and the plates were incubated at 20 °C. The number of colonies was determined after moisture correction of the soil. AODC slides were created using the 10^2 dilution of soil in the presence of acridine orange stain. This mixture was filtered (Poreties Corporation) using vacuum filtration. The bacteria present on the filter were viewed using a fluorescent light microscope.

Table 1. Properties of the characterized soil.

Organic Carbon	Particle Density	pH	Particle size Distribution	Moisture Content
4 %	1.01 g/cm^3	7.49	81 % medium sand 14 % fine sand 4.5 % very fine sand 0.6 % silt	4 %

Nutrient Addition. After the soil stabilized to room temperature, it was left undisturbed for a week to allow the microbes to repopulate. After a week, nutrient addition was carried out with the help of a distributor assembly that pumps in the nutrient at a pre-determined rate. The soil was flushed with the nutrient until a solution equivalent to one residence period was collected through the drainpipe at the bottom of the box. Organic (Nutrient Agar), Inorganic (0.1 N M9 buffer) and millipore water have been used as nutrients. The sample retrieval continued for another eight days after nutrient addition.

RESULTS AND DISCUSSION

Figures 4-6 show the results obtained. The temperatures at various locations in the boxes were averaged to give a mean temperature distribution at each time interval. Data points from the AODC curves represent the individual sampling points at temperatures along the resistive heating, stagnation and cooling curves. The data after nutrient addition spanning from Day 0 to Day 7 is also plotted in the figures. The microbial counts are reported in units of Microorganisms/g at 10^2 dilution of soil.

For the organic treatments, increase in temperature results in increase in microbial counts in Figure 4,left. However, as the stagnation phase is approached (Temperatures around 100 °C) there is a substantial decrease in the microbial counts from 5500 microbes/g to 10 microbes/g. This indicates the thermal stress prevailing during the stagnation period. The cool down period is accompanied by an increase in microbial count to 1500 microbes/g. The organic Nutrient augments the growth of microorganisms to levels around 4000 microbes/g on Day 4. The microbial counts for Run 2 are higher during the heat up phase because of the relatively low temperatures in Run 2 compared to Run 1.

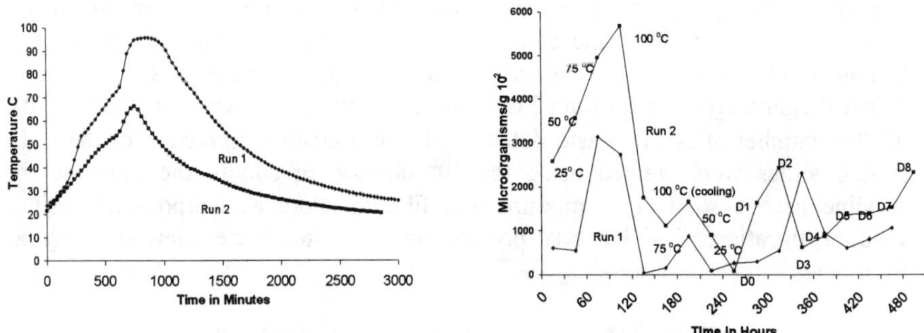

Figure 4. Comparison of Temperature-Time distribution and Microbial counts (AODC) for organic nutrient treatment.

For the inorganic treatments, microbial counts in Runs 1&2 increase with an initial increase in temperature. At temperatures near 100 °C the counts fall considerably to levels of 300 microorganisms/g. There is a favorable response to the addition of the inorganic nutrient as Runs 1&2 show an increase in the microbial population during the nutrient addition phase. The microbial counts for

Run 2 is smaller than Run 1 because of the extended nature of the temperature curve for Run 2. (Figure 5, left).

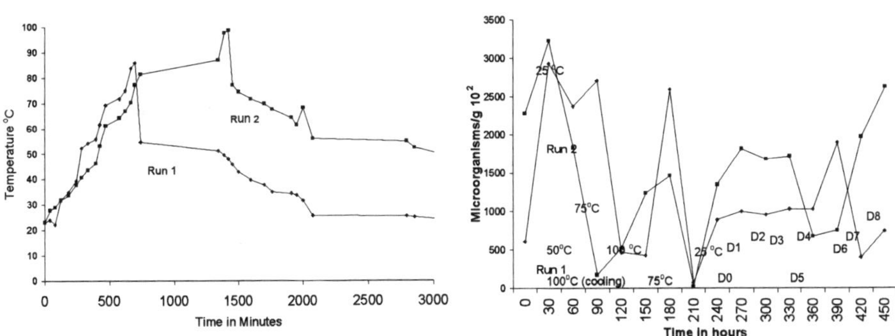

Figure 5. Comparison of Temperature-Time distribution and Microbial counts (AODC) for inorganic buffer treatment.

For the Millipore water treatment, Runs 1&2 in Figure 6 start out with a small microbial count of 400 microbes/g. However, the heating favors growth and around 50 °C a maximum count of 2500 microbes/g is reached. Higher temperatures are marked by a reduction in microbes either due to microbial death or spore formation. In the cooling phase when ambient conditions are restored the microorganisms repopulate and reach values of 1000 microbes/ g.

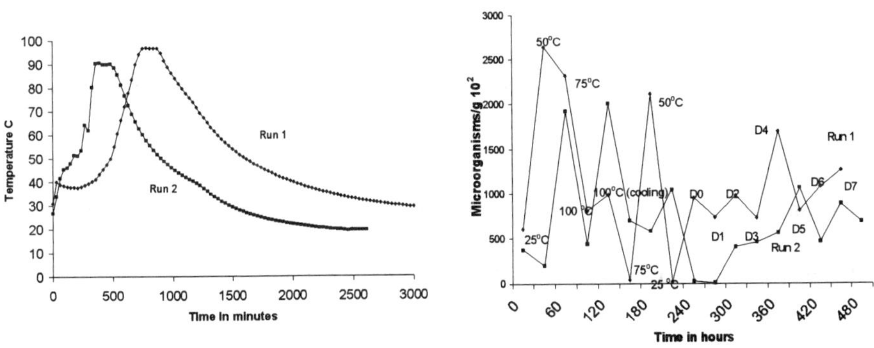

Figure 6. Comparison of Temperature-Time distribution and microbial counts (AODC) for Millipore water.

CONCLUSIONS

Results of the preliminary experiments in ovens indicated that heating decreases microbial counts as determined by the plate counts, MPN and AODC. Slight increases in temperature (up to 30 °C) were associated with an increase in the number of detectable microorganisms, but temperatures higher than 40 °C were associated with a decrease in detectable microbial populations. Decreases in the microbial populations at high temperatures tend to be larger when the soils were incubated with a high water content as compared to dry soils. These results indicate that microbial populations in heated soils may be damaged by resistive heating technologies, and if it is assumed that residual contamination remaining after heating will be removed by natural attenuation, the subsurface may need to be amended with nutrients to increase microbial populations and activity. This hypothesis was corroborated in the box experiments where a distinct increase in microbial activity was observed after nutrient addition. To determine more about the nature of the microbial community changes during heating and cooling, Phospholipid Fatty Acid Analysis (PLFA) is being undertaken with the assistance of Oak Ridge National laboratories.

ACKNOWLEDGEMENTS

The aid of Tameka Terry and Devang Panchigar in analyzing samples and the use of the facilities of the Air Force FAST center for environmental Remediation, Fate and Transport of Hazardous chemicals (Grant Number F49620-95-1-0514) are gratefully acknowledged. Effort sponsored by the AFRL/MLQE, Air Force material command, USAF, under the grant number F41624-98-1-0003. The U.S. government is authorized to reproduce and distribute reprints for governmental purposes notwithstanding any copyright notation thereon. The views and conclusions contained herein are those of the authors and should not be interpreted as necessarily representing the official policies or endorsements, either expressed or implied, of the AFRL/MLQE or the U.S. government.

REFERENCES

Bergsman, T.M., and L.M. Peurrung. 1997. *Applications Analysis Report: Six Phase Soil Heating of the saturated zone; Dover Airforce base, Delaware.* Pacific Northwest National Laboratory; Richland, WA.

Heron, G. M., Van Zutphen, T.H. Christensen and C.G. Enfield. 1998. "Soil Heating for Enhanced Remediation of Chlorinated Solvents: A laboratory study on resistive heating and Vapor extraction in a silty, low-permeable soil contaminated with Trichloroethylene." *Environ. Sci. Technology. 32*: 1474-1481.

DIESEL REMOVAL FROM A CONTAMINATED SOIL BY NATURAL HYDROCARBON-DEGRADING MICROORGANISMS

Facundo J. Márquez-Rocha (CICESE, Ensenda, B.C., México)
Vanessa Hernández-Rodríguez and Rafael Vázquez-Duhalt (IBT-UNAM, Cuernavaca, Mor., México)

ABSTRACT: A bacterial community was used for a biotreatability test of a diesel-contaminated soil. The consortium was obtained from a complex microbial community isolated from a hydrocarbon-contaminated site. Transferring this community to a diesel-containing liquid medium as sole carbon source, more than 90 % of the diesel added (7%: v/v) was consumed to produce bacterial biomass as analyzed by gas chromatography and protein determinations. The biomass produced was lyophilized and used as a bacterial inoculum for bioremediation of a diesel-contaminated soil, at both the laboratory- and semi-pilot-scale levels. Bacterial cells were reactivated on liquid medium containing diesel and inoculated on a diesel-containing soil (20 % v/p of diesel). Diesel absorbed to soil was removed by more than 60 % during five weeks of treatment at both levels. Diesel removal was improved (> 90 %) by an additional inoculation of the white rot fungus *Pleurotus ostreatus* at laboratory-scale. Biodegradation does not appear to be limited by microbial metabolic activity, since high molecular weight hydrocarbons (> C16) were biodegraded at the same rate. The lack of degradation of the remaining hydrocarbons seems to be due to mass transfer limitations in which hydrocarbons are not available to the microorganisms.

INTRODUCTION

Bioremediation is being used or proposed as a treatment option at many hydrocarbon-contaminated sites (Braddock et al., 1997). The effectiveness of bioremediation is often a function of the extent to which a microbial population or consortium can be enriched and maintained in an environment. Microorganisms with the ability to degrade crude oil are ubiquitously distributed in soil and marine environments (Venkateswaran and Harayama, 1995). However, when few or no indigenous degradative microorganisms exist in a contaminated area or when there is not time for the natural enrichment of a suitable population, inoculation may be a realistic option. Inoculation may sometimes shorten the acclimation period prior to the onset of biodegradation. The same ecological principles that influence biodegradation in general will also govern the effectiveness of inocula, regardless of whether they are natural isolates or genetically engineered microorganisms (Liu and Suflita, 1993).

Various authors have reported that inoculation had no positive, or only marginal, effects on oil biodegradation rates. Microorganisms able to degrade organic pollutants in culture may fail to function when inoculated into natural environments because they may be susceptible to toxins or predators in the environment, they may use other organic compounds in preference to the pollutant, or they may be unable to move through the soil to the area containing the contaminant. The successful use of microbial inocula in soils requires the microorganisms to contact the contaminant. Physical adsorption to soil particles or filtration by small pores may inhibit the transport of organisms (Margensin and Schinner, 1997).

Although bacteria are probably responsible for most hydrocarbon biodegradation in the environment, fungi, yeast, and even some cyanobacteria and algae have some ability to degrade hydrocarbons, albeit at very slow rates. A major question in hydrocarbon biodegradation is how the microbes actually contact substrates. Three main mechanisms are often involved, and different bacterial species may use one or more of them: interactions with low levels of dissolved material, direct contact with drops or surfaces of the insoluble phase, and interactions with hydrocarbon "solubilized" by interaction with surfactants (Prince, 1993). In this work, we utilized a bioaugmentation procedure for bioremediation experiments of a diesel-contaminated soil at laboratory and semi-pilot-scale levels.

Objective. Evaluate the possibility to use a natural bacterial community to scale up a bioremediation process to remove diesel from contaminated soil.

MATERIALS AND METHODS

Bacterial community and fungal strain. A bacterial consortium was obtained from the Mexican Institute of Petroleum (IMP, Mexico City). It was acclimated to grow on crude oil prior to diesel acclimation. The mineral medium consisted as followed (g/L): K_2HPO_4 1.71, KH_2PO_4 1.32, NH_4Cl 1.26, $MgCl_2 \cdot 6H_2O$ 0.011, $CaCl_2$ 0.02, trace metals 1 mL, and 7 mL of diesel per 100 mL of medium. The bacterial consortium was acclimated to growth in the mineral medium with diesel as sole carbon source. A dense culture was lyophilized to keep the bacterial community. An inoculum of the consortium was activated and used to prepare the inoculum for soil treatment. A strain of the white rot fungus *Pleurotus ostreatus* ECS-110 (IE-8, Intituto de Ecología, Xalapa, Ver., Mex.) was used for laboratory experiments.

Soil preparation and physicochemical analyses. A 5 g sample of soil was dried at 80°C over night (this may serve to eliminate indigenous bacteria from the soil). Once dried, 1 mL of diesel was uniformly placed on the soil and allowed to adsorb for 30 minutes. Humidity was adjusted to approximately 43%. Physicochemical analyses of soil were done by the Environmental Biotechnology Laboratory of the University of Morelos, Morelos. The physicochemical characteristics are summarized in Table 1.

Biodegradation conditions. (1) Laboratory-scale level. Soil containing diesel was treated with 1 mL of bacterial consortia previously grown in liquid medium. A control test was carried out without addition of any microorganisms. Experiments were carried out in triplicate and sampling was programmed once a week. (2) Semi-pilot-scale level. Two unit cells of 5 m^2 and 20 cm deep were used for these experiments. One of these units was used as a control or non-contaminated soil. Inocula for this level were prepared on a small cement mixer during one week.

TABLE 1. Soil physicochemical characteristics

Parameter	Value
pH	5.64
Electrical conductivity (µohms/cm)	94.70
Organic matter (%)	6.22
Humidity (%)	1.76
Texture	
Clay (%)	36.0
Sand (%)	54.0
Slime (%)	10.0
Nitrogen (%)	0.214
Available phosphate (mg/kg)	26.50
Available cations (solubles and exchangeables (meq/100 g)	
Calcium	35.70
Magnesium	6.10
Sodium	6.62
Potassium	1.58

Bacterial analyses. Lyophilized isolates were classified to their generic level on the basis of gram stain, catalase reaction, oxidase activity, and ability to utilize glucose and lactose under aerobic and anaerobic conditions.

Analysis. In liquid, diesel was extracted with dichloromethane (1:1) by using 5 mL of sample and mixing vigorously. In soil, one flask or 5 g of soil (semi-pilot) was used every sampling time; the diesel was extracted utilizing 50 mL of solvent. The organic phase was passed through $NaSO_4$ and concentrated to 0.2 mL and analyzed by gas chromatography using a Varian model 3700 equipped with a flame detector and capillary column (30 m long, I.D. 0.25 mm, SPB-20, Supelco).

RESULTS AND DISCUSSION

Bacterial community growth on diesel. Figure 1 shows the time-course of a liquid culture of the bacterial consortium using diesel as the sole carbon source. The growth followed a typical Monod-type curve, where hydrocarbon was consumed and biomass produced. Diesel was consumed more than 90%. At initiation, the chromatographic signal was typical for high molecular saturated hydrocarbons and at the conclusion, diesel was almost consumed.

Diesel removal from soil. The bacterial consortium was used to remove diesel adsorbed in soil. Bioaugmentation is one of the strategies used for bioremediation. Here we confirmed that native organisms acclimated to grow on diesel can be produced ex situ and reinoculated to a contaminated soil for effective treatment to remove diesel from the soil. Figure 2 shows the biodegradation of diesel adsorbed in soil by the bacterial consortium (C) and *P. ostreatus* (P) or both (PC). Diesel oil consists mostly of linear and branched alkanes with different chain lengths and contains a variety of aromatic compounds. Many of these compounds, especially linear alkanes, are known to be easily biodegradable. Due to low water solubility however, the biodegradation of these compounds is often limited by slow rates of dissolution, desorption, or transport. In general, the bioavailability of hydrophobic compound is determined by their sorption characteristics and dissolution or partitioning rates and by transport processes into the microbial cell (Sticher et al., 1997). Biodegradation does not appear to be limited by the aqueous solubility of hydrocarbon molecules since a significant fraction of high molecular weight (> C16) saturates were biodegraded (>60%). However, the degradation activity of the bacterial community may be limited by its movement, since *P. ostreatus* achieved the removal of diesel more efficiently. Four bacterial genuses were identified in the bacterial community, *Pseudomona*, *Serratia*, *Acinetobacter* and *Flavobacterium*.

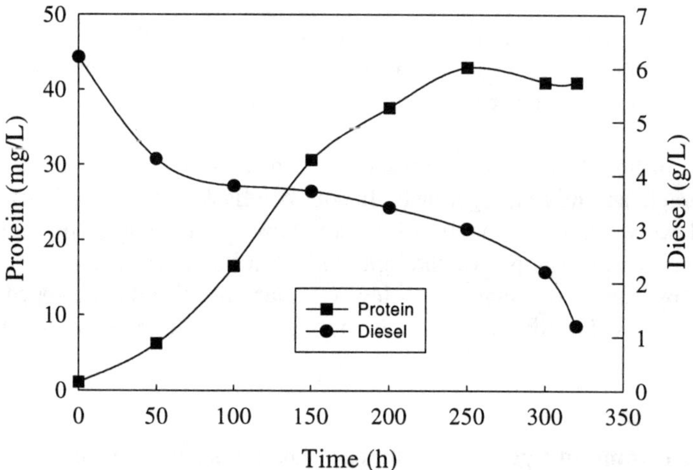

FIGURE 1. Time course of the growth of the bacterial consortium grown on diesel as sole carbon source.

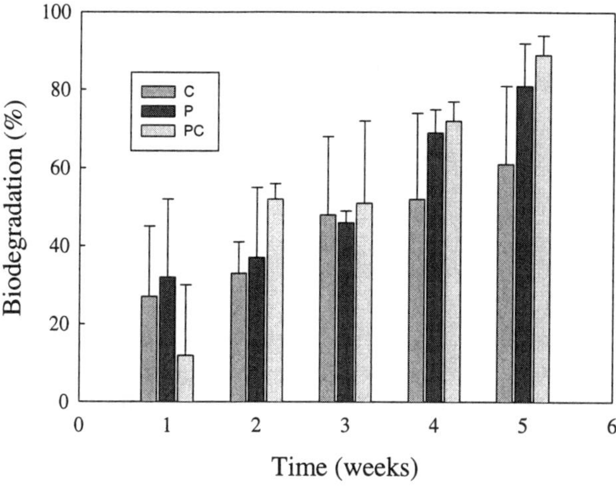

FIGURE 2. Time course of diesel removal from soil (laboratory-scale level) C: microbial community; P: *P. ostreatus*; PC: *P. ostreatus* + C.

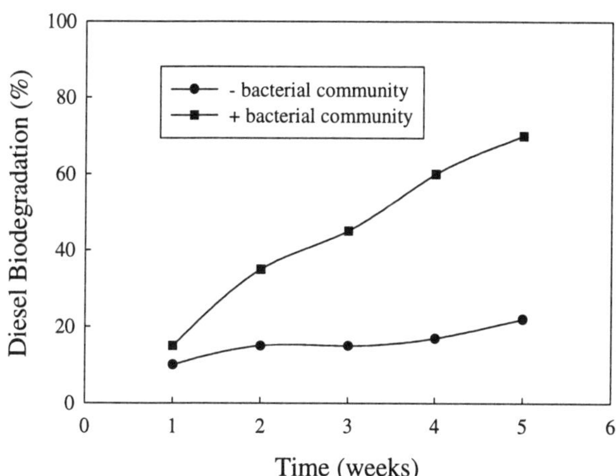

FIGURE 3. Diesel biodegradation at semi-pilot-scale level.

Semi-pilot-scale experiments inoculated with the same bacterial inocula were able to remove more than 70% of the diesel absorbed (Figure 3). Non-inoculated soil eliminated only a little more than 20% of the diesel, possibly due to the presence of some hydrocarbonoclastic bacterial activity.

For optimization of bioremediation, it is necessary to understand the factors that control microbial activity at a specific site. One of the important factors is the substrate concentration. By analyzing diesel removal by the exponential growth and high substrate concentration kinetic model ($-ds/dt = ke^{rt}$; where [s] >> Km), we obtained a degradation rate of 0.091 h^{-1} for liquid degradation and <0.04 d^{-1} in soil at the laboratory-scale with the microbial community. At the semi-pilot-scale level, the biodegradation rate increased to 0.051 d^{-1}. This may be due to indigenous microbial activity. In laboratory experiments, with the presence of *P. ostreatus,* the biodegradation rate was 0.062 d^{-1}.

Enhancement of microbial activity at a hydrocarbon-contaminated site may be used in combination with microbial addition. The semi-pilot-scale level based results support the concept of bioaugmentation by increasing the biodegradative activity that may be improved by inoculum stimulation.

REFERENCES

Braddock J. F., M. L. Ruth, P. H. Catterall, J. L. Walworth, and K. A. McCarthy. 1997. "Enhancement and inhibition of microbial activity in hydrocarbon contaminated arctic soils: implications for nutrient-amended bioremediation". Environ. Sci. Technol. 31(7): 2078-2084.

Huesemann M. H. 1995. "Predictive model for estimating the extent of petroleum hydrocarbon biodegradation in contaminated soils". Environ. Sci. Technol. 29(1): 7-18.

Liu S. And Suflita J. M. 1993. "Ecology and evolution of microbial populations for bioremediation". TIBTECH 11(8): 344-352.

Margensin R. and F. Schinner. 1997. "Efficiency of indigenous and inoculated cold-adapted soil microorganisms for bioremediation of diesel oil in Alpine soils". Appl. Environ. Biotechnol. 63(7): 2660-2664.

Prince R. C. 1993. "Petroleum spill bioremediation in marine environments". Crit. Rev. Microbiol. 19(4): 217-242.

Sticher P., Jaspers M. C. M., Stemmler K., Harms H., Zehnder A. J. B., and van der Meer J. R. 1997. "Development and characterization of a whole-cell bioluminescent sensor for bioavailable middle-chain alkanes in contaminated groundwater samples". Appl. Environ. Microbiol. 63(10): 4053-4060.

Venkateswaran K. and Harayama S. 1995. "Sequential enrichment of microbial populations exhibiting enhanced biodegradation of crude oil". Cand. J. Microbiol. 41: 767-775

FAILURE OF NON-INDIGENOUS MICROORGANISMS TO REMEDIATE PETROLEUM HYDROCARBON CONTAMINATED SITES

Lisa S. Dernbach
(California Regional Water Quality Control Board,
Lahontan Region, South Lake Tahoe, CA)

ABSTRACT: Researchers promote non-indigenous microorganisms as a viable method to remediate soils contaminated with petroleum hydrocarbons and to restore groundwater quality. Suppliers claim that microaerophilic bacteria are engineered to degrade a wide range of hydrocarbon contaminants, even in cold weather areas. However, staff with the California Regional Water Quality Control Board, Lahontan Region (Regional Board), found that non-indigenous microorganisms failed to achieve target cleanup levels in groundwater at gasoline contaminated sites where applied in the Lake Tahoe Basin.

Between 1993 and 1995, Regional Board staff approved three in situ field-scale bioremediation proposals using non-indigenous bacteria. These proposals were made with the knowledge that temperatures below freezing (0° C) can exist in the Lake Tahoe Basin for five months of the year or more. The geology at the three sites consisted of alluvial and/or lacustrian sediments. Ground water varied from 1 to 12 ft (0.3 to 3.7 m) below ground surface. Batch solutions of bacteria and water were infiltrated into the subsurface at each site via a series of temporary boreholes, on two or three occasions.

The results indicated that injected non-indigenous microorganisms conclusively failed to remediate gasoline contaminated sites. Monitoring well data revealed that total petroleum hydrocarbon (TPH) and ethylbenzene concentrations did not change, or showed increased concentrations in groundwater after the inoculations. Benzene, toluene, and xylenes concentrations increased in one monitoring well but decreased in other monitoring wells. None of the three sites came close to reaching target cleanup levels for groundwater. Since no exact explanation was provided as to why the biocultures failed to restore groundwater quality, Regional Board staff no longer approves proposals for in situ remediation using non-indigenous microbes in cold weather areas of the Lahontan Region.

INTRODUCTION

Regulators saw an increasing number of cases with petroleum contamination in soil and ground water during the 1980's as more underground storage tanks were removed to comply with federal and state regulations. Responsible parties cried out for timely and cost effective remedial options that would cause little disturbance to impacted properties. In response, federal agencies encouraged states to allow the implementation of alternate remedial technologies, even when a track record for a particular method had not been established. In situ bioremediation, consequently, became an increasingly popular choice for responsible parties trying to comply with regulatory requirements and restore water quality.

Between 1993 and 1995, staff with the Lahontan Regional Water Quality Control Board (Regional Board) approved in situ bioremediation proposals using non-indigenous microorganisms at three gasoline impacted sites. The sites were located in the Lake Tahoe Basin, in the Sierra Nevada. The proposals stated that biocultures were naturally occurring psuedamoma and bacillus which were engineered to degrade a wide range of hydrocarbons (Bio-Tec, 1993). Microaerophilic organisms would be mixed on site with micronutrients prior to placement in the plume. Following several bioculture inoculations, each site should achieve target cleanup levels in groundwater in about ten months. Regional Board staff was provided assurances by the supplier that high altitude, 6,229 ft (1900 m), and extreme cold temperatures (< 0° C) for at least five months of the year would not impair the effectiveness of non-indigenous microbes.

The bioremediation approach undertaken at each of the three sites was similar. Therefore, the implementation method and results are discussed below in detail for only one gasoline contaminated site.

CAR WASH PROPERTY

Site Description. Extensive soil and groundwater contamination were discovered in 1991 following the removal of three 10,000 gal (37,850 L) gasoline underground storage tanks (Fig. 1). The 18-year old tanks had been part of a gas station and car wash in South Lake Tahoe. The area of impacted soil was estimated at 100 ft x 100 ft (30.5 m x 30.5 m). Up to 3,300 mg/kg as Total Petroleum Hydrocarbons (TPH) as gasoline and 50 mg/kg as benzene existed in the soil. Because the tanks had not been in operation since the mid-1980's, MTBE was not a contaminant of concern.

The site is located less than 800 ft (244 m) from Lake Tahoe, a drinking water source. The car wash property is underlain with fine sand having silt interbeds. The geology is a mixture of alluvial sediments eroded from surrounding mountains intertongued with lacustrine deposits. Groundwater ranged from 2 to 6 ft (0.6 to 1.8 m) below ground surface year round. The groundwater cleanup goal was set to background levels (i.e. non-detect) to protect Lake Tahoe and restore water quality for future beneficial use. The Regional Board does not set soil cleanup goals. However, in this instance due to shallow groundwater, goals for groundwater cleanup implies similar goals for soil cleanup.

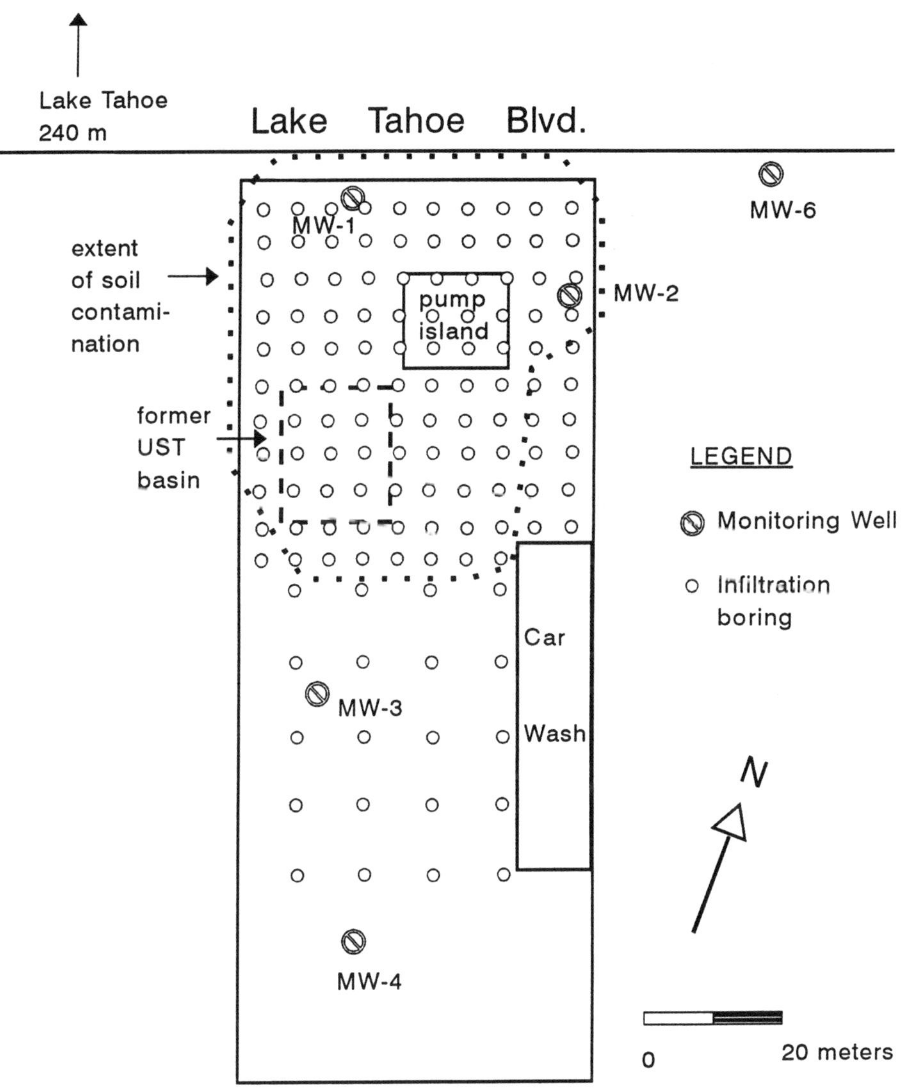

Figure 1. Site plan for car wash showing area of soil contamination and infiltration borehole locations.

Method. A hollow stem auger drill rig was used to install 135 infiltration boreholes at the car wash, upgradient and in the area of soil contamination. The borings, arranged in a grid pattern on 10 ft (3.1 m) centers, were 4-in (10-cm) in diameter, 10 ft (3.1 m) deep, and backfilled with pea gravel. The non-indigenous microbes were mixed on site with natural source water and micronutrients. After micronutrients were consumed, the batch solution was infiltrated in the boreholes. The car wash received two inoculations of biocultures of approximately 6,000 gal (22,710 L) each during October 1993. A third inoculation occurred in June 1994.

RESULTS

Table 1 below shows hydrocarbon concentrations in groundwater from monitoring events prior to the initial inoculation through after the third inoculation. In monitoring well MW-1, TPH concentrations and benzene, toluene, ethylbenzene, and xylenes (BTEX) concentrations increased after the first and second inoculation and decreased after the third inoculation. One year after the first inoculation, gasoline constituents in groundwater at MW-1 were actually higher over that from initial concentrations. In monitoring wells MW-2 and MW-6, the overall TPH and BTEX concentrations had decreased slightly after the first two inoculations and showed a greater decrease after the third inoculation. Benzene and toluene concentrations, one year after the first inoculation, were at least half of their initial concentrations. Xylenes showed only a slight decrease in concentrations. TPH and ethylbenzene concentrations, however, increased from initial concentrations one year before.

TABLE 1. Groundwater analytical results

Monitoring Well	Date	TPH-gas (µg/L)	Benzene (µg/L)	Toluene (µg/L)	Ethylbenzene (µg/L)	Xylenes (µg/L)
MW-1	8/93	86,000	8,600	18,000	2,300	7,800
	12/93	100,000	17,000	22,000	2,000	8,800
	3/94	71,000	22,000	15,000	1,600	4,000
	7/94	11,000	1,700	2,800	240	1,200
	10/94	160,000	12,000	32,000	4,600	8,100
MW-2	8/93	65,000	7,800	11,000	3,200	8,800
	12/93	66,000	7,700	9,200	2,800	7,600
	3/94	79,000	5,800	10,000	2,900	7,600
	7/94	32,000	4,100	5,500	430	6,100
	10/94	81,000	3,600	4,800	5,200	7,700
MW-6	8/93	49,000	2,600	8,400	2,200	5,100
	12/93	47,000	2,400	6,100	2,500	3,600
	3/94	72,000	3,200	14,000	3,400	8,000
	7/94	33,000	1,900	5,900	320	4,600
	10/94	59,000	1,300	2,000	2,200	4,400

DISCUSSION

Groundwater monitoring after the completion of bioculture inoculation showed that target cleanup levels were not achieved at the car wash property. The results indicate that non-indigenous microorganisms failed to have a significant effect on gasoline constituents. This finding was similar to those at the other two gasoline sites that received inoculations with non-indigenous microbes (Fugro, 1994 and Hydro-Search, 1996).

The consultants that implemented the bioremediation projects provided conflicting explanations why biocultures failed to restore water quality. The explanations included: excess soil moisture during springtime snowmelt; insufficient inoculation events; higher gasoline levels in soil and groundwater existed; and lower air and groundwater temperatures existed.

Regional Board staff believes that the latter explanation played a strong role in the failure of non-indigenous microorganisms' ability to consume hydrocarbons and achieve cleanup levels. Besides having cold air temperatures, the Lake Tahoe Basin groundwater is also cold, typically about 8° C. These conditions contrast with ideal in situ bioremediation settings, such as warm air temperatures and groundwater temperatures of at least 12.5° C (Quigley, 1998). Non-indigenous microorganisms applied to the subsurface in the Lake Tahoe Basin were likely unable to adapt to local environmental conditions.

CONCLUSIONS Regional Board staff has taken the position that in situ bioremediation using non-indigenous microorganisms is not an effective cleanup strategy at gasoline contaminated sites in cold weather areas. The costs to implement in situ bioremediation using non-indigenous microorganisms were reimbursed to the responsible parties by the California Underground Storage Tank Cleanup Fund. The total remedial costs for the gasoline sites amounted to nearly half a million dollars. State regulations and policies require regulators to approve only cost effective remedial methods that can likely achieve target cleanup levels. Thus, Board staff no longer approves such projects in these settings in the Lahontan Region.

REFERENCES

Apex Envirotech, Inc. 1994. *Third Quarter 1994 Groundwater Monitoring Report, Jet-Thru Car Wash.* Fair Oaks, CA.

Bio-Tech, Inc. 1993. *A Remedial/Corrective Action Plan for the Bioremediation of Hydrocarbon Contaminated Soil and Water.* Colorado Springs, CO.

Fugro West, Inc. 1994. *Quarterly Groundwater Monitoring and Remediation Progress Report.* Reno, NV.

Hydro-Search, Inc. 1996. *Monitoring Report, Sierra Boat Company.* Reno, NV

Quigley, J. T. 1998. "Engineered microbe effectiveness, bioaugmentation in lab vs. field and transgenic microorganisms." *Underground Tank Technology Update*. 12(3): 5-7.

LABORATORY STUDY OF CRUDE OIL REMEDIATION BY BIOAUGMENTATION

Paulo N. Seabra, (Petrobras' R&D Center, Rio de Janeiro, Brazil)
Monica M. Linhares and Lidia M. Santa Anna (Petrobras' R&D Center, Rio de Janeiro, Brazil)

ABSTRACT: Bioremediation can be carried out through the addition of nutrients or optimization of soil environmental conditions, often referred as biostimulation, using native microorganisms. Another strategy, termed bioaugmentation, consists of the addition of microorganisms capable of readily degrading specified contaminants to a contaminated site. Petrobras' Research and Development Center – CENPES – isolated a consortium of natural soil microorganisms from a contaminated site. Continuous 45-day laboratory remediation tests were carried out in two unsaturated sand Plexiglas columns (6.0 cm ID × 50.0 cm long), one containing the microbiological consortium. These studies indicated that bioaugmentation improved biodegradation of light Arabian crude oil (28.4%) as compared to results without added microorganisms (20.4%). After the tests, the hydrocarbon compounds in the residual oil trapped in the sand showed a higher biotransformation in the bioaugmentation column.

INTRODUCTION

Each year huge quantities of chemicals are released into terrestrial and aquatic environments as a result of industrial activities. In the petroleum industry, some releases are accidental crude oil spills. In the first half of the 90s, an average of 7.9 million gallons of oil was spilled each year in the USA (API, 1997). Bioremediation is a cost-effective alternative to physical and chemical methods for removing pollutants from the environment and restoring contaminated sites. Bioremediation technologies were used to remediate more than 10% of the Superfund sites treated between 1982 and 1995 (Betts, 1998). The ability of microorganisms to degrade and stabilize hydrocarbons as sources of energy and carbon from crude oil spills is well known (Atlas, 1988). Bioremediation can be carried out by the addition of nutrients and optimization of other soil environmental conditions, often referred to as biostimulation, or by the addition of microorganisms that have the capability to degrade specific contaminants readily, called bioaugmentation.

The utility of bioaugmentation is a source of controversy. Some say that bioaugmentation is capable of solving most site remediation problems, while others believe that adding microorganisms to a site can confound existing engineering problems (e.g., well plugging) and contribute to higher costs with minimal proven benefit. Many commercial microbiological products are available to increase the remediation rate in sites polluted by petroleum hydrocarbons. Most of these products have been tested in contaminated Brazilian soils. However, results have not been promising, probably due to specific environmental factors that are not well understood.

Objective. The objective of this study was to evaluate the biodegradation of light Arabian crude oil attained with the addition of a consortium of natural soil microorganisms, isolated from a contaminated site by Petrobras' Research and Development Center (CENPES).

MATERIALS AND METHODS

Continuous 45-day tests were carried out in two unsaturated sand Plexiglas columns (6.0 cm ID × 50.0 cm long) each one packed with 1.7 kg sand contaminated by light Arabian crude oil (Figure 1). The sandy material was obtained from Pontal Beach, Angra dos Reis, in southeast Brazil, and possessed a particle mass density of 2.73 g/cm^3. In column A, a mixture of oil-degrading bacteria (Linhares and Seabra, 1997) was injected from the top after the light Arabian crude oil contamination. This consortium of microorganisms was passed continuously through the sand column 2 times a week, during 6 hours for each run. No inoculum was added to the other column (B). The porous medium porosity was 19.0% and 22.4%, respectively for columns A and B. An aqueous solution containing mineral nutrients was passed continuously through the sand column for 45 days at a flowrate of 1 porous volume (VP)/24 h (nearly 0.16 L/h). The nutrient solution consisted of $(NH_4)_2SO_4$ at 0.3 g/L, NH_4NO_3 at 0.15 g/L, KH_2PO_4 at 0.15 g/L, K_2HPO_4 at 1.0 g/L, $MgSO_4 \cdot 7H_2O$ at 0.05 g/L, $CaCl_2$ at 0.05 g/L and micronutrient solution at 1 mL/L. The micronutrient solution was made up of 100 mg $FeSO_4 \cdot 7H_2O$, 40 mg $(NH_4)_6Mo_7O_{22}.4H_2O$, 100 mg $MnSO_4 \cdot H_2O$, 30 mg $Na_2B_4O_7 \cdot 10H_2O$, 25 mg $Co(NO_3)_2 \cdot 6H_2O$, 25 mg $CuCl_2$, 25 mg $ZnCl_2$, and 15 mg NH_4VO_3 in 1 liter of deionized water.

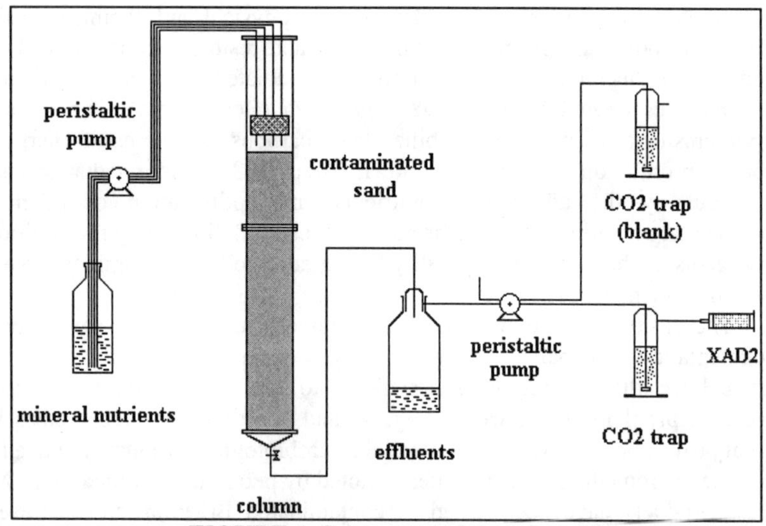

FIGURE 1. Schematic of the laboratory test.

Analytical Procedures. At the end of the tests, concentration and composition of oil in sand and effluents were determined (Ducreux et al., 1994). Organics from soil and effluent samples were released using a Soxhlet dichloromethane extraction system (Zymark). The organic extract was separated into aliphatic, aromatic, polar, resin, and asphaltene fractions using a column chromatography system (Knauer) as described elsewhere (Linhares and Seabra, 1997). A gas chromatograph (HP 6890) with a flame-ionization detector (CG/FID) was utilized for saturate fraction analysis using a DB-1 fused non-polar silica capillary column with 0.25 mm inside diameter, 30 m long, and 250 μm film thickness. The microbiological enumeration on soil was quantified by standard plate count. Two types of culture media, TSA and McConkey, were used.

RESULTS AND DISCUSSION

Soil Residual Oil. Table 1 shows the amount of oil released from each one of the two columns and the residual oil in the sandy material.

TABLE 1. Total amount of oil released from the column and biodegraded.

Parameter	Column A	Column B
Initial residual oil (g/kg of dry sand) – t_0	28.8	26.5
Final residual oil (g/kg of dry sand) – t_{45}	20.1	19.7
Total crude oil released from column (%)	30.2	25.7
Total oil biodegraded (%)	28.4	20.4
Amount of oil released biodegraded (%)	94.0	79.4
Degradation rate constant (day^{-1}) – k	0.0074	0.0051

In column A, where the consortium of microorganisms was added, 30.2 percent of the initial residual oil was released, against 25.7 percent for column B, where only the natural microorganism were stimulated by the mineral medium. From the total crude oil released, 94.0 and 79.4 percent have been reported to be biodegraded in columns A and B, respectively. It was assumed here that the substrate removal follows first-order kinetics. The degradation rate constant (k) obtained from column A assay was 0.0074 day^{-1}, which is higher than from column B (0.0051 day^{-1}).

At the end of the test, the sandy material of each column was divided in two parts, bottom and top. The gas chromatograms of the soil samples used for evaluating biodegradation are set forth in Figure 2. Figures 2a and 2b show the aliphatic hydrocarbon chromatograms for top and bottom of column A, respectively. For the top of column A the aliphatic peaks are smaller than for the bottom. For column B (Figures 2c and 2d), the chromatograms of the top and bottom exhibit the same aspect. These results are explained by a better disposal of oxygen and nutrients in the top of the columns, improving the oil biodegradation. In both columns the residual oils trapped in sand have smaller peaks compared to the chromatogram of original crude oil (Figure 2e). Chromatograms from column A present smaller peaks than those from column B, showing a better biodegradation of the trapped residual oil.

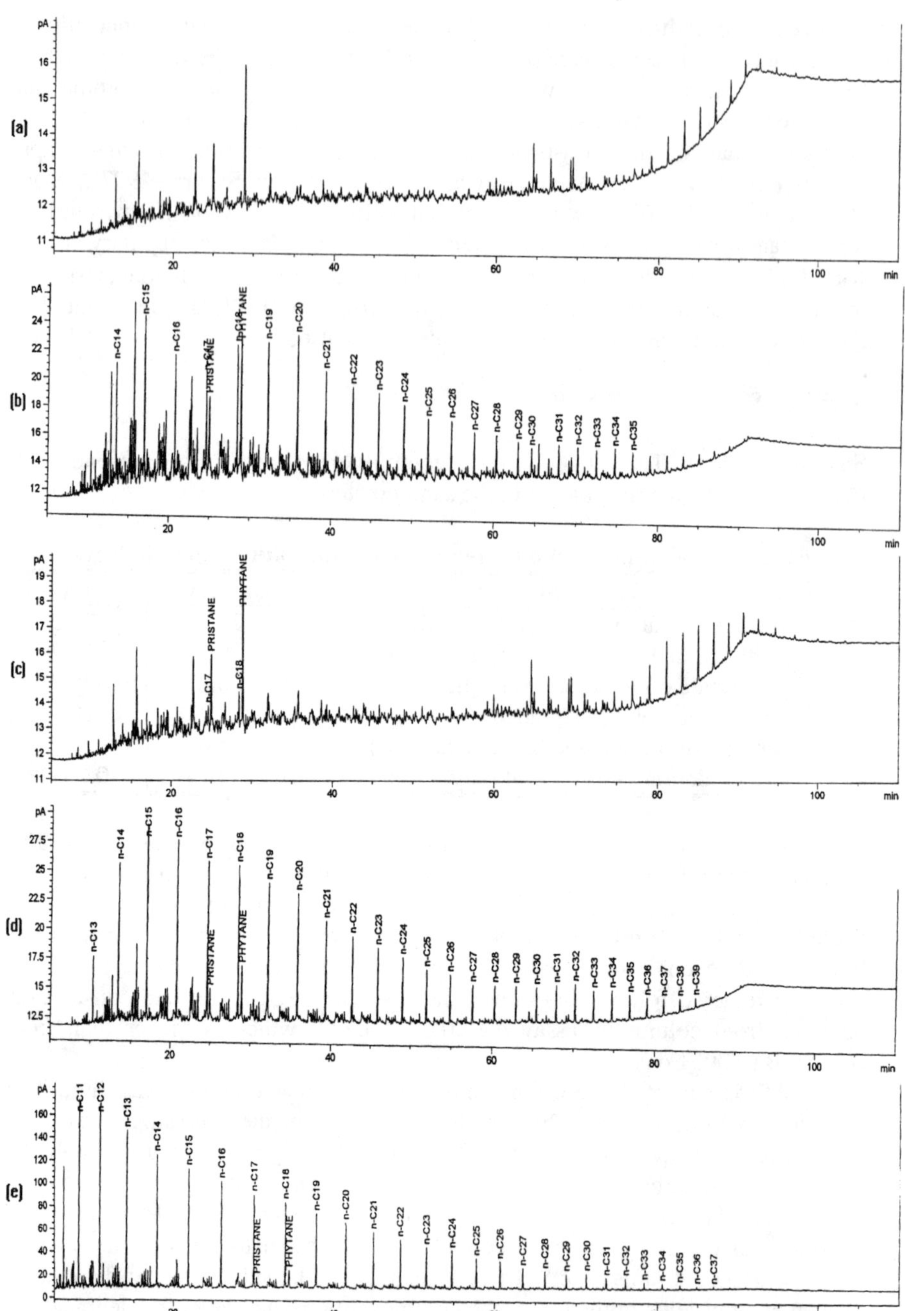

FIGURE 2. Gas chromatogram of the soil samples from columns A and B. (a) Column A top part. (b) Column A bottom part. (c) Column B top part. (d) Column B bottom part. (e) Original crude oil.

The selective disappearance of microbiologically labile isomers is another way to evaluate the biodegradation level. The microbiological resistant branched-chain isomer, phytane, served as an internal tracer which persisted relative to octadecane (C_{18}). This study focused on C_{17}/pristane and C_{18}/phytane ratios. The C_{17}/pristane and C_{18}/phytane ratios from column A are smaller than those from column B. These results indicate that in column A, using a bioaugmentation technique, the residual oil trapped in sand was more susceptible to microbial attack. So, the use of bioaugmentation has shown an increase in the biodegradation levels.

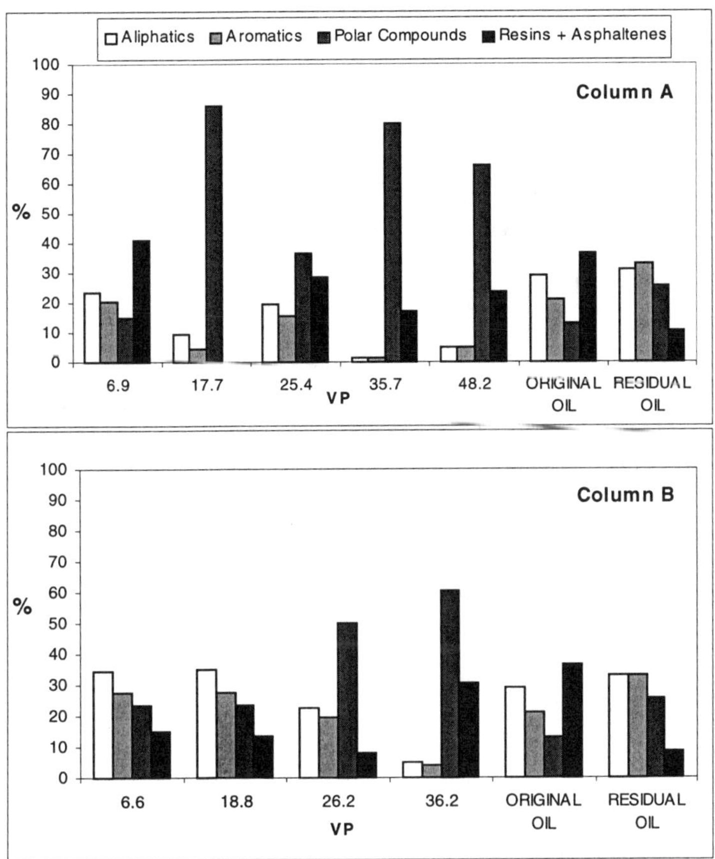

FIGURE 3. Distribution of the effluents of columns A and B by chemical families.

Effluents. The concentration and composition of oil in effluents were analyzed during the assays. The effluents were collected and divided in 4 or 5 different fractions based on the number of pore volumes that passed through the columns. Aliphatic and aromatic biodegradation yields a certain amount of polar compounds. Effluents from column A showed a polar compound percentage larger than those from column B (Figure 3), suggesting better microbial activity in column A.

Microorganism Count. After the assays, the genera of the microorganisms and a count of those entrapped in sand were determined. The microbial consortium used was mainly of the gram-negative *Pseudomonas* genus. McConkey medium is selective for gram-negative bacteria, while the TSA medium is used for total bacteria count. Table 2 lists the results obtained. In column A, nearly 75% (equal to the McConkey/TSA ratio) and 90% of all bacteria are gram-negative in the top and bottom parts, respectively. For column B the gram-negative bacteria represent 63% and 12% in the top and bottom parts, respectively.

TABLE 2. Microorganisms count entrapped in sand bacteria.

Column	Level	TSA (cfu/g of soil)	McConkey – gram (-) (cfu/g of soil)
A	Top	$2,8 \times 10^{10}$	$2,1 \times 10^{10}$
	Bottom	$2,9 \times 10^{10}$	$2,6 \times 10^{10}$
B	Top	$2,4 \times 10^{10}$	$1,5 \times 10^{10}$
	Bottom	$4,2 \times 10^{10}$	$5,0 \times 10^{9}$

CONCLUSION

The addition of a consortium of oil-degrading bacteria – a bioaugmentation process – to a sand contaminated with light Arabian crude oil improved its biodegradation.

REFERENCES

American Petroleum Institute. 1997. *Petroleum Industry Environmental Performance - Fifth Annual Report*. No: N10040.

Atlas, R.M. 1988. "Biodegradation of Hydrocarbons in the Environment." *Basic Life Science*, *45*: 211-222.

Betts, K.S. 1998. " New Cleanup Technologies Battle Credibility Gap". *Environ. Sci. Technol. 32(11)*: 266A-270A.

Ducreux, J., D. Ballerini, and C. Bocard. 1994. "The Role of Surfactants in Enhanced In Situ Bioremediation". In R.E Hinchee, B.C. Alleman, R.E. Hoeppel, and R.N. Miller (Eds.), *Hydrocarbon Bioremediation*, pp. 237-242. Lewis Publishers, Ann Arbor, MI.

Linhares, M.M. and P.N. Seabra. 1997. *Efficiency Evaluation of Oil Spills Treatment Using Microorganisms in Soil - Final Report* (in Portuguese). Petrobras/CENPES/SEBIO, Rio de Janeiro, Brazil.

ACCELERATED BIOREMEDIATION AS AN ALTERNATIVE TO CONVENTIONAL REMEDIAL TECHNOLOGIES

Dan C. Buzea and Edward J. DeStefanis (Leggette, Brashears & Graham, Inc. White Plains, New York)

ABSTRACT: Application of bioremediation as an alternative technology to expedite soil and ground-water remediation has been effective in reducing the concentration of hydrocarbons at an affordable cost. Bioremediation processes can be achieved by the addition of biological organisms into the soil and ground water which are contaminated and are microorganism deficient. The microorganism's activities can also be increased by adding nutrients. A combination of microorganisms and nutrients was injected in low permeability soil contaminated with hydrocarbons using Geoprobe® drilling equipment. Ground-water samples collected prior to the injection and five months after the first application of microorganisms (Micro-Bac®, International, Inc.) and Oxygen Release Compounds (ORC®, Regenesis, Inc.) showed a significant decrease in dissolved hydrocarbon concentrations in most of the monitoring points. Additional ground-water sampling conducted after seven months of application showed, in most of the areas, that the bioremediation process was still active. The reduction in hydrocarbon concentrations was obtained at a low cost if compared with other remedial technologies which are currently used for similar projects.

INTRODUCTION

Hydrocarbon components from releases of petroleum products into the subsurface are in general degraded by the native microorganisms located in soil and ground water. An increase of biological activity by introducing microorganisms and nutrients can accelerate the biodegradation process of hydrocarbons and remediate the soil and ground water. The following is a description of bioremediation treatment using microorganisms and ORC® in order to accelerate the biodegradation process of hydrocarbon compounds at a significant reduction in cost.

SITE DESCRIPTION/MATERIALS AND METHODS

The site (105' X 45') is a former gasoline in New York, NY. Following the removal of the underground storage tanks (USTs), a soil-vapor extraction (SVE) system was installed in the former UST excavation which was backfilled, paved and used as a parking lot. Additional subsurface investigations were conducted and concluded that the fine grained sediments mixed with building debris encountered beneath the site continued to exhibit high levels of hydrocarbons and that the ground water which was found at 22 ft bg (feet below grade) also contained dissolved hydrocarbons. The data suggested that the SVE reached the point of diminishing return and for that reason was removed from the site.

After analysis of several additional remedial technologies, bioremediation was selected for this site because it has the potential for significant reduction of hydrocarbon concentrations while causing minimal site disturbance. Soil samples was analyzed in the field for pH and moisture percentage and submitted to a laboratory for analysis of volatile organics, by EPA Method 8020 modified to include MTBE as well as heterotrophic plate counts (HPC) and total organic carbon (TOC). Ground-water samples were analyzed in the field for dissolved oxygen, eH, pH, conductivity, dissolved solids and then analyzed for contaminants as described for soil In order to determine the bioremediation applicability each ground-water sample was also analyzed for nitrate, ammonium, orthophosphate, chemical oxygen demand, sulfate and RCRA metals.

Two types of bioremediation treatment were used at the site. The first treatment consisted of injecting microorganisms into the vadose zone using the Geoprobe® drilling unit The second bioremediation treatment consisted of an ORC® slurry injection beneath the ground-water level also using the Geoprobe® borings. The application of ORC® into the ground-water plume was done in order to increase the aerobic biodegradation of the contaminants and to provide additional impact of oxygen to the microorganisms which are the engine of bioremediation process.

The first application of microorganisms and ORC® at the site was conducted in April 1997. Eight test borings drilled by Geoprobe® method were used and a liquid containing microorganisms supplied by Micro-Bac® International, Inc. was injected in four borings above the ground-water level; a slurry of ORC® manufactured by Regenesis® was injected below the ground-water level using four different borings. Data was collected at five months and seven months after the first bioremediation treatment was applied.

RESULTS

Field Results: Bioremediation results are presented in Table 1 and Figure 1. With respect to Table 1, a comparison between the application of microorganisms versus ORC® was made in order to determine which application has the potential to accelerate the bioremediation process. The data indicates that following microbe application at Injection Point TB-1 the benzene concentration in ground water decreased in five months by 54% and at seven months to 84%. Values for total BTEX were 44% after five months with a small increase to 40% by Month 7. Similarly, for injection point, TB-6, there was a benzene reduction of 14% after five months and a total of 21% after seven months. The total BTEX was reduced 41% after five months and a small increase to 41% after two more months. The data suggested that the microbe application have been successful in accelerating the bioremediation process during the first five months.

The ORC® application shows a similar pattern for benzene (TB-3 and TB-5 injection points) and a higher reduction of total BTEX. For example, at Injection Point TB-3, the total BTEX reduction after five months of ORC®

application was 80% and increased to 95% after two more months. The data from a second ORC® application point (TB-5) showed a smaller reduction of total BTEX (34%) after 5 months but continued to indicate that the bioremediation process was still active after 7 months (51% reduction) following the ORC® application.

The data suggest that the ORC® application was more efficient then the microbe application. Because the results are completely based on ground-water samples and the ORC® application was conducted below the ground-water level while the microorganisms were injected in the vadose zone the effect of ORC® application is higher to the ground water. The injection of microorganisms and ORC® in two adjacent Geoprobe® borings at the site stimulated biodegradation and finally resulted in a reduction of the hydrocarbon concentrations in ground water. In addition, the bioremediation treatment reduced the distance where the plume migrates at concentrations that pose a risk to the environment.

A contour map illustrating the overall impact of the treatments on total BTEX at baseline, Month 5 and Month 7 is presented in Figure 1. The highest BTEX concentration prior to the first application was in the TB-3 area. The extent of the plume showing BTEX concentrations between 30,000 and 200,000 ug/l was approximately 40 feet long. After five months, following the application of bioremediation treatment in eight injection points the 30,000 ug/l BTEX concentration plume was reduced to 30 feet. The concentration of dissolved BTEX was reduced substantially in most of the injection points and the plume started to show that the application process was very efficient in the core of the contamination plume.

By November 1997, after seven months of the accelerated bioremediation treatment, the configuration of the dissolved plume indicates that while the concentration of the plume remained in the same location a significant reduction of dissolved BTEX was obtained. The results of ground-water analysis showed that the bioremediation treatment was efficient and a significant reduction of hydrocarbon plume was obtained beneath the site.

Cost Comparison: The cost of the bioremediation treatment was compared with the cost for other remediation technologies which were determined to be feasible for these site conditions. Remediation techniques like soil excavation and disposal, ground-water pumping and treat, and on-site thermal treatment are not feasible because of site conditions (i.e., access, low aquifer yield, low ground-water levels, high cost) and were eliminated. The costs for implementation of remedial technologies as soil-vapor extraction, air sparging/soil-vapor extraction (AS/SVE) and high vacuum extraction were evaluated in comparison with the bioremediation treatment.

Table 2 lists the capital and operations and maintenance costs for each of these remediation technologies. Capital cost for bioremediation treatment includes drilling of Geoprobe® borings and application of microbe and ORC® treatment for six times during a three-year period. However, based on results obtained at the Manhattan site, it is expected that the bioremediation treatment will take less than three years.

The operation and maintenance cost for the other remediation technologies include monthly visits and sampling of remediation system and quarterly ground-water sampling for a five-year period. The operation and maintenance cost for bioremediation treatment is related to quarterly ground-water sampling and analysis for five years.

CONCLUSIONS

The results of bioremediation treatment indicate that this technology was very effective for remediation of soil and ground water with hydrocarbon components at this site. Significant reduction of BTEX in ground water was obtained by using Geoprobe® points for injecting microorganisms and Oxygen Release Compound above or below the ground-water level. The bioremediation technology has the potential to reduce hydrocarbon concentrations from soil and ground water while causing minimal site disturbance. This remediation technology has a low cost, and depending on subsurface conditions, in some cases is more efficient than other available technologies.

TABLE 1. Water Quality Summary.

Sample ID	Date	Benzene (ug/l)	Toluene (ug/l)	E-benzene (ug/l)	Xylenes (ug/l)	BTEX (ug/l)	MTBE (ug/l)
TB-1	4/2/97	7,000	23,000	4,000	20,100	54,100	1,300
TB-1	8/27/97	3,200	14,000	2,000	11,000	30,200	<500
TB-1	11/17/97	1,100	11,000	3,200	17,000	32,300	180
TB-3	4/2/97	6,200	52,000	26,000	122,000	206,200	10,000
TB-3	8/27/97	3,200	5,100	6,000	25,000	39,300	6,900
TB-3	11/17/97	3,100	1,600	2,300	2,900	9,900	4,800
TB-5	4/2/97	20,000	43,000	5,600	24,500	93,100	33,000
TB-5	8/27/97	14,000	21,000	5,100	21,000	61,100	4,900
TB-5	11/17/97	13,000	17,000	3,400	12,000	45,400	5,600
TB-6	4/2/97	14,000	47,000	6,400	31,600	99,000	22,000
TB-6	8/27/97	22,000	22,000	3,300	15,000	52,300	3,800
TB-6	11/17/97	22,000	22,000	4,300	21,000	58,300	2,400
TB-7	4/2/97	670	6,600	2,400	10,900	20,570	540
TB-7	8/27/97	230	490	150	490	1,360	76
TB-7	11/17/97	1,200	14,000	3,200	16,000	34,400	300
TB-8	4/2/97	<1	270	1,200	3,800	5,270	1,000
TB-8	8/27/97	6	25	330	850	1,211	590
TB-8	11/17/97	<1	8	2	11	21	3,300
TB-11	4/2/97	<1	2	<1	5	7	<1
TB-11	8/27/97	1	21	15	120	157	1
TB-11	11/17/97	42	240	72	380	734	4
TB-13	4/2/97	<1	<1	<1	<1	<1	7,100
TB-13	8/27/97	2	4	2	12	20	4,900
TB-13	11/17/97	50	910	3,100	13,000	17,060	580

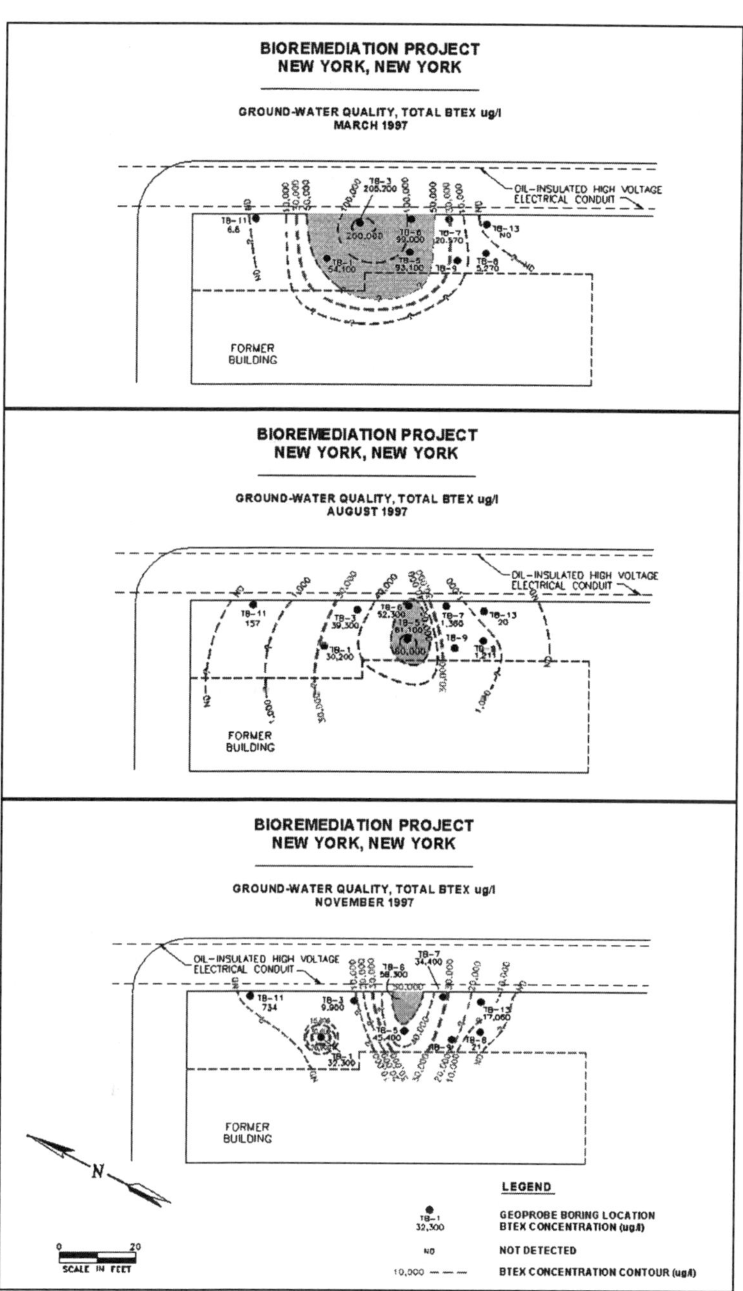

BIOREMEDIATION PROJECT
NEW YORK, NEW YORK
GROUND-WATER QUALITY, TOTAL BTEX ug/l
MARCH 1997
OIL-INSULATED HIGH VOLTAGE ELECTRICAL CONDUIT
FORMER BUILDING
BIOREMEDIATION PROJECT
NEW YORK, NEW YORK
GROUND-WATER QUALITY, TOTAL BTEX ug/l
AUGUST 1997
OIL-INSULATED HIGH VOLTAGE ELECTRICAL CONDUIT
FORMER BUILDING
BIOREMEDIATION PROJECT
NEW YORK, NEW YORK
GROUND-WATER QUALITY, TOTAL BTEX ug/l
NOVEMBER 1997
OIL-INSULATED HIGH VOLTAGE ELECTRICAL CONDUIT
FORMER BUILDING
N
0
20
SCALE IN FEET
LEGEND
GEOPROBE BORING LOCATION BTEX CONCENTRATION (ug/l)
ND NOT DETECTED
BTEX CONCENTRATION CONTOUR (ug/l)

FIGURE 1.

TABLE 2. Cost comparison of remediation technology.

Remediation Technology	Capital Cost			Total Capital Cost
	Wells, Trenching and Piping	Equipment, Material, Permits and Air Treatment	Bioremediation Treatment	
Soil-Vapor Extraction (SVE)	$27,000	$19,000	--	$46,000
Air Sparging/ SVE	$31,000	$25,000	--	$53,000
High Vacuum Extraction (HVE)	$27,000	$51,500	--	$78,500
Bioremediation Treatment[1/]	--	--	$48,000	$48,000

Remediation Technology	Operation and Maintenance			Total O&M
	Monthly O&M	Quarterly Ground-Water Monitoring	Years	
Soil-Vapor Extraction (SVE)	$ 1,000	$ 2,000	5	$100,000
Air Sparging/ SVE	$ 1,500	$ 2,000	5	$130,000
High Vacuum Extraction (HVE)	$ 2,000	$ 2,000	5	$160,000
Bioremediation Treatment	$ 0	$ 2,000	3	$ 24,000[2/]

[1/] 6 applications in 3 years [2/] Based on only 3 years O&M

<u>TOTAL COST - CAPITAL AND O&M</u>

Soil-Vapor Extraction (SVE):	$146,000
Air Sparging/SVE:	$183,000
High Vacuum Extraction (HVE):	$238,500
Bioremediation Treatment:	$ 72,000

ORGANOMERCURIAL RESISTANT BACTERIA: TRANSPORT AND DEGRADATION POTENTIALS ESTIMATION IN LABORATORY

Valérie A. E. Guérin, Michel A. Buès
Laboratoire Environnement géomécanique et ouvrages
Vandoeuvre-lès-Nancy, France

ABSTRACT

Laboratory column tests are used to evaluate transport of phenyl mercuric acetate (PMA) and *Pseudomonas fluorescens* bacterial strain. The measured retardation factor and the outlet concentration gave access to the retention capacity of sand towards bacterial strain, as well as the ability of degradation to take place under flow conditions. These tests also enabled us to evaluate the effect on the breakthrough curves of different environmental factors (bacterial density, injection method, ionic strength, and value of flux). In all cases, the bacterial strain proved to be fixable as well as transportable, which is a required condition for bioremediation use.

The bacterial strain studied, even in non-growing conditions, is able to capture pollutants, and to act as a vector of transport. In growing conditions, bacteria colonized along the length of the column before reaching an equilibrium growth rate. PMA affects growth rate for concentrations over 2 mg/L. The degradation in column is constant up to 2.5 mg/L, and is consistent with degradation capacities observed in a batch reactor.

INTRODUCTION

The recent interest in the actions of microorganisms in porous media is motivated by the role of microorganisms in the biorestoration of contaminated aquifers. However, the persistence of pollutants that are or were accidentally spilled in the environment depends on both the mobility and retention capacity of the bacterial strain in the media and its degradation capacity.

Our study took place following the pollution of the Rhine River by pesticides, including organomercurials such as phenyl mercuric acetate (PMA). After this pollution, a bacterial strain emerged in Rhine sediments: a *Pseudomonas fluorescens* strain repertoried PL IV.

Objective. Tests were conducted to evaluate the possible use of this bacterial strain in bioremediation.

EXPERIMENTAL PROCEDURE

Laboratory column testing was conducted in carefully controlled conditions.

Materials. The pollutant considered is PMA, an organomercurial compound which is used as a pesticide in seed culture. The resistant bacterial strain selected is a subclone of strain *Pseudomonas fluorescens,* isolated in 1986 from the bulk mercury contaminated sediments of the Rhine river (Mirgain, 1992). This strain, named PL IV, is a non-spore-forming, gram-negative rod (approximately 0.5 x 2 micrometers), which develops a resistance to PMA (Robinson and Tuovinen, 1984). The nutrient growth medium used is called complete medium (CM) and contains Casamino-Acids, Casein Hydrolysat, and some minerals and trace elements (Meyer and Abdallah, 1978; Veron, 1975).

The solid phase used for column experiments is a quartz sand from the Rhine river, which contains a little pyrite and clay. Before use in the column, the sand was washed using hydrochloride acid and sodium polyacrilate. The d_{60} of the sand is 0.63 and its uniform coefficient U is 2.3.

Analysis. PMA measurements were performed according to a method modified by Marcandella and Bues (1995) taken from the Hatch-Ott procedure (1968). Mercury-light absorption was measured with a Perkin-Elmer MAS 50 B spectrophotometer. Bacterial concentration was measured using the turbidity at 640 nm.

Experimental setup. The system used is a Pyrex chromatography column (inside diameter 2.6 cm, length 15 cm). A variable-flow peristaltic pump (Figure 1) regulates the flow of the eluate in each type of experiment. The sand is poured into the column to minimize stratification.

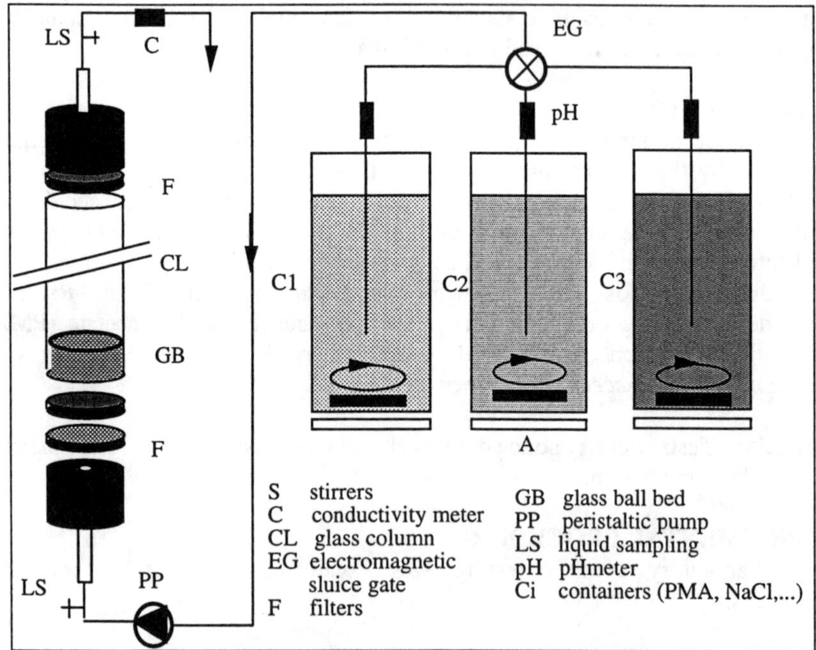

Figure 1. Experimental setup

The solution (PMA or bacteria) is substituted for distilled water as a constant flow rate is established. For column experiments, distilled water was used to test adsorption of PMA and initial bacterial adhesion on solid phase; a dilution of the CM was used to test biodegradation in flow conditions.

A pulse of 0.1 M NaCl was introduced into the column after the experiments to determine the hydrodynamic characteristics (e.g.: kinematic porosity of 0.38 +/-0.02). All tubing, glassware, and aquifer sand were autoclaved to avoid external contamination. The experiments were performed at an average temperature of 30°C.

RESULTS AND DISCUSSIONS

Interaction of bacterial cells with solid phase: Bacterial adhesion in dynamic condition. To ensure effectiveness of in site detoxification, two conditions must be verified: the bacterial strain must be both transportable and fixable. So we were initially interested in evaluating the movement of strain PL IV in a porous medium constitute of sand in distilled water. The role of a variety of parameters on bacterial retention was also considered.

Bacterial retention. The growth and mortality rates of bacteria are assumed to be negligible as suggested by Lindqvist et al. (1994). And the size of our bacterial strain is small enough to assume that sedimentation is insignificant.

As well as for pollutant (Bicheron et al., 1991) it was also possible to draw a breakthrough curve in reduced concentration (Figure 2). In the case of bacteria, breakthrough takes place earlier than in the case of PMA, and the increase of the outlet bacterial density is regular up to reaching equilibrium (Nb/Nb_0= 1). Colonization of the medium is gradual.

Influence of environmental conditions. The effect of a number of parameters such as value of flow, ionic strength and bacterial density of the solution on bacterial retention was evaluated (Table 1).

Table 1. Influence of environmental conditions on retardation factor

	Flux (ml/min)			Ionic Strength (Molarity)			Bacterial density		
	0.54	0.9	1.9	0	10^{-2}	10^{-1}	0.05	0.2	0.4
R	4.5	4.7	1.8	4.7	7	3	4.8	3.9	3.2

Increase in flow tends to decrease the total number of retained bacteria by not allowing some adsorption mechanisms that occur in slower flows, a result that indicates the existence of kinetic limitations. At the same time, the decrease of R with increasing concentrations indicates that the number of fixation sites is limited (Figure 3).

The influence of ionic strength is more complex. An increase in ionic strength from 0 to 10^{-2} M enhances fixation and increases the retardation factor by reducing of the thickness of the double layer at the interface. But an increase in ionic strength up to a molarity of 0.1 decreases the retardation factor by causing flocculation of the bacteria: bacterial flocs no longer have the same retention capacity. At the end of each percolation, run measurements show that fixed bacteria are distributed throughout the length of the column.

Biodegradation in flow conditions. For the study of the transport of PMA and bacteria, two approaches are available. First we consider a porous medium that contains bacteria, and the solution is considered as being equivalent to the growth medium. Pollution is then injected. This situation is called natural degradation. Alternatively we suppose that pollution is already present in the column, and then we inject bacteria with growth medium.

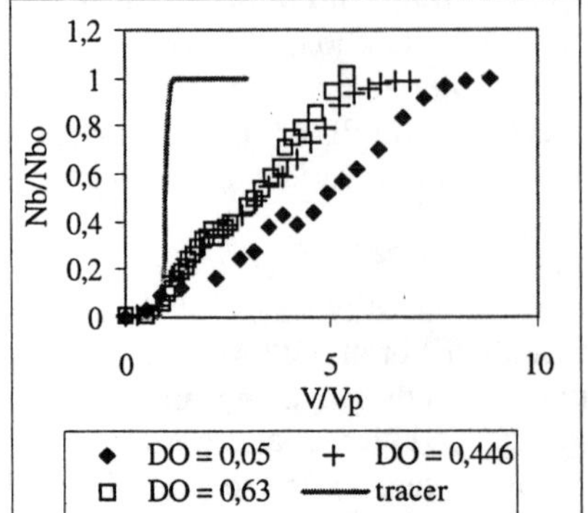

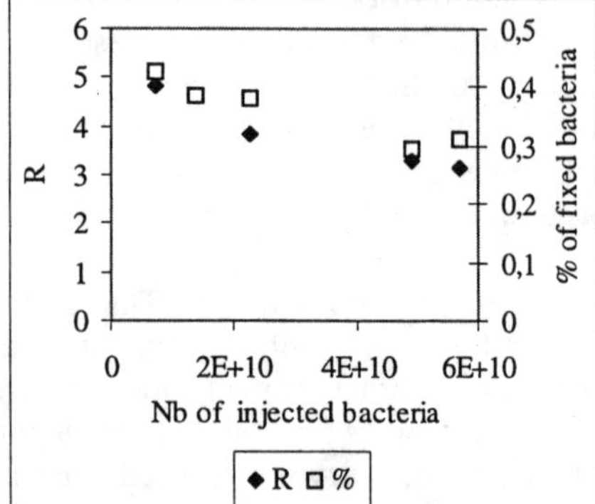

Figure 2. Bacterial retardation factor with bacterial density

Figure 3. Retardation factor and percentage fixed with injected bacteria

Natural degradation. It corresponds to the case of a porous medium, containing bacteria with a circulating solution equivalent to the growth medium. PMA is then injected as reported on Figure 4. When bacteria were injected by a dirac (10 min DO = 0.4) in porous media, we observed a peak of bacteria corresponding to the first phase of colonization: the fixation of inlet bacteria and of the growing bacteria already fixed. The outlet concentration of bacteria then reached an equilibrium, which matched the growth and death of fixed bacteria.

At the beginning, the elution of PMA is similar to that in the column without bacteria. But after 7.5 pore volumes, PMA is only forty percent of the injected bacteria, so degradation in the column is efficient. Study at different PMA concentrations (0.5 to 2.5 mg/L) indicates that if for low concentration (< 1.5 mg/L) the depletion on the growth rate is temporary, an increase in PMA concentration affects the growth rate is durably. So, the toxicity of PMA on bacterial growth is of the same order of magnitude as in a batch test.

In flow conditions, degradation is efficient up to 2.5 mg/L; that is, fixed bacteria have equivalent properties compared to free living bacteria: fixation does not alter bacterial capacity. When a mass balance of PMA concentration in column is drawn up, it is clear that degradation of PMA gives metallic mercury that still remains partially in porous medium, so there is no real cleansing of the medium.

Bioremediation. Alternatively we suppose that pollution is already present in the column, then we inject bacteria with growth medium (Figure 5). The high level of PMA leaching is due to three mechanisms: the change in solution composition (ionic strength), the degradation of PMA by bacteria, and the fixation of pollutant on bacterial walls. Compared to the case of leaching by growth medium only, the presence of bacteria enhanced departure of PMA from the porous medium.

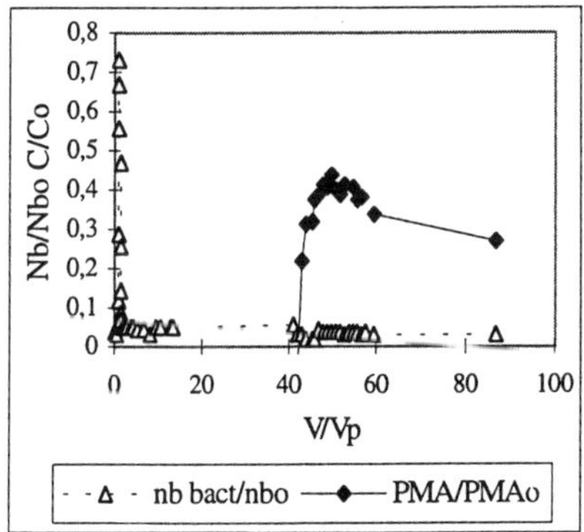

Figure 4 Natural degradation [PMA] = 0.59 g/l

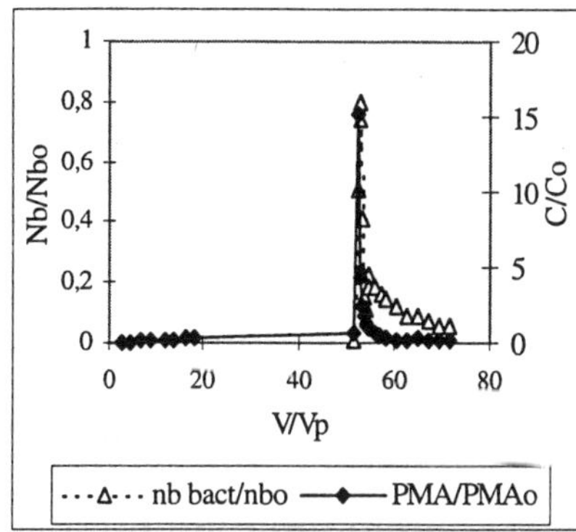

Figure 5 Bioremediation [PMA] = 0.59 g/l

CONCLUSIONS AND PROSPECTS

Dynamic studies have shown that bacteria could be partially fixed by porous media, but are sensitive to environmental conditions, and mechanisms have not yet been well listed.

The results of concomitant transport of bacteria and pollutant show that the best use of this bacterial strain is as a bioreactor. The aerobic metabolism of PMA has been well documented elsewhere (Marcandella and Bues, 1995; Guerin, 1998). However, degradation results in dynamic conditions with in place bacteria as well as injected ones, which indicates that the aerobic volatilization of PMA can yield a good level of detoxification. The potential for in site transformation of PMA nevertheless needs to provide a complete mineral supplement and excess electron donors.

REFERENCES

Bicheron, C., J.M. Strauss, and M.A. Bues. 1991. "Comparison of the behavior of two mercurial compounds during their transport through a natural saturated porous medium". *Computer Methods and Water Resources, Vol. 1, Ground Water Modeling and Pressure Flow*. Ed., D. Ben Sari et al. Springer Verlag, 185-196.

Guerin V.A.E., 1998. "Réactivité Bio-physico-chimique d'un micropolluant organomercuriel en milieu poreux saturé". thèse de l'I.N.P.L ., 260 p.

Hatch, W.R., and W.L. Ott. 1968. "Determination of sub-microgram quantities of mercury by atomic absorption spectrophotometer." *Annal. Chem. 40*: 2085-2087.

Lindqvist, R., J. Soo Cho, and C.G. Enfield. 1994. "A kinetic model for cell density dependent bacterial transport in porous media." *Water Resources Research 30*(12): 3291-3299.

Marcandella, E. and M.A. Bues. 1995. "Caractérisation par des essais en réacteurs fermé de la réactivité physico-chimique et de la biodégradation bactérienne d'un organomercuriel : application à la dépollution de sédiments contaminés". Hydrogéologie/Hydrogeology, n°1, p.79-87.

Marcandella, E., C. Bicheron, and M.A. Bues. 1995. "Influence of bacterial degradation on the transport of a micropollutant through a natural saturated porous medium", In Microbial Processes for Bioremediation, Ed. by R.E. Hinchee et al., Batelle Press, p. 281-287.

Meyer, J.M., and M.A. Abdallah. 1978. "The fluorescent pigment of Pseudomonas fluorescens: Biosynthesis, purification, and physicochemical properties". *J. Gen. Microbiol.* 107: 319-328.

Mirgain I. 1992. "Dégradation bactérienne de deux pesticides (acétate de phényl mercure et atrazine) dans l'environnement naturel". Thèse de l'Univervité Louis Pasteur, Strasbourg, 178 p.

Robinson, J.B. and O.H. Tuovinen. 1984. "Mechanisms of microbial resistance and detoxification of mercury and organomercury compounds : physiological, biochemical, and genetic analyses". *Microbiological Reviews*, 48, (2), p. 95-124.

Veron, M. 1975. "Nutrition et taxonomie des Enterobacteriaceae et bactéries voisines". *Anna. Microbiol.* (Institut Pasteur), 126A, n°3.

BIODEGRADATION OF HYDROCARBONS IN A BIOBARRIER UNDER OXYGEN-LIMITED CONDITIONS

L. Yerushalmi, A. G. Beard, A. Al-Hakak and S. R. Guiot
Biotechnology Research Institute, Montreal, Quebec, Canada

ABSTRACT: Continuous bioremediation of gasoline- and benzene-contaminated water in a packed-bed biobarrier system under oxygen-limited condition is discussed. Protruded stainless steel pieces and granulated peat moss were used as packing material to support microbial growth in two biobarriers. An enrichment culture of an indigenous microbial population from a soil sample was used as the inoculum. The biobarriers' inlet hydrocarbon concentration and the linear liquid velocity were similar to those commonly found at *in situ* conditions. Overall removal efficiencies in the stainless steel-packed biobarrier ranged from 94% to 99.9% for gasoline, and from 63.9% to 99.9% for benzene. Benzene concentrations below drinking water standards (0.002 mg/L) were obtained at inlet concentrations below 1.76 mg/L. In the peat moss-packed biobarrier, the removal efficiencies ranged from 86.6% to 99.6% for gasoline and from 70.4% to 97.2% for benzene. Both biobarrier packings supported near complete removal of the most soluble aromatic hydrocarbons of gasoline (BTEX). The consumption of sulfate and the presence of anaerobic sulfate-reducing bacteria suggested the involvement of anaerobic metabolism during the biodegradation of hydrocarbons.

INTRODUCTION

The biobarrier is an alternative technology for the *in situ* bioremediation of contaminated groundwater. It consists of a permeable bioreactive zone or wall placed across the path of a contaminated plume. As the contaminated groundwater moves under natural hydraulic gradient through the bioreactive zone, the contaminants are removed or biodegraded, leaving uncontaminated groundwater to emerge from the downgradient side (Figure 1).

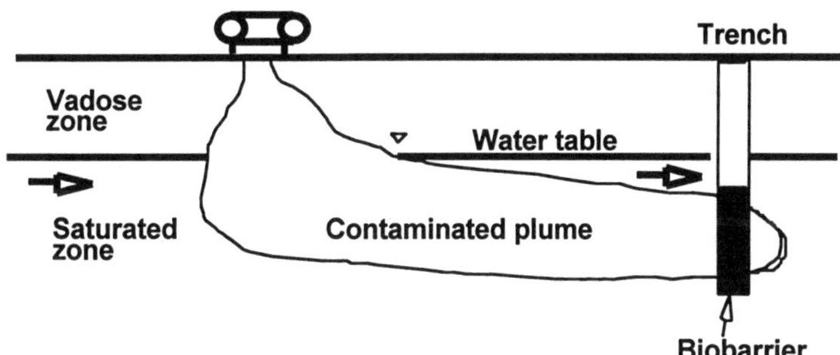

FIGURE 1. On-site and *in situ* remediation of contaminated groundwater by a permeable biobarrier.

Compared to conventional treatment techniques such as pump and treat or air stripping, the biobarrier offers many advantages such as low maintenance requirements, low energy consumption due to the passive nature of the system using natural gravity for water transport, and the lack of on-site manipulation of the contaminated soil or groundwater, thus eliminating the risk of phase transfer or volatilisation of contaminants to surrounding properties.

The developed biobarrier consists of a bioreactor filled with a permeable packing material to support microbial growth and biofilm formation. The performance of a laboratory-scale biobarrier, packed with protruded stainless steel pieces or granulated peat moss, in bioremediation of gasoline- and pure benzene-contaminated water under oxygen-limited conditions was examined in the present work. Remediation of groundwater under anoxic or oxygen-limited condition is very important due to the limited supply or complete lack of oxygen under *in situ* conditions. In fact, oxygen limitation is one of the major problems affecting the performance of *in situ* biological treatment systems. Also, supplying alternative electron acceptors such as sulfate and nitrate is more economical and more convenient. The developed biobarrier provides a promising alternative technique for the treatment of contaminated groundwater.

MATERIALS AND METHODS

Culture Isolation and Conditions. The microbial culture used as the inoculum for the biobarriers was isolated by enrichment techniques from the top layers of a gasoline-contaminated soil sample from Montreal, Canada. During the enrichment, gasoline with a concentration of 74 mg/L obtained from a commercial gasoline station was used as the source of carbon and energy for the microorganisms. A minimal salts medium (MSM) was used for culture enrichment as well as the biobarriers' feed (Yerushalmi and Guiot, 1998).

Analytical Techniques. Concentration of hydrocarbons in the gas and liquid phase was determined by the headspace and the gas phase chromatographic techniques, respectively, as described before (Yerushalmi and Guiot, 1998). The concentration of dissolved oxygen was determined by a glass polarographic DO probe Model E05643 (Cole Parmer, Chicago, Illinois, USA) and monitored on a DO controller Model 01972 (Cole Parmer, Chicago, Illinois, USA).

Biobarrier Packing. The Pro-Pak, protruded stainless steel (SS packing, Cannon Instrument Company, Pa., USA) and granulated peat moss (PM packing, Produits Recyclable Bioforet, Quebec, Canada) were used as the packing material and support for the microbial biofilm in the biobarriers. The Pro-Pak S.S. packings were 0.6 x 0.6 cm in size and had an average particle density of 5.13 g/mL with a free space of 93%. The granulated peat moss had an average diameter of 0.4 to 0.7 cm, a particle density of 0.64 g/mL and a free space of 57%.

Biobarrier Operation. The biobarriers consisted of a packed-bed permeable bioreactor with a stainless steel body (25 cm x 20 cm x 10 cm), a rectangular cross sectional area and a total volume of 5 liters. They were equipped with two glass windows and several ports on the sides as well the top for feeding, liquid recirculation during the initial batch operation and sampling. The nutrient solution (MSM) was continuously agitated and sparged with sterile air in order to provide oxygen for the process. A four-channel peristaltic pump model Gilson minipuls 2 (Gilson Medical Electronics, WI, USA) was used to deliver feed to the four inlet ports on the side of the biobarriers. Two peristaltic pumps model Masterflex (Cole Palmer Instrument Company, Illinois, USA) were used to introduce hydrocarbon solution containing 370 mg/L gasoline or benzene in MSM medium into the biobarriers. The hydrocarbon solution was constantly mixed throughout the course of the experiments.

Experiments were performed at room temperature (~25° C). Overall gasoline removal was evaluated by analyzing gas and liquid samples of the effluent. Gas samples were taken from a side tower located on the outlet port of the biobarriers. Hydrocarbon concentration inside the biobarriers was determined by analyzing liquid samples taken from the sampling ports along the biobarrier.

During the continuous operation of biobarriers, the loading rate was increased step-wise by keeping a constant hydraulic retention time (HRT) while increasing the inlet hydrocarbon concentration. The efficiency, and the overall capacity of biobarriers in the removal of hydrocarbons was evaluated according to their removal efficiency (RE) and elimination rate (ER) as defined by the following equations:

$$\text{Removal Efficiency (RE)} = (S_i - S_e) * 100 / S_i \quad (1)$$

$$\text{Elimination Rate (ER)} = Q\,(S_i - S_e)/V_f \quad (2)$$

Where S_i is the inlet concentration of hydrocarbon (mg/L), S_e is the outlet concentration of hydrocarbon (mg/L), Q is the inlet feed flow rate (l/d) and V_f is the volume of the biobarriers occupied by the packing and liquid (L_f).

RESULTS AND DISCUSSION

Overall Performance of Biobarriers. The effluent concentration of hydrocarbons along with the biobarriers' operating conditions are presented in Tables 1 and 2. While in the stainless steel-packed biobarrier biodegradation was the major mechanism of hydrocarbon removal, in the peat moss-packed biobarrier hydrocarbon disappearance was also due to physical-chemical adsorption by the support material. Both biobarrier packings supported high hydrocarbon removal efficiencies under the conditions examined. Gasoline Removal efficiencies ranged from 94.0% ± 2.0% to 99.9% ± 0.02% in the stainless steel-packed biobarrier, and from 86.6% ± 3.5% to 99.6% ± 0.11% in the peat moss-packed biobarrier. The higher free space in the stainless steel-packed biobarrier (93%) compared to that in the peat moss-packed biobarrier (57%) caused a higher liquid flow rate at any given operating condition in terms of the linear liquid velocity and hydrocarbon concentration. This resulted in an elevated hydrocarbon loading rate and a higher elimination rate in the stainless steel biobarrier as calculated from Equation 2. The overall gasoline elimination rate ranged from 1.1 to 64.8 ± 1.4 mg/L_f.d in the stainless-steel-packed biobarrier and from 0.6 ± 0.02 to 40.8 ± 0.37 mg/L_f.d in the peat moss-packed biobarrier.

Similarly, high benzene removal efficiencies, ranging from 63.9% ± 3.8% to 99.9% ± 0.02% in the stainless steel-packed biobarrier and from 70.4% ± 3.4% to 97.2% ± 0.05% in the peat moss-packed biobarrier, were obtained. Under these conditions benzene elimination rate ranged from 0.1 to 3.7 ± 0.2 mg/L_f.d in the peat moss-packed biobarrier and from 0.2 to 10.4 ± 0.6 mg/L_f.d in the stainless steel-packed biobarrier.

TABLE 1. Operating parameters and gasoline removal performance of the biobarriers.

Linear Liquid Velocity (cm/d)	Inlet Concentration (mg/L)	Effluent Concentration (mg/L)	
		SS Packing	PM Packing
25	3.7	0.02	0.18
25	14.8	0.13	0.55
25	37.0	0.27	0.14
25	74.0	4.44	2.5
12.5	3.7	0.01	0.35
12.5	37.0	0.30	0.21
12.5	74.0	0.29	2.3
8.3	74.0	0.30	0.59

TABLE 2. Operating parameters and benzene removal performance of the biobarriers.

Linear Liquid Velocity (cm/d)	Inlet Concentration (mg/L)	Effluent Concentration (mg/L)	
		SS Packing	PM Packing
12.5	0.44	0.002	0.03
12.5	1.76	0.002	0.05
12.5	5.3	0.02	0.17
12.5	8.8	2.6	2.6
12.5	13.2	3.5	3.0
12.5	17.6	-	4.6
12.5	26.4	7.6	-
12.5	35.1	12.7	-

The dependence of hydrocarbon elimination rate on the biobarriers' loading rate is presented in Figure 2. Overall gasoline and BTEX removal efficiencies of 98% and 99.9% were obtained from the slope of the lines (Figure 2a), signifying the potential *in situ* applications of the biobarrier since the water-soluble BTEX is the major fraction of gasoline that causes groundwater contamination. During the removal of benzene, increase of the inlet benzene concentration above 5.3 mg/L (Figure 2b), corresponding to the loading rates of

1.5 mg/L.d in the peat moss-packed biobarrier and 2.4 mg/L.d in the stainless steel-packed biobarrier, resulted in the reduction of benzene removal efficiency below 77.2 %.

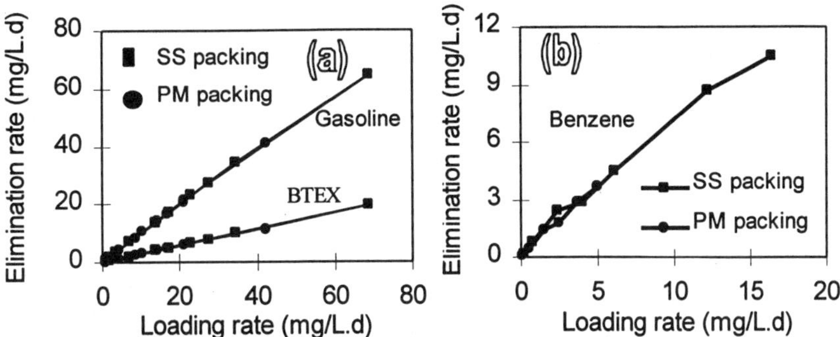

FIGURE 2. Dependence of the biobarriers' elimination rate on hydrocarbon loading rate. (a) Gasoline treatment, (b) Benzene treatment.

Mechanism of Hydrocarbon Biodegradation. Measurements of the dissolved oxygen (DO) concentration in the effluent as well as at the first sampling port of the biobarriers showed values of less than 0.3 mg/L under most of the conditions examined in this work. This indicates that the supplied oxygen was quickly consumed leading to a decrease in the DO concentration. The inlet feed stream was always saturated with oxygen at concentrations above 8.0 mg/L. However, as the inlet hydrocarbon concentration increased, the supplied oxygen was not sufficient to support complete aerobic biodegradation of hydrocarbons, leading to the establishment of a microaerophilic condition in the biobarriers.

The high hydrocarbon removal efficiencies of up to 99.9% obtained with limited supply of oxygen imply that the microaerophilic condition did not prevent the degradation of hydrocarbons. This is in contrast with previous investigations that demonstrated the requirement for a minimum concentration of DO in the range of 1.0-1.5 mg/L to initiate petroleum hydrocarbon biodegradation (Wilson and Bouwer, 1997; Chiang et al., 1989).

The occasional analysis of remaining sulfate in the biobarriers' effluents showed the consumption of sulfate during the removal of gasoline and benzene. Evidence for the presence of sulfate-reducing bacteria was provided by the growth of microorganisms, withdrawn from the biobarriers, on anaerobic media specifically designed for growth of sulfate-reducing bacteria followed by the formation of black FeS precipitation in the presence of ferrous sulfate. These findings suggest that anaerobic metabolism was present in the biobarriers during the degradation of hydrocarbons. They further suggest that hydrocarbons were degraded under mixed electron-accepting conditions as reported previously (Barbaro et al., 1997; Su and Kafkewitz, 1996). Sulfate has been recognized as a terminal electron acceptor in the biodegradation of hydrocarbons (Edwards and Grbic-Galic, 1992; Lovely et al., 1995). It is also possible that some gasoline constituents, particularly those that are more recalcitrant under anaerobic condition, were not completely mineralized but were partly transformed to intermediate metabolites.

The high hydrocarbon removal efficiencies of greater than 99% obtained at low dissolved oxygen concentrations and with limited supply of molecular oxygen is a notable advantage of the biobarrier, exhibiting its potential for *in situ* applications and under DO conditions normally found underground. Research is currently in progress to evaluate the performance of biobarriers in the removal of polycyclic aromatic hydrocarbons under various operating and culture conditions.

REFERENCES

Chiang, C. Y., J. P. Salanitro, E.Y. Chai, J. D. Colthart and C. L. Klein. 1989. "Aerobic Biodegradation of Benzene, Toluene, and Xylene in a Sandy Aquifer-Data Analysis and Computer Modeling." *Groundwater* 27: 823-834.

Edwards E. A. and D. Grbic-Galic. 1992. "Complete Mineralization of Benzene by Aquifer Microorganisms under Strictly Anaerobic Conditions." *Appl. Environ. Microbiol.* 58: 2663-2666.

Lovely D. R., J. D. Coates, J. C. Woodward and E. J. P. Phillips. 1995. "Benzene Oxidation Coupled to Sulfate Reduction." *Appl. Environ. Microbiol.* 61: 953-958.

Su J. J. and D. Kafkewitz. 1996. "Toluene and Xylene Degradation by a Denitrifying Strain of *Xanthomonas maltophilia* with Limited or no Oxygen." *Chemosphere* 32: 1843-1850.

Wilson L. P. and E. J. Bouwer. 1997. "Biodegradation of Aromatic Compounds under Mixed Oxygen/Denitrifying Conditions: A Review." *J. Ind. Microbiol. Biotechnol.* 18: 116-130.

Yerushalmi, L and S. R. Guiot. 1998. " Kinetics of Biodegradation of Gasoline and its Hydrocarbon Constituents." *Appl. Microbiol. Biotechnol.* 49: 475-481.

DESIGN AND OPERATION OF A FULL-SCALE AEROBIC BIOBARRIER

H.Tonnaer, H.S. Michelberger, C.G.J.M. Pijls
(Tauw bv, P.O. Box 133, NL-7400 AC Deventer, Netherlands),
F. De Palma (Ecodeco S.p.A., Loc. Cassinazza di Baselica,
27010 Giussago (PV), Italy)

ABSTRACT: An aerobic biobarrier has been installed at an industrial site in northern Italy in order to prevent contaminant migration beyond the site borders. The full-scale biobarrier, which is based on biosparging, has been in operation since November 1998. Prior to designing and installing the full-scale system, the effectivity of the system has been tested in a pilot plant. Results showed that in the first four months of full-scale operation the total mass flux of contaminants leaving the site via the groundwater dropped by 98%. Expectations are that within a few more months, the quality of the groundwater leaving the site will be below the total aromatic hydrocarbons target value of 10 ppb, which is the national standard for drinking water quality.

INTRODUCTION

Compared to more traditional containment techniques (e.g. pump and treat), aerobic biobarriers based on air sparging can be a good alternative for the containment of hydrocarbons contamination plumes. Although the potential applicability of biobarriers is widely accepted, the number of full-scale systems actually installed and operated is limited (Pankow et all., 1993; Bass and Brown, 1997; Kershner and Theoret, 1997). In this paper a full-scale application of an aerobic biobarrier is presented.

SITE DESCRIPTION

The site in question accommodates a paint production facility and is located in a residential area in northern Italy. Industrial activities were taken up on site in 1962. The production facility is still in use at present. The geohydrological situation on site can be characterized as shown in Table 1.

TABLE 1: Geohydrological situation

Depth (m –gl)	Soil structure	Geohydrological unit
0 – 10	Medium coarse sand, gravel and pebbles	Phreatic groundwater
10 – 20	Fine sand/silty sand	Idem
20 – 30	Medium coarse sand with fine gravel	Idem
> 30	Fine sand/clay	First aquitard

The depth of the groundwater table ranges from 3 to 4 m –gl. A pumping test was performed in order to determine the hydrological conductivity. From its results and the leveling data available, a groundwater velocity of approx. 280 m/year was derived.

Due to spills that occurred in the past, the soil and groundwater of the site are contaminated with aromatic solvents, mainly toluene, ethylbenzene, xylenes and solvent naphta. Solvent naphta is a mixture of 1-ethenyl, 2-methylbenzene, isomers of trimethylbenzene, propylbenzene, styrene and other aromatic compounds with various alkyl groups.

Prior to carrying out the soil and groundwater investigations on site, four risk areas were determined, which are shown in Figure 1. *Area 1* accommodates a total of 25 underground storage tanks (USTs), all of them presently in use for the storage of several types of solvents. *Area 2* is the most heavily contaminated area; it accommodates a building for the storage of raw materials. In the south of this area shelves for the storage of liquid raw materials are still in use. *Area 3* and *Area 4a* and *4b* are less severely contaminated and will not be discussed any further in this paper.

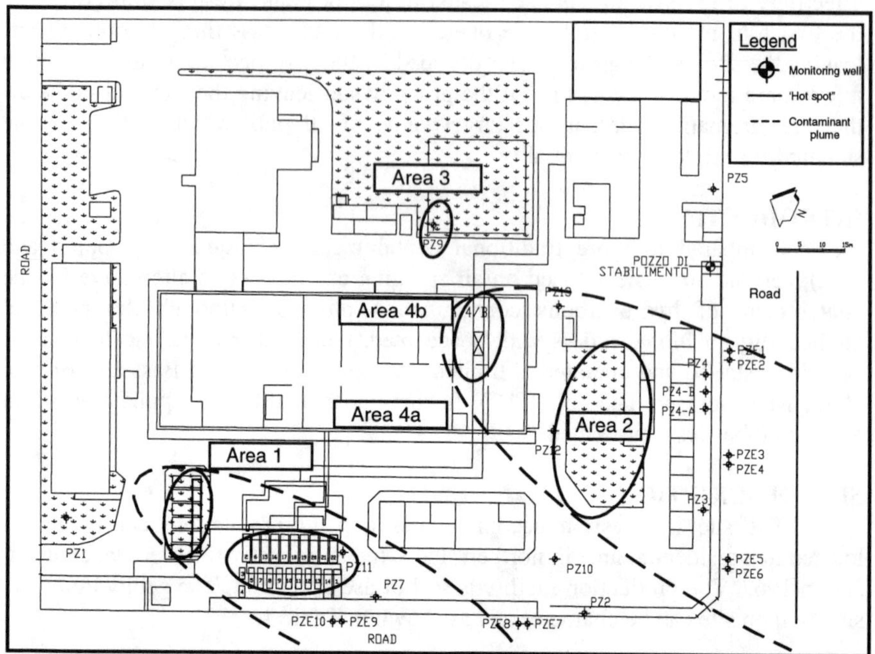

FIGURE 1: Contaminant distribution

The total hydrocarbon levels encountered in the groundwater varied from 100 to 1,000 ppb in the Areas 1, 3, 4a and 4b. In Area 2, they were as high as 100,000 µg/l, indicating the presence of non dissolved pure product. The maximum depth of the groundwater contamination was 10 m –gl.

OBJECTIVES OF THE REMEDIAL ACTIONS

The objectives of the remedial actions taken on site can be split up in short- and long-term objectives. In the short term, the remediation is to prevent further contaminant migration in the groundwater to areas beyond the site borders; this is to protect and restore the groundwater quality off site, a requirement imposed by the local Italian authorities. In the long term, the soil and groundwater quality on site is to be restored; free from any governmental pressure, the site owner intends to find a definite solution for the soil and groundwater problem on site.

REMEDIAL OPTIONS

Two options were considered to meet the short-term objective. They consisted of geohydrological containment (pump and treat) and the installation of an aerobic biobarrier based on air sparging on site, the latter in combination with natural attenuation of the plume off site (Figure 2).

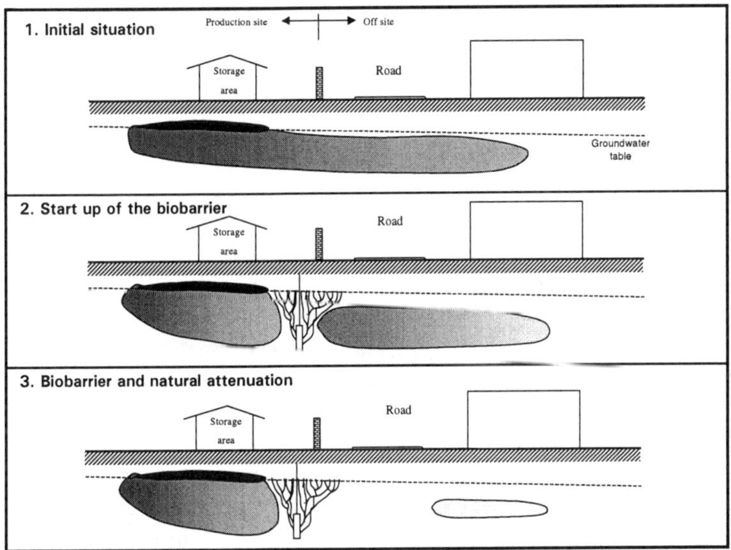

FIGURE 2: Aerobic biobarrier in combination with natural attenuation

Geohydrological containment proved to be far less cost-effective than a sparging biobarrier. The high containment cost were primarily caused by the groundwater treatment required, as there was no adequate sewage system on the site. Therefore, the technical feasibility of an aerobic biobarrier in combination with natural attenuation was investigated, following which a full-scale design could be made.

The feasibility study consisted of two parts: 1) a pilot trial to determine the effectiveness of air sparging and the key design parameters (effective radius of influence (ROI) and injection flow and pressure), and 2) model calculations to determine whether and how quickly the plume would be attenuated following the start-up of the full-scale biobarrier.

RESULTS OF THE FEASIBILITY STUDY

Pilot trial. The pilot trial was carried out in the heavily contaminated zone downstream of Area 2. One air sparging well was installed at a depth of 10 m –gl.

In total, 8 monitoring wells were installed to collect the relevant data, 3 of them perpendicular to the groundwater flow and the remaining 5 downstream of the in air sparging well.

The pilot plant was operated for a period of 11 weeks. Throughout this period, the development of the contaminant concentrations and the oxygen content in the groundwater within the ROI was measured. Figure 3 shows the development of the groundwater quality during the pilot trial in one monitoring well. Other monitoring wells gave similar results.

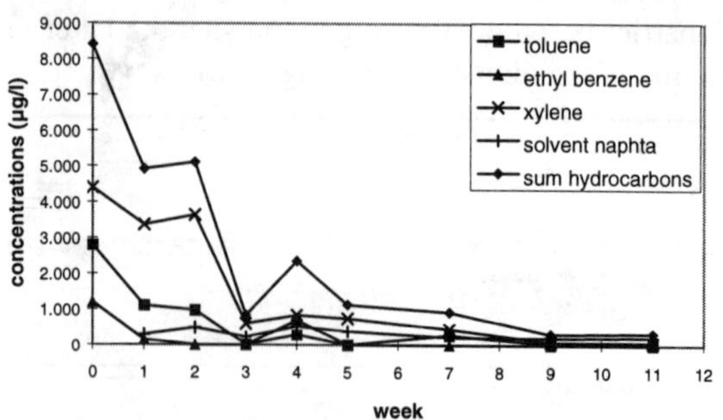

FIGURE 3: Development of groundwater quality during pilot trial

As shown in Figure 3, air sparging reduced the level of total hydrocarbons in the groundwater from 8,400 ppb to 300 ppb (mainly solvent naphtha) within 11 weeks, equaling a reduction by more than 95%. The radius of influence achieved in the pilot trial was about 4 meters, judging from the oxygen level in the groundwater and the reduction in the level of contaminants.

The hydrocarbon levels measured in the soil vapor in the unsaturated zone remained below 10 ppm. This indicated that no SVE-system to capture stripped contaminants is required for the full-scale biobarrier.

Furthermore, a microbiological count was performed in groundwater samples from the monitoring wells of the pilot plant. In all samples, toluene, xylene as well as naphta and TMB-degraders were found. The total number of bacteria capable of degrading the relevant aromatic compounds was found to be 10^4-10^6 M.P.N./ml (M.P.N. = Most Probable Number).

It was concluded from the results of the pilot trial that the aerobic biobarrier should be able to reduce groundwater contaminations to an acceptable level within a manageable period of time.

Model calculations. An on-site field investigation into natural attenuation showed that the groundwater entering the site was aerobic and contained approx. 5 mg/l of oxygen. Using this data and the groundwater velocity derived, the plume development was modeled using MODFLOW/RT3D. The calculations showed that intercepting the plume off site using an aerobic biobarrier would result in the degradation of the off-site plume within 15 years.

DESIGN AND OPERATION OF THE FULL-SCALE BIOBARRIER

On the basis of the feasibility study results it was decided to proceed with the design and installation of the full-scale aerobic biobarrier. At the southwestern site border, a single row of air sparging wells is installed (11 wells in total)At the southeastern site border, a double row of air sparging wells is installed (18 wells in total). This will reduce the risk of 'leakages' in the biobarrier. The distance between the sparging wells is 8 meters (Figure 4). For the supply of air, a compressor with a capacity of 100 m^3/hr is used.

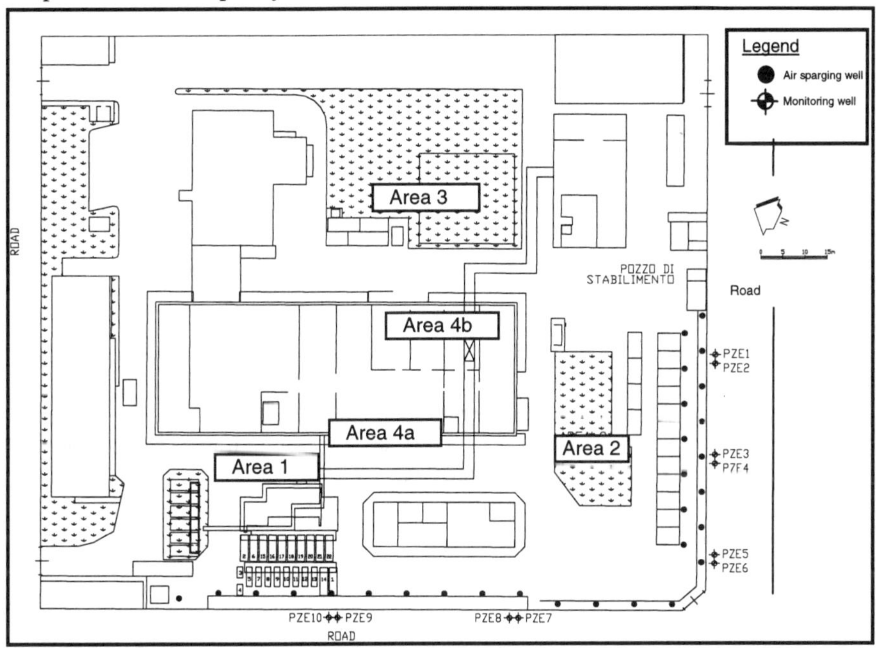

FIGURE 4: Full-scale aerobic biobarrier

At present, the air sparging system is operated intensively in order to quickly reduce the concentrations in the biobarrier. The injection pressure is 1.2-1.5 bar, which results in a flow of approx. 70 m^3/hr. The intermittent injection period is 2 minutes per well, following which injection proceeds to the next well. Each well is in operation once an hour. Once the drinking water limits will be reached in the groundwater leaving the site, the intermittent operation of the air sparging system will be reduced to a less intensive setting.

In order to achieve the long-term objective, "hot spot" remediation systems are installed in the Areas 1 to 4, consisting of air sparging in combination with soil vapor extraction. These systems are also in operation now.

RESULTS OF THE FULL SCALE BIOBARRIER

The development of the groundwater quality beyond the site border is monitored using a total of 10 monitoring wells (Figure 4). The results are summarized in Table 2.

TABLE 2: Total hydrocarbon levels during initial period of operation

		Total hydrocarbon levels (ppb)			
Well	**Depth (m –gl)**	**week 0**	**week 7**	**week 12**	**week 17**
PZE1	4 – 5	85	66	35	10
PZE2	9 – 10	16	15	8	4
PZE3	4 – 5	3,523	461	2	6
PZE4	9 – 10	870	253	1	10
PZE5	4 – 5	143	63	11	9
PZE6	9 – 10	0	0	0	2
PZE7	4 – 5	217	105	2	5
PZE8	9 – 10	4	6	1	4
PZE9	4 – 5	39	34	19	7
PZE10	9 – 10	18	12	8	6

The monitoring results clearly demonstrate the effectiveness of the biobarrier. The levels of hydrocarbons in all 10 monitoring wells are equal to or below the target value of 10 ppb (the national drinking water standard). The total mass flux of contaminants leaving the site has decreased from approx. 28 kg/year to 0,3 kg/year since the start-up of the operation (> 98 % reduction).

CONCLUSIONS

The feasibility study and the preliminary results of the full scale system have shown that a sparging biobarrier can effectively contain groundwater contaminations. Within 4 months contaminant levels in groundwater within the biobarrier were reduced to the target levels. Contaminant emissions by groundwater migrating beyond the site borders were reduced by more than 98 %.

REFERENCES

Bass, D.H., Brown, R.A., 1997, "Performance of Sparging Systems – a Review of Case Studies", In B.C. Alleman and A. Leeson, *4th International In situ and On-Site Bioremediation Symposium*, Vol 1: pp.117-122, Battelle Press, Columbus OH.

Kershner, M.A., Theoret, D.R., "Horizontal Air Sparging utilized as a Site Control Measure", ", In B.C. Alleman and A. Leeson, *4th International In situ and On-Site Bioremediation Symposium*, Vol 1: pp.215-220, Battelle Press, Columbus OH.

Pankow, J.F., Johnson, R.L., Cherry, J.A., 1993, "Air Sparging in Gate Wells in Cutoff Walls and Trenches for Control of Plumes of Volatile Organic Compounds (VOC's)", *Groundwater*, July-August 1993 31(4): pp. 654-659.

Orrigi, G., Colombo, M., De Palma, F., Rivolta, M., Rossi, P., Andreoni, V., 1997, "Bioventing of hydrocarbon – contaminated soil and biofiltration of the gas-off: results of a field scale investigation", *J. Environ. Sci. Health,* A 32(8): pp. 2289-2310.

DEVELOPMENT OF A BIOFILTRATION SYSTEM FOR *IN SITU* REMEDIATION OF BTEX-IMPACTED GROUNDWATER

Geneviève Martineau, Annick Tétreault, Josée Gagnon, Louise Deschênes and ***Réjean Samson***. (NSERC Industrial Chair for Site Bioremediation, École Polytechnique de Montréal, Montréal (Québec) Canada. H3C 3A7)

ABSTRACT : Based on a study of different filtering media, granular peat moss was chosen for the treatment of BTEX-contaminated groundwater. Laboratory and field studies were conducted using two 200-L pilot scale biofilters : a percolating bed and a saturated bed with up flow circulation. Biological activity and hydrodynamic properties of the beds were studied in order to estimate the effect of aging on the biofilters. Aerobic microcosms were used to characterize the biological activity of the granular peat moss at 10°C. Results showed a quick adaptation of the microbial population after 14 weeks, and more than 60 % mineralization was obtained for all BTEX. Acclimation was also observed for *m*- and *p*-xylenes isomers. After 19 weeks of operation, an analysis of the liquid residence time distribution indicated an important deterioration of physical properties and hydrodynamic behavior. These poor hydrodynamic characteristics combined with low dissolved oxygen and high iron concentrations resulted in a low BTEX removal. Addition of a prefilter designed for iron precipitation is expected to improve the removal efficiency.

INTRODUCTION

Groundwater contamination by petroleum hydrocarbons from leaking underground storage tanks and other spills is now a well known problem. To limit contaminant dispersion, novel approaches using bio-barriers or *in situ* filters were developed. For example, Barker *et al.* (1993) developed the funnel-and-gate system, in which the contaminated plume is confined by underground cutoff walls and treated within *in situ* reactors located in the saturated zone. This technology provides interesting results, but is time-, money- and space-consuming. To alleviate these problems, a biofiltration system combining hydraulic confinement and installation in the vadose zone is proposed. Such a system will also have to meet the requirements listed in Table 1. Following a column study, granular peat moss was chosen for its good physical and hydrodynamic properties and for its capacity to treat toluene- and benzene-contaminated water (Forget *et al.* 1997). Two pilot scale biofilters were then designed and laboratory and field-tested for their hydrodynamic properties and BTEX-treatment capacities. The objective of this paper is to present the results obtained after six months of field operation.

TABLE 1. Basic Requirements for In *Situ* Groundwater Filtration Systems.

Requirements
Physical requirements for the filtering medium
✓ important permeability
✓ no compaction
✓ stable over long periods
Hydrodynamic requirements
✓ low channeling and dead zones
✓ able to operate in plug flow mode
Treatment requirements
✓ support important mineralization activity
✓ able to operate at 10°C
✓ able to operate at low dissolved oxygen concentrations

MATERIALS AND METHODS

Filtering Media Characteristics. Granular peat moss (GPM) was provided by Premier Tech (Rivière-du-Loup, Canada). Granules are cubical with widths ranging from 1 to 2,5 cm, a bulk density ranging from 0,14 to 0,81 g/cm³ and a high porosity of 90 % (v/v), meaning a high peat/water contact area, which should increase mass transfer between the filter bed and the contaminant, and subsequently improve the biodegradation rates. When completely saturated, the water content of GPM is 595 % (w/w) and the hydraulic conductivity is $8,1 \times 10^{-1}$ cm/s. Solid, static liquid and dynamic liquid retention were determined to be respectively 13 ± 1 %, 46 ± 9 % and 41 ± 6 % (v/v). These properties are excellent for a biofiltration medium since high flow rates may be imposed to the bed without reaching the drowning point (superficial velocity higher than the medium hydraulic conductivity).

Biofilters Development. Two pilot scale biofilters were tested : a percolating bed in which a distribution system feeds the affluent over the bed section and a saturated bed with up flow circulation (Figure 1). In the percolation design, water is fed from the top of the unit via a flow distribution system. Water trickles freely through the filtering bed, allowing air/water exchanges to help maintain high dissolved oxygen concentrations. At the bottom of the biofilter, water exits via a pierced tube placed in the gravel. In the saturated up flow design, water enters the unit by the bottom, in the gravel, and leaves via an overflow connected to a drain. A perforated Plexiglas plate is placed between the filtering medium and the overflow to prevent the lines from clogging with GPM fragments.

Field Experiments. Biofilters were operated on a contaminated site located in Trois-Rivières (Canada). Both designs were studied in duplicate. During 20 weeks, BTEX-contaminated groundwater was pumped from the aquifer and fed to the treatment units. Since no dissolved oxygen was naturally present in water, a preliminary aeration was made in a 200-L holding tank. Groundwater contamination at the entrance of the biofilters was (in mg/L) : total BTEX: 5,0 (B: 1.8 ; T: 0.8 ; E: <0.015 ; X: 2.4); total C_{10}-C_{50}: 1.1; N-NH_4: 2; total iron: 37; sulfate: 290; chlorine: 100; calcium: 120; suspended solids: 50.

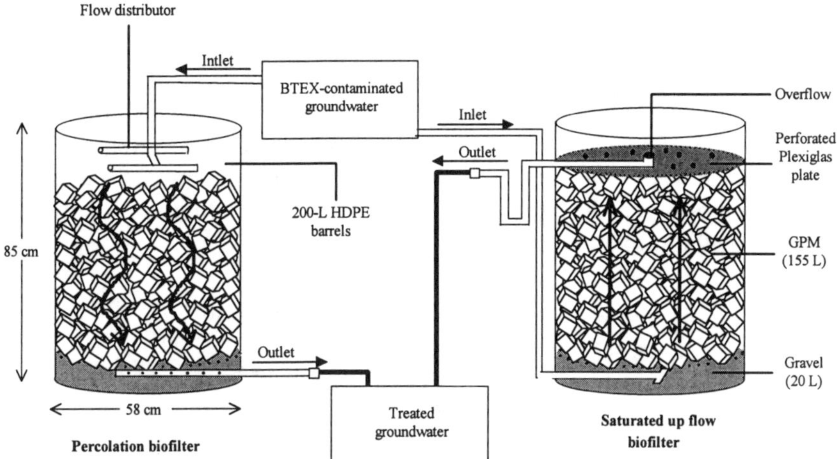

FIGURE 1. Schematic illustration of the biofilters design.

Biological Activity in Granular Peat Moss. Mineralization of ^{14}C-BTEX in GPM was studied in 120-mL serologic bottles under aerobic conditions at 10 °C. Two grams of filtering medium was introduced in each bottle with 80 mL of groundwater. An injection of 0,4 µL (250 kdpm/µL) of radiolabeled contaminant (either ^{14}C-benzene, toluene, ethylbenzene, *o*-, *m*- or *p*-xylene) was then made in every microcosm. The sampling method was described previously (Forget *et al.*, 1997). For each contaminant (B, T, E, *o*-, *m*- or *p*-X), three microcosm series were studied. The first two consisted of fresh GPM in synthetic groundwater (Ross *et al.*, 1997) with and without addition of gasoline (total BTEX equivalent of 23.9 and 4.3 mg/L). The last series contained acclimated GPM (taken from a biofilter that had been filtering contaminated groundwater for 14 weeks) and contaminated groundwater (from Trois-Rivières, Canada) containing a total BTEX concentration of 9.3 mg/L.

Hydrodynamic Study. The flow pattern of water in the two types of biofilters was evaluated by determining the residence time distribution (RTD) of a non-reactive tracer (tritium ^3H_2O, 40 kdpm/µL). At a hydraulic loading rate of 10 m³/m².d, a spike of tracer was induced at the entrance of the biofilters. The water leaving the units was sampled at regular interval, and counted for tritium isotope with a liquid scintillation counter (Wallac model 1409, Turku, Finland).

RESULTS AND DISCUSSION

From preliminary essays, it has been shown that GPM meets all physical requirements determined in Table 1.

Characterization of Biological Activity in Granular Peat Moss. Treatment efficiency was characterized for all BTEX in the fresh medium and after 14 weeks of biofiltration on field. It was observed that for all BTEX, mineralization curves were almost similar showing no lag phase and a plateau at 70 ± 10 %

of biofiltration on field. It was observed that for all BTEX, mineralization curves were almost similar showing no lag phase and a plateau at 70 ± 10 % mineralization after approximately 30 days. Similar results were obtained with fresh GPM for benzene and *o*-xylene at low concentration (4.3 mg/L). Mineralization of toluene and ethylbenzene reached the same level, but a lag phase of 5 and 10 days was observed. Thus, it seems that for low concentration of benzene and *o*-xylene, the intrinsic biological activity of GPM is not significantly enhanced by the adaptation period.

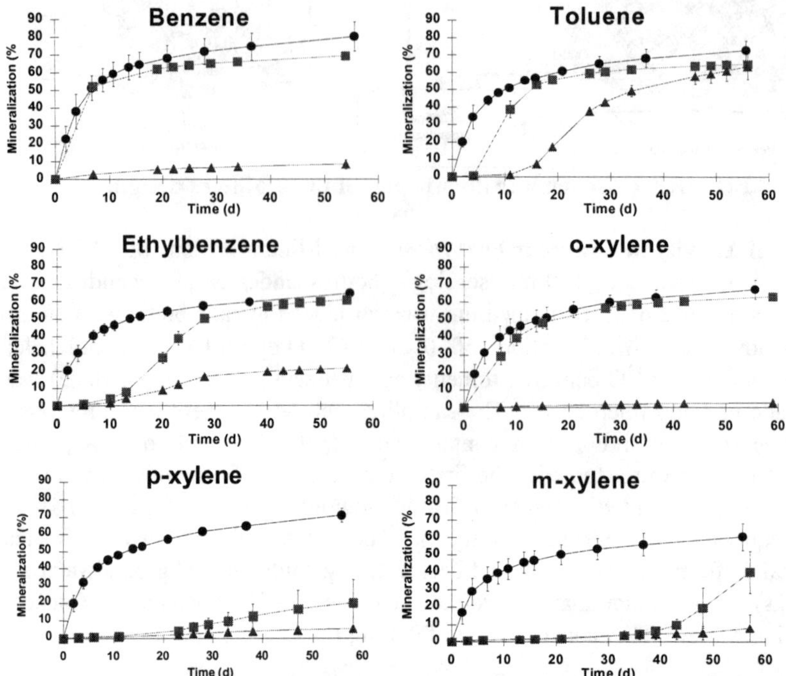

FIGURE 2. BTEX mineralization at 10°C in aerobic conditions by fresh and acclimated GPM. (●)Acclimated GPM + 9.3 mg/L BTEX; (■)Fresh GPM + 4.3 mg/L BTEX; (▲)Fresh GPM + 23.9 mg/L BTEX.

The influence of adaptation is more obvious with *m*- and *p* - xylenes. For the fresh GPM, no mineralization was observed until 40 days while 60% mineralization was observed within the acclimated GPM. Moreover, when a high BTEX concentration was present in water (23.9 mg/L), no significant mineralization was observed with fresh GPM, except for toluene (66 %) and ethylbenzene (20 %), however, a 10 days lag phase was observed.

Hydrodynamic Study. To estimate the effect of GPM aging on the RTD, a tracer study was performed on both type of biofilters. After 19 weeks of field operation, a new tracer study was made on all pilot units (two designs in

presence of channeling and dead zone but not to a large extent (Levenspiel, 1972). After 19 weeks of operation, the RTD curves appaears much more irregular, even for the duplicates. In percolation mode though, it is clear that channeling had intensified, leaving a very short residence time for the biodegradation to take place. In saturated up flow mode, duplicates were quite different, showing channeling for no 1 and recirculation for no 2.

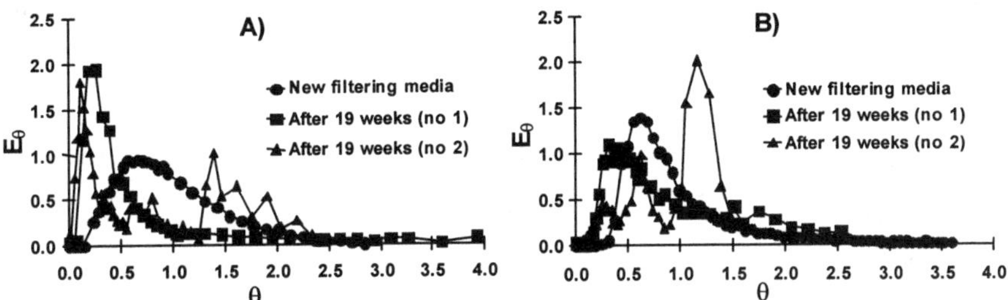

FIGURE 3. Effect of aging on the RTD in GPM biofilters operated at 10 m³/m²d. A) percolation design; B) saturated up flow design, where θ and E_θ are the dimensionless residence time and RTD function.

Hence, it appears that for both designs, aging had a negative impact on the RTDs. The compaction of the filtering material and its clogging with biomass and iron deposits was responsible for the bad hydrodynamic performances obtained. Fortunately, it has been demonstrated that no compaction occurred in a 0,5 L GPM column after 4 months of operation Forget *et al.* (1997). In this experiment, the groundwater was free of iron.

Field operational problems. During the twenty weeks of field experimentation, others difficulties were encountered. First, the very low dissolved oxygen (DO) in groundwater made impossible the aerobic biodegradation to perform correctly. To solve this problem, an aeration system was installed and DO reached 6 mg/L. However, the aeration produces some side effects : loss of volatile hydrocarbons in the thank; oxidation of the iron to ferrous oxide; production of a biofilm in the system; and clogging of the aerator, lines and liquid distributors with ferrous deposits and biomass. Because of the important biomass production within the lines, the DO was consumed before the entrance of the biofilters, leaving intact the problem of DO absence. GPM porosity was also affected since the medium acted as a filter and kept all the suspended solids (SS) (150 mg SS/L after aeration in comparison with 50 mg SS/L before). Because of these problems were principally linked to the chemical characteristics of groundwater and the low DO concentration dissolved oxygen, no significant degradation was noted in the continuously fed biofilters.

CONCLUSION

From preliminary essays, granular peat moss appears extremely promising as a new filtering media because it offers good physical and microbial properties.

Unfortunately, results obtained with two 200-L pilot scale biofilters tested in the field were not very good. A poor BTEX degradation was observed in the continuously fed biofilters and hydrodynamic characteristics indicated that such system is not suitable to treat groundwater containing high concentration of iron. On the positive side, characterization of biological activity in aerobic microcosms at 10 °C showed adaptation of indigenous GPM microbial population after 14 weeks. After some modifications, the system should provide interesting results since the specific BTEX-degraders are present and active in the GPM.

ACKNOWLEDGMENTS

The authors acknowledge the financial support of the Chair partners : Alcan, Bodycote-Analex, Bell Canada, Browning-Ferris Industries (BFI), Cambior, Hydro-Québec, Petro-Canada, SNC-Lavalin, Centre québécois de valorisation de la biomasse et des biotechnologies and the Natural Science and Engineering Research Council (NSERC).

REFERENCES

Barker, J.F., J. Weber, J.S. Fyfe, J.F. Devlin, D.M. McKay, J.A. Cherry. 1993. "Efficient Addition Technology for In Situ Bioremediation. " In *Third Annual Symposium on Groundwater and Soil Remediation*, pp. 285-296. Quebec city, Canada.

Forget, D., L. Deschênes, D. Karamanev, and R. Samson. 1997. "Characterization of a New Support Media for In Situ Biofiltration of BTEX-Contaminated Ground Water." In *Fourth International In Situ and On-Site Bioremediation Symposium,* pp. 221-226. New Orleans, LA.

Levenspiel, O. 1972. "Non Ideal Flow. " In John Wiley & Sons (Eds), *Chemical Reaction Engineering, 2nd Ed.*, pp. 253-315, New York,.NY.

Ross, N., L. Deschênes, J. Bureau, B. Clément, Y. Comeau, and R. Samson. 1998. "Ecotoxicological Assessment and Effect of Physicochemical Factors on Biofilm Development in Groundwater Conditions." *Environ. Sci. Technol.*, 32(8) : 1105-1111.

DIESEL OIL: A SOURCE FOR THE PRODUCTION OF BIOSURFACTANTS

Carvalho, D. F.; Corsi, F. K.; Furquim, F. E.; Leal, P. & Durrant, L. R.
DCA/FEA - UNICAMP (Campinas State University), Campinas-SP, Brazil

Abstract: Fifteen bacterial strains were isolated from soil samples collected near Paulinia's petroleum refinery, Campinas, Brazil, and grown on 1.5% of diesel oil. The emulsification activities of the cell free extracts were determined by measuring either the increase in absorbance of the oil-in-water emulsions at 610 nm or the height (cm) of the water-in-oil emulsion layers (halo) formed, using diesel oil as the test substrate. Six bacteria were selected and identified and their emulsification activities were determined following growth on 3.0%, 5.0%, 15.0% or 30.0% of diesel oil. The strains *Actinobacillus lignieresii, Moraxella nonliquefaciens, Acinetobacter calcoaceticus* and *Flavobacterium* sp showed high emulsification activities (halos higher than 1.5 cm) after growth for 120 hours in all the concentrations of diesel oil used. *Bacillus megaterium* produced oil-in-water emulsions after growth for 120 hours only at the diesel oil concentration of 1.5%. *Planococcus citreus* was able to emulsify all of the diesel oil present in the growth medium, regardeless of the concentrations and of the incubation period, but no emulsification activities were detected in its culture supernatants.

INTRODUCTION

Biosurfactants are a diverse group of surface-active chemical compounds mainly produced by hydrocarbon-utilizing microorganisms, which can enhance the degradation of oils in the environment [1].

Microbial derived surfactants have special advantages over their chemically manufactured counterparts because of their lower toxicity, biodegradable nature and effectiveness at extreme temperatures and pH values [2].

Hydrocarbons, such as oil products, petroleum products and halogenated compounds, form an important class of pollutants on a global scale. The presence of hydrocarbons in the environment is of considerable public health and ecological concern due to their persistence, ability to be bioaccumulated and toxicity to a wide variety of biological systems [3].

Diesel fuel is one of the major pollutants of soil and ground water near petrol stations [4]. The addition of biologically produced emulsifier, or the stimulation of microbial production of biosurfactants and bioemulsifiers in these contamianted sites can aid in the biorremediation of oil-contaminated environment [5].

Biosurfactants produced by oil-degrading bacteria may be useful for cleaning oil tanks, in the clean-up of oil spills and in the biodegradation of oils present in industrial wastewater [6].

The influence of a large variety of carbon sources, such as glucose, kerosene or vaseline to biosurfactant production have been reported, but diesel oil is little studied [4].

MATERIALS AND METHODS

ISOLATION OF MICROBIAL STRAINS: Samples from soil contaminated with diesel oil were collected near Paulinia's petroleum refinery in Campinas (SP, Brazil), and used for the isolation of microorganisms as follows: homogenization of soil samples, surface plating, incubation at 30 ^{0}C and isolation of pure colonies.

IDENTIFICATION OF THE BACTERIAL STRAINS: For the tentative identification of the bacteria, various biochemical tests were undertaken and observation was made of all morphological characteristics [7].

CULTURE CONDITIONS: The bacteria were cultivated in a medium containing either 1.5%, 3.0%, 5.0%, 15.0% or 30.0% (vol/vol), of diesel oil (TEXACO) as the carbon source plus 0.5 g $MgSO_4$, 3.0 g $NaNO_3$, 1.0 g KH_2PO_4, 1.0 g yeast extract and 0,3 g peptone per liter of medium. Cultures were grown in 100 mL Erlenmeyer flasks with 50 mL of medium and incubated with shaking (150 rpm) at 30 ^{0}C. Duplicate flasks were collected at 48, 72 and 120 hours of incubation and the emulsification activities were determined as described below. Non-inoculated flscks were also incubated under the same conditions and were used as control.

EMULSIFICATION ACTIVITIES MEASUREMENTS: Culture broth was made cell free by centrifugation. 3.5 ml of the cell free broth was vigorously shaken with 2.0 ml of diesel oil on a vortex shaker and left undisturbed. After one hour, optical density of the oil-in-water emulsion phase was measured at 610 nm. The O.D. was reported as emulsification activity[8]. After 24 hours the height of the emulsion layer (water-in-oil) was measured and emulsification activity was expressed in cm [9]. The test tubes utilized for activities measurements have diameters of 1.0 cm, producing a 2.0 cm substrate layer which corresponds to 2 ml of diesel oil.

RESULTS AND DISCUSSION

All fifteen isolated bacterial strains were tested following growth in 1.5% of diesel oil as the carbon source. Six strains were selected because they showed high emulsification activities (halos higher than 1.0 cm or absorbance increases greater than 0.5). They were identified according to the results obtained after the biochemical tests plus observation of all morphological characteristics, as shown in table 1.

CODE	CULTURE	CODE	CULTURE	CODE	CULTURE
18	*Planococcus citreus*	5BII	*Actinobacillus lignieresii*	5B4	*Flavobacterium sp*
5C2	*Acinetobacter calcoaceticus*	5D2	*Bacillus megaterium*	5E2	*Moraxella nonliquefaciens*

TABLE 1: Bacterial strains identification.

Table 2 shows the emulsification activities expressed as optical density (Abs sample – Abs control = Abs). With the exception of strain 5E2 grown on 1.5% and 5.0% of diesel oil, all the others significant results (absorbance higher than 0.5) were obtained following growth of the strains for 120 hours.

Emulsification Activities (Abs)						
Diesel oil concentration/ Growth time	**Bacterial strains**					
	18	**5BII**	**5B4**	**5C2**	**5D2**	**5E2**
1.5%/48 hours	ND	0.02	0.05	0.07	0.05	0.75
3.0%/48 hours	ND	0.06	ND	0.08	ND	0.01
5.0%/48 hours	0.12	0.12	0.09	0.26	0.06	0.87
15.0%/48 hours	0.08	0.05	0.16	0.04	ND	0.28
30.0%/48 hours	0.18	0.17	0.08	0.11	ND	0.32
5.0%/72 hours	0.13	0.06	0.035	0.22	ND	0.03
15.0%/72 hours	0.06	0.08	0.15	0.21	ND	0.06
30.0%/72 hours	0.15	0.06	0.04	0.24	ND	0.07
1.5%/120 hours	ND	0.62	0.30	0.35	0.96	0.23
3.0%/120 hours	ND	0.27	0.23	0.65	0.38	0.70
5.0%/120 hours	0.1	0.51	0.65	0.53	0.14	0.47
15.0%/120 hours	0.66	0.11	0.03	0.02	ND	0.02
30.0%/120 hours	0.45	0.06	0.05	0.09	ND	0.04

TABLE 2: Oil-in-water emulsification activities in cell-free culture fluids of the bacterial strains. ND=not detected

Acinetobacter calcoaceticus was the only strain able to produce a significant amount of biosurfactants detected by water-in-oil emulsions, when grown for 48 hours on 3.0% of diesel oil, resulting in a halo formation superior to 1.5 cm. This strain showed good results (halos higher than 2.0 cm), when grown for 120 hours in diesel oil concentrations up to 5.0%.

The bacterium *Moraxella nonliquefaciens* showed good emulsification activities when grown on 3.0%, 5.0%, 15.0% and 30.0% of diesel oil for 72 and 120 hours.

When 1.5% and 5.0% of diesel oil was used for growth of *Actinobacillus lignieresii*, halos greater than 2.0 cm were observed following 120 hours cultivation. But, when higher concentrations of diesel oil were utilized, *Actinobacillus lignieresii* did not show any emulsification activity.

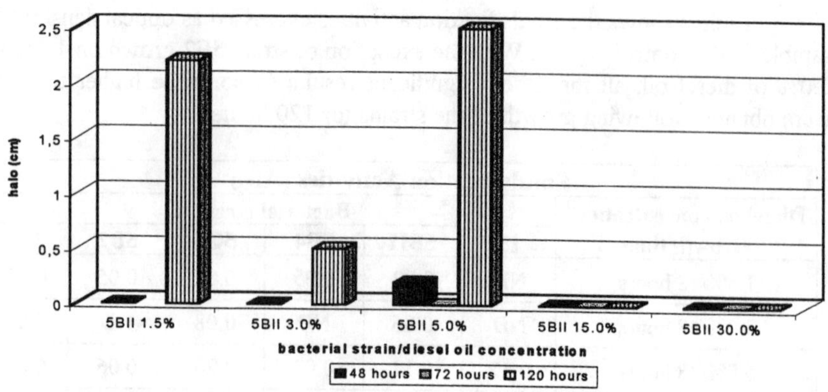

FIGURE 1: Water-in-oil emulsification activities in cell-free culture fluids of the strain *Actinobacillus lignieresii*

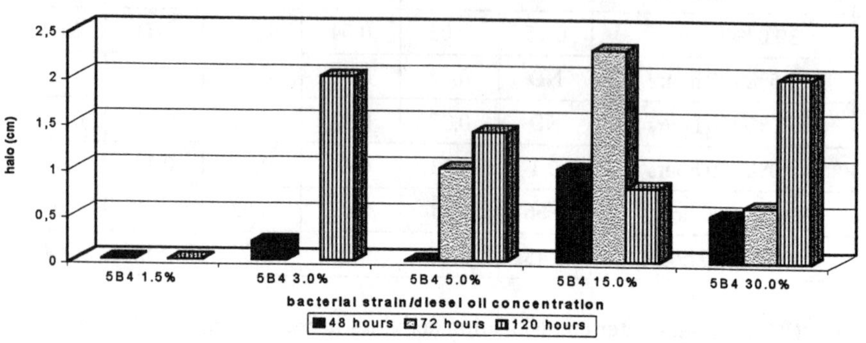

FIGURE 2: Water-in-oil emulsification activities of the strain *Flavobacterium sp*

The concentrations of 3.0%, 15.0% and 30.0% of diesel oil, for 72 and 120 hours of growth, were the best conditions for biosurfactants production by *Flavobacterium sp*.

Apparently, the high diesel oil concentration stimulated the biosurfactant production by the strains *Flavobacterium sp* and *Moraxella nonliquefaciens*, not showing a toxic effect to these bacterial cells.

Planococcus citreus and *Bacillus megaterium* did not show any emulsification activities by halo formation.

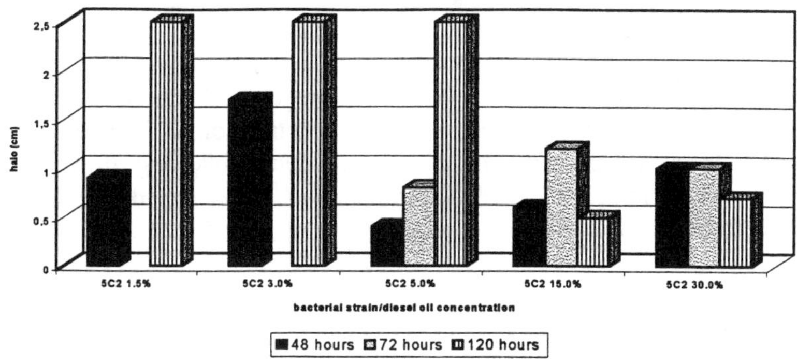

FIGURE 3: Water-in-oil emulsification activities of the strain *Acinetobacter calcoaceticus*

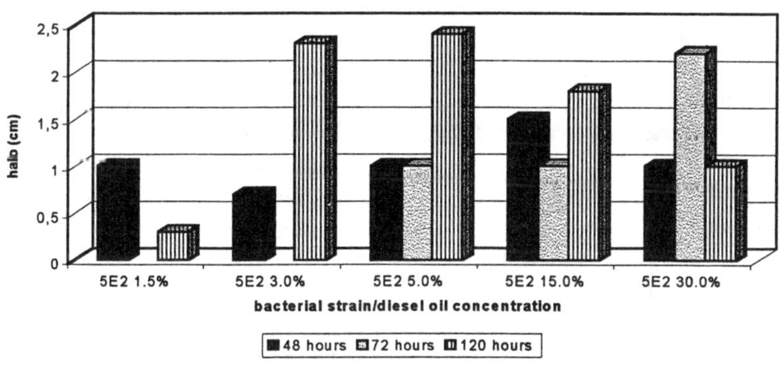

FIGURE 4: Water-in-water emulsification activities of the strain *M. nonliquefaciens*

Planococcus citreus emulsified all the diesel oil present in the growth medium, regardless of the concentration used and of the incubation period. Unexpectedly, low emulsification activities were obtained from this strain at incubation time tested. Probably, the greater amount of biosurfactant produced by *Planococcus citreus* was released to culture medium during the first hours of incubation, being utilized to emulsify the diesel oil present in the culture medium.

With the exception of *Bacillus megaterium* and *Planococcus citreus,* each bacterium was able to produce biosurfactants depending on the diesel oil concentration and growth time.

Actinobacillus lignieresii was able to produce a significant amount of biosurfactant, detected by a 2.5 cm halo formation when grown on 5% of diesel oil, but the introduction of this strain in a *in situ* bioremediation should be done with care because this strains produces hydrogen sulfide. In this case, the aplication of the isolated biosurfactant could be more profitable.

The best results were obtained by cultures that produced water-in-oil emulsions. Values greater than 2.0 cm indicate that all the oil substrate was emulsified. In these cases, the diesel oil was totally transformed into a white foamy layer. The biosurfactants and emulsification produced following growth of these bacterial strains in diesel oil concentrations up to 30.0% demonstrate their potential to be used in biorremediation process.

ACKNOWLEDGEMENTS

We thank the CNPq and FAPESP for financial support.

REFERENCES

1. Banat, I. M. (1995) Characterization of Biosurfactants and their use in Pollution Removal - State of the Art (Review) **Acta Biotechnol.** 15 (3), 251-267
2. Pruthi, V. & Cameotra, S. S. (1997) Short Communication: Production of a biosurfactant exhibiting excellent emulsification and surface active properties by *Serratia marcescens* **World Journal of Microbiology & Biotechnology** 13, 133-135
3. Jain, D. K., Lee, H. & Trevors, J. T. (1992) Effect of addition of *Pseudomonas aeruginosa* UG2 inocula or biosurfactants on biodegradation of selected hydrocarbons in soil **Journal of Industrial Microbiology** 10, 87-93
4. Geerdink, M. J., van Loosdrecht, M. C. M. & Luyben, K. Ch. A. M. (1996) Biodegradability of diesel oil **Biodegradation** 7, 73-81
5. Carrillo, P. G., Mardaraz, C., Pita,-Alvarez, S. I. and Giulietti, A M. (1996) Isolation and selection of biosurfactant-producing bacteria. **World J. Micorbiol. & Biotechnol**., 12, 82-84
6. Van Dyke, M. I., Lee, H. & Trevors, J. T. (1991) Applications of Microbial surfactants **Biotech. Adv.** 9, 241-252
7. Holt, J. G., Krieg, N. R., Sneath, P. H. A., Staley, J. T. & Williams, S. T. (1994) Bergey's Manual of Determinative Bacteriology. 9, USA.
8. Johnson, V., Singh, M., Saini, V. S., Adhikari, D. K., Sista, V. and Yadav, N. K. (1992). Bioemulsifier prodution by an oleaginous yeast *Rhodotorula glutinis* IIP-30. **Biotechnology Letters**, 6, 487-490
9. Cooper, D. and Goldenberg, B. G. (1987). Surface active agents from two Bacillus species. **Applied and Environmental Microbiology**, 53, 224-229

OXYGEN CONSUMPTION OF NATURAL REDUCTANTS IN AQUIFER SEDIMENT RELATED TO IN SITU BIOREMEDIATION

Jasper Griffioen, Bas van der Grift, Alice Buijs (Netherlands Institute of Applied Geoscience TNO, Delft, The Netherlands)
Niels Hartog (Institute of Earth Sciences, Utrecht University, Utrecht, The Netherlands)

ABSTRACT: Biogeochemical experiments were performed to characterize the reactivity of natural reductants versus that of benzene, toluene, ethylbenzene, and xylenes (BTEX) compounds. Two types of experiments were performed: (1) fluidized-bed experiments in which competition between BEX compounds and natural reductants for incoming oxidants was characterized and (2) microoxymax experiments in which O_2 reduction by natural reductants in sediment was characterized. Oxidation of bulk organic matter (BOM) is distinguished from that of Fe sulfides by the production of CO_2. Oxidation of BEX compounds is generally fast compared to that of BOM or Fe sulfide. However, oxidation of BOM or Fe sulfide is a slow but ongoing process. Natural reductants thus consume oxidants that are introduced naturally or human-induced and need to be considered in planning of in situ bioremediation strategies.

INTRODUCTION

Sandy aquifer sediment may contain natural reductants such as bulk organic matter (BOM) and pyrite (FeS_2). Natural reductants will compete with contaminant reductants, such as petroleum-type hydrocarbons, for the oxidants available. These oxidants may be naturally available (natural attenuation strategy) or introduced by man (stimulated in situ bioremediation strategy). The success or failure of in situ bioremediation strategies is determined partly by the biogeochemical reactivity of the natural reductants present.

The objective of our study was to relate the reactivity of natural reductants to that of BTEX compounds. We have performed two types of experiments: fluidized-bed experiments in which the competition between the two types of reductants can be studied, and microoxymax experiments in which O_2 consumption and CO_2 production of natural reductants is characterized as function of time. Toluene was not studied, because field investigations have shown that toluene was always degraded naturally and is thus of less environmental concern.

MATERIALS AND METHODS

Aquifer sediment was sampled anaerobically using Akkerman sampling tubes. The samples were stored anaerobically in glass jars. The samples were analyzed for carbonate, organic carbon, and total sulfur content by combustion at increasing temperature and infrared (IR) detection using a Stroehlein C-mat 5500 analyzer. Groundwater was sampled with a peristaltic pump and stored in 5- or

10- L glass bottles. The groundwater samples were used immediately in the fluidized-bed experiments, except for one experiment in which artificial groundwater was used.

Fluidized-bed experiments were performed using a 100-cm-long glass tube in which 50 g of sediment was placed (Fig. 1). The glass tube was filled with native groundwater from below. A peristaltic circulation pump was used to fluidize the sediment. A 0.45-µm filter in a stainless steel filter holder was placed following the glass tube to prevent migration of fine particles. Another P-50 Hiload pump injected (treated) groundwater at a flowrate of 30 mL/hr. All materials in contact with sediment and water were made of glass or stainless steel, except a short tube in the peristaltic pump. This was composed of tygon, which is a gastight plastic. The groundwater was flushed with a N_2/CO_2 gas mixture to maintain the anaerobic status and fix the pH. A syringe pump injected a BEX solution at a flowrate of 0.7 mL/hr; Br was added as an inert tracer to calculate the dilution with injected (ground)water. Dilution of the BEX solution with (ground)water resulted in BEX concentrations that are representative for the sites studied. The fluidized-bed tube and the groundwater bottle were cooled at 12°C. Influent and effluent samples were analyzed using routine techniques.

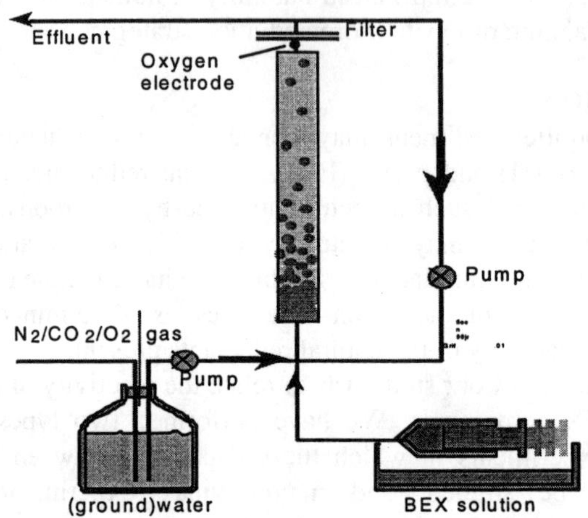

FIGURE 1. Setup of fluidized-bed experiments.

Respiration experiments were performed in duplicate at 22°C using the microoxymax, which is a fully automated respirometer that measures the O_2 consumption and CO_2 production of batch incubations as a function of time. Two experimental series were performed in duplicate. In the first series, 15 grams of sediment was incubated with 50 mL of tapwater in 100-mL glass bottles for 27 days. The supernatant was analyzed afterwards on pH, alkalinity, major cations, and anions using routine techniques. Several sediments were decalcified by extraction with 1.0 M HCl during 16 hours and subsequently repeated flushing with demineralized water.

In the second series, we studied four samples with organic C content ranging from 0.15 to 0.30%. Twenty grams of sediment was incubated in 100-mL glass bottles. A 50-mL solution containing 25 mg/L of trace elements and 0.5 mg/L of vitamins, was added to ensure substrate-limiting conditions. We checked this by running the same experiments with doubled concentrations. Phosphate buffer was added to keep the pH constant at 7.5 to 8.0. The series ended after 107 days.

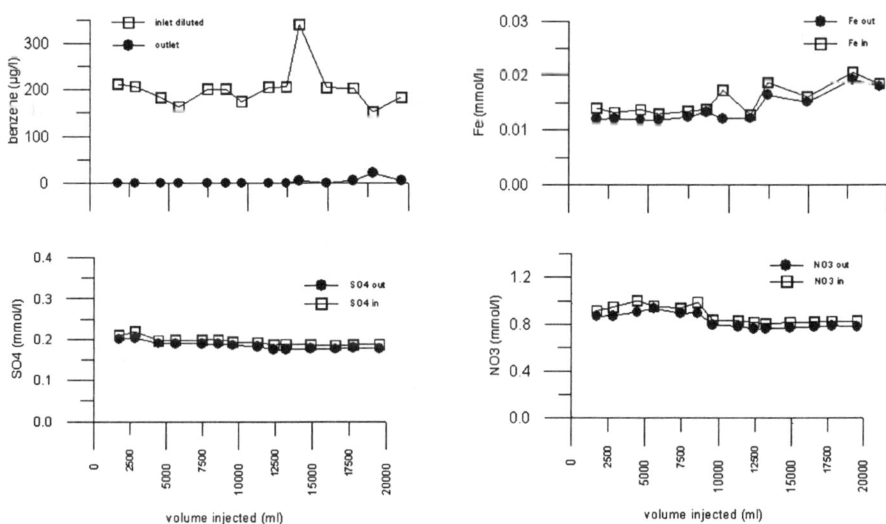

FIGURE 2. Time series of influent and effluent concentration for fluidized-bed experiment with injection of NO_3-enriched tapwater using Flebo sediment.

RESULTS

The research focused on three sites for which the potential for stimulated in situ bioremediation was studied. We will illustrate the results for the Flebo and Slochteren sites. The first has a deep anaerobic status where dissolved H_2S and CH_4 are present above 0.1 mg/L. The Flebo site is contaminated with benzene at concentrations of at most 220 μg/L. The Slochteren site is an Fe-anaerobic site contaminated with gas condensate. Figure 2 shows the results of a fluidized-bed experiment for Flebo in which tapwater saturated with air and 50 mg NO_3/L added was injected together with a benzene-containing solution. Comparison of the influent concentrations with the effluent ones reveals that benzene is degraded during most of the experiment. The oxygen concentration of the effluent was about 0.20 mmol O_2/L, which is significantly below saturation with air at 12°C. No significant changes for the redox-sensitive species (NO_3, SO_4, and Fe) were observed. Aerobic degradation of benzene thus occurred. The first-order degradation constant was minimal at 11.0 day^{-1}, based on the residence time in the fluidized bed and the difference between the influent and effluent concentrations. No consumption of natural reductants was observed in these experiments.

Figure 3 shows the results of the microoxymax experiments for Flebo. An experiment with untreated sediment and another with decalcified sediment were performed. Degradation of BOM was the major process for the untreated sediment. This is indicated by the consumption of O_2 together with the production of CO_2. The molar ratio of CO_2 produced to O_2 consumed was equal to 1.0; degrading BOM can thus be represented by CH_2O. The results show that oxidation of BOM occurred at a rate of 1.3 μmol O_2/g_{sed}.day initially and 0.6 μmol O_2/g_{sed}.day after 27 days. For the decalcified sediment, O_2 was consumed at a much higher rate of 4.2 μmol/g_{sed}.day initially and 2.9 μmol/g_{sed}.day after 27 days. No CO_2 production was observed after 2 days, which means that another redox process must have taken place. CO_2 production in the initial stage is attributed to degassing. Analyses of the supernatant after the incubations gave SO_4 concentrations of 28.7 and 30.3 mmol/L and Fe concentrations of 12.6 and 14.7 mmol/L. Sulfides thus were oxidized in the second experiment. The extraction by HCl had removed a coating that prevented oxidation of pyrite. This coating was most likely either Ca carbonate or Fe oxyhydroxide. The pHs decreased to 2.15 and to 2.17 in these experiments, whereas the pHs for the untreated sediment were 7.57 and 7.90 (and for tapwater 8.0). Pyrite oxidation causes strong acidification and may change the oxidation mechanism (Evangelou & Zhang, 1995). We have not further investigated this.

The fluidized-bed experiments for Flebo indicate that benzene is more reactive than naturally present reductants. An experiment with sediment from Slochteren, in which artificial suboxic NO_3-containing water was injected, shows the following. Initial Fe sulfide oxidation by pyrite was observed together with degradation of ethylbenzene and the three xylenes, but little or no degradation of benzene. The small increase in SO_4 concentration (except for one initial and

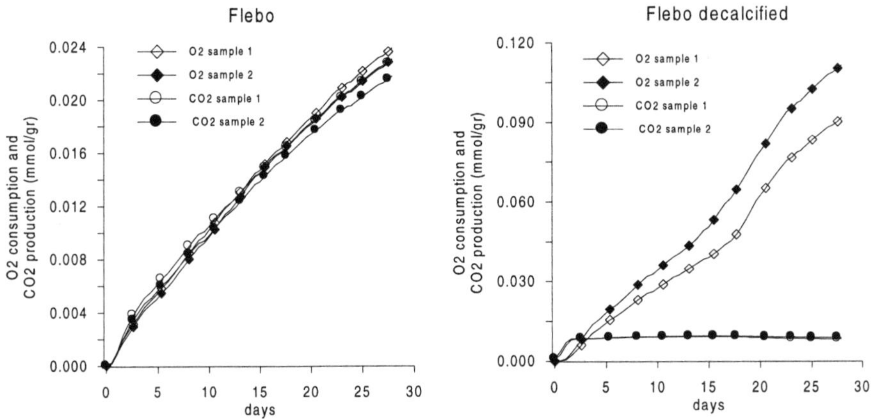

FIGURE 3. Cumulative O_2 consumption and CO_2 production for microoxymax experiment with untreated and decalcified Flebo sediment.

another measurement) for the first 7000 ml was less than the decrease in NO_3 concentration, when corrected for reaction stoichiometry. Additional reduction of NO_3 thus happens. Dissolved Fe was below the detection limit of 1.0 μmol/L and the pH remained near-neutral, which means that complete Fe sulfide oxidation to Fe oxyhydroxide had occurred. The amount of SO_4 produced corresponds to a S content of the sediment of 190 ppm, which is an order of magnitude smaller than the total elemental S content of 0.4%. The microoxymax experiments for the Slochteren sediment also indicate Fe sulfide oxidation for both the untreated and decalcified sediment (not shown here). The total O_2 consumption rate varied from 0.6 to 2.1 μmoL/g_{sed}.day for the untreated sediment, and it was two times lower for the decalcified sediment. Little BOM oxidation was observed despite an organic C content of 0.8%.

The reactivity of BOM was studied in more detail in series 2 for a set of aquifer sediments from 't Klooster, Hengelo. The sediments contained less than 0.1% Fe sulfides. The low SO_4 concentrations in solution (< 0.8 mmol/l) at the end of the experiment indicate that no significant Fe sulfide oxidation had occurred. The results of the duplicate experiment are comparable, and the identical rates for both low and high nutritional conditions confirmed C_{org}-limiting conditions. In all samples the molar production of CO_2 equalled the molar O_2 consumption and thus indicated the degradation of BOM. The remaining amount of initial BOM ranged from 91% to 80% at the end of the experiment (Fig. 4). The decreasing degradation rates during the experiment suggest that a significant fraction of BOM will be relatively stable under the oxidizing conditions applied. The first-order rate constants ranging from 0.0008 yr^{-1} to 0.002 yr^{-1} are, however, intermediate to high rates in marine sediments in the broad range of oxic degradation rates (Henrichs, 1993).

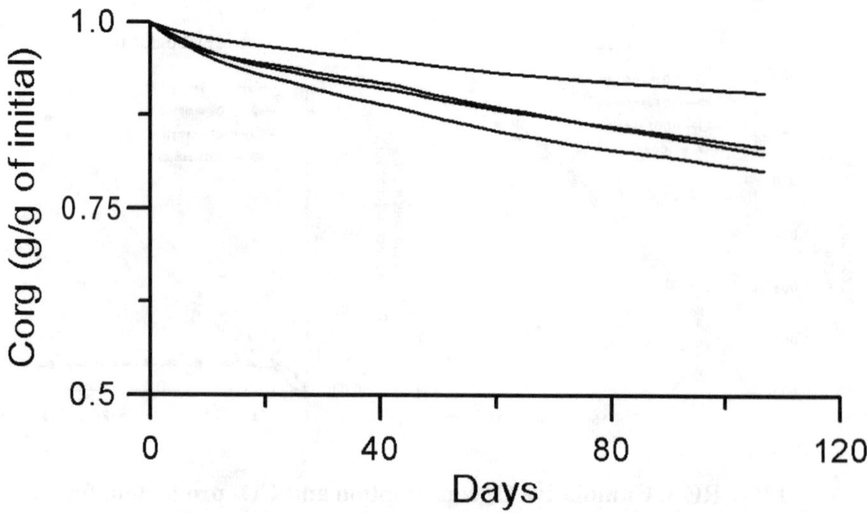

FIGURE 4. Calculated fraction of bulk organic matter remaining during microoxymax incubation with sandy sediments from 't Klooster, Hengelo.

DISCUSSION AND CONCLUSIONS

The general results show that natural reductants are an important sink for either natural or human-induced oxidants. Although the reactive amount is in general low for sandy sediments, it is sufficient to contribute significantly to the reduction capacity of sediments. The slow kinetics sets a limit to stimulated bioremediation of, for example, petroleum-type hydrocarbons by means of injecting oxidants such as oxygen or nitrate. The time period during which the natural reductant reacts will be on the order of several weeks to months. A significant loss of oxidants introduced will occur if sufficient reaction time is available.

ACKNOWLEDGMENTS

We would like to thank Hugo van Buijssen for his assistance in the microoxymax experiments.

REFERENCES

Evangelou, V.P., and Y.L. Zhang. 1995. "Pyrite oxidation mechanisms and acid mine drainage prevention." *Crit. Rev. Env. Sci. & Technol. 25*: 141-199.

Henrichs, S. M. 1993. "Early diagenesis of organic matter: The dynamics (rates) of cycling of organic compounds." In M.H. Engel and S.A. Macko (Eds.), *Organic Geochemistry,* pp. 101-117. Plenum, New York.

THE EFFECT OF ELECTRON ACCEPTORS ON THE NITRATE UTILIZATION EFFICIENCY IN GROUNDWATERS

Chih-Jen Lu, Chi Mei Lee, Shung-Hung Liu, and Yen-Han Du
National Chung Hsing University, Taichung, Taiwan, TAIWAN, ROC

ABSTRACT: This study focused on the nitrate utilization efficiency for the biodegradation of toluene in simulated groundwaters. The nitrate utilization efficiency is defined as the ratio of the total amount of toluene removed to the total amount of nitrate consumed. The experiments were conducted with a series of batch reactors. The reactors were seeded with only denitrifers or with both denitrifers and aerobic toluene degrading cells. The reactors contained only nitrate or both nitrate and dissolved oxygen as the electron acceptors. The experimental results indicated that toluene could be effectively biodegraded in the aerobic and anoxic conditions. When nitrate was used as both the nitrogen source and electron acceptor, the nitrate utilization efficiency was 0.22 mg-toluene-removed/mg-nitrate-consumed in the reactors seeded with only denitrifers. When reactors were seeded with both aerobic toluene degraders and denitrifers in the presence of both nitrate and limited dissolved oxygen, toluene could be simultaneously removed by aerobic cells and denitrifers. The presence of a small amount of dissolved oxygen and sufficient nitrate resulted in a better nitrate utilization efficiency, 0.31 mg-toluene-removed/mg-nitrate-consumed compared to the reactor contained only nitrate. The results suggested that the presence of both limited dissolved oxygen and nitrate as well as both aerobic cells and denitrifers enhanced toluene removal and reduced nitrate consumption.

INTRODUCTION

Aromatic hydrocarbons can be readily biodegraded in the aerobic environment. Aerobic biodegradation in the subsurface, however, usually is hindered after the depletion of dissolved oxygen (DO), because of the limited oxygen supply and its low solubility (Aggarwal *et al.*, 1991). When the environment is in an oxygen limited condition, hydrogen peroxide can be used as an alternative oxygen source to enhance the biodegradation of aromatic compounds (Lu, 1994). However, the utilization efficiency of hydrogen peroxide was very low and decreased with an increase in its initial concentration (Lu *et al.*, 1996). Therefore, other alternative electron acceptor is considered to replace DO to overcome the limitation of DO. Nitrate is the most acceptable alternative electron acceptor because of its high solubility, low adsorption potential, and high energy yield for cells (Hunchins and Wilson, 1994). Barenschee *et al.* suggested that the addition of hydrogen peroxide resulted in diesel fuel degradation rate four to seven times higher than that with nitrate, because of the higher bacterial activity and higher total cell counts (Barenschee *et al.*, 1991). However, the addition of hydrogen peroxide caused less utilization efficiency of electron acceptor. For example, the removal of 1 mg hydrocarbon required 4.1 to 5.7 mg hydrogen peroxide compared to 2.7 to 4.7 mg nitrate (Barenschee *et al.*, 1991). Therefore, using DO or nitrate as the only electron acceptor in the bioremediation system has different limitations. Especially, it is very impractical to keep the bioremediation system in a complete aerobic or anoxic condition in the field. The bioremediation system using nitrate as the major electron acceptor usually also contained limited DO (Schreiber *et al.*, 1997; Barbaro *et al.*, 1997). Therefore, this paper focuses on the biodegradation of an aromatic compound (toluene) in the presence of both nitrate and DO or only nitrate to study the nitrate utilization efficiency in the aerobic and anoxic conditions.

MATERIALS AND METHODS

Chemostat. The major purpose of this work was to compare the nitrate utilization efficiency in the toluene removal using both nitrate and DO or only nitrate as the electron acceptor. Therefore, two chemostats fed with toluene as the only substrate were operated to acclimate toluene-degrading cells. One chemostat was operated in the aerobic condition using nitrate as the nitrogen source. The other was operated in the anoxic condition using nitrate as both the nitrogen source and electron acceptor. These two chemostats were continuously operated for about 6 months. Then, the acclimated cells were ready for use as the cell source in this study.

Batch Reactors. The removal of toluene and the consumption of nitrate were conducted with a series of batch reactors. In addition to micronutrients, the reactor contained enriched cells, toluene, and nitrate in the adequate initial concentrations. The reactors were shaken at 100 rpm with a temperature-controlled water bath (25°C) in dark. The batch reactor was sampled to analyze the remaining concentrations of toluene and nitrate with a gas chromatography (GC/FID) and ion chromatography (IC), respectively. DO and pH were also measured.

RESULTS AND DISSCUSION

Toluene Removal Using Nitrate as the Electron Acceptor. The theoretical stoichiometry of toluene biodegradation using nitrate as both the electron acceptor and nitrogen source can be expressed as following:

$$C_7H_8+5.67\ NO_3^-+5.67\ H^+ \longrightarrow 0.333\ C_5H_7O_2N+5.33\ CO_2+5.67\ H_2O+2.67\ N_2 \quad (1)$$

Theoretically, the consumption of 1 mg nitrate results 0.26 mg toluene removal (0.26 mg-toluene/mg-nitrate). Table 1 shows the removal of toluene using nitrate as both the electron acceptor and nitrogen source for cell growth in a series of batch reactors seeded with a mixture of denitrifers. The initial toluene and nitrate concentrations were 19 and 102 mg/L, respectively. The initial DO concentration was 0.95 mg/L. The DO concentration was very stable during the experimental period, 6 days. The stable DO concentration indicated that DO was not used as the electron acceptor for toluene removal. After 6 days of operation, toluene was completely removed with a decrease in nitrate concentration from 102 mg/L to 17 mg/L (85 mg/L nitrate consumption). The results suggested that the removal of 1 mg/L toluene required 4.47 mg/L nitrate. The nitrate requirement (4.47 mg-NO_3^-/mg-toluene) for toluene removal was higher than that calculated from the theoretical equation (3.82 mg- NO_3^-/mg-toluene) (equation 1). The experimental results can be used to modify the stoichiometry equation of anoxic toluene removal. The modified equation can be expressed as following:

$$C_7H_8+6.63\ NO_3^-+6.63\ H^+ \ -->0.14\ C_5H_7O_2N+6.30\ CO_2+6.99\ H_2O+3.25\ N_2 \quad (2)$$

The modified stoichiometry equation of toluene removal (equation 2) indicates that the practical nitrate requirement is higher than the theoretical value. In other words, equation (2) indicates the practical nitrate utilization efficiency (0.22 mg-toluene/mg-nitrate) is less than the theoretical value (0.26 mg-toluene/mg-nitrate). In here, the nitrate utilization efficiency is defined as the ratio of the total amount of toluene removed to the total amount of nitrate consumed.

TABLE 1. Toluene removal and nitrate consumption in reactors using nitrate as both the nitrogen source and electron acceptor by denitrifers.

	Nitrate (mg/L)	Toluene (mg/L)	DO (mg/L)	pH
Initial Conc.	102	19	0.95	7.1
Final Conc. (after 6 days)	17	0	0.90	7.5
Removal	85	19	----	----

※ nitrate utilization efficiency: 19/85 = 0.22 (mg-toluene/mg-nitrate)

Toluene Removal in the Presence of both Nitrate and DO. In this study, the batch reactors were seeded with both aerobic toluene degraders and denitrifers. The reactors also contained nitrate and dissolved oxygen at initial concentrations of 124.8 mg/L and 4.6 mg/L, respectively. The stoichiometry equation of toluene removal using nitrate as the nitrogen source and oxygen as the electron acceptor can be expressed as following:

$$C_7H_8+0.446\ NO_3^-+0.446\ H^++5.82\ O_2->0.446\ C_5H_7O_2N+4.71\ CO_2+2.64\ H_2O \quad (3)$$

Table 2 shows the toluene removal in the presence of both nitrate and DO. Toluene removal can be divided into two stages. During the first stage (day 0 to day 2), the concentration of dissolved oxygen (DO) was 4.6 mg/L. Therefore, DO was available to be primarily used as the electron acceptor. The stoichiometry of toluene removal (equation 3) indicates that the removal of 1 mg/L toluene requires 2.02 mg/L DO and 0.3 mg/L nitrate. During the first stage, toluene concentration decreased from 32.1 to 24.7 mg/L (7.4 mg/L removal) and nitrate concentration decreased from 124.8 to 102.3 mg/L (22.5 mg/L consumption). At the same time, DO concentration decreased from 4.6 to 1.8 mg/L (2.8 mg/L consumption). According to the aerobic toluene biodegradation using nitrate as the nitrogen source, the removal of 1 mole toluene required 5.82 mole oxygen and 0.446 mole nitrate (equation 3). Therefore, the consumption of 2.8 mg/L DO should have resulted in the removals of toluene and nitrate at 1.4 and 0.4 mg/L, respectively. However, the results indicate that the removals of toluene and nitrate were 7.4 and 22.5 mg/L, respectively. The removal of toluene and the consumption of nitrate were higher than the calculated stoichiometry values. In addition to the aerobic toluene removal, the results suggested that toluene and nitrate should be simultaneously removed by other mechanism. Because the reactors were seeded with both aerobic toluene degraders and denitrifers with the presence of both DO and nitrate, it might be possible that toluene was simultaneously removed by aerobic toluene degraders and denitrifers. The detachment of nitrate consumption used as nitrogen source during the stage of aerobic toluene removal (0.4 mg/L) from the total consumption of nitrate (22.5 mg/L) was the nitrate being used as electron acceptor by denitrifers (22.1 mg/L). According to the stoichiometry of denitrification, the consumption of nitrate at 22.1 mg/L should result in the toluene removal of 5.8 mg/L. Therefore, the aerobic toluene removal (1.4 mg/L) and denitrifying toluene removal (5.8 mg/L) were the theoretical total toluene removal (7.2 mg/L) during the first stage. The theoretical value (7.2 mg/L) was very close to the experimental result (7.4 mg/L). The results suggested that toluene was simultaneously removed by aerobic toluene degraders and denitrifers during the first stage with the presence of both DO and nitrate.

TABLE 2. Toluene removal and nitrate consumption in reactors using both nitrate and DO as the electron acceptor.

	Nitrate (mg/L)	Toluene (mg/L)	DO (mg/L)	pH
First Stage				
Initial Conc.	124.8	32.1	4.6	6.8
Final Conc. (after 2 days)	102.3	24.7	1.8	6.8
Removal	22.5	7.4	2.8	----
Second Stage				
Initial Conc.	102	24.7	1.7	6.8
Final Conc. (after 4 days)	21	<0.1	1.7	7.3
Removal	81	24.7	----	----
Overall Removal	103.8	32.1	----	----

※ overall nitrate utilization efficiency: 32.1/103.8 = 0.31

At the end of the first stage, the DO concentration was about 1.8 mg/L and was not further consumed. The lower DO could not effectively support the aerobic removal of toluene. Therefore, nitrate was the next available electron acceptor. When nitrate was the electron acceptor, toluene concentration significantly decreased from 24.7 mg/L to less than 0.1 mg/L. At the same time, nitrate concentration decreased from 102 to 21 mg/L. The denitrification process also could be indirectly confirmed from the increase in pH, from 6.8 to 7.3. The results showed that the consumption of 1 mg nitrate resulted in 0.29 mg toluene removal. The theoretical stoichiometry (equation 1) showed that the consumption of 1 mg nitrate caused 0.26 mg toluene removal. The experimental result (0.29 mg-toluene/mg-nitrate) was close to the calculated theoretical stoichiometry value (0.26 mg-toluene/mg-nitrate) (equation 1). However, this value (0.29 mg-toluene/mg-nitrate) is relatively higher than the value calculated from the modified stoichiometry equation (0.22 mg-toluene/mg-nitrate) (equation 2).

When nitrate was the only available electron acceptor during the denitrification stage using toluene as the substrate (Table 1), the results showed that the consumption of 1 mg nitrate resulted in 0.22 mg toluene removal. However, when the reactors contained a small amount of oxygen and abundant nitrate, the nitrate utilization efficiency was relatively better. According to the results in Table 2, the total nitrate consumption and toluene removal in the first and second stages were 103.8 mg/L and 32.1 mg/L, respectively. The presence of both a small amount of DO and sufficient nitrate resulted a better overall nitrate utilization efficiency (0.31 mg-toluene/mg-nitrate). However, the actual anoxic nitrate utilization efficiency during the second stage was 0.29 mg-toluene/mg-nitrate, which was still better than that obtained from the anoxic condition (equation 2) with the presence of only nitrate (0.22 mg-toluene/mg-nitrate). During the second stage, nitrate might not be the only available nitrogen source. During the denitrification stage, nitrogen released from the decayed aerobic cells could be used to support cell growth to reduce nitrate consumption.

SUMMARY

This work focused on the nitrate utilization efficiency in simulated groundwaters using nitrate as the only electron acceptor or simultaneously using both limited DO and nitrate as the electron acceptors. The theoretical stoichiometry equation indicated that the removal of 1 mg toluene required 3.82 mg nitrate, when nitrate was both the electron acceptor and nitrogen source. The results of denitrification study showed that the removal of 1 mg toluene consumed 4.47 mg nitrate by denitrifers. The experimental results suggested that the practical nitrate requirement was higher than the theoretical value. When the reactors contained both aerobic toluene degraders and denitrifers using both DO and nitrate as the electron acceptors, the removal of 1 mg toluene consumed 3.1 mg nitrate. Therefore, the presence of both DO and nitrate resulted in a better nitrate utilization efficiency, reducing nitrate consumption. After the depletion of DO in reactors containing both DO and nitrate, the removal of 1 mg toluene only consumed 3.28 mg nitrate. The results suggested that toluene could be simultaneously removed by aerobic toluene degraders and denitrifers using limited DO and nitrate as the electron acceptors. The nitrate was more effectively utilized in the reactors containing both aerobic cells and denitrifers in the presence of both limited DO and nitrate. When the reactor contained both limited DO and nitrate, the nitrate utilization efficiency was 0.31 mg-toluene/mg-nitrate compared to 0.22 mg-toluene/mg-nitrate with the presence of only nitrate.

REFERENCES

Aggarwal, P. K., J. L. Means, R. E. Hinchee. 1991. "Formulation of Nutrient Solutions for In Situ Bioremediation." In R. E. Hinchee and R. F. Olfenbuttel (Eds.), *In Situ Bioreclamation: Application and Investigations for Hydrocarbon and Contaminated Site Remediation*, pp. 51-66. Butterworth-Heinemann, Stoneham, MA.

Barbaro, J. R., J. F. Barker, and B. J. Butler. 1997. "In Situ Bioremediation of Gasoline Residuals Under Mixed Electron-Acceptor Conditions." In B C. Alleman and A. Leeson (Eds.), *In Situ and On-Site Bioremediation: Volume 5, The Fourth International In Situ and On-Site Bioremediation Symposium*, pp. 21-26. Battelle Press, Columbus, Ohio.

Barenschee, E. R., P. Bochem, O. Helmling, and P. Weppen. 1991. "Effectiveness and Kinetics of Hydrogen Peroxide and Nitrate-Enhanced Biodegradation of Hydrocarbons." In R. E. Hinchee and R. F. Olfenbuttel (Eds.), *In Situ Bioreclamation: Application and Investigations for Hydrocarbon and Contaminated Site Remediation*, pp. 103-124. Butterworth-Heinemann, Stoneham, MA.

Hutchins, S. R. and J. T. Wilson. 1994. "Nitrate-Based Bioremediation of Petroleum-Contaminated Aquifer at Park City, Kansas: Site Characterization and Treatability Study." In R. E. Hinchee, B. C. Alleman, R. E. Hoeppel, and R. N. Miller (Eds.), Hydrocarbon Bioremediation, pp. 80-92. Lewis Publishers, Boca Raton, FL.

Lu, C. J. 1994. "Effects of Hydrogen Peroxide on the In Situ Biodegradation of Organic Chemicals in a Simulated Groundwater System." In R. E. Hinchee, B. C. Alleman, R. E. Hoeppel, and R. N. Miller (Eds.), Hydrocarbon Bioremediation, pp. 140-147. Lewis Publishers, Boca Raton, FL.

Lu, C. J., L. C. Fan, and C. M. Lee. 1996. "The Utilization Efficiency of Hydrogen Peroxide on the Removal of Volatile Organic Acids in Sand Columns." *Wat. Sci. Tech. 34*(7/8):359-364.

Schreiber, M. E., J. M. Bahr, M. Zwolinski, Y. Shi, and W. J. Hickey. 1997. "Field and Laboratory Studies of BTEX Bioremediation Under Denitrifying Conditions." In B C. Alleman and A. Leeson (Eds.), *In Situ and On-Site Bioremediation: Volume 5, The Fourth International In Situ and On-Site Bioremediation Symposium*, pp. 13-18. Battelle Press, Columbus, Ohio.

IN SITU REMEDIATION OF A BTEX-CONTAMINATED AQUIFER BY ADDITION OF NITRATE

Hans-Peter Rohns (Stadtwerke Düsseldorf AG, Düsseldorf, Germany)
Paul Eckert, Jürgen Schubert (Stadtwerke Düsseldorf AG, Düsseldorf, Germany)
Harald Strauss, Frank Wisotzky, Peter Obermann (University of Bochum, Bochum, Germany)

ABSTRACT : A multitracer test using the conservative tracer pyranin and the reactive tracer nitrate was performed to investigate the reactivity of nitrate in a contaminated aquifer under field conditions. To start the field study the tracer solution (1,100 l) containing 25 kg KNO_3 and 1 kg pyranin had been injected simultaneously. Taking into account of the infiltrated tracer mass and the detected concentrations of pyranin the breakthrough curve of a conservative transported nitrate was calculated. The comparison of the calculated and the monitored breakthrough curve revealed that 85 % of the injected nitrate was reduced on the flow path. In the meantime the sulfate concentration rose up to 20 mg l^{-1}. Theoretically an increase of sulfate up to 50 mg l^{-1} had been expected, if nitrate had been totally consumed by pyrite oxidation. The predominant part of the injected nitrate was used by microorganisms for the degradation of the contaminants (BTEX).

INTRODUCTION

Aromatic hydrocarbons such as benzene, toluene, ethylbenzene and xylene (BTEX) and polyaromtic hydrocarbons (PAH) are biodegradable under aerobic and anerobic conditions. High degradation rates may be achieved in the presence of oxygen and oxidized iron (Lovely 1994). Under nitrate and sulfate reducing conditions the degradation rates are slower, particular substances reveal to be persistent.

At a former gasworks plant in Düsseldorf (Germany) massive soil and groundwater contaminations with BTEX were detected in a shallow unconfined aquifer. The study presents the results of investigations concerning natural assimilation processes and field studies, which were conducted in order to stimulate natural biodegradation processes by adding nitrate. A thorough analysis of the hydrochemical conditions (Schmitt et al. 1997, Wisotzky et al. 1998) proved essential for a successful approach, focussing especially on the possible oxidation of pyrite due to addition of nitrate. Therefore a multitracer test with pyranin (conservative tracer) and nitrate (reactive tracer) was performed.

NATURAL BIODEGRADATION PROCESSES IN THE PLUME

The contaminant plume extend about 600 m in the direction of the groundwater flow. A balance of electron acceptors upgradient and downgradient of the source area together with isotope investigations showed that the main part of the contaminants were degraded under sulfate reducing conditions (Wisotzky et al. 1998). Oxygen and nitrate are not detectable in the contaminated plume while the Fe(II) concentration increases up to 25 mg l^{-1}. The depth oriented sampling verified (Figure 1) that sulfate is consumed by anaerobic microorganisms. The microbial breakdown of the BTEX-aromatics resulted in an increase of dissolved organic carbon (DIC), organic acids (Schmitt et al. 1997) and δ^{34}S-sulfate in the remaining sulfate. The dissolved ferrous iron precipitates with sulfide and remains as a sulfur source in the aquifer.

Microbial sulfate reduction is associated with a distinct change in isotopic composition of sulfur and oxygen in the residues, unreacted sulfate towards more positive values. Residual sulfate concentrations about 50 mg/l are characterized by sulfur and oxygen isotope values as high as +41‰ and +15‰, respectively. This indicates an advanced level of sulfate reduction under closed system conditions with respect to sulfate availibility. In contrast, aquifer sulfate which is unaffected by bacterial reduction displays an average concentration of 175 mg/l and isotope values around +6‰ (sulfur) and +9‰ (oxygen), respectively.

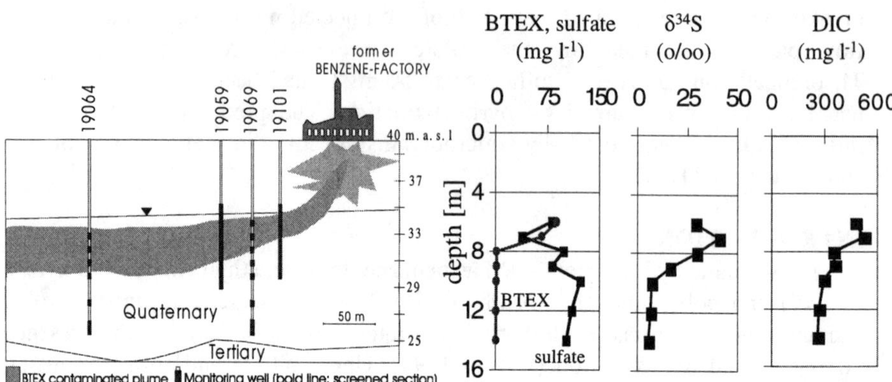

FIGURE 1. Cross section showing the extent of the contaminant plume and the vertical distribution of BTEX, sulfate, δ^{34}S-sulfate and DIC at the multilevel well 19,069

STIMULATION OF INTRINISIC BIOREMEDIATION

Monoaromatic hydrocarbons are easily biodegradable under aerobic conditions. Nitrate can also been applied for the remediation of BTEX aromatics in a contaminated aquifer. Introducing nitrate in an anaerobic aquifer to enhance the microbial activity may be negatively influenced by the oxidation of inorganic reduced species. In several field studies the oxidation of pyrite due to nitrate degradation (equation 1) was observed (Kölle et al. 1983; Postma et al. 1991).

$$5\,FeS_2 + 14\,NO_3^- + 4H^+ \rightarrow 7\,N_2 + 10\,SO_4^{2-} + 5\,Fe^{2+} + 2H_2O \quad (1)$$

A multitracer test using the conservative tracer pyranin and the reactive tracer nitrate was performed to investigate the reactivity of nitrate under field conditions. The test field (Figure 1) is located between the monitoring well 19059 and the injection well 19101. Initially 1.1 m³ of the tracer solution containing 25 kg KNO_3 and 1 kg pyranin had been injected within 30 minutes. The mass measurement of the infiltrated tracer (M_{pyr}, M_{NO3}) and the concentration of pyranin $C_{pyr}(x,y,z,t)$ in the field experiment were used to calculate the breakthrough curve of the conservatively transported nitrate c by equation 2.

$$C_{NO_3}^{kons}(x,y,z,t) = [C_{pyr}(x,y,z,t) / M_{pyr}] * M_{NO3} \qquad (2)$$

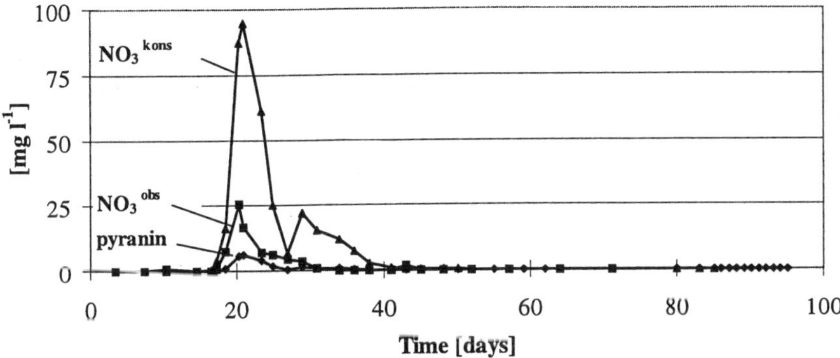

FIGURE 2. Measured pyranin, measured nitrate (NO_3^{obs}) and calculated conservative nitrate (NO_3^{kons}) breakthrough curves at the multilevel sampling point 19069/10m.

Figure 2 shows the calculated breakthrough curve and the measured nitrate concentrations $C_{NO_3}^{obs}(x,y,z,t)$ in the monitoring well 19069 at a depth of 10 m. The comparison of the concentrations reveals that 85 % of the injected nitrate was reduced on the flow path. Equation 3 describes the relation of the sulfate concentration and the amount of nitrate used for the oxidation of pyrite.

$$dC_{SO4}^{max} = 10/14*96.1/62 * (C_{NO_3}^{obs}(x,y,z,t) - C_{NO_3}^{obs}(x,y,z,t)) \qquad (3)$$

Figure 3 shows the layer 19069 at a depth of 9 m with the most significant mobilization of sulfate up to 20 mg l^{-1}. Theoretically an increase of the sulfate concentration approximately 50 mg l^{-1} had been expected, if nitrate had been consumed only by the oxidation of pyrite. However, FeS_2-Oxidation could also be observed in various layers. The predominant part of the injected nitrate was used by microorganisms for the degradation of the contaminants (BTEX) in the aquifer (Figure 4).

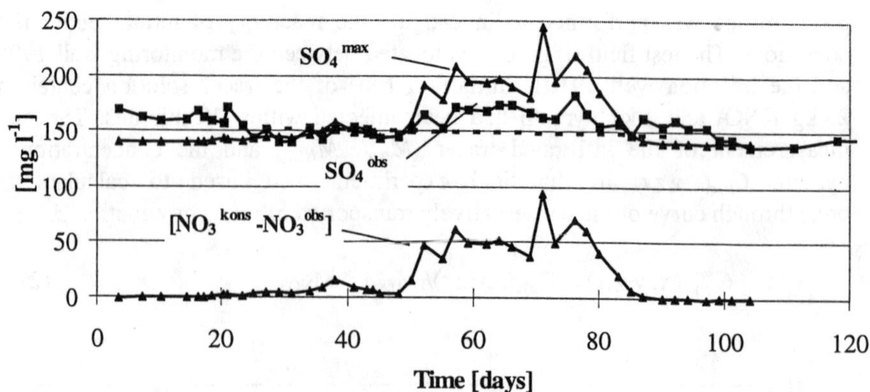

FIGURE 3. Reduced nitrate (NO_3^{kons}- NO_3^{obs}) , observed sulfate (SO_4^{obs}) and calculated sulfate (SO_4^{max}) breakthrough curves at the monitoring well 19069/9m.

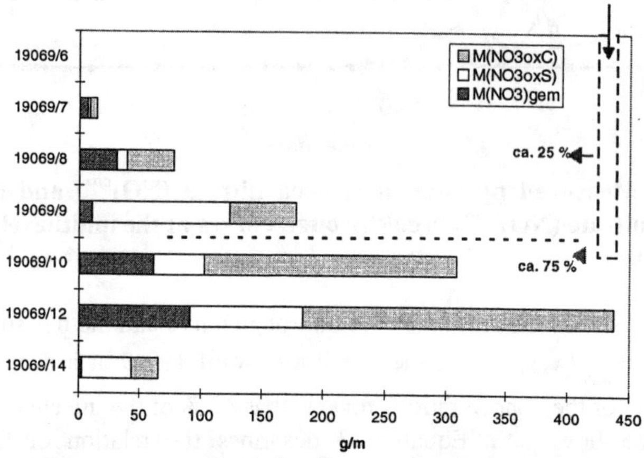

FIGURE 4. Mass balance of Nitrate used for the microbial breakdown of BTEX and the oxidation of sulfide

REFERENCES

Lovley, DR.; Woodward, JC.; and FH. Chapelle 1994. „Stimulated anoxic biodegradation of aromatic hydrocarbons using Fe (III) ligands." Nature 370, 128 - 131.

Kölle, W.; Werner, P.; Strebel, O.; and J. Böttcher 1983. „Denitrifikation in einem reduzierten Grundwasserleiter". Vom Wasser 61, 125 - 147.

Postma, D.; Boesen, C.; Kristiansen, H.; and F. Larsen 1991. „Nitrate reduction in an unconfind sandy aquifer: water chemistry, reduction processes, and geochemical modelling." Wat. Resour. Res. 27, 2027 - 2045.

Schmitt, R.; Langguth, H.R.; Püttmann, W.; Rohns, HP.; Eckert, P.; and J. Schubert 1997. „Biodegradation of aromatic hydrocarbons under anoxic conditions in a shallow sand and gravel aquifer pf the Lower Rhine Valley". Org. Geochem. 1(2), 41 - 50.

Wisotzky, F.; Eckert, P.; and P. Obermann 1998. „Hydrochemische Reaktionen im Bereich einer Grundwasserbelastung mit monoaromatischen Kohlenwasserstoffen und deren Modellierung." Neue Jb. Geol. Paläon. 208, 1 - 18.

INTRINSIC AND ENHANCED BIODEGRADATION OF BENZENE IN STRONGLY REDUCED AQUIFERS

W.N.M. van Heiningen[1], A.A.M. Nipshagen[2], J. Griffioen[3], A.G. Veltkamp[4], A.A.M. Langenhoff[1], and ***H.H.M. Rijnaarts***[1]

[1] TNO Institute of Environmental Sciences, Energy Research and Process Innovation, P.O.Box 342, 7300 AH, Apeldoorn, The Netherlands
[2] IWACO, Consultants for Water and Environment, Branch Office North, Groningen, The Netherlands
[3] NITG-TNO, the Netherlands Institute of Applied Geoscience
[4] NAM, Nederlandse Aardolie Maatschappij, Assen, The Netherlands

ABSTRACT. Intrinsic and enhanced benzene bioremediation has been investigated for three deeply anaerobic aquifers located in the northern part of the Netherlands. Field data could not evidence intrinsic benzene degradation. Laboratory microcosm studies with sediment and groundwater samples taken at five different locations were performed. Spontaneous intrinsic anaerobic benzene biodegradation does not occur in any of these samples during incubation times up to 450 days. The effect of addition of nitrate, sulphate, limited amounts of oxygen and combinations of these electron acceptors was also studied. In one out of the five series of samples, addition of nitrate initiated benzene biodegradation after lag-times greater than 70 days. Benzene degradation could be initiated in all samples by spiking stepwise small amounts of oxygen (25% of the reduction capacity of the aquifer materials). The level of oxygen required to initiate benzene degradation increases with the reduction capacity of the aquifer materials. Under these conditions, initial aerobic oxidation of the benzene appears to be followed by anaerobic processes that further degrade the formed oxidized intermediates. Sulphate and/or iron are most likely the electron acceptors used in this second step. Anaerobic bioremediation of strongly reduced benzene contaminated aquifers often seems to be limited by the absence of a microbial community adapted to anoxic benzene degradation. Bioaugmentation with natural benzene degrading microbial consortia from other sites may be an interesting option. Alternatively, aerated bioactivated zones may be another feasible remediation approach.

INTRODUCTION

Many sites exist that are contaminated with petroleum hydrocarbons, organic solvents or other organic mixtures. At such sites benzene, ethylbenzene, toluene and xylenes, are often the dominant groundwater contaminants. Benzene is in many cases the risk determining chemical due to its high mobility, its carcinogenicity and its recalcitrance towards degradation under anoxic conditions (Nipshagen *et al.*, 1996).

Many reports exist on natural attenuation of BTEX plumes in naturally oxic aquifers. In general, such plumes are found to be relatively small and often in a shrinking stage. For deeply reduced aquifers the situation is different. Benzene biodegradation under such conditions has been reported to occur and not to occur (Nales *et al.*, 1998). Often the autochtonous microbial community of such systems is not (yet) adapted towards anaerobic benzene degradation (Anderson *et al.*, 1998). This paper reports a microcosm investigation of three strongly reduced aquifers in the Netherlands contaminated with benzene. The purpose of the study is to select the optimal intrinsic or enhanced bioremediation approach for these sites.

MATERIALS AND METHODS

Groundwater chemistry and contaminant situation have been reported in another paper (Nipshagen *et al.*, 1997). Sample locations and characteristics are described in Table 1. The groundwater samples were collected in nitrogen flushed glass bottles with butylrubber/Teflon sealed screw caps. Sediment samples were taken after flushing the borehole with nitrogen. Sample cores were sealed and immediately transported to an anaerobic glovebox where further sample manipulation occurred. The reduction capacity (amount of oxygen per gram aquifer material required to completely oxidize the material) of the sediment and groundwater samples was determined using oxymax™ equipment. Anoxic laboratory microcosms were made with 120 cm^3 flasks crimp-sealed with 1 cm thick Viton-stoppers, containing 30 g of aquifer solids supplemented with 90 cm^3 groundwater leaving no headspace. Before capping, the batches were flushed with nitrogen. After capping, benzene and other TEX concentrations corresponding to the field situation were applied (Table 1). During the incubation period, liquid samples were taken and analyzed by GC-PID using an adsorptive fiber sampler (SPME). The triplicate microcosms were used for each condition tested, i.e. no additions (intrinsic degradation), nitrate addition (concentrations of 50, 100, and 150 mg/l), sulfate addition (concentrations of 50, 100, and 150 mg/l), and oxygen addition in steps of 25% of the reduction capacity.

Table 1. Sample characteristics and BTEX-concentrations in groundwater

Location	depth [m]	Benzene [µg/l]	Toluene [µg/l]	Ethylbenzene [µg/l]	Sum xylenes [µg/l]
1. Slochteren	18-19	11000	200	180	600
2. Slochteren	20-21	60	2	<0.2	4
3. Schoonebeek	12.2-13	700	5	110	40
4. Schoonebeek	3.5-4.5	120	100	390	480
5. Hoogezand	7.3-7.7	222	<0.1	<0.1	<0.1

RESULTS AND DISCUSSION

Typical results obtained with these microcosm studies are depicted in Figure 1. For all samples tested intrinsic benzene degradation was insignificant within incubation periods up to 450 days. Apparently the microbial communities in these sulfate-reducing to methanogenic sediments are not adapted to benzene degradation. In one out of five samples tested, an anaerobic benzene degradation could be induced by adding nitrate (sample 5). A lag phase of 70 days or more was observed in the three individual batches (Figure 1b) and other batches (not shown), and indicated the involvement of a microbial process. During the first part of the lag period, nitrate concentrations decreased with a few mg/l (data not shown), due to nitrate reducing conversion of organics other than benzene. After stabile nitrate levels were achieved, benzene degradation was observed. Possibly, easily degradable organics needed to be converted before benzene degradation could start up. In the other four samples tested, such a nitrate induced benzene degrading microbial activity appears to be absent. Addition of sulphate did not induce an anoxic benzene degradation, but promoted ethylbenzene degradation in sample 4 (not shown).

The possibility to initiate benzene biodegradation by adding limited amounts of oxygen was also tested. A typical result of such a test, shown in Figure 1C, demonstrates that a portion of the benzene is being degraded after each spike. In all sediment samples except one (sample 2) a similar result was obtained, indicating that aerobic benzene degradation can readily occur in aquifers that have been anaerobic for thousands of years. From reaction stoichiometry it follows that per mg oxygen applied, 0.26 mg benzene can be mineralized or 2.8 mg benzene can be oxidized to catechol via dioxygenase mediated reactions. The observed ratio's of amounts of benzene converted to oxygen added varied between 0.9 and 2.8 mg benzene/mg oxygen. This suggests that at least a part of the benzene is being converted to oxidized aromatics that may be further degraded through anoxic degradation processes. The contribution of anaerobic processes is even more likely when considering that significant amounts of oxygen are being used for the oxidation of the aquifer material and not for benzene degradation. Competition between benzene biodegradation and aquifer material oxidation is also indicated by the finding that the level of oxygen required to initiate benzene degradation increases with increasing reduction capacity of the aquifer materials.

Anaerobic bioremediation of strongly reduced aquifers contaminated with BTEX is for technical and cost-effectiveness reasons favorable over aerobic applications. The results of this study show that such an anaerobic bioremediation of these aquifers cannot easily be achieved by adding alternative electron acceptors like nitrate and/or sulphate. It appears that the autochtonous microbial consortia present are often not adapted to such anaerobic benzene degradation processes.

Enhanced adaptation, i.e. by inoculation with natural anaerobic benzene degrading microbial consortia from other sites, may be an interesting option. In such a way biologically activated benzene degrading zones may be created in naturally unadapted aquifers. The feasibility of such an approach still needs to be tested in the laboratory and in the field. When a completely anaerobic application is not feasible, an aerobic benzene degrading activated zone in which benzene is partially degraded followed by further anaerobic degradation may be an alternative approach. This requires a thorough investigation and quantification of the competition between benzene biodegradation versus sediment and groundwater oxidation for the applied oxygen. Reduction capacities of sediment and groundwater are crucial parameters and are needed for the engineering of aerobic pilots or full-scale applications.

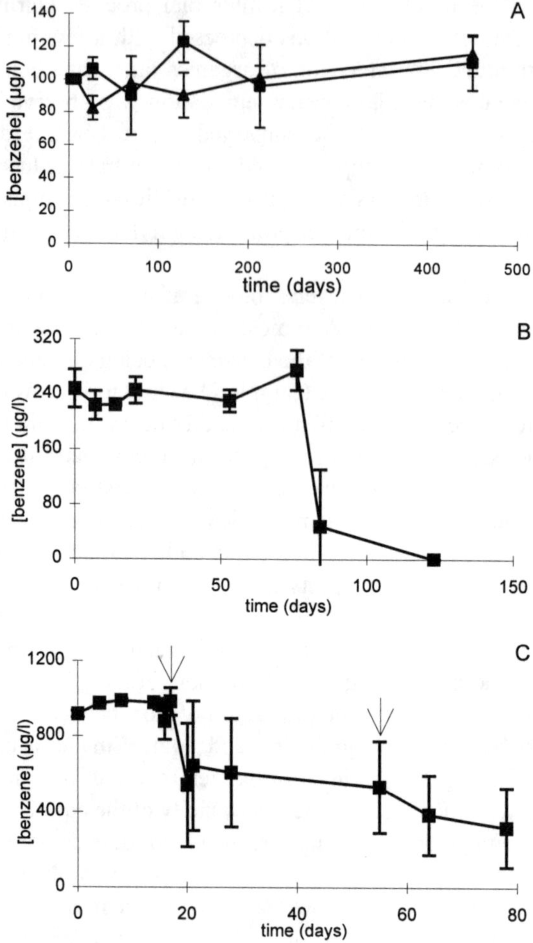

FIGURE 1. Examples of benzene biodegradation performance in anoxic triplicate microcosms: A) no addition (intrinsic degradation test;) versus sterile control, sample 1; B) nitrate addition, 50 mg/l nitrate, sample 5; and C) oxygen addition (arrows), in portions of 25% of the reduction capacity of the aquifer material, sample 1.

ACKNOWLEDGMENTS

Major parts of the work presented have been financed by NOBIS, the Netherlands Research Program on *In-Situ* Bioremediation, The NAM, and the Province of Groningen. Arcadis-Heidemij Consultancies and Oranjewoud Consultancies, the Netherlands, have contributed by providing groundwater and aquifer samples from some of the sites.

REFERENCES

Anderson, R. T., Rooney-Varga, J. N., Gaw, C. V. and Lovley, D. R. (1998). Anaerobic benzene oxidation in Fe(III) reduction zone of petroleum hydrocarbons. *Environmental Science and Technology* **32**, 1222-1229.

Nales, M., Butler, B. J. and Edwards, E. A. (1998). Anaerobic Benzene Biodegradation: a Microcosm Survey. *Bioremediation Journal* **2**, 125-144.

Nipshagen, A., Keuning, S., Baten, H.H.M., Kersten, R. H. B., Rijnaarts, H. H. M., Griffioen, j., and Doelman, p. (1997). Anaerobic bioremediation of BTEX contaminated sites, phase 1.1 NOBIS report 95-1-43 (In Dutch)

ACKNOWLEDGMENTS

Major parts of the work presented here were financed by [illegible] Research Program on In-Situ Bioremediation. The NAM and the [illegible] Arcadis Heidemij Consultancy and Oranjewoud Consultancy [illegible] have contributed by providing groundwater and aquifer samples from some of the sites.

REFERENCES

Anderson, R. T., [illegible] Rooney-Varga, J. N., Gaw, C. V. and Lovley, D. R. (1998) Anaerobic benzene oxidation in the Fe(III) reduction zone of petroleum contaminated aquifers. *Environ. Sci. Technol.* 32, 1222-[illegible]

[illegible], M., [illegible], B. J. and Edwards, E. A. (1998) Anaerobic benzene biodegradation: A microcosm survey. *Bioremediation Journal* 2, 135-144.

[illegible] Gerritse, J., [illegible] Rijnaarts, H. H. M. [illegible] (1996) Anaerobic biodegradation of BTEX contaminated sites [illegible] NOBIS report 96-1-04 (in Dutch).

ENHANCEMENT ON IN SITU AND ON-SITE BIOREMEDATION BY RADIOWAVE FREQUENCY SOIL HEATING

S. Hüttmann (Groth & Co., Itzehoe, Germany)
B. Angelmi (ARBES-Umwelt GmbH, Berlin, Germany)
H. Peters (Groth & Co., Itzehoe, Germany)
P. Jütterschenke (ARBES-Umwelt GmbH, Berlin, Germany)
L. Beyer (Institute of Soil Science, University of Kiel, Germany)

ABSTRACT: Temperature is a very important parameter affecting microbial activity in the soil. The application of radiowave frequency electromagnetic fields (RF) at 6.78 Mhz for soil heating to promote hydrocarbon degradation was studied in the laboratory as well as in small scale field studies. Microbial respiration, microbial biomass, viability of soil microorganisms, total petroleum hydrocarbon (TPH), and soil physical conditions were monitored.

The effect of the electromagnetic field on microbial respiration and viable cell counts largely depended on the duration and level of energy input. Soil respiration was generally enhanced in the radiowave frequency electromagnetic field at temperatures of up to 35°C. Temperatures above 45°C caused severe damage of the microbial biomass within a few hours. Long term RF-applications reduced microbial biomass by up to 80% and viable cell counts by up to 99% even if soil temperatures were kept below 40°C. Changes in soil physics were caused by water, contaminant and electrolyte movement from the exciter electrode towards the grounded electrode.

In order to prevent these unfavorable impacts on soil microorganisms and soil physical properties, the radiowave-frequency soil heating systems was applied in intervalls. Daily heating and cooling cycles induced a strong respiratory response, viable cell counts maintained fairly constant and higher hydrocarbon degradation rates were monitored. Energy consumption was reduced by up to 60 % compared to energy consumption with continuous heating.

Applied carefully, the radiowave frequency soil heating system accelerated biodegradation processes. In addition it promoted further breakdown of strongly degraded hydrocarbons in soils. The results suggest that the applied soil heating system is an appropriate tool to maintain favorable soil temperatures for high microbial activity even in cold regions in order to reach bioremediation aims quickly and economically.

INTRODUCTION

Temperature has a strong ecological impact on microbial processes (Brock and Madigan, 1991). In a range between 5 – 30°C biological respiration rates increase 2-4 fold with each 10°-temperature step. For this reason temperature is the minimum factor controlling TPH-degradation rates in bioremediation systems if all other parameters are set at their optimum. This is a typical situation in in situ and on-site bioremediation under cold climate conditions.

One way to heat up soil material homogeneously is to use radiowave frequency (RF) electromagnetic fields. Soil material within the range of the electromagnetic field is heated up by transformation of mechanical friction into thermal energy, a process called dielectric heating (Edelstein et al., 1994).

A major advantage of this method is the homogeneous temperature distribution within the range of the electromagnetic field compared to external heating systems.

Objective. The objective of this project was (a) to evaluate the possibilities of an application of the RF-technology for bioremediation purposes, (b) to study the impacts on soil microbiology and contaminant chemistry and (c) to optimize the efficiency of the system.

MATERIALS AND METHODS

For the RF-treatments in laboratory scale soil columns (outer diameter:200mm; inner diameter: 80mm; height: 40mm) were prepared and aerated with 2 – 3 l h^{-1} kg^{-1} of carbondioxid-free air. For the RF-electromagnetic fields a frequency of 6.78 MHz was chosen, which is reserved for industrual, scientific and medical applications (ISM-frequency). They were generated by a SEG 100 RF-generator at 0.1 kW (former VEB Funkwerk Köpenick, Berlin). The exciter electrode was located in the column center, the grounded electrode surrounded the outside of the glass column. Both electrodes were water-cooled to allow longer RF-incubation times for the soil materials.

In technical scale the triangular-shaped biopiles had the dimensions 1.8 x 1.0 m (footlength) and 0.45 m in height. The RF-electromagnetic field was generated with a Hüttinger PFG 5000 RF generator at 0.5 – 2.5 kW with different electrode configurations.

Microbial respiration was measured using the Isermeyer-method based on CO_2-absorbtion for soil samples in the lab(Alef, 1994) or a portable infrared-based CO_2-detection system for in-situ-measurements of soil respiration in the biopile (Blanke and Bacher, 1997). Microbial biomass was calculated using the substrate-induced respiration (Anderson and Domsch, 1978) and viable cell counts determined with the microdrop method (Drews, 1974) in a nutrient broth and a mineral oil hydrocarbon-containing broth..

Total Petroleum Hydrocarbons (TPH) were measured in a soil extract with 1,1,2-trichlorotrifluoroethane by FTIR.

RESULTS AND DISCUSSION

Short term incubation experiments. In 2-hour-RF-incubation experiments hydrocarbon-contaminated soil material was heated up to temperatures of 25, 35, 45 and 55°C. Soil respiration rates were generally positively correlated with increasing temperature up to 35°C and declined gradually at higher temperatures. There were no detectable negative effects on a 10%-level for viable cell counts of heterotrophic bacteria at temperatures up to 35°C (table 1, experiment 1 – 4).

However, an exposure of soil material in a RF-electromagnetic field at 55°C for 2 hours lead to a microbial survival rate of less than 5% according to viable cell counts. The reduction of microbial biomass measured by substrate induced respiration was less severe.

Long term incubation experiments. The effects of a RF-exposure of different soil material on soil respiration, microbial biomass and viable cell counts for various soil materials and different incubation times are shown in table 1, experiment 5 - 12. In most experiments there was an increase in soil respiration rate at 35-38°C under radiowave exposure compared to respiration at 25°C. With incubation times exceeding 100 hours there was a considerable reduction of the microbial biomass determined by SIR of up to 81% and a drastic reduction of viable cell counts of up to 99% (table 1). This effect was also observed in soil material incubated at 35°C in the oven for comparable periods of time. This is possibly due to a lack of temperature-adapted microorganisms.
In the incubation experiments exceeding more than a few hours there was a considerable movement of water, salts and mineral oil components towards the grounded electrode caused by a side effect called steam distillation. Water losses within the soil column induced a change in the impedance of the soil material. This made a constant adjustment of the RF-electromagnetic field necessary.

TABLE 1: Effects of RF-electromagnetic fields on soil respiration, microbial biomass and viable cell counts for various soil materials (soil type: s. = sandy, w. = weak, l. = loamy, si. = silty; RF-application: c = constant RF-electromagnetic field; i = RF-electromagnetic field in daily intervalls; n.d. = not determined)

No. of Exp.	Soil material*	Incub. time [h]	RF-energy input/soil [kWh/kg]	RF-applic.	Max. temp. [°C]	change in soil resp. [%]	change in biomass [%]	change in viable cell cts. [%]
1	s. loam	2	0.04	c	25	40	0	-10
2	s. loam	2	0.08	c	35	125	0	17
3	w.l.sand	2	0.04	c	25	-13	-5	66
4	w.l.sand	2	0.08	c	35	17	-20	37
5	l. sand	140	0.32	c	37	100	-10	-41
6	s. loam	72.5	0.57	c	37	0	-2	-71
7	s. loam	164	0.43	c	37	11	-18	**-95**
8	sand	170	0.29	c	38	0	-81	**-99**
9	si. sand	528	1.40	c	38	43	-53	**-96**
10	sand	430	0.24	i	38	n.d.	-67	-59
11	l. sand	386	0.22	i	35	n.d.	-20	-26
12	si. sand	264	0.25	i	35	n.d.	3	-44

TPH-degradation tests in the RF-electromagnetic field had to take into account the uneven spatial distribution of TPH in the soil column resulting from steam distillation processes.TPH-degradation could therefore only be calculated by a hydrocarbon mass balance of the whole soil column. The loss of volatile TPHs was neglectable. TPH-degradation rates as measured in the RF-electromagnetic field and compared to degradation rates at 25°C and at 35°C oven temperature are shown in table 2. TPH degradation rates were generally higher at an incubation temperature of 35°C compared to 25°C. In experiment 12, which represents a strongly aged petroleum hydrocarbon contamination, microbial degradation processes were only detected at temperatures above 25°C. Comparative studies with soil material incubated at a constant temperature of 35°C in the oven and at daily heating cycles by RF-electromagnetic fields up to 38°C (table 2, experiment 10) showed comparable TPH degradation rates. The temperature-enhanced degradation process seems to be independent of the heat source and of the duration of daily heating cycles.

TABLE 2: Total petroleum hydrocarbon degradation rates in different soil materials at different incubation temperatures (* = see table 1; n.d. = not determined)

No. of experi.	Soil material*	incub. time [d]	TPH degradation rate [mg/kg x d] at 25°C	at 35°C in oven	at 25 - 35°C RF-intervalls
10	sand	18	19	32	30
11	l. sand	16	0	n.d.	11
12	si. sand	14	15	n.d.	21

Technical scale experiments. For the experiments in technical scale the commonly used triangular-shaped biopile formation was chosen. An adequate electrode configuration in the biopile is required to reduce spatial variability of temperature in the soils. For this reason the primarily applied stick electrodes were replaced by surface electrodes. These were installed in form of a metallic plate on the biopile bottom and in form of a metallic textile on the top of the biopile. The resulting spatial temperature distribution after 3 hours of continuous heating is shown in figure 1.

The slightly inhomogenous electromagnetic field on the edges of the exciter electrode lead to a considerable loss of 2.4% water (25% total loss) in the adjacent soil material leading to unfavourable soil conditions for microbial activity.

The resulting impacts on basal respiration, microbial biomass and viable cell counts were less severe than in the laboratory scale soil column experiments of comparable RF-exposure. Following a 4.25 hour RF-exposure soil respiration measured in situ increased from undetectable values (<0.0 g CO_2 m^{-2} h^{-1}) to values between 0.7 to 1.4 g CO_2 m^{-2} h^{-1} in the center of the biopile. Basal respiration of soil samples measured at constant temperature (22°C) before and after RF-exposure increased by 50%. Microbial biomass (SIR) decreased by 25% and no significant change was observed in the viable cell counts.

Thus it is technically possible to heat even unfavourably shaped biopiles more or less homogeneously. Nevertheless, special attention must be paid to the soil physical effects of drying out as well as salt and contaminant movement within the biopile towards the grounded electrode.

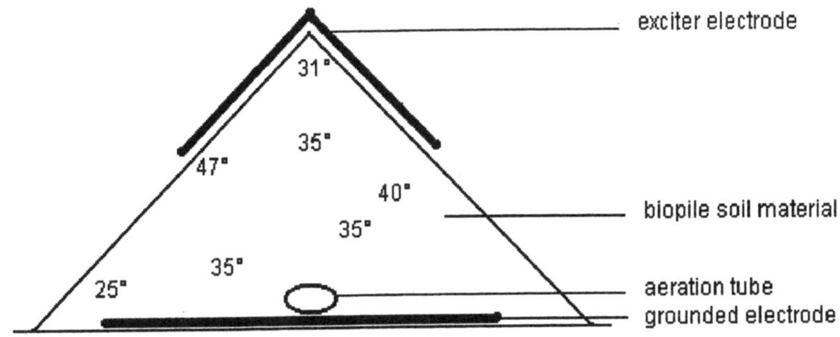

FIGURE 1: Temperature distribution in a technical scale biopile with top-bottom electrode configuration

Conclusion and future aspects. Radiofrequency-electromagnetic fields can successfully support temperature-dependent biological TPH-degradation processes in soil materials. However, expert knowledge is required to establish a homogeneous electromagnetic field by choosing an adequate electrode configuration. This is necessary due to a high thermal sensitivity of soil microorganisms with respect to the small degree between promoting and killing of microbes. The application of the RF-technology for other biotechnological systems located outdoors, such as in situ bioremediation, biofilters, biobarriers, and biological wastewater treatment systems looks promising and will have to be studied in the near future.

REFERENCES

Alef, K. 1994. Biologische Bodensanierung: Methodenhandbuch. VCH-Verlagsgesellschaft mbH, Weinheim, Germany.

Anderson, J.P.E. and K.H. Domsch.1978. „A Physiological Method for the Quantitative Measurement of Microbial Biomass in Soil". *Soil Biol. Biochem.* 10: 215-221.

Blanke, M.M. and W. Bacher. 1997. Atmung eines Bodens im Gemüsebau zu Beginn der Vegetationsperiode. *Zeitschrift für Pflanzenernährung und Bodenkunde.* 160: 52-55.

Brock, T.D. and M.T. Madigan. 1991. Biology of Microorganisms. 6th ed. Prentice Hall, Englewood Cliffs, New Jersey.

Drews, G. 1974. Mikrobiologisches Praktikum. 2nd edition. Springer-Verlag, Berlin, New York.

Edelstein, W.A., I.E.T. Iben, O.M. Mueller, E.E. Uzgiris, H.R. Philipp, and P.B. Roemer. 1994. Radiofrequency Ground Heating for Soil Remediation: Science and Engineering. *Environmental Progress.* 13. 247-252.

BIODEGRADATION OF FUEL OIL UNDER LABORATORY AND ARCTIC MARINE CONDITIONS

Robert M. Garrett, Copper E. Haith, and Roger C. Prince.
(Exxon Research and Engineering Co., Annandale, New Jersey)

ABSTRACT: An international consortium carried out a field trial of bioremediation on an Arctic beach on Spitsbergen (78°N, 17'E). An intermediate fuel oil was applied to a beach, and slow release and soluble fertilizers were added to stimulate microbial degradation of the oil. Laboratory degradation studies using the same oil and a marine sediment inoculum from Spitsbergen were carried out in parallel.

Oil biodegradation by microorganisms from the Spitsbergen marine environment follows a similar pattern to that seen with microorganisms from temperate Atlantic and Pacific environments. n-Alkanes are degraded before iso-alkanes, smaller aromatic molecules are degraded before larger ones, and alkylated polycyclic aromatic hydrocarbons are degraded more slowly than the parent compounds. These patterns allow us to identify biodegradation in samples collected from the field, and to make estimates of the extent of biodegradation that occurred in the field trial.

INTRODUCTION

Oil that enters the marine environment is subject to a variety of natural processes. Photooxidation may play a small role in removing oil from the environment (Garrett *et al.*, 1998) but the majority of the oil is eventually removed by biodegradation (National Academy of Sciences, 1985). Bioremediation by adding fertilizers has proven to be an environmentally acceptable and cost-effective way of stimulating this natural process in temperate climates (Prince, 1993, Lee *et al.*, 1995, Swannell *et al.*, 1996, Venosa *et al.*, 1996, Prince and Bragg, 1997), and there have been strong indications that similar results can be expected in the Arctic (Sendstad *et al.*, 1982, 1984; Sveum, 1987a,b; Sveum and Ladousse, 1989).

A multinational collaborative project entitled *In Situ Treatment Of Oiled Shoreline Sediments* (Sergy *et al.*, 1998) has addressed the potential that bioremediation by adding fertilizer stimulates oil biodegradation on marine shorelines in Arctic climates (Prince *et al.*, this meeting). As a corollary of this work, we have carried out laboratory studies with the same oil using microbial inocula from shorelines that had been oiled and treated as part of this project. We have monitored the progress of biodegradation over 12 weeks, analyzing the residual oil at various times by gas chromatography coupled with mass spectrometry (GC/MS, see Garrett *et al.*, 1998). The experiment used saline

Bushnell-Haas medium (Bushnell and Haas, 1941) to ensure that nitrogen and phosphorus nutrients were not limiting these laboratory experiments. Eight flasks were used, and duplicates were completely extracted after 1, 3, 6 and 12 weeks (see Garrett *et al.*, 1998). These laboratory experiments were done at room temperature since most marine Arctic microorganisms are known to be mesophilic, and this allowed extensive biodegradation in a reasonable time frame. Other experiments, where the temperatures were maintained at 6°C have shown that low temperatures slow the rate of biodegradation, but not its final extent.

RESULTS AND DISCUSSION

Figure 1 shows the total ion chromatograms of the initial oil, and oils extracted after 1, 3, 6 and 12 weeks of incubation.

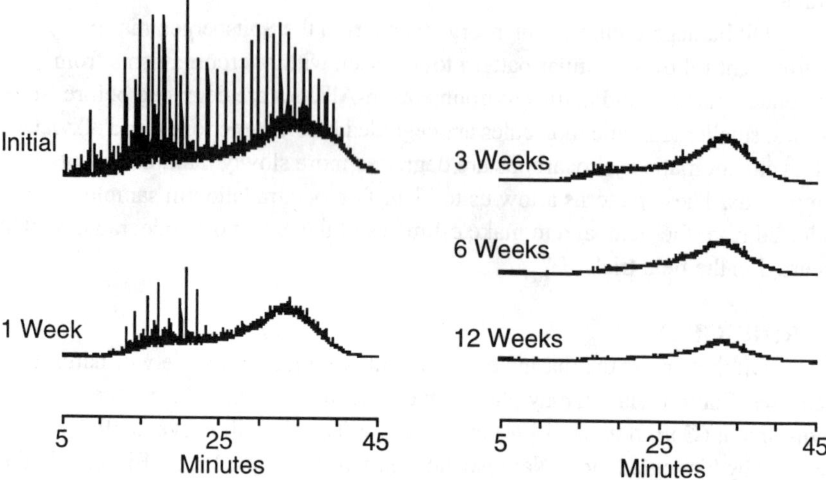

FIGURE 1. Total ion chromatograms of initial oil, and oil extracted from cultures incubated for the indicated time.

The resolved peaks in the initial oil are the n-alkanes and iso-alkanes, and the former have been completely degraded within 1 week under these conditions. The iso-alkanes are essentially completely degraded within 3 weeks.

Aromatic hydrocarbons are not seen as resolved peaks in the total ion scans of Figure 1, but they are readily resolved by selective ion monitoring. Figure 2 shows the progress of biodegradation of phenanthrene (a three-ring aromatic hydrocarbon) and its alkylated derivatives; as expected (e.g. Elmendorf *et al.*, 1994) the parent compound is degraded most readily, followed by the methyl-phenanthrenes, dimethyl- and ethyl-phenanthrenes and trimethyl-, methyl-ethyl- and propyl-phenanthrenes. Quantitation of biodegradation was achieved using hopane as a conserved internal marker that remained unchanged during the

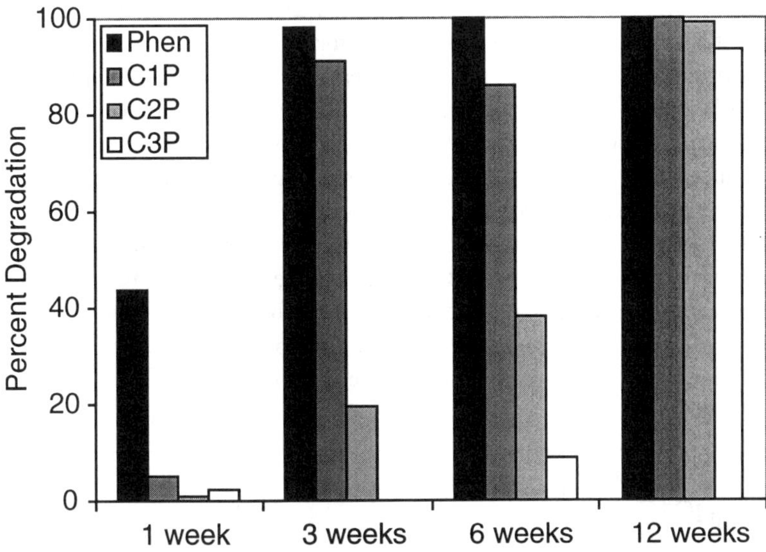

FIGURE 2. Biodegradation of phenanthrene (Phen), methyl-phenanthrenes (C1P), dimethyl- and ethyl-phenanthrenes (C2P) and trimethyl, methyl-ethyl- and propyl-phenanthrenes (C3P), based on hopane as a conserved internal marker.

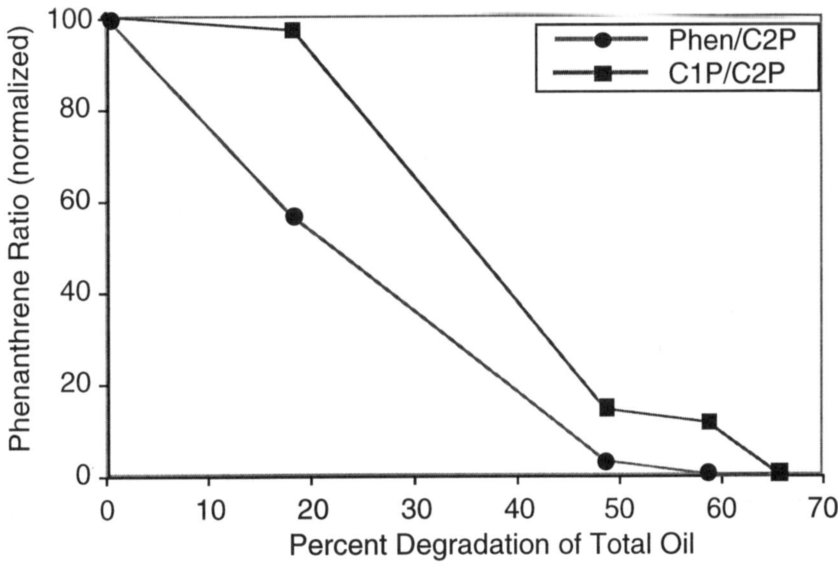

FIGURE 3. Changes in the ratio of phenanthrene to the dimethyl and ethyl phenanthrenes (Phen/C2P) and methyl phenanthrenes to the dimethyl and ethyl phenanthrenes (C1P/C2P) as a function of total oil degradation based on hopane as a conserved internal marker.

experiment (Prince *et al.*, 1994). A very similar pattern of degradation was seen for dibenzothiophene and its alkyl-substituted forms.

As discussed in the paper on the field part of the Svalbard experiment (Prince *et al.*, this meeting), the oil applied to the shoreline was not the same for all the plots., and the concentration of hopane in the oils on different parts of the plot were quite varied. The ratio of phenanthrene to dimethyl- and ethyl-phenanthrenes was, however, similar in all the oils. We thus investigated the possibility of using changes in this ratio as a surrogate for monitoring biodegradation. Figure 3 shows how this ratio, and the ratio of methyl-phenanthrenes to dimethl- and ethyl-phenanthrenes, changes as biodegradation proceeds. It is clear that the two ratios provide a reasonably sensitive indicator of biodegradation up to 50% loss of total GC-detectable oil.

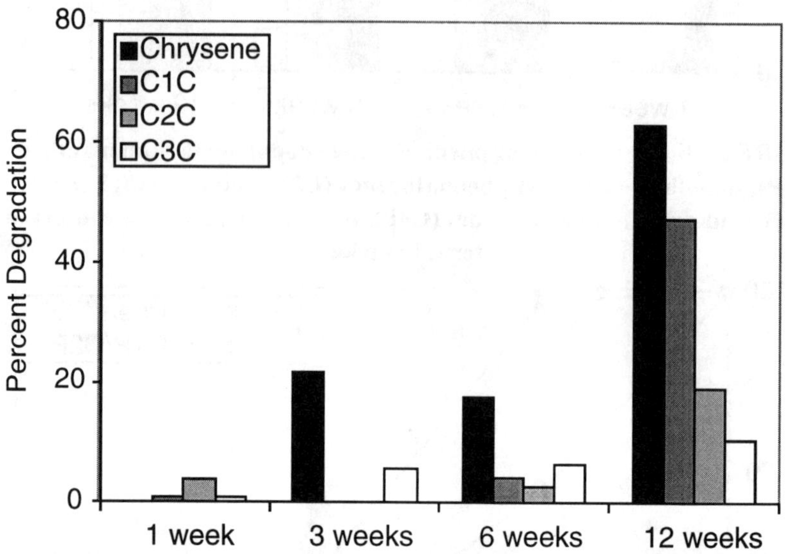

FIGURE 4. Biodegradation of chrysene, methyl- chrysenes (C1C), dimethyl- and ethyl- chrysenes (C2C) and trimethyl, methyl-ethyl- and propyl-p chrysenes (C3C) based on hopane as a conserved internal marker.

As expected, the biodegradation of chrysene (a four-ring aromatic hydrocarbon) lagged behind that of phenanthrene and dibenzothiophene. Nevertheless, once biodegradation began, it followed a very similar pattern (Figure 4).

CONCLUSIONS

The pattern of oil biodegradation by indigenous microorganisms on an Arctic shoreline is very similar to that exhibited by microorganisms from temperate Atlantic and Pacific marine environments (e.g. Elmendorf *et al*, 1994). n-Alkanes are degraded before iso-alkanes, smaller aromatic molecules are degraded before larger ones, and alkylated polycyclic aromatic hydrocarbons are degraded more slowly than the parent compounds. These patterns allow us to identify biodegradation in samples collected from the field, and to make estimates of the extent of biodegradation that has occurred. In particular, changes in the ratio of phenanthrene to dimethyl- and ethyl-phenanthrenes are useful indicators of the extent of biodegradation of an intermediate fuel oil up to 50% removal of the total GC-detectable hydrocarbon.

ACKNOWLEDGMENTS

The *In situ* Treatment of Oiled Sediment Shorelines Program was sponsored by an international partnership of spill response and research agencies composed of; Canadian Coast Guard, Environment Canada, Exxon Research and Engineering Co., Fisheries and Oceans Canada, Imperial Oil Canada, Marine Pollution Control Unit (UK), Minerals Management Service (USA), Norwegian Pollution Control Authority, Swedish Rescue Service Agency and the Texas General Land Office. We are indebted to these organizations, and the field crews who helped apply the oil.

REFERENCES

Bushnell, L. D. and Haas, H. F. 1941. "The utilization of certain hydrocarbons by microorganisms." *J. Bacteriol.* **41**, 653-673.

Elmendorf, D. L., Haith, C. E., Douglas, G. S. and Prince, R. C. 1994. "Relative rates of biodegradation of substituted polycyclic aromatic hydrocarbons." In *Bioremediation of Chlorinated and Polycyclic Aromatic Hydrocarbon Compounds.* (R. E. Hinchee, A. Leeson, L. Semprini and S. K. Ong, eds.) Lewis Publishers, Boca Raton, FL. pp. 188-202.

Garrett, R. M., Pickering, I. J., Haith, C. E. and Prince, R. C. 1998. "Photo-oxidation of crude oils." *Environ. Sci. Technol.* **32**, 3719-3723.

Lee, K., Tremblay, G. H. and Cobanli, S. E. 1995. "Bioremediation of oiled beach sediments: assessment of inorganic and organic fertilizers." In *Proceedings of the 1995 International Oil Spill Conference*, American Petroleum Institute, Washington DC, pp. 107-113.

National Research Council 1985. *Oil in the Sea: Inputs, Fates and Effects*, National Academy of Sciences, Washington DC

Prince, R. C. 1993. "Petroleum spill bioremediation in marine environments." *Critical Reviews Microbiology 19*, 217-242.

Prince, R. C., Elmendorf, D. L., Lute, J. R., Hsu, C. S., Haith, C. E., Senius, J. D., Dechert, G. J., Douglas, G. S. and Butler, E. L. 1994 17α(H),21β(H)-hopane as a conserved internal marker for estimating the biodegradation of crude oil. *Environ. Sci. Technol.* **28**, 142-145.

Prince, R. C. and Bragg, J. R. 1997. "Shoreline bioremediation following the *Exxon Valdez* oil spill in Alaska." *Bioremediation J.* **1**, 97-104.

Sendstad, E., Hoddo, T., Sveum, P., Eimhjellen, K., Josefsen, K., Nilsen, O. and Sommer, T. 1982. "Enhanced oil degradation on an arctic shoreline." In *Proceedings of the Fifth Arctic Marine Oilspill Program Technical Seminar*, Edmonton, Alberta, pp. 331-340.

Sendstad, E., Sveum, P., Endal, L. J., Brattbakk, Y. and Ronning, O. I. 1984. "Studies on a seven-year old seashore crude oil spill on Spitsbergen." In *Proceedings of the Seventh Arctic Marine Oilspill Program Technical Seminar*, Edmonton, Alberta, pp. 60-74.

Sergy, G. A., Guénette, C. C., Owens, E. H., Prince, R. C. and Lee, K. 1998. "The Svalbard shoreline oilspill field trials." In *Proc. 21st Arctic and Marine Oil Spill Program Seminar*. Environment Canada, pp. 873-889.

Sveum, P. 1987a. "Fate and Effects of Dispersed and Non Dispersed Oil on Arctic Mud Flats." In - *Proceedings of the Tenth Arctic and Marine Oilspill Program Technical Seminar,* Edmonton, Alberta, pp. 149-167.

Sveum, P. 1987b. "Accidentally spilled gas oil in a shoreline sediment on Spitsbergen: natural fate and enhancement of biodegradation." In *Proceedings of the Tenth Arctic Marine Oilspill Program Technical Seminar*, Edmonton, Alberta, pp. 177-192.

Sveum, P. and Ladousse, A. 1989. "Biodegradation of oil in the Arctic: enhancement by oil-soluble fertilizer application." In *Proceedings of the 1989 International Oil Spill Conference*, American Petroleum Institute, Washington DC, pp. 439-446.

Swannell, R. P. J., Lee, K. and McDonagh, M. 1996. "Field evaluations of marine oil spill bioremediation." *Microbiol. Rev. 60*, 342-365.

Swannell, R. P. J., Lee, K., Basseres, A. and Merlin, F. X. 1994. "A direct respirometric method for in-situ determination of bioremediation efficiency." In *Proceedings of the Seventeenth Arctic Marine Oilspill Program Technical Seminar*, Environment Canada, Ottawa, pp. 1273-1286.

Venosa, A. D., Suidan, M. T., Wrenn, B. A., Strohmeier, K. L., Haines, J. R., Eberhart, B. L., King, D. and Holder, E. 1996. "Bioremediation of an experimental oil spill on the shoreline of Delaware Bay." *Environ. Sci. Technol. 30*, 1764-1775

EFFECTS OF NUTRIENTS ON MICROBIAL ACTIVITY IN CONTAMINATED ARCTIC GRAVEL PADS

Vikram N. Chaobal (University of Alaska Anchorage, Anchorage, AK)
Craig R.Woolard (University of Alaska Anchorage, Anchorage, AK)
Daniel M. White (University of Alaska Fairbanks, Fairbanks, AK)

ABSTRACT: Experiments were conducted to evaluate the effect of nutrient addition on the oxygen uptake rates (OUR) of petroleum contaminated, nutrient-limited, cold-region cryic soils. The effects of nitrogen and phosphorus addition to soils at 5°, 10°, and 15°C increased the OUR of soils collected from a hydrocarbon contaminated gravel pad on the North Slope of Alaska. Laboratory studies indicate that application of 20-25 mg/kg N and 2-2.5 mg/kg P result in a several-fold increase in soil OUR. Nutrient application early in the treatment season, when soil temperatures average 5° C, can increase the OUR when soils warm later in the treatment season.

INTRODUCTION

Alaska has many of gravel pads on the North Slope, some of which are contaminated with petroleum hydrocarbons. Remediation of these contaminated sites is expensive due to remote logistics associated with mobilizing equipment and materials. Passive remediation systems can result in substantial savings. However, there are concerns about the effectiveness of these treatment technologies under relatively harsh arctic conditions. Environmental parameters such as cold temperatures, nutrient-limited soils, and short treatment seasons contribute to the low biodegradation rates typically observed at these sites. Many of these environmental conditions cannot easily be controlled. Soil characteristics such as inorganic nitrogen concentration, however, can be manipulated at a relatively low cost. In this research, studies were conducted to evaluate the effectiveness of nutrient amendments on biodegradation rates in petroleum contaminated, nutrient-limited, cold-region gravel soils. This paper presents a portion of the experiments conducted to determine the effect of low concentrations of nutrients added to soils incubated at typical field temperatures.

MATERIALS AND METHODS

Soils used in the laboratory experiments were collected from Child's Pad, a typical North Slope gravel pad in Deadhorse, Alaska. Soils from Child's Pad were amended with varying concentrations of ammonium nitrate (NH_4NO_3) and triple superphosphate ($Ca(H_2PO_4)_2$. Oxygen uptake rates in soil samples from Child's Pad were measured using a batch reactor system. Between 9 and 10 kg of soil were added to a 6-inch diameter, 5.5 L plastic cylinder or cell. Each cell was equipped with a Figaro Model KE-50 oxygen sensor connected to a data logger that recorded soil gas oxygen levels (as % oxygen) on regular intervals (usually every 0.25 to 1 hour).

When the oxygen concentration in the reactor was depleted to approximately 10% by microbial activity compressed air was passed through the soil cells to restore oxygen levels to 17 to 21%. Oxygen uptake tests resumed were resumed after reaeration. During oxygen uptake tests, a linear reduction in oxygen concentration was generally observed. The oxygen utilization rate was determined by taking the slope of the oxygen concentration data. The oxygen uptake rate (OUR) was calculated in mg of oxygen consumed per kg of dry soil per hour following the method of Leeson and Hinchee (1996).

Oxygen uptake tests were conducted in 3 separate experiments. In experiment #1, two cells were prepared. One cell was amended with nutrients (25 mg N/kg dry soil and 2.5 mg P/kg dry soil), and one cell served as the control (no additional N or P added). Nutrients were added in distilled water. The initial DRO concentration was 2,000 mg/kg and the moisture content 6%. These cells were incubated at 5°C for six days and the oxygen concentration monitored continuously at 1 hr intervals. After 6 days, the cells were incubated at 10°C for another 4 days. After ten days, the cells were incubated at 15°C for 6 days. The total experimental time was 16 days.

In experiment #2, two soil reactors were prepared and incubated at 5°C. One cell served as a control in which no nutrients were added. A second cell was amended with 25 mg N/kg dry soil and 2.5 mg P/kg dry soil. Nutrients were added in distilled water, and the control received an equal volume of distilled water (without nutrients) as the amended cell. The initial DRO concentration of the soils was 6,100 mg/kg, the soil moisture content 6%, and the pH 7.5. The oxygen uptake rate was monitored over a 22-day period on each cell.

In experiment #3, three soil reactors were incubated at 10°C. A soil reactor with no nutrient amendments served as a control and two duplicate cells contained 21 mg N/kg dry soil and 2 mg P/kg dry soil. The moisture content of the soils was 7% and the DRO concentration approximately 1,000 mg/kg. Soils were incubated for a 30-day period in which the oxygen uptake rate was monitored.

RESULTS

The contaminated soil at Child's Pad was characterized as a well-graded gravel with sand. A chemical analysis of the material indicated that the soil contained 1.7% total carbon, 0.025% total nitrogen, and less than 1 mg/kg of mineralizable nitrogen and available phosphorous. These results indicate that the soils are nutrient poor and that the organic nitrogen present in the site is largely unavailable for microbial use. As shown in Figure 1, soil temperatures at Child's Pad remained relatively low, ranging from 4.5° to 11.7° C, throughout most of the remediation season in 1997.

The results of experiment #1 are shown in Figure 2. OURs of less than 10 mg O_2/kg dry soil/day were observed at 5°C for both the control and nutrient amended soil. An increase of the soil temperature to 10°C resulted in an OUR increase of 6-8 mg O_2/kg dry soil/day in the cell amended with nutrients, but did not result in an increase in the control. When the soil temperature was increased to 15°C, the OUR of the cell amended with nutrients increased 3-fold to 30-33 mg O_2/kg dry soil/day.

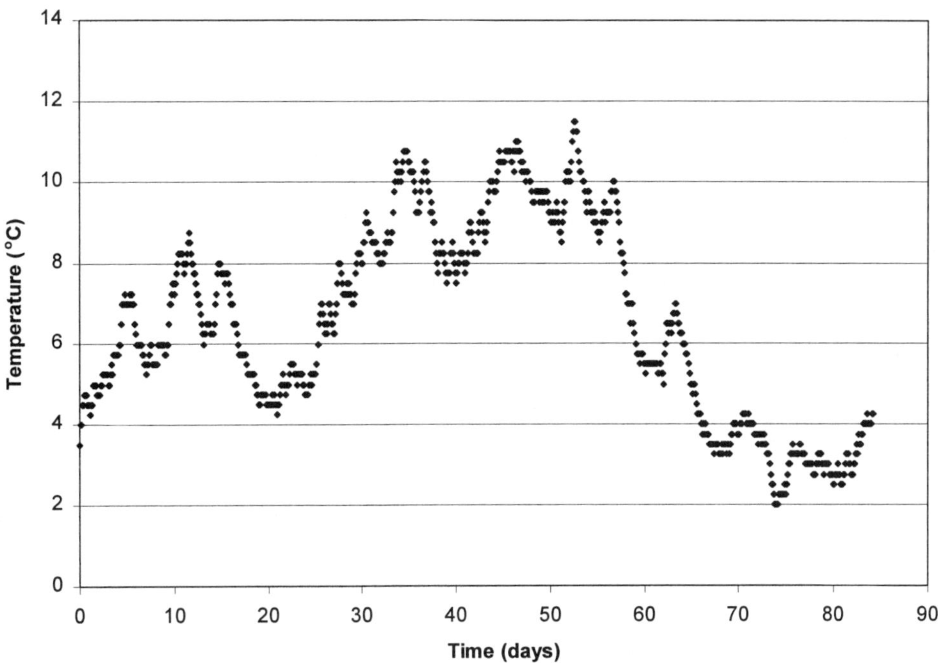

Figure 1. Soil Temperatures at Child's Pad (July-September, 1997)

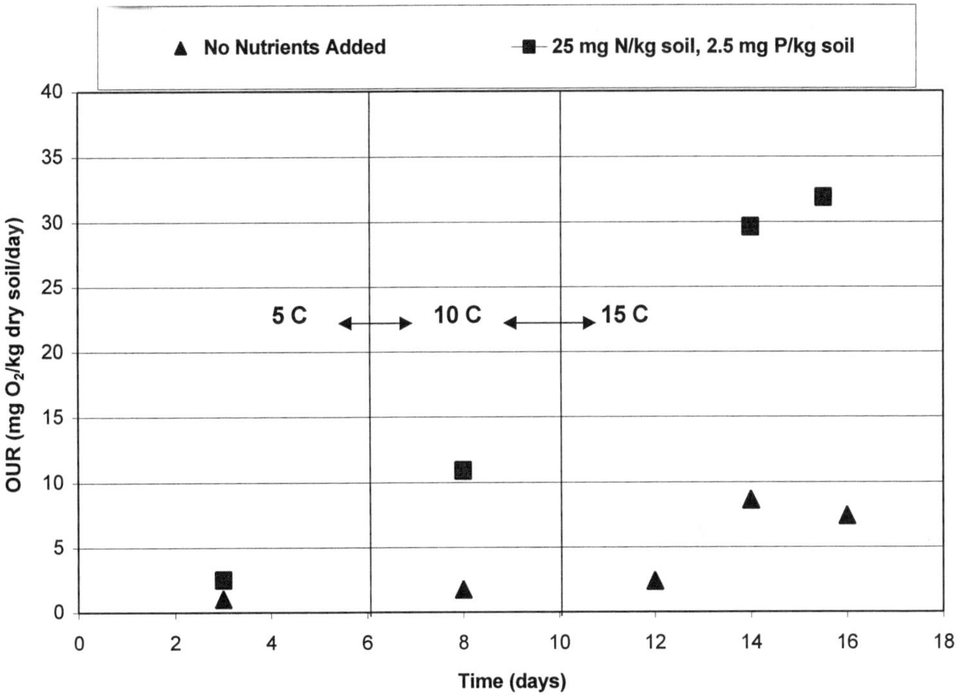

▲ No Nutrients Added ■ 25 mg N/kg soil, 2.5 mg P/kg soil

OUR (mg O_2/kg dry soil/day)

5 C ↔ 10 C ↔ 15 C

Time (days)

Figure 2. Increase in OUR in Response to Temperature Increase in Experiment #1

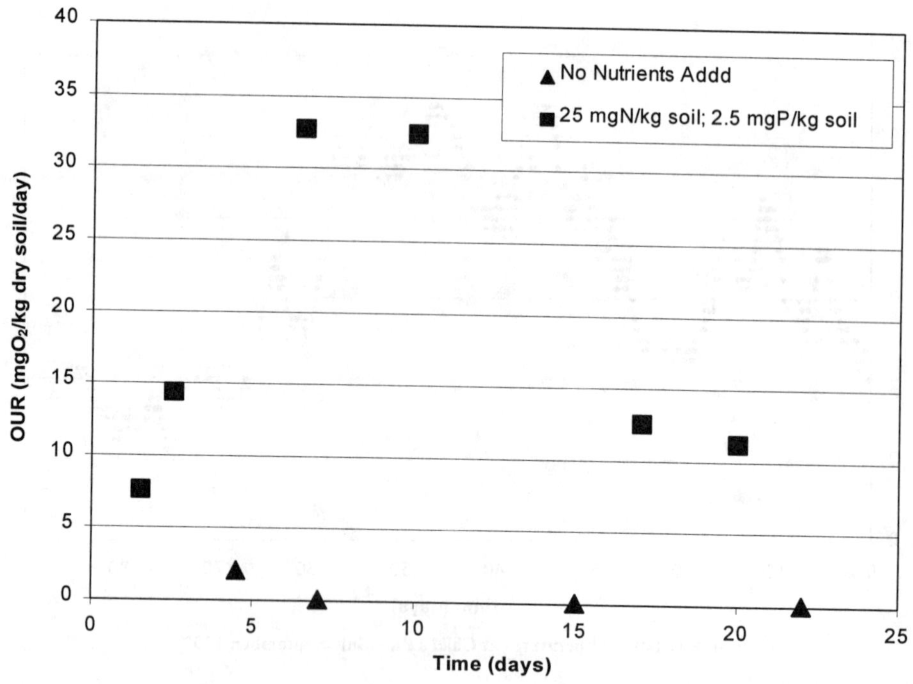

Figure 3. Effect of N and P on OUR in Soils Containg 6,100 mg/kg DRO Incubated at 5°C

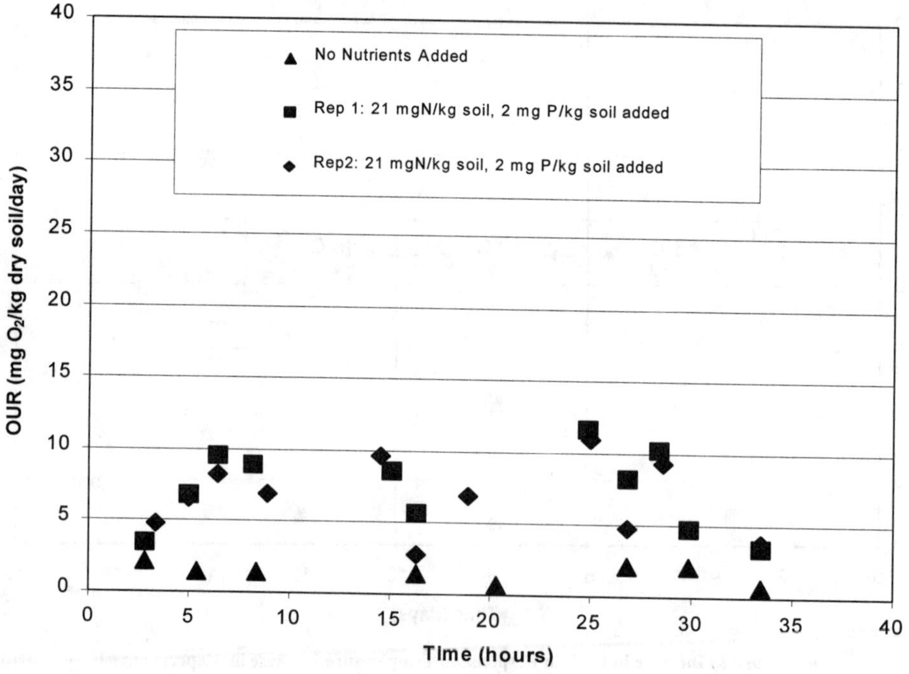

Figure 4. Effect of N and P Addition on Soil Incubated at 10°C.

Much larger increases in OUR with nutrient addition were observed in experiment #2 which was conducted on soil containing 6,100 mg/kg of DRO at 5°C (see Figure 3). Addition of 25 mg/kg N and 2.5 mg/kg P increased the OUR to approximately 32 mg/kg/day. The maximum OUR was obtained within 1 week of nutrient application, and a subsequent decrease in the OUR after 22 days.

In experiment #3 which was conducted at a temperature of 10° C, the addition of 21 mg N/kg soil and 2 mg P/kg soil increased the OUR to approximately 8-10 mg/kg/day (see Figure 4). The OUR rates measured in the control remained at less than 5 mg/kg/day for the duration of the study.

DISCUSSION

The experimental results presented in this paper indicate that even small additions of ammonium nitrate (25 mg N/kg soil) and phosphate (2.5 mg P/kg dry soil) salts to nutrient limited gravel pad can substantially increase microbial activity relative to soils which receive no nutrients. Furthermore, the application of nutrients at cold soil temperatures (5°C) which occur in the gravel pads at the beginning of the season have a beneficial effect as soil temperatures increase during the remediation season. Substrate concentration (DRO in these experiments), also appears to influence the response to nutrient addition. The OUR increase observed in soils with a 1,000 and 2,000 mg/kg DRO and nutrients were 10-12 mg/kg /day. However, when the substrate concentration was 6,100 mg/kg addition of nutrient increased the OUR to over 30 mg/kg/day.

REFERENCES

Leeson, A., Hinchee, R.E. (1996) Principles and Practices of Bioventing: Volume I: Bioventing Principals, Battelle Press, Columbus, OH.

NUTRIENT & TEMPERATURE INTERACTIONS IN BIOREMEDIATION OF PETROLEUM-CONTAMINATED CRYIC SOIL

Jim Walworth (University of Arizona, Tucson, AZ)
Craig Woolard (University of Alaska Anchorage, Anchorage, AK)
Larry Acomb (Geosphere, Inc., Anchorage, AK)
Vikram Chaobal (University of Alaska Anchorage, Anchorage, AK)
Mark Wallace (US Army Corp. of Engineers, Anchorage, AK)

ABSTRACT: A factorial study was conducted with a sandy, petroleum-contaminated subsoil from Ft. Wainwright, Alaska to evaluate the effect of temperature, moisture content, nutrient levels and diesel range hydrocarbon concentration on microbial activity at a cold climate bioremediation site. In soils with low hydrocarbon concentrations (DRO = 500 mg/kg) soil heating and nutrient addition increased biodegradation rates only slightly, suggesting a substrate limitation. In more highly contaminated soil (DRO = 8,100 mg/kg) soil warming increased microbial activity, indicating low soil temperature is the primary factor limiting biodegradation rates at the Ft. Wainwright site. Maximum microbial activity was observed at about 21°C, with little or no additional increase at 31°C, whereas heating the soil to 41°C reduced biodegradation rates. Addition of nitrogen and phosphorus increased biodegradation when the soils were heated above 1°C. Of these two, soil nitrogen concentration was the more important factor in maximizing biodegradation rates.

INTRODUCTION

At U.S. Air Force AFCEE bioventing sites with diesel and jet fuel contamination, typical biodegradation rates range from about 2 to 20 mg/kg/day. At some cold-region bioventing sites, however, biodegradation rates of less than 0.5 mg/kg/day are often measured. These low biodegradation rates extend the required remediation time and increase the overall cost of the project.

Low soil temperatures, soil moisture contents, nutrient availability have all been hypothesized as reasons for low observed biodegradation rates at certain Alaskan bioventing sites. While it is possible that a single rate-limiting factor may control biodegradation rates at a site, it is more likely that several environmental hierarchical factors control biological activity. Typically one variable is the primary limiting factor in a given situation, and when that limitation is alleviated the next most limiting factor controls activity at a new but higher level. Increasing a factor that is not the primary limiting factor will have a limited effect on biological activity until the primary limiting factor is increased.

The results presented in this paper were collected from a factorial experiment conducted to evaluate the interactions between soil moisture content, temperature, nutrient amendments, and hydrocarbon concentration on the aerobic biodegradation rates of diesel range organic hydrocarbons in nutrient-poor soils from Fort Wainwright, Alaska.

MATERIALS AND METHODS

Contaminated sand (88.6% sand, 8.6% silt, 2.8% clay) was collected from a site at Ft. Wainwright, Alaska. Gasoline range organics (GRO) ranged from about 130 to 2,600 mg/kg and diesel range organic (DRO) concentrations ranged from 410 to 3,400 mg/kg. The soil was low in total carbon (0.32%), total nitrogen (200 mg/kg), mineralizable nitrogen (<1 mg/kg), and available phosphorus (4 mg/kg).

Oxygen uptake tests were conducted at 5 different temperatures in 18 experimental reactors. Each reactor had a different combination of soil moisture, nitrogen, and phosphorus levels. Three soil moisture levels (5, 7.5, and 10% by weight), two phosphorus levels (0 and 75 mg P/kg soil), and three nitrogen levels (0, 100, and 200 mg N/kg soil) were set up in a factorial experimental design. Phosphorus was applied as $Ca(H_2PO_4)_2 \cdot H_2O$ (24.5% P) and nitrogen was added as NH_4Cl (26.2% N). This experiment was repeated with two hydrocarbon concentrations (500 mg/kg DRO and 8,100 mg/kg DRO). Diesel fuel was added to the soil to achieve the higher contaminant concentration. All of the reactors were subjected to a temperature cycle starting at approximately 1° C and then increasing to approximately 11, 21, 31, and finally 41° C.

Experiments were conducted in batch-type reactors constructed of 10 cm diameter by 25 cm long ABS plastic pipe, with a total volume of 1.838 L. Each reactor was equipped with two gas-tight ports and a Figaro Model KE-50 oxygen sensor. A Campbell CR10 data logger recorded oxygen sensor measurements ($\%O_2$) every hour. The reactors were loaded with a predetermined mass of soil, sealed to maintain an airtight system, and placed in temperature-controlled incubators.

When the oxygen concentration in the reactor was depleted to approximately10% by microbial activity, passing compressed air through the soil column purged the reactors, restoring oxygen levels to 17 to 21%. Oxygen uptake tests resumed were resumed after purging. During oxygen uptake tests, a linear reduction in oxygen concentration was generally observed. The oxygen utilization rate was determined by taking the slope of the linear uptake rate. The biodegradation rate was calculated from the oxygen utilization rate data using the method of Leeson and Hinchee (1996) using the stoichiometry for the oxidation of hexane.

Statistical analysis of the data was performed using biodegradation rate data from the various treatments (nitrogen levels, phosphorus levels, soil water contents, and temperatures) as the independent variable data for a standard analysis of variance (ANOVA). In the absence of true replication, the interaction between all variables was used as the experiment-wise error term (Cochran and Cox, 1957). SAS computer software was used to run ANOVA analysis (SAS Institute, 1996).

RESULTS AND DISCUSSION

Low Hydrocarbon Concentration (500 mg/kg DRO) Results. In the tests conducted with soil containing 500 mg/kg DRO, the following one and two-way effects were statistically significant at the 95% level: nitrogen, water, temperature, nitrogen by phosphorus, phosphorus by water, nitrogen by temperature, phosphorus by temperature, and water by temperature. Although the experiment shows that warming and adding nutrients to soils with low hydrocarbon concentrations will increase biodegradation rates, the biodegradation rates measured in the study remained below 2.5 mg/kg/day. Selected results are shown in Figures 1 through 3.

High Hydrocarbon Concentration (8,100 mg/kg DRO) Results. Soils with high hydrocarbon concentrations (initial DRO = 8,100 mg/kg) showed changes in biodegradation rates of a magnitude to be of interest for field applications.

Temperature Main Effect. The effect of temperature on microbial activity is shown in Figure 1. For the conditions tested, maximum microbial activity was

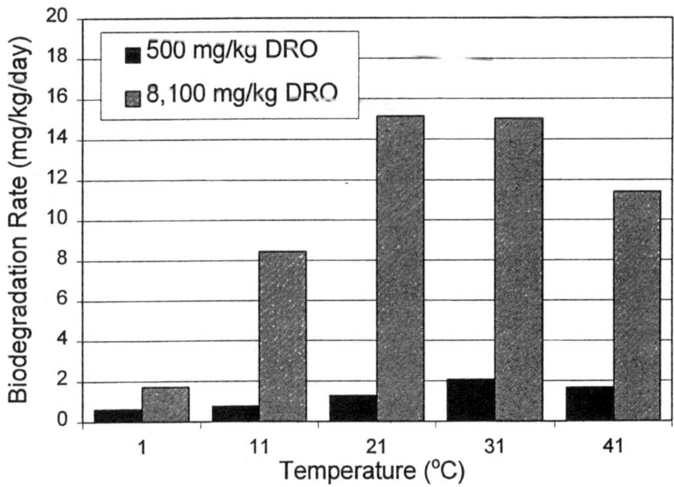

Figure 1. The Effects of Temperature on Soil Biodegradation Rates.

observed between 21°C and 31°C. The average biodegradation rates increased from 1.7 mg/kg/day at 1°C to 8.2 mg/kg/day at 11°C, and to 15.1 mg/kg/day at 21°C (a total increase of 790 percent). Average biodegradation rates were similar at 21°C and 31°C, and decreased slightly when the temperature was increased to 41°C. The greatest rate of biodegradation rate increase with increasing temperature appeared to occur between 1°C and 11°C. The effect of temperature on biodegradation rates was significant at the 95 % level. (Note that the average biodegradation rates listed here include results from reactors amended with nutrients.)

Moisture Main Effect. Over the range of moisture contents tested, the effect of moisture content was small compared to the other factors. Average biodegradation rates of 9.6, 11.2, and 10.2 mg/kg/day were measured at moisture contents of 5, 7.5 and 10%, respectively (data not shown). Moisture content was significant at the 90% level.

Nitrogen Main Effect. As shown in Figure 2, the addition of 100 and 200 mg/kg nitrogen increased the average biodegradation rate from 7.4 mg/kg/day to 11.5 and 12.1 mg/kg/day, respectively (a 55% to 64% increase). Nitrogen was significant at the 95% level.

Phosphorus Main Effect. The main effect observed with was similar to the results obtained with nitrogen addition (data not shown). However, the 25% increase in average biodegradation rate due to phosphorus addition (from 9.2 to 11.4 mg/kg/day) is less than half of the increase observed when nitrogen was added. Phosphorus was significant at the 95% level.

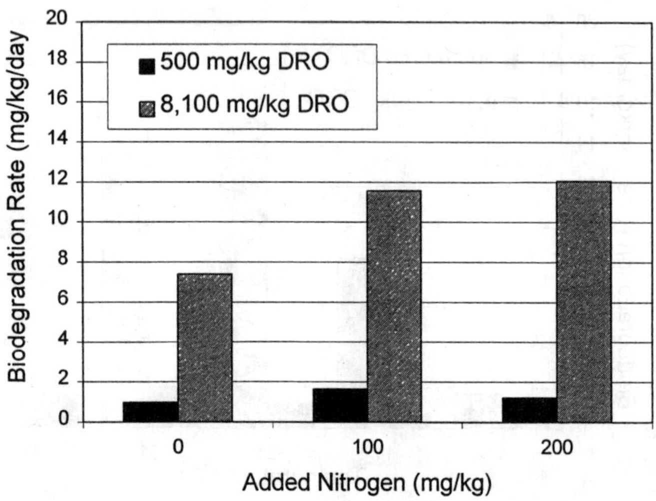

Figure 2. The Effects of Nitrogen on Soil Biodegradation Rates.

Nitrogen by Phosphorus Interaction. Additional phosphorus can enhance microbial activity at some nitrogen levels. Biological activity was enhanced when phosphorus was added to soil containing 0 or 100 mg/kg N, but the response disappeared when 200 mg/kg N was applied (data not shown). The nitrogen by phosphorus interaction was significant at the 95% level.

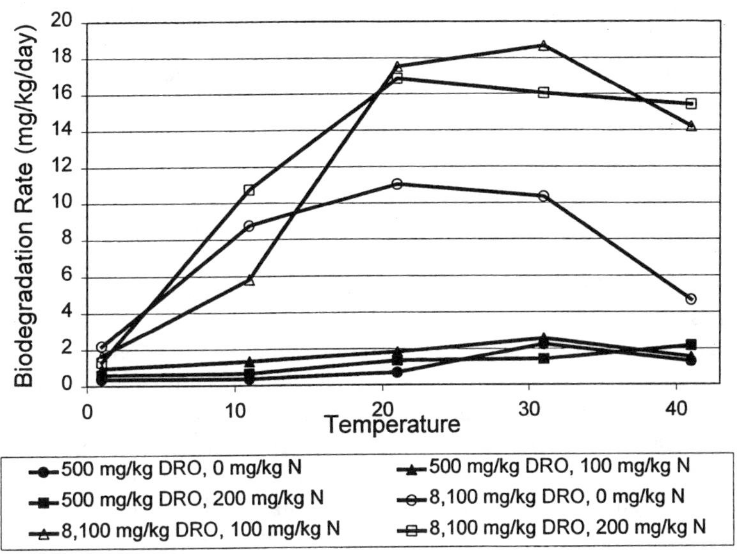

Figure 3. Combined Effects of Temperature and Nitrogen on Soil Respiration.

Nitrogen by Temperature Interaction. The data in Figure 3 indicate that nitrogen addition can enhance the microbial response to soil warming. Without the addition of nitrogen a maximum average biodegradation rate of 11 mg/kg/day was observed at 21°C. At 21°C, the addition of 100 mg/kg and 200 mg/kg N increased the maximum biodegradation rate to approximately 17.5 and 16.8 mg/kg/day, respectively. At all three nitrogen levels, the maximum microbial activity was observed at temperatures between 21°C and 31°C for the conditions tested. At temperatures of 1 and 11°C, beneficial effects of nitrogen addition were not clearly demonstrated. This interaction was significant at the 95% level.

Phosphorus by Temperature Interaction. The beneficial effects of soil warming can also be enhanced with the addition of phosphorus, although the effects are not as great as with nitrogen addition. The maximum average biodegradation rates were increased from 13.7 to approximately 17 mg/kg/day by the addition of phosphorus (data not shown). The beneficial effect of phosphorus addition was not clearly demonstrated at soil temperatures below 21°C. The phosphorus by temperature interaction was significant at the 95% level.

CONCLUSION

The results of the first experiment indicate that in soils with low hydrocarbon concentrations (DRO = 500 mg/kg) soil heating and nutrient addition may increase biodegradation rates slightly. However, peak biodegradation rates may still be relatively low (less than about 2.5 mg/kg/day), and are likely limited by the low hydrocarbon concentration. The results of the second experiment (DRO = 8,100 mg/kg) suggest that soil warming can increase microbial activity, and that low soil temperatures may be the primary factor limiting biodegradation rates at the currently operated Fort Wainwright sites. For the Fort Wainwright soils tested, the maximum microbial activity was observed at about 21°C. Little or no additional increase in biodegradation rate occurred by heating the soil to 31°C, whereas heating the soil to 41°C reduced biodegradation rates. The experimental results also indicate that addition of nitrogen and phosphorus can increase biodegradation when the soils are heated above 1°C (at 1°C nutrient addition did not significantly increase biodegradation rates). Of these two nutrients, soil nitrogen concentration appears to be the most important factor in maximizing biodegradation rates. These data indicate that temperature is the primary rate-limiting factor at Fort Wainwright.

The effects of soil temperature on biodegradation rates has been addressed in the literature and the increase in biodegradation rates observed in this study by adding heat alone are consistent with these studies. For example, 1) laboratory column studies using jet-fuel-contaminated soils from Eielson Air Force Base found that biodegradation rates at 10°C and 20°C were 6 and 10 times greater than the biodegradation rate observed at 2°C (Smith & Hinchee, 1993); 2) Sayles et al. (1995), in pilot scale field studies, measured a doubling or tripling of biodegradation rates when soils at Eielson Air Force Base were heated from an average annual temperature of 3.5°C to an average temperature of approximately 15°C; and 3) Simpkin et al. (1995) found that biodegradation rates in Prudhoe Bay soils doubled or tripled as soil temperatures were increased from about 2°C to 7°C.

The literature is not as clear on the effects of nutrient addition at bioventing sites. Based on the data collected as part of the Air Force Bioventing Initiative, Leeson and Hinchee (1996) concluded that the addition of nutrients does not necessarily improve biodegradation rates. However, numerous other studies have demonstrated the positive effect of providing supplemental nutrients. One reason for these apparent inconsistencies is that the complex interactions between nutrient addition, temperature, moisture content and hydrocarbon concentration are rarely considered in published studies. The factorial experiment conducted as part of this work demonstrates that these interactions must be considered to understand the true effect of adding nutrients. If nutrients were added, for example, to a system where biological activity is limited by temperature or hydrocarbon concentration, then no positive effect would be realized.

REFERENCES

Cochran, W.G. and G.M. Cox (1957) Experimental Designs. John Wiley & Sons, NY

Leeson, A., Hinchee, R.E. (1996) Principles and Practices of Bioventing: Volume I: Bioventing Principals, Battelle Press, Columbus, OH.

SAS Institute. 1989-1996. SAS Institute Inc., Cary, NC, USA.

COLOMBO: A NEW MODEL FOR THE SIMULATION OF SOIL BIOREMEDIATION

Marco Villani, Mariangela Mazzanti, Massimo Andretta, and
Roberto Serra
(Centro Ricerche Ambientali Montecatini, Marina di Ravenna, RA, Italy)
Salvatore Di Gregorio (Università della Calabria, Rende, CS, Italy)

ABSTRACT: In this paper we present the first results of the EU-ESPRIT Project Colombo focused on the application of parallel computers to the simulation of the complex phenomena occurring in the bioremediation of contaminated soils. In Colombo, discrete models (i.e. Cellular Automata) are used, instead of the more classical approach based on partial differential equations. In fact, cellular automata are intrinsically parallel and they can be easily and "naturally" implemented on parallel computers of the MIMD type. This approach has been already applied to describe bioremediation of a phenol contaminated soil in a previous ESPRIT project, named Caboto. In this paper we present the main theoretical characteristics of the Caboto/Colombo simulation models and the results of some comparisons between experimental and simulation data.

INTRODUCTION

In-situ bioremediation is a complex phenomenon, where the behavior of the macroscopic observables is the outcome of the interaction of different "microscopic" processes, which take place on a smaller scale, and which are themselves only partly known. It is therefore very difficult to develop reliable simulation models of these phenomena. Actually, while in other engineering disciplines the use of mathematical models is widespread, bioremediation is presently in a peculiar situation: a recent, albeit limited, survey among companies which operate in Italy has shown that mathematical models play no role in determining the decision of choosing bioremediation instead of alternative technologies. However, the same survey has shown that the difficulty of forecasting the duration and therefore the costs of the operations is the topic of greatest concern for decision makers [Andretta et al, 1996, Serra et al., 1998].

It seems therefore that the bioremediation field would need simulation models, but that present-day models are still perceived as insufficient for practical purposes. It is therefore very important to be able to show that models can accurately describe experimental data. A number of initiatives are under way, and we will describe in detail one of them which receives partial support from the UE Esprit program. The modeling framework described below was initially developed within the UE Esprit project Capri-Caboto, and major model developments are presently under way within another UE Esprit project, Colombo. We will collectively refer to the common modeling framework of the two projects as the Caboto/Colombo (shortly, CabCol) approach.

Most existing simulation models [see e.g. , Molz et al, 1986, Jennings & Manocha, 1994 and further references cited therein] are based upon a continuum formalism and, as they are not amenable to analytical treatment, numerical integration is needed.

In CabCol a different approach was explored, based upon direct discrete modeling of the system with macroscopic cellular automata [Di Gregorio et al, 1998, Serra et al., 1998]. Macroscopic cellular automata are described in Section 2, while Section 3 describes the main features of the CabCol model which has been developed.

The model includes a number of parameters which cannot be either directly determined or deduced from first principles, but whose values must be chosen so as to match experimental data; the parameter values have been determined by comparing the model simulations with the results of some experiments on suitably equipped pilot plants, using genetic algorithms [Goldberg, 1989]. Some results regarding the comparison between model and experiments are shown in Section 4. Finally, critical discussion and indications for further work are given in Section 5.

MACROSCOPIC CELLULAR AUTOMATA

Cellular automata are "artificial universes" divided into discrete cells; one or more state variables are associated to each cell, and they can take only discrete values [se e.g. Serra & Zanarini, 1990 and further references cited therein]. The state variables change in time: time itself is discrete, and there is a unique global clock for the whole automaton. In each cell, a transition function determines the value of the state variable at time t+1, from the values taken at time t by the variables of the cells belonging to a given neighborhood.

As in our case we are interested in modeling phenomena which take place on a large space scale, it is appropriate to choose macroscopic variables as state variables. This amounts to choosing a cell dimension which comprises several pores, so that it is meaningful to consider the value of variables like water saturation, concentration of chemicals, biomass density, etc. The size of the automaton cell in CabCol models must be large enough to comprise several pores (in order to make a macroscopic approach possible) but still much smaller than the length scale of appreciable variations in the system properties, so as to make it possible to consider them as constant within each cell.

The choice of the topology of the CA should be driven by physical insight and by some simplicity criterion. For this purpose, we have relied (in the 3D case) upon a fairly well established choice of a rectangular cell, with a neighborhood composed of the cell itself, its four neighbors in the same horizontal plane (E,N,W,S), and by its upper and lower cell in the vertical direction.

THE MODEL

The model has a layered structure, where the lower layer describes the fluid dynamical properties (multiphase flow in a porous medium), the second layer describes the fate of the solute (advection, dispersion, adsorption/desorption, chemical reactions) and the third layer describes the growth of the biomass and its interaction with nutrients and contaminants.

A further remark concerns the use of kinetic coefficients. The bioremediation model is fairly ambitious, as it aims at describing a complex set of interacting phenomena, which take place on different time scales. Usually, in simple CA applications, where one has to deal with a single phenomenon, the CA clock is adjusted so to match the observed kinetics; however, as we need to deal with several phenomena, it is necessary to introduce kinetic coefficients which describe the different velocities involved. Therefore, a CA algorithm is applied at each time step to compute the maximum possible value of a given flow from one cell to another, let us call it MF(t). Then the actual flow AF(t) between these two cells is obtained by multiplying MF(t) times an appropriate kinetic coefficient $\eta \leq (1)$: AF(t) = ηMF(t).

The basic equations of the model are obtained by considering conservation laws, coupled with suitable constitutive relationships, based on phenomenological evidence. The formalism of macroscopic CA can incorporate the empirical knowledge available about a given field.

As far as the fluid dynamical layer is concerned (multiphase flow), it is assumed that mass flow is driven by potential differences [Helmig, 1997]and that, once a reference phase is chosen, the behavior of the other phases can be reproduced as a function the reference phase. Therefore, the algorithm first computes the potential of the reference phase and, from this, the potential of the other phases. Then mass flows, proportional to the gradient of potential and corrected by appropriate kinetic coefficients, are computed.

The computation of the hydraulic head (from which the pressure is trivially derived) is based on a method of successive approximations in order to reach an equilibrium situation between a cell and its neighbors in all the cellular space; this is obtained when, for each cell of the CA, the total amount of mass incoming from the neighbor equals the total outflow. If this is not verified, the condition of mass flow conservation is superimposed, the potential of the cells is corrected and flows are recomputed until the equilibrium is reached; the rate is proportional to the hydraulic conductivities of the cell and to the hydraulic head difference from the cell and the its neighbor.

When the steady state is achieved, the procedure for the calculation of the pressure of the reference phase stops and the pressures of the other phases are computed. Then the flows of the phases can be calculated, for each CA step, by using the methods of multiphase flow theory (e.g., relative permeability coefficients). An example of pressure distribution in a water saturated system computed according to the above procedure is given in Figure 1.

As far as chemicals are concerned, solutes are transported by the phases flowing in the system. The adsorption/desorption process has been modeled by first order kinetic equations. It should also be recalled that, when there is an inflow of a given phase (e.g. water) in a cell at time t, and the phase carries a certain chemical, it may require some time before the concentration of that chemical reaches a homogeneous value in that cell, because some water was already present, before the inflow. According to the spirit of the CA approach, only two kinds of water are considered, *new water* and *old water*; for example, in a contamination experiment, new water is contaminated, while old water is "clean". As time goes on, a growing fraction of the water in the cell becomes contaminated, while of course the contamination level diminishes, in order to guarantee mass conservation. The introduction of chemical reactions is straightforward, as these reactions take place within each cell and do not involve transport among neighbors.

The third layer of the model describes biomass dynamics. It can been assumed (taking a so called "Occam razor approach") that bacteria are immobile, and that their access to the contaminant and the nutrients does not suffer from mass transfer limitations, or else, in a more general case, it can be taken into account the passive bacterial transport, which may be carried by the water flow.

If bacterial transport may be ignored, one is left with bacterial kinetics within each cell. It has been assumed in CabCol that Monod kinetics, which can describe a wide range of laboratory data regarding single bacterial strains, can be applied. It has also been assumed that the growth of all kinds of bacteria are ruled by Volterra dynamics, in the absence of nutrients and contaminants. Nutrients and contaminants may induce growth and/or death: for some compounds (e.g. phenol) both effects can be observed and modeled. The effect of nutrients and other physical or chemical conditions can also be modeled in a simplified way by letting the kinetic parameters of the growth equation vary.

COMPARISON WITH EXPERIMENTAL RESULTS

Several comparisons between model results and experimental data have been performed. Some experiments have been selected for tuning adjustable parameters, and it is worth stressing that good results have been achieved in cases different from those used for adjusting parameters (Di Gregorio et al, 1998).

In fig. 1 it is shown how the procedure for computing the pressure of the water phase accounts for the case where a pumping and an extracting well are at work (steady state).

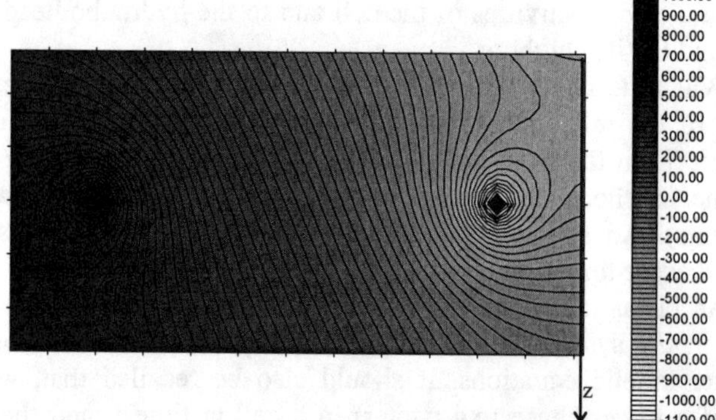

FIGURE 1. The pressure distribution (arbitrary units) generated by two constant opposite pumping sources in a water saturated environment.

In order to test the performance of the chemical layer, an example is given in Figure 2 of the concentration of phenol detected in the percolated water, during an experiment where a pilot plant was contaminated with phenol (see Serra et al, 1998, for further details).

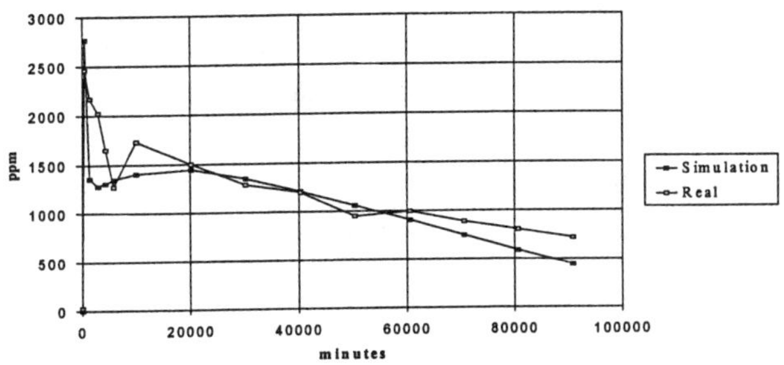

FIGURE 2. Concentration of phenol in drainage water.

A test of the model for bacterial dynamics is given in Figure 3, also referring to contamination by phenol.

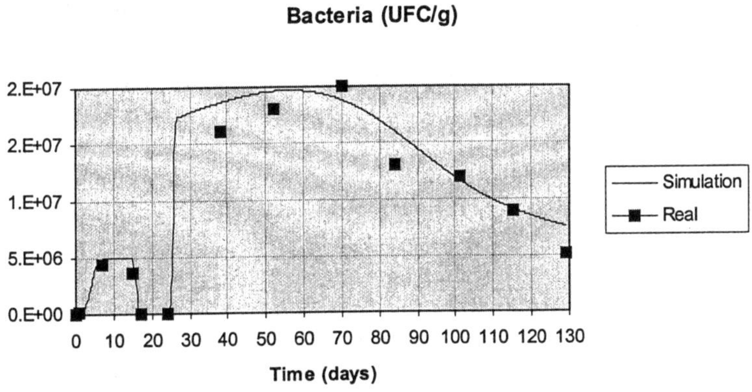

FIGURE 3. Comparison of actual and estimated density of phenol-degrading bacteria (acclimatization, contamination, bioremediation events).

CONCLUSIONS

The agreement between experimental data and simulation results is remarkable, taking into account the complexity of the system under study, the fact that the phenomena are fully non stationary and the different time scales of the involved phenomena. This agreement provides an *a posteriori* test of the validity of the cellular automata approach.

Several tests, on a limited scale, are under way within the Colombo project to give the experimental data required to test the model in multi-phase systems.

The major issue which is still open concerns the accuracy of the model scale up to the field test, which requires, among other considerations, expensive and accurate large scale measurements. This issue is also addressed in the ongoing Colombo project.

We believe that these results, which show how even such a complex phenomenon can be precisely described by a simulation model, can contribute to promote the diffusion of model based techniques in the sector of *in situ* bioremediation, which is still largely dominated by a highly empirical approach.

ACKNOWLEDGEMENTS

This work has been partially funded by the UE Esprit project # 24907 Colombo. We gratefully acknowledge the contribution of many friends and colleagues: Federica Abbondanzi, Darinn Cam, Antonella Iacondini, Nello Lombardi, Rocco Rongo, William Spataro, Giandomenico Spezzano, Domenico Talia.

REFERENCES

Andretta, M., R. Baroncelli, M. Matteuzzi, R. Serra, G. Spezzano, and D. Talia. 1996. *Industrial exploitation.* Deliverable D8 of the UE-Esprit project Capri-Caboto, Centro Ricerche Ambientali Montecatini, Ravenna.

Di Gregorio, S., R. Serra, and M. Villani. 1998. "A cellular automata model of soil bioremediation". *Complex Systems. 11*(1): 31-54.

Goldberg, D. E. 1989. *Genetic Algorithms in Search, Optimization, and Machine Learning.* Addison-Wesley.

Helmig, R. 1997. *Multiphase Flow and Transport Processes in the Subsurface.* Springer Verlag, Heidelberg.

Jennings, A. A., and A. Manocha. 1994. "Modeling soil bioremediation". In D. L. Wise and D. J. Trantolo (Eds.), *Remediation of hazardous waste contaminated soils,* pp. 645-680. Marcel Dekker, New York.

Molz, F. J., M.A. Widdowson, and L. D. Benefield. 1986. "Simulation of Microbial Growth Dynamics Coupled to Nutrient and Oxigen Transport in Porous Media". *Water Resour. Res. 20*: 1207-1216.

Serra, R., and G. Zanarini. 1990. *Complex systems and cognitive processes.* Springer Verlag, Heidelberg.

Serra, R., S. Di Gregorio, M. Villani, and M. Andretta. 1998. "Bioremediation simulation models". In R. Serra (Ed.), *Biotechnology for soil remediation*, pp. 125-154. Cipa Editore, Milano.

RHEINGOLD PROJECT: BIOREMEDIATION OPTIMISATION THROUGH ON-LINE MONITORING OF BIOACTIVITY INDICATORS

Bruce N. Anderson (RMIT University, Melbourne, Australia)
Rolf Greis (Institute for Bioremediation, Rösrath, Germany)
Eckart Bär (Environment Expert Office, Rösrath, Germany)
Christoph Henkler (King's College, London, United Kingdom)
Rolf D. Henkler (ICI Paints, Slough, United Kingdom)

ABSTRACT: The feasibility of using bioremediation to treat localised areas of solvent hydrocarbon contamination of soil and groundwater at an operating paint factory in Germany was investigated. Existing natural hydrocarbon attenuation levels of 0.27 mg/kg soil/day were demonstrated, and enhanced biodegradation rates of up to 10 mg/kg soil/day were achieved in bioventing and biosparging pilot trials. A full-scale bioremediation system was installed during 1997 to remediate solvent hydrocarbon contamination in 20,000 m^3 of unsaturated zone soils and shallow groundwater to a depth of 12m below grade. The system is comprised of groundwater air injection sparge points (biosparging), unsaturated soil zone air injection points (bioventing) and a soil vacuum extraction (SVE) system to collect injected air for treatment through compost filters (biofiltration) prior to discharge. The above-ground services are of a modular layout to facilitate re-deployment for the bioremediation of contamination plumes elsewhere on the factory site. A feature of the system is the extensive use of on-line monitoring to achieve real-time quantification of contaminant biodegradation, and to enable operational control for optimisation of the biodegradation processes. More than 4,000 kg of diffuse contamination was biodegraded in the first year of operation and target remediation levels are expected to be achieved.

INTRODUCTION: The Rheingold Project site is located in the Rhine valley in the state of Nord Rhein Westfalen, Germany. The natural sediments in the area range from medium to coarse sands, and fine gravels, underlain by Quaternary sediments. An unconfined aquifer extends from the groundwater surface (5 - 7 m below grade) to an aquitard layer of fine, silty sands approximately 40 m below grade. The average groundwater velocity through the unconfined aquifer is 100 m per annum. The site has been used for the manufacture of solvent-based paints, lacquers and varnishes for over 140 years, and has a number of underground storage tank farms, several of which have been out of service for decades. Solvents used in traditional paint manufacture included "solvent naphtha" and "Stoddard solvent", a paraffinic hydrocarbon mixture containing up to 35% aromatic hydrocarbons including benzene, toluene, ethylbenzene, xylenes (BTEX) and trimethylbenzenes. Areas of soil and groundwater contamination resulting from the historical operations have been identified beneath the tank farms, transfer pipelines and recycling and waste treatment facilities.

The strategy adopted within the Rheingold project included measures to prevent any further infiltration of contaminants into soil and groundwater, physical recovery of free-phase material (LNAPL recovery), soil vacuum extraction (SVE) of gross vadose-zone contamination, followed by bioremediation of diffuse contamination wherever possible. Implicit in the strategy is the use of "natural attenuation" wherever possible rather than active remedial intervention.

BIOREMEDIATION FEASIBILITY ASSESSMENT: Initial desktop and laboratory-based feasibility assessments followed the guidelines outlined by the National Research Council (NRC) in 1993. The results obtained, together with standard chemical analysis of groundwater, fulfilled the requirement to document the loss of contaminants at field scale, to incorporate chemical analytical data into mass balance calculations, and to demonstrate biodegradation of the contaminants in laboratory microcosms seeded with samples from the site (Anderson, 1995). Pilot-scale *in situ* soil gas injection and respiration studies were undertaken to assess the level of hydrocarbon solvent biodegradation that could be achieved in the vadose zone through aerobic bioventing (Peck et al., 1996) and in the saturated zone through biosparging. The extent of on-going natural attenuation of the solvent hydrocarbon contamination at the study site was assessed essentially as described by Wiedemeier et al.,1994. Expressed Assimilative Capacities of up to 47mg/L benzene equivalent were demonstrated for localised regions of the site (McLaughlan et al., 1996). More commonly, expressed capacities of around 20 - 30 mg/L benzene equivalent are found with typical daily natural attenuation rates of 0.2 - 0.3 mg/L. Potential constraints to the successful implementation of bioremediation were assessed using *lux*-modified bacterial toxicity biosensors and the constraint characterisation protocol described by Sousa et al, 1998a, 1998b.

MONITORING STRATEGY: Barcelona (1994) suggested that the parameters used in site characterisation should provide the basis for monitoring throughout the life of a remediation project, with technically sound monitoring networks that provide reasonable estimates of the *in situ* distribution of contaminants over time. Barcelona's principle has been adopted within the Rheingold project through the use of on-line monitoring systems where possible, recognising that prevailing trends are often more useful indicators of remediation progress. The traditional off-line chemical analyses for site assessment, monitoring of remediation progress, and determination of active remediation "end-points" are augmented through the use of toxicity biosensors, DNA probes to determine biodegradation potential, mineralisation tests to monitor biodegradation activity, and the measurement of changes in biogeochemical parameters. The overall Rheingold project bioremediation assessment and monitoring strategy was described by Anderson and Henkler (1997). On-line monitoring parameters during active remediation include, as appropriate, oxygen, carbon dioxide, temperature, flow, pressure, methane and volatile organic compounds (VOCs), together with calculated or derived parameters including inferred biodegradation rates.

FULL-SCALE REMEDIATION: Based upon the experience gained through laboratory and pilot-scale investigations, a full-scale bioremediation system was designed and commissioned to treat diffuse soil and groundwater contamination in an area of the site formerly occupied by production buildings A & B (A/B Area). The levels of contamination with benzene and total BTEX are shown in Table 1, together with the applicable intervention values. The A/B Area is close to the site boundary (<20m) and adjacent private housing.

TABLE 1: Contamination levels in the A/B area at the Rheingold project site.

Contaminant	Maximum Level [μg/L]	Intervention Value [μg/L]	Excess
Benzene	850	5	170 fold
Total BTEX	5,234	100	52 fold
Chlorinated Hydrocarbons	495	70	7 fold

The installed remediation system consists of 14 soil vapour extraction wells, three vadose zone air injection wells, seven saturated zone air sparging wells and four multi-level groundwater sampling points. Ancillary monitoring systems and mechanical equipment are housed in "portable" containers. A biofiltration system (two parallel trains each of two biofilters in series) has been installed for treatment of the air captured by the SVE system. The design performance parameters for the system are shown in Table 2.

TABLE 2: Design performance parameters for A/B area bioremediation.

	Biosparging (7 Wells)	Bioventing (3 Wells)	SVE (14 Wells)
Volumetric Flow Rate	200 m^3/h	200 m^3/h	440 m^3/h
Depth	18.5m	5.5 m	5 m
Porous Section	17.5 – 18.5 m	4 – 5 m	2 – 5 m
Expected ROI*	8 – 10m at 4 hPa	20 m at 1 hPa	8 – 16 m at 0.2 hPa

*ROI = Radius of Influence.

The biofiltration system was designed to accommodate the full volumetric flow from the SVE system (440 m^3/h) with a maximum BTEX loading of 900 mg/m^3. The installed remediation system was commissioned in mid-1997 and has operated almost continuously since start-up, with on-line monitoring and logging of 138 data items versus time, as summarised in Table 3. The total cumulative *in situ* attenuation of hydrocarbons and the daily average attenuation rates (g/h) since start-up are shown in Figure 1.

TABLE 3: On-line data acquisition for A/B area bioremediation system.

On-Line Items (Measured)	Bio-sparging	Bio-venting	SVE	Bio-filtration	Total
Flow (m^3/h)	7	3	13	1	24
Pressure (mbar)	7	3	13	1	24
Temperature (°C)	7	3	13	1	24
Gas Analysis: [CO_2, O_2 (Vol. %)]	2	2	39	9	52
Gas Analysis: [BTEX, CH_4 (mg/m^3)]			13	3	16
Total Measured Items					**140**

On-Line Items (Calculated)	Bio-sparging	Bio-venting	SVE	Bio-filtration	Total
Current Mass Flow BTEX (g/h)				3	3
Cumulative Mass Flow [BTEX (kg)]				1	1
Current In situ Attenuation [BTEX (g/h)]	2	2	2		6
Cumulative *In situ* Attenuation [BTEX (kg)]					1
Total Calculated Items					**11**
TOTAL DATA ITEMS					**151**

*The inferred biodegradation rates calculated from oxygen consumption and carbon dioxide evolution [McCarty (1988)], as described by Peck et al, (1996).

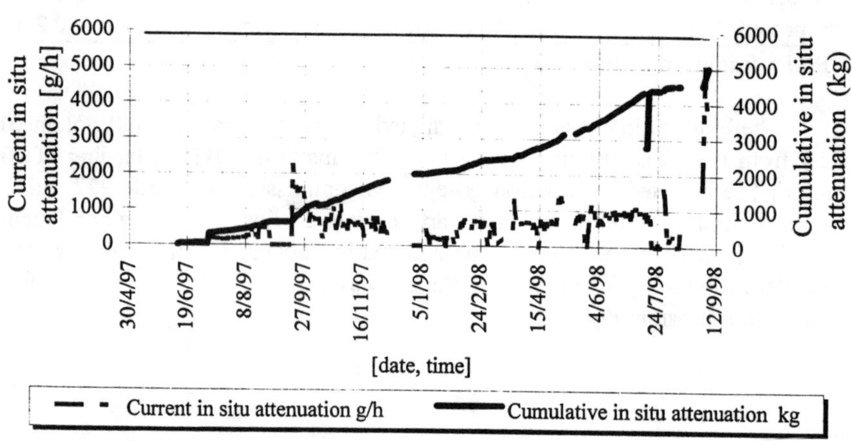

FIGURE 3: Cumulative total and average hourly hydrocarbon attenuation.

SUMMARY: A bioremediation system was installed in the A/B Area at the Rheingold project site after careful feasibility assessment and pilot-scale testing, and with implementation of an on-line monitoring strategy that could be used throughout the remediation project until completion. On-line monitoring is augmented, as appropriate, with chemical analyses, laboratory-based testing with toxicity biosensors, DNA probing to assess biodegradation potential, and mineralisation tests for biodegradation activity. More than four tonnes of diffuse solvent contamination have been mineralised to date in the A/B area, and consent levels for the contaminants of concern should be achieved in the near future. The ready availability of trend data from the on-line monitoring systems has enabled optimisation of the system in terms of sparge pulse cycles and air flow rates to an extent not possible when reliant upon off-line and off-site laboratory analysis.

REFERENCES:

Anderson, B. N. (1995). "Recent Developments in Contaminated Site Investigation-The Assessment of Indigenous Microflora and Natural Attenuation." In Bioremediation: The Tokyo '94 Workshop, pp.189-194. OECD, Paris.

Anderson, B. N. and R-D. Henkler (1997). "A Novel Bioremediation Feasibility Assessment and Implementation Strategy." Australasian Biotechnol. 7, 355-362.

Barcelona. M. J. (1994). "Site Characterisation: What Should We Measure, Where (When) and Why?" In US EPA Symposium on Intrinsic Bioremediation of Ground Water, Denver, pp. 1 - 9. US EPA/540/R-94/515.

National Research Council (1993). "In situ Bioremediation - When does it work?" National Academy Press, Washington, DC., USA.

McCarty, P. L. (1988). "Bioengineering Issues Related to *In Situ* Remediation of Contaminated Soils and Groundwater." In Environmental Biotechnology: Reducing Risks from Environmental Chemicals through Biotechnology, G.S. Omenn (Ed.), , pp. 143-162. Plenum Press, New York.

McLaughlan, R. G., K. P. Walsh, R-D. Henkler and B. N. Anderson. (1996). "Identification of Biological Processes in a Mixed Hydrocarbon Plume at a Paint Manufacturing Facility." In The 5th World Congress of Chemical Engineering, Volume III, pp.767-771.

Peck, P. C., S. H. Rhodes, B. N. Anderson and R-D. Henkler. (1996). "Bioremediation of Solvent Hydrocarbons: Laboratory and *In Situ* Field Studies." In The 5th World Congress of Chemical Engineering, Volume III, pp.737-742.

Sousa, S., C. Duffy, K. Killham, H. Weitz, L. A. Glover, R. Henkler, E. Bär, B. N. Anderson and J. R. Mason. (1998a). "Contribution of lux-marked bacterial biosensors to remediation of BTEX contaminated Land." In Contaminated Soil, pp. 839-840. Thomas Telford, London.

Sousa, S., C. Duffy, H. Weitz, L. A. Glover, E. Bär, R. Henkler and K. Killham (1998b). "Use of a lux-modified bacterial biosensor to identify constraints to bioremediation of BTEX contaminated sites." Environmental Toxicology and Chemistry, 17, 1039-1045.

Wiedemeier, T.H., J.T. Wilson, D.H. Kampbell, R.N. Miller and J.E. Hansen (1994). Technical Protocol for implementing intrinsic remediation with long term monitoring for natural attenuation of fuel contamination dissolved in groundwater. US Air Force Centre for Environmental Excellence, Brooks Air Force Base, San Antonio, Texas, USA.

BIOGENIC INTERFERENCE WITH PETROLEUM ANALYSIS IN ORGANIC SOILS FROM ALAKSA

Craig Woolard (University of Alaska Anchorage, Anchorage, AK)
Daniel White (University of Alaska Fairbanks, Fairbanks, AK)
James Walworth (University of Arizona, Tucson, AZ)
Marty Hannah (University of Alaska Anchorage, Anchorage, AK)

ABSTRACT

Analysis of the uncontaminated soil samples show that organic soils can have several hundred to several thousand mg/kg of extractable natural organic material that could be inadvertently quantified as contamination. The concentration of NOM that was recorded as contamination varied with sample location and depth. Even at the relatively small 15'x15' sampling site, biogenic interference levels varied by an order of magnitude.

The variability observed in the uncontaminated samples has several implications for the remediation of contaminated soils prone to biogenic interference. Large variations in biogenic interference levels with sample location will make it difficult to establish a "background" value at a given site unless a large number of samples are collected. In addition, highly variable background levels will make it difficult to determine if remediation strategies are effective and when remediation is complete.

INTRODUCTION

Standard methods for testing contaminant concentration in soil are well established and effective in most soils where the principle constituents are inorganic gravels, sands, silts or clays. However, in organic soils the interpretation of contaminant analyses is more difficult, particularly when the contaminant is a petroleum product. Since many of the same compounds are found in petroleum and natural organic material (NOM), true contamination and natural compounds become indistinguishable when analyzed with standard test methods. "Biogenic interference" is the term used to describe the NOM quantified as "petroleum" during a standard test for soil contamination.

Because the extraction methods are used in establishing cleanup limits, biogenic interference can prompt inaccurate cleanup standards, perhaps lower than the actual contamination warrants. Industries and government agencies responsible for cleanup want to direct resources at treating contamination and not "remediating" biogenic material. In addition, the uncertain accuracy of extraction and analytical methods can discourage the use of alternative, cost effective technologies. Responsible parties are reluctant to try these technologies in highly organic soils where low cleanup limits require "treatment" of biogenic material. NOM interference can also make it difficult to determine if natural attenuation of hydrocarbon contamination is occurring.

The basic problem is that many similar compounds appear in both highly organic soils and petroleum products. Petroleum products are complex mixtures containing both non-polar (e.g., n-triacontane) and polar (e.g., acetic acid) compounds (Dragun, 1988). Non-polar materials in crude oil include the straight chain, branched and cyclic alkanes, and polynuclear aromatic hydrocarbons. Similar natural petroleum-like compounds are also present as part of the soil bitumen, or lipid-derived organic material.

In standard test methods for soil contamination in Alaska, the contaminant is extracted from the soil with methlylene chloride. The solvent extracts compounds from the natural organic material (NOM) and petroleum with the same elemental composition and those with similar degrees of polarity. Once the contaminant is extracted from the soil, it can be quantified. Certain constituents in NOM, such as sterols, can be differentiated from petroleum compounds using gas chromatography/mass spectrometry (GC/MS). Others, such as the straight chain alkanes, cannot. When biogenic interference cannot be quantified independent of petroleum contamination, it is difficult to determine the true contamination level and set appropriate cleanup limits.

Although it is well established that the current extraction and analytical methods are complicated by biogenic interference, no systematic studies have been published that document the magnitude and variability of this interference in highly organic soils. This paper summarizes the results of a study conducted at one site to determine background concentrations and variability of extractable NOM at one location in Arctic Alaska.

MATERIALS AND METHODS

A total of 37 samples were collecteed to evaluate the variability of "natural" DRO (NDRO) and "natural" RRO (NRRO) levels at a single site. All of the samples were taken from a poorly defined polygon near Betty Pingo in the Prudhoe Bay Unit, Alaska (70°16'45.5"N; 148°53'30.0"W). The polygon, which is shown schematically in Figure 1, was divided into 3 transects (A, B and C) and each transect was sampled at 5 different locations using a 4" coring tool. Sub-samples from distinct soil horizons of each core were separated and analyzed for NDRO and NRRO.

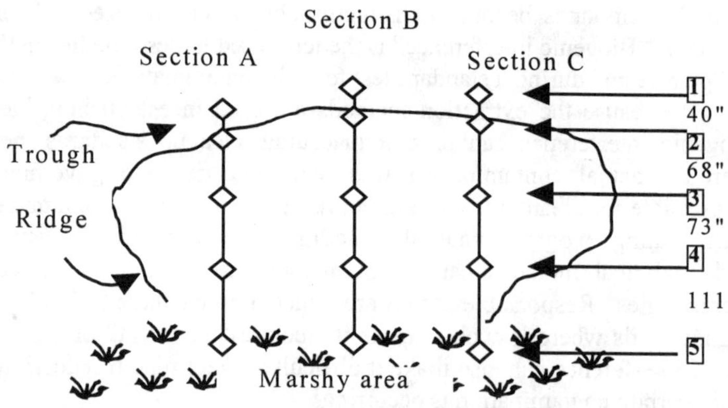

Figure 1. Schematic of Betty Pingo Sampling Locations

Extractable natural organic material from the pristine soil samples was quantified as either naturally occurring diesel range organics (NDRO) which quantifies compounds in the C_{10}-C_{25} range or as naturally occurring residual range organics (NRRO) which quantifies compounds in the C_{25}-C_{36} range. Each sample was extracted with methylene chloride and analyzed using a gas chromatograph equipped with a flame ionization detector (EPA, 1986).

RESULTS

The pristine soil samples collected at the Betty Pingo site were characterized by variable concentrations of NDRO and NRRO. As shown in Table 1, the NDRO and NRRO concentrations in the surface layer ranged from 201-591 and 365-1670 mg/kg, respectively. If all of the samples from Betty Pingo are considered, NDRO and NRRO concentrations varied by an order of magnitude within the site. Although the amount of NDRO and NRRO varied significantly with sampling location and depth at the Betty Pingo site, far less variation was observed in the NRRO/NDRO ratio.

Table 1. Summary Statistics for Samples Collected from Betty Pingo

Parameter	Average	Coefficient of Variation (%)	Low and High Values
Surface Horizon			
NDRO	399	29.6	201-591
NRRO	1009	34.2	365-1670
NRRO/NDRO	2.51	11.6	1.95-3.31
All Horizons			
NDRO	601	55.2	159-1590
NRRO	1685	66.4	365-4740
NRRO/NDRO	2.69	16.4	1.95-3.65

A set of chromatograms from soil core 2A is shown in Figure 2. At this location, the deeper horizon contained a greater total mass of NDRO and NRRO as well as a greater number of different compounds. Although this soil core showed an increasing concentration of extractable natural organic material with depth, this was not the case in all soil cores. No distinct pattern of NDRO or NRRO was observed with vertical position at the Betty Pingo site.

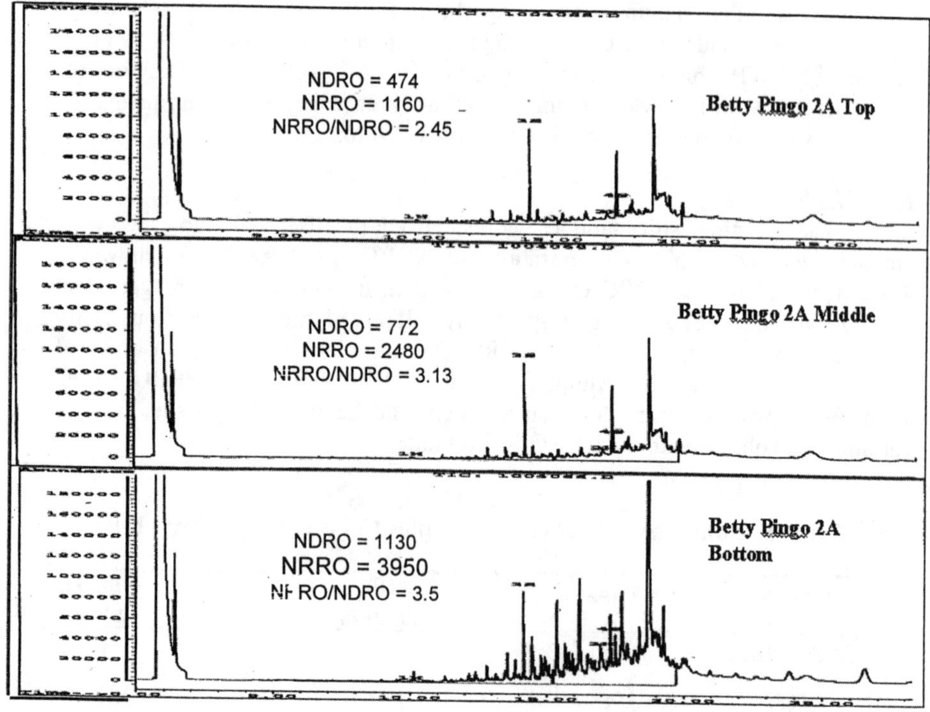

Figure 2. Chromatograms of Extractable Natural Organic Material from Location 2A at the Betty Pingo Site.

DISCUSSION

The results of analysis of the 37 samples from the Betty Pingo site shown that uncontaminated organic soils can have several hundred to several thousand mg/kg of extractable natural organic material that could be inadvertently quantified as contamination. The concentration of NOM that was recorded as contamination varied with sample location and depth. Even at this relatively small 15'x15' sampling site, biogenic interference levels varied by an order of magnitude.

Organic soils are generally formed from environmental conditions that favor primary production over decomposition. In soils on the North Slope of Alaska and other circumpolar regions, decomposition is hindered by cold temperatures, low oxygen transfer rates in the commonly saturated soils and frequently acidic conditions. The variety of controlling conditions in different soils results in many unique mixtures of organic material in different stages of decomposition. In addition, frost heaving may mix the soil horizons causing situations where fresh organic material is buried beneath layers of much older soil. Since the molecular composition of these materials will change with age and environmental conditions, the NDRO and NRRO fractions should also vary vertically throughout the soil profile.

The variability observed in the uncontaminated samples has several implications for the remediation of contaminated soils prone to biogenic interference. Large variations in NDRO and NRRO concentrations with sample location will make it difficult to establish a true "background" NDRO and NRRO value at a given site unless a large number of samples are collected. For example, consider a site where the background concentration is to be estimated to within +/- 100 mg/kg of the true background average concentration at the 95% confidence level. A simple random sampling design can be prepared assuming the Student's "t" probability density function applies and that the NDRO measurements made at the Betty Pingo are representative of the variance that could be expected at the site. With these assumptions, approximately 43 samples would be required to determine the average background concentration at the specified confidence level. Based on the analysis of the Betty Pingo samples, it will be difficult to determine the background biogenic interference level if just a few background samples are collected.

CONCLUSIONS

Results from 37 samples collected at one location in Arctic Alaska show that organic soils can have several hundred to several thousand mg/kg of extractable natural organic material that could be inadvertently quantified as contamination. The concentration of NOM that was recorded as contamination varied with sample location and depth. Even at the relatively small 15'x15' sampling site, biogenic interference levels varied by an order of magnitude.

REFERENCES

Dragun, J. (1988) The Soil Chemistry of Hazardous Materials. Silver Spring, MD: Hazardous Materials Control Research Institute, pp. 325-445.

White, D.M., Luong, H., and Irvine, R.L. (1998) "Pyrolysis-GC/MS analysis of contaminated soils in Alaska", *ASCE Journal of Cold Regions Engineering*, March, pp. 1-10.

MONITORING OF BIOVENTING AND INTRINSIC REMEDIATION OF KEROSENE AT SEMI-FIELD SCALE

C.-S. Hwu and ***J.T.C.Grotenhuis*** (Wageningen Agricultural University, The Netherlands)
A.L. Smit and J. Groenwold (DLO-Research Institute for Agrobiology and Soil Fertility, Wageningen, The Netherlands)

ABSTRACT: Process monitoring of transport and biodegradation of kerosene at semi-field scale was conducted using three 2.5 m^3 compartments homogeneously filled with a sandy soil. Two compartments were identically contaminated as a point contamination with 3 l kerosene at 50 cm below the soil surface. In one of the two compartments the effect of bioventing on biodegradation was investigated and in the other a situation of intrinsic remediation was simulated. The third compartment functioned as a control. After a period of rapid infiltration both spills stabilized more or less. Hereafter compositions as well as concentrations of the volatile components of kerosene in the soil atmosphere were measured with time. At termination of the experiments, kerosene residues in soil were identified and quantified.

Three-dimensional figures showed a remarkable difference regarding the soil gas concentrations when bioventing was compared to intrinsic remediation. However, soil extraction at the end of the experiment showed comparable distribution of the kerosene concentrations in both compartments. Still bioventing removed 20% more kerosene than did intrinsic remediation. The kerosene constituents with boiling-points (BP) lower than decane (C_{10}) were removed more by bioventing. Data of soil extraction showed that the concentrations of constituents higher than decane in the two contaminated compartments were very similar. With regards to transport phenomena, the volatile components of kerosene migrated fast in the vertical downwards direction, however, also an upward flow was monitored. It was also found that the low BP constituents in the gas phase migrated a longer distance in all directions.

INTRODUCTION

Kerosene contains mobile and immobile components with carbon numbers ranging from 6 to 16. Especially the mobile components in kerosene can threaten the quality of soil and groundwater if a spill occurs. Kerosene-contaminated sites are commonly found in areas of air force bases and refinery plants.

Process monitoring of transport and biodegradation of contaminants such as kerosene in the field is often difficult to realize, as in most cases the amount of measuring points is limited. Also monitoring processes at laboratory or bench scale is often not possible due to limiting boundary conditions of these small systems. To overcome both restrictions we performed an experiment at semi-field scale in the AB-DLO facility Wageningen Rhizolab using three 2.5 m^3 compartments homogeneously filled with a sandy soil.

Soil gas composition (with respect to O_2, CO_2 and volatile hydrocarbons) of the three compartments was followed during a 135 days period. The results concerning monitoring of transport and biodegradation of kerosene during bioventing and intrinsic remediation are discussed.

MATERIALS AND METHODS

At the AB-DLO research facility Wageningen Rhizolab, The Netherlands (van de Geijn et al., 1994), three compartments (dimensions: 1.25 m x 1.25 m x 1.65 m) were filled homogeneously with a sandy soil. The soil had a bulk density of 1400 kg/m^3, water content of 10% (w/w, ca. 56% of field capacity) and organic matter content of 4%. In two of the three compartments a point contamination of kerosene was applied at the center at 50 cm below the soil surface. Compartment 1 mimicked a bioventing treatment (a homogeneous gas flow introduced at the bottom of the compartment at a rate of 600l/h) whereas compartment 2 functioned as intrinsic remediation by flushing only the headspace with the same amount of air. Compartment 3 functioned as a control that received the same treatment as compartment 1 except for the kerosene spill.

The flushing of air was applied when low concentrations of O_2 were detected in the soil gas. The on-line oxygen measurement throughout the soil profile was used to determine the moment to resume the gas flow. To avoid inhibition of biodegradation, due to oxygen limitation, a trigger value of 5% oxygen was chosen, this value was also confirmed by laboratory experiments (Malina et al., 1997). The total time course of the experiments, started at kerosene spill and ended at taking soil samples, was 3225 h (135 d). Figure 1 shows the important time instants at which bioventing in compartment 1 and headspace (HS) flushing in compartment 2 was implemented.

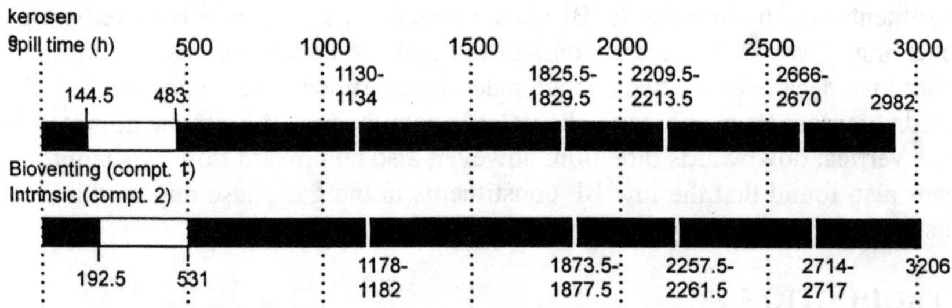

FIGURE 1. Schematic diagram showing the time course (in hours) of the experiment. The white bands indicate the time intervals during which venting in compartment 1 (bioventing) and headspace flushing in compartment 2 (intrinsic remediation) was implemented.

Kerosene used was a gift from Arcadis (Waalwijk, the Netherlands) and was introduced to the two compartments at a rate of on average 4 l/h. The contaminant (3 l kerosene) was infiltrated at an area of 64 cm^2 using a pressure head of about 5 cm. Within 1 h a sampling program was started for the soil gas by using the CH_4 channel of a BK1311 apparatus (Brüel & Kjær, Denmark) for on-line measurement of kerosene (as CH_4 equivalents). To determine the composition of the soil gas 48 gas sampling cells, connected to Viton™ tubing, were placed in each compartment. Gas samples were taken both for gas chromatography (GC, HP 5890A) analysis and for photoacoustic spectroscopy to determine compositions and concentrations of the contaminant vapor, oxygen and carbon dioxide. Methods for GC analyses and soil extraction by CS_2 have been described elsewhere (Hwu et al., 1998a,b). Mapping of data to make three-dimensional (3-D) graphs was made by use of the program Surfer (Golden Software, Colorado).

RESULTS AND DISCUSSION

Transport Phenomena. The on-line measurement by BK1311 apparatus showed that at all sampling cells the change of kerosene concentrations became insignificant after 120 h time. Thus we assumed that the kerosene spills stabilized more or less within 6-8 days time. Previously we reported that infiltration of the spills in both contaminated compartments agrees very well with each other in the Wageningen Rhizolab. This shows that the procedure of contamination can be reproduced. The infiltration rates in all directions measured were in the order: vertical downwards > diagonal downwards > horizontal > vertical upwards (Hwu et al., 1998a).

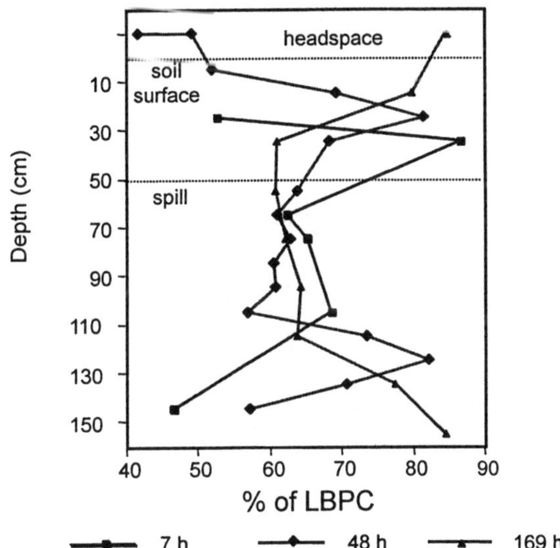

Figure 2. Changes of LBPC percentage in the center of compartments at different depths. Results presented are merged data of compartments 1 and 2.

Since kerosene is a complex mixture, the infiltration of each individual component might behave different from each other due to the complicated reactions with soil particles, e.g., adsorption. However, it is very difficult to differentiate this between components. Therefore we divided kerosene into two major groups by selecting a cut-off at retention time (RT) after xylene peak (ca. 0.75 min) in the GC chromatograms. The migration of the low BP components (LBPC, i.e., RT ≤ xylene) in the two soil compartments, before venting started, with the elapsed time is given in Figure 2.

The LBPC migrated farther and faster, leading to an increasing percentage LBPC with time at a longer distance from the spill. This, however, can be expected since these components are relatively more volatile and thus more mobile. When in compartment 1 venting occurred on day 7 the kerosene concentrations in soil gas instantly increased in the upper part of the compartment and a strong decrease in the lower part to even nil below a certain depth was found. Figure 3 shows the typical contrast between compartment 1 and 2.

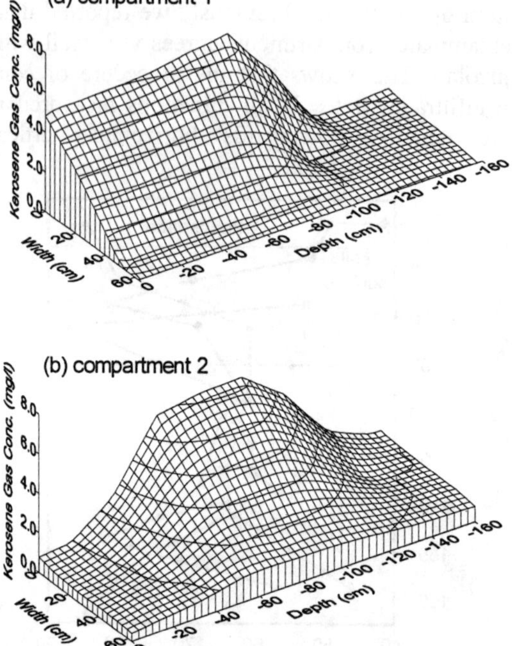

FIGURE 3. The 3-D presentation of the profiles of kerosene gas concentration in the soil compartments during the first period of venting. (a) Bioventing, 217 h after spill, air-vented for 72.5 h, (b) Intrinsic remediation, 243 h after spill, headspace air-flushed for 50.5 h. (See also Figure 1.)

The monitoring of the kerosene in the soil gas showed quite a different behavior in the two compartments respectively. GC chromatograms revealed that the soil gas mainly consisted of components with retention times shorter than decane. However, the amount of these components occupies only a relatively small fraction in the kerosene applied to the compartments (GC-MS results, data not shown).

The effectiveness of venting on removal of the total kerosene was determined by soil extraction of the total amount of kerosene at the end of the experiment. Figure 4 showed that the shapes of the kerosene profiles in both compartments were more or less comparable; indicating that venting had limited effect on the removal of total kerosene. Comparison of the residual concentrations at the end of the experiment with the high concentrations in the top of compartment 1 during venting (i.e., Figure 3a vs. 4a) shows that the kerosene in the soil gas did not adsorb to the soil.

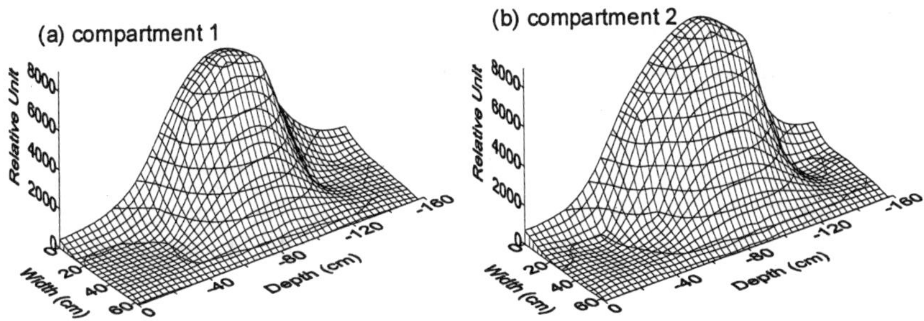

FIGURE 4. Profiles of kerosene residue in soil solid phase at termination of the experiment.

Biodegradation of Kerosene. By integration of the volumes of the 3-D maps as shown in Figure 4 representing kerosene residual amount we obtained the relative difference of the effectiveness between bioventing and intrinsic remediation. Bioventing removed 20% more kerosene than did intrinsic remediation under the semi-field conditions. The higher removal in the bioventing compartment is most probably due to the higher amount of O_2 introduced to this compartment. On-line soil gas monitoring showed a ca. 20% higher CO_2 production for the bioventing compartment compared to the compartment with intrinsic remediation. Therefore it becomes clear that the increased removal is caused by biodegradation and that ventilation of kerosene plays a minor role.

CONCLUSIONS

The transport and biodegradation processes in the soil gas of both bioventing and intrinsic remediation could be well-monitored with the Wageningen Rhizolab facility.

Monitoring of the soil gas led to a clear difference in the kerosene spatial 3D distribution at semi-field scale. For bioventing, however, this information may not reflect the genuine distribution of kerosene adsorbed to the solid phase.

Bioventing compared to intrinsic remediation showed 20% higher removal efficiency by biodegradation that was stimulated by the input of oxygen.

REFERENCES

Hwu, C.-S., J. T. C. Grotenhuis, G. Malina, A. L. Smit, and J. Groenwold. 1998a. "Monitoring of Kerosene Infiltration in an Unsaturated Sandy Soil at Semifield Scale in the Wageningen Rhizolab." In. *Proc. 6th Intl. FZK/TNO Conf. on Contaminated Soil, ConSoil '98*. Vol. 2. May 1998, Edinburgh, pp. 989-990.

Hwu, C.-S., J. T. C. Grotenhuis, G. Malina, A. L. Smit, and J. Groenwold. 1998b. "Gas-Solid Partitioning and Infiltration of Kerosene in an Unsaturated Sandy Soil." In. *Procesmonitoring van Bodemluchtventilatie onder Gecontroleerde Omstandigheden op Semi-veldschaal. Tussenrapport Fase 1*. A. L. Smit (Ed.), Dutch Research Programme of Biotechnological In-Situ Remediation (NOBIS), Gouda, The Netherlands.

Malina, G., J. T. C. Grotenhuis, and W. H. Rilkens. 1997. "Effective Use of Air During Bioventing of Model Hydrocarbons." In. *In Situ and On Site Bioremediation.* Vol. 1. B. C. Alleman and A. Leeson (Eds.). Bioremediation 4 (1), Battle Press, pp. 367-372.

Van de Geijn, S. C., J. Vos, J. Groenwold, J. Goudriaan, and P. A. Leffelaar. 1994. "The Wageningen Rhizolab—A Facility to Study Soil-Root-Shoot-Atmosphere Interactions in Crops. I. Description of the Main Functions." *Plant and Soil 161*: 275-287.

CATABOLIC PATHWAY CHARACTERIZATION FOR THE BIOTRANSFORMATION OF AROMATIC HYDROCARBONS

M. J. IQBAL and J. R. Mason (King's College London, UK.)

ABSTRACT; Benzene dioxygenase catalyses the conversion of benzene to *cis*-benzene dihydrodiol. The genes *bed*C_1C_2BA, which encode the broad specificity enzyme have been cloned and expressed in *Escherichia coli.* In this study, the *bed*D gene from *P. putida* ML2, encoding the enzyme *cis*-benzene dihydrodiol dehydrogenase was cloned 3' to the *bed*C_1C_2BA , transformed and expressed in *E.coli.* The *E.coli* cells containing the recombinant, quantitatively transformed benzene to catechol. The strain could also degrade toluene and *p*-xylene to catechol without the accumulation of any metabolic intermediates.The new recombinant strain could also attack a broad range of aromatic hydrocarbons to produce their respective catechols, demonstrating the broad substrate specificity of the benzene dioxygenase and *cis*-benzene dihydrodiol dehydrogenase enzymes.

INTRODUCTION

Extensive use of chlorinated aromatic compounds has led to widespread industrial pollution. Among these compounds, the substituted groups of benzene, toluene and xylene are major environmental pollutants. The accumulation of these compounds in the environment would indicate that the rate of the evolutionary process is inadequate to protect the biosphere from industrial pollution (Mason et al., 1997). Many of the hydrocarbon degrading microorganisms contain plasmids which code for catabolic genes. By understanding the biochemistry and genetics of plasmid-borne degradation and by using recombinant DNA techniques, it is possible to characterize the appropriate genes and transfer them to construct improved strains with enhanced ability for the degradation of toxic compounds.

In the aerobic biodegradation pathway, benzene is converted to *cis*-benzene dihydrodiol by the action of benzene dioxygenase. Subsequently the enzyme *cis*-benzene dihydrodiol dehydrogenase is able to dehydrogenate the *cis*-benzene dihydrodiol to form catechol.We have employed such an approach to manipulate the broad substrate specificity enzyme benzene dioxygenase from *Pseudomonas putida* ML2 which initiates the oxidation of aromatic compounds. The genes *bed*C_1C_2BA, which encode the enzyme have been cloned into the vector pKK223-3 to generate the plasmid pJRM501, and expressed in *Escherichia coli.* (Tan & Mason, 1990; Tan et al., 1993).

Objectives. In this study, the *bed*D gene was cloned and expressed in *E. coli.*The activities of benzene dioxygenase and *cis*-benzene dihydrodiol dehydrogenase were compared and the ability of the two enzymes to transform a broad range of aromatic compounds into their respective diols and catechols were examined.

MATERIALS AND METHODS

The bacterial strains and plasmids used in this study are listed in Table 1. Cultures of *E. coli* transformed with recombinant plasmids were grown and maintained on Luria-Bertani (LB) medium at 37°C. *Pseudomonas putida* ML2 (NCIB 12190) was maintained at 30°C on Hutner's minimal medium (Zamanian and Mason, 1987) with succinate or benzene as the carbon source. The primers for *bed*D gene were designed using the MacVector programme (Oxford Molecular Limited, UK.).

TABLE 1. BACTERIAL STRAINS AND PLASMIDS

Strain / Plasmid	Relevant Characteristics	Source / Reference
P.putida ML2	*bed*$^+$ prototroph bearing pHMT112	Axcell & Geary, 1975(NCIB 12190)
E.coli JM109	end A1 rec A1 gyr A96 thi hsd R17 (r_k^- m_k^+) rel A1 supE44£$^-$ Δ (lac-pro AB) [F' tra D36 pro AB lacIqZ ΔM15]	Promega
PCR^{TM} 11	Ap^r TA Cloning Vector	Invitrogen
pJI-100	1.259 kb EcoR 1 *bed*D fragment in PCR^{TM}11	This study
pKK223-3	Expression vector in *E. coli*	Pharmacia
pJRM501	3.7 kb EcoR 1 *bed* C_1C_2BA fragment in pKK223-3	Tan & Mason, 1990
pJRM500	5.2 kb BamH 1 *tac* promoter *bed*C1C2BAD fragment in pKK223-3.	This study

PCR analysis and cloning of the *bed*D gene. The *bed*D gene encoding for *cis*-benzene dihydrodiol dehydrogenase was amplified by means of PCR using pHMT112 plasmid DNA as a template.The PCR product was directly cloned into PCR II vector and transformed into *E. coli* JM109. The *bed*D gene was subcloned into the *EcoR*1 site of a plasmid pJRM501 expressing *bed*C_1C_2BA to produce a plasmid pJRM500 coexpressing *bed*C_1C_2BAD.

Preparation of cell extracts and protein measurement. Cell extracts were prepared as described by Zamanian and Mason (1987). Protein concentration was determined by the method of Lowry et al., (1951), using bovine serum albumin as a standard.

SDS-PAGE. Sodium dodecyl sulphate polyacrylamide gel electrophoresis was performed at room temperature in a 12.5% (w/v) resolving gel as described by Zamanian & Mason (1987).

Enzyme assay. *Cis*-benzene dihydrodiol dehydrogenase was assayed spectrophotometrically (Mason and Geary, 1990).

Biotransformation of aromatic compounds. *E. coli* cells were grown in 400ml M9+succinate +ampicillin medium in two litre conical flasks (Sambrook et al., 1989). IPTG was added when the cells were in mid-exponential growth phase. After three hours of induction, the cells were then harvested, washed, and resuspended in 90 ml of sodium succinate M9 medium. These were then used for the transformation of aromatic hydrocarbons. Samples were taken after 30, 60, 120 and 180 minutes. Dihydrodiols and catechols were measured spectroscopically at 260 nm and 276 nm respectively using a Beckman DU-650 spectrophotometer.

RESULTS AND DISCUSSION

The catabolic pathway utilized by *Pseudomonas putida* ML2 for the degradation of benzene involves an initial dioxygenation step followed by the dehydrogenation reaction to yield a catechol. In this study, the *bed*D gene was first cloned into the PCR II vector and then subcloned in the expression vector pKK223-3:*bed*C_1C_2BA (pJRM501) expressing the genes (*bed*C_1C_2BAD) under the control of the IPTG induced *tac* promotor.

The *cis*-benzene dihydrodiol dehydrogenase gene was shown to be functional using expression studies and enzyme activity assays. An SDS-PAGE analysis of the gene products revealed a polypeptide band with an estimated molecular size of 39 kDa.Similar results were found by Fong et al., 1996. Almost all of the *cis*-benzene dihydrodiol dehydrogenases reported so far are homotetramers with a 28 kDa polypeptide and similar with regard to their specificity for *cis*-dihydrodiols and their absolute requirement for NAD^+ as their primary electron acceptor (Patel & Gibson,1974; Rogers & Gibson, 1977). The enzyme also exhibited a high degree of similarity to glycerol dehydrogenases from *E. coli, C. freundii, and B. stearothermophilus*. The dehydrogenase reported here, however, has biochemical properties which differ significantly from those of the *cis*-benzene dihydrodiol dehydrogenase described earlier (Axcell & Geary, 1973). The distinguishing feature of the other enzyme are its subunit molecular weight (110,000) and requirement for ferrous ions and glutathione for maximum activity . This indicates the presence of two isoenzymes within the *Pseudomonas putida* ML2.

Benzene dioxygenase activity was expressed in both recombinant strains (*E. coli*:pJRM501 and *E. coli*:pJRM500), however, the *cis*-benzene dihydrodiol dehydrogenase activity was present only in the strain harbouring pJRM500 (Table 2). The specific activity of *cis*-benzene dihydrodiol dehydrogenase was twenty times greater than that of benzene dioxygenase. The Km^{app} values of the enzymes for *cis*-benzene dihydrodiol and NAD^+(159 μM and 81 μM respectively), determined from double reciprocal plots were comparable to those reported previously for the purified enzyme (Mason & Geary, 1990).

TABLE 2. Benzene dioxygenase and *cis*-benzene dihydrodiol dehydrogenase activities in cell extracts of *E.coli* recombinant strains.

Plasmid	Specific activity (n moles/min/mg of protien) [a] Benzene dioxygenase	*cis*-benzene dihydrodiol dehydrogenase
pJRM501	21.2 ±5.7	0.00
pJRM500	20.2 ±3.4	431.71 ±18.0

a Values are Means±SD of duplicates

Benzene dioxygenase activity was measured polarographically using oxygen electrode and *cis*-benzene dihydrodiol dehydrogenase activity was measured spectroscopically at 340 nm based on NAD^+ reduction.

Benzene dioxygenase and *cis*-benzene dihydrodiol dehydrogenase were also able to transform toluene and xylene (Table 3). In the case of *E.coli*:pJRM500, there was no accumulation of metabolic intermediate compound which confirmed the higher activity of *cis*-benzene dihydrodiol dehydrogenase than benzene dioxygenase. The new recombinant strain (*E.coli*:pJRM500) could also transform a broad range of chlorinated aromatic compounds to produce their respective catechols (Table 3), demonstrating the broad substrate specificity of the combined benzene dioxygenase and *cis*-benzene dihydrodiol dehydrogenase.

TABLE 3. Biotransformation of aromatic hydrocarbons to diols and catechols by *E.coli* recombinant strains.

Substrate	pJRM501		pJRM500	
	Diol	catechol	Diol	catechol
	(nmol/min/mg)		(nmol/min/mg)	
Benzene	17.0 ±0.6	0.0	0.0	20.3 ±4.0
Toluene	13.9 ±0.5	0.0	0.0	15.6 ±0.3
Xylene	7.5±0.3	0.0	0.0	8.3±0.5

Dihydrodiols and catechols were measured spectroscopically at 260 nm and 276 nm respectively by using the DU-650 spectrophotometer.

TABLE 4. Substrate specificity of benzene dioxygenase and *cis*-benzene dihydrodiol dehydrogenase in *E.coli*:pJRM500.

Substrate	Catechol production
Chlorobenzene	+
Fluorobenzene	+
Trifluorotoluene	-
1,2-dichlorobenzene	+
1,3-dichlorobenzene	+
1,4-dichlorobenzene	+
1,2,4-trichlorobenzene	+
1,3,5-trichlorobenzene	-
1,2,4,5-tetrachlorobenzene	-
2-chlorotoluene	+
4-chlorotoluene	+
2,4-D	-
2,4,5-T	

\+ Catechol production
_ No catechol production

Recent work by Beil et al., (1998) has shown that the ability of class II B dioxygenase such as benzene dioxygenase to dechlorinate 1,2,4,5-tetrachlorobenzene reside with a single amino acid in the active site of the enzyme. Current work in our laboratory is in progress to both extent, the substrate specificity of benzene dioxygenase and to increase its turnover numbers.

REFERENCES

Axcell, B. C and P. J. Geary. 1973. "The metabolism of benzene by bacteria: Purification and some properties of enzyme *cis*-1, 2-dihydrocyclohexa-3, 5-diene (nicotinamide adenine dinucleotide) oxidoreductase(*cis*-benzene glycol dehydrogenase)". *Biochem J 136*: 927-934.

Beil, S., J. R. Mason, K. N. Timmis and D. H. Peiper. 1998. "Identification of chlorobenzene dioxygenase sequence element involved in dechlorination of 1,2,4,5-tetrachlorobenzene". *J Bacteriol 180*(21): 5520-5528.

Fong, K. P. Y., C. B. H. Goh and H.-M. Tan. 1996. "Characterization and expression of the plasmid-borne *bed*D gene from *Pseudomonas putida* ML2, which codes for a NAD^+-dependent *cis*-benzene dihydrodiol dehydrogenase". *J Bacteriol 178*: 5592-5601.

Lowry, O. H., N. J. Rosebrough, A. L. Farr and R. J. Randall. 1951. "Protein measurement with the Folin phenol reagent". *J. Biol. Chem. 193*: 265-275.

Mason, J. R., F. Briganti and J. R.Wild . 1997. "Protein engineering for improved biodegradation of recalcitrant pollutants", *pp.* 107-118. In *Perspectives in bioremediation, Technologies for environmental improvement.* Edited by J. R. Wild, S. D. Varfolomeyev and A. Scozzafa. Kluwer Academic Publishers.

Mason, J. R. and J. Geary. 1990. "*Cis*-1, 2-Dihydroxycyclohexa-3, 5-diene (NAD) Oxidoreductase (*cis*-Benzene Dihydrodiol Dehydrogenase) from *Pseudomonas putida* NCIB 12190", *pp.* 134-137. In *Methods in Enzymology. vol. 188*. Edited by M. E. Lidstrom. Academic press, Inc. London.

Patel, T. R., & Gibson. 1974. Purification properties of (+) *cis*-naphthalene dihydrodiol dehydrogenase of *Pseudomonas putida. J Bacteriol 119*: 879-888.

Rogers, J. E. and D. T. Gibson. 1977. "Purification and properties of *cis*-toluene dihydrodiol dehydrogenase from *Pseudomonas putida*". *J Bacteriol 130*: 1117-1124.

Sambrook, J., E. F. Fritsch and T. Maniatis. 1989. *Molecular cloning.* A Laboratory Manual. Cold Spring Harbor, New York: Cold Spring Harbor Laboratory Press.

Tan, H.-M. and J. R. Mason. 1990. "Cloning and expression of the plasmid-encoded benzene dioxygenase genes from *Pseudomonas putida* ML2". *FEMS Microbiol Lett 72*: 259-264.

Tan, H.-M., H.-Y. Tang, C. L. Joannou, N. H. Abdel-Wahab and J. R. Mason. 1993."The *Pseudomonas putida* ML2 plasmid-encoded genes for benzene dioxygenase are unusual in codon usage and low in G+C content".*Gene 130*: 33-39.

Tan, H.-M., C. L. Joannou, C. E. Cooper, C. S. Butler, R. Cammack and J. R. Mason. 1994. "The effect of $ferredoxin_{BED}$ overexpression on benzene dioxygenase activity in *Pseudomonas putida* ML2". *J Bacteriol 176*: 2507-2512.

Zamanian, M. and J. R. Mason. 1987. "Benzene dioxygenase in *Pseudomonas putida.* Subunit composition and immuno-cross-reactivity with other aromatic dioxygenases".*Biochem J 244:*611-616.

ENVIRONMENTAL/ECONOMIC EVALUATION AND OPTIMISING OF REMEDIATION AT CONTAMINATED SITES. LIFE PROJECT 96ENV/DK/16

Lars Deigaard & Søren Toft Nielsen (Scanrail Consult, Copenhagen, Denmark)

ABSTRACT: Site remediation is often initiated without thinking about the environmental effects resulting in resource scarcity, energy consumption, acidification and greenhouse effect. This article presents the preliminary results for developing a decision-making model for optimal restoration of soil and ground water. The decision-making model operates with a program phase, a proposal phase, a project phase and a construction and operation phase. The model is based on practical experience from four in situ methods. The four in situ methods are Biosparging. Bioventing, Reactive Wall and Methane Stimulated Biological Degradation. The in situ methods are tested at five sites contaminated with oil or chlorinated solvents.

INTRODUCTION

The Danish National Railway Agency has initiated a development project to achieve effective and environmentally optimal remediation of soil and ground water at sites contaminated with oil and halogenated solvents. A decision-making model for use in choosing the best method of remediation for a given situation will be developed and verified. Furthermore the clean-up methods applied will be optimised. The project is a part of the Danish National Railway Agency's general environmental action plan.

The development project consists of the following main activities:

- Development of a decision-making model for optimal restoration of soil and ground water, including preparation of guidelines for an all-inclusive environmental accounting.
- Demonstration and optimisation of in-situ remedial methods at five railroad localities in Denmark.

Status. The project was initiated in the spring of 1997, and the final results are expected in the year 2000.

The prototype of the decision-making model has been tested on the demonstration projects during the design and construction phases.

The demonstration projects have been in operation since the summer of 1998 and the first results from the operating phase are achieved.

THE DECISION-MAKING MODEL

The decision-making model integrates the environmental costs and benefits in choosing and optimising the clean up strategy/method at a site. The decision-making model will provide a useful tool for any public or private organisation faced with the problem of cleaning up contaminated sites.

The regular practice is to base the selection and optimising of remediation method on demands corresponding to function, economy and time, in relation to the desired environmental benefit.

The decision-making model is a method of involving an assessment of the environmental costs and benefits on equal terms with the other decision parameters (function, economy and time). In the decision-making model it is possible to effect an environmental weighing in all the phases of a remediation project.

A remediation project can typically be divided into the following four phases; Program, Proposal, Project and Construction and operation.

The program phase. Identification of possible remediation concepts. Qualitative descriptions of environmental costs and benefits and economy of each possible concept in a remediation program. The descriptions are based upon key descriptions and general experience. Key descriptions of the methods used in the demonstration projects will be provided.

Selection of remediation strategy. It is possible to select environmental parameters for further description in the following phases.

The proposal phase. Preliminary design of possible methods. Quantitative and qualitative descriptions of environmental costs and benefits and economy of the selected methods in a project proposal. The descriptions are based upon key figures and rough estimates. In this project key figures for methods used in the demonstration projects will be provided. Selection of one method for the project phase.

The project phase. Detailed design of selected method. Selection of materials and processes for the remediation project. Environmental and economic budget for selected method. The environmental budget is based upon key numbers and calculations.

The construction and operation phase. Construction and operation of the selected method. Environmental and economic accounts will be made. The environmental account is based upon the budget and measurements during the construction and operating phase.

Tools. The environmental benefits must be calculated in accordance with the established practice for risk analysis. The degree of detail for the benefits depends on the phase. In the early phases benefits typically are described generally using a risk assessment for the site and key descriptions for the remediation concept. In the later phases where physical and chemical conditions are fully described, it will be possible to elucidate benefits in much greater detail.

The environmental cost will be estimated in accordance with the principles for life-cycle assessments(LCA). Normally LCA is used on a single product or process. In this method LCA is used on remediation projects containing a multiplicity of components and processes. This provide a lesser amount of detail than normally used in LCA. Therefore it is important to create a perspective of a complicated and complex system with many sub-components.

For the early phases use is made of key descriptions and figures of the remediation concepts used in the demonstration projects. The descriptions are intended to be a tool to create a general view of the advantages and disadvantages of different remediation concepts.

For the later phases a spreadsheet is used. The spreadsheet enumerates material and resources use. The quantities can be used to calculate the environmental effect of the activities in a specific LCA-software. In the spreadsheet default values is used for the most common processes, machinery, parts etc.

THE DEMONSTRATION PROJECTS

The project utilises a comparative analysis in which various remedial methods are tested and optimised. A full account of the environmental cost of each remedial method will be included in each analysis, and the remedial method chosen will be optimised in each case.

The chosen remediation methods are Biosparging, Bioventing combined with drainage, Reactive wall and Methane stimulated degradation. The methods are used at five sites where three are contaminated with oil and two with chlorinated solvents.

Methods. The Biosparging method is used for contaminants biodegradable under aerobic conditions. Two geographically different sites with almost the same size and concentration level are treated by Biosparging. Optimising will be carried out throughout the entire clean-up sequence. The optimising parameters include reduced air flow and pulsed sparging in optimised intervals to reduce the energy consumption.

Bioventing combined with drainage is used for contaminants biodegradable under aerobic conditions. The method is used at a site situated next to one of the Biosparging sites making a total comparison possible. Optimising of the technique includes reduced air flow, pulsed sparging and reduced drainage pumping.

The Reactive Wall is used at a site contaminated with chlorinated solvents that degrade catalytically in the presence of iron. No optimisation is planned, but if ground water pumping is required the pumping rate can be optimised.

The Methane Stimulated Biological Degradation method is used for contaminants that are degradable in the presence of a primary carbon source and oxygen. We use the method at a site contaminated with chlorinated solvents. A biological treatment zone is established across the plume. Optimising includes reduced air flow, pulsed sparging and controlled addition of carbon.

Excavation is the most common method to eliminate contamination of soil and ground water. The method is used as a reference at a site contaminated with oil.

Environmental Costs and Benefits. Figure 1 provides an example of selected effects of environmental costs. The graph shows the environmental costs during the phases of installation, operation and removal. Five effects have been chosen: crude oil / natural gas (because it is a limited resource); waste, which is both hazardous waste, volume waste, cinders and fly-ash; toxicity which consists of human toxicity and eco-toxicity; acidification and green house effect are general effects that affect the environment.

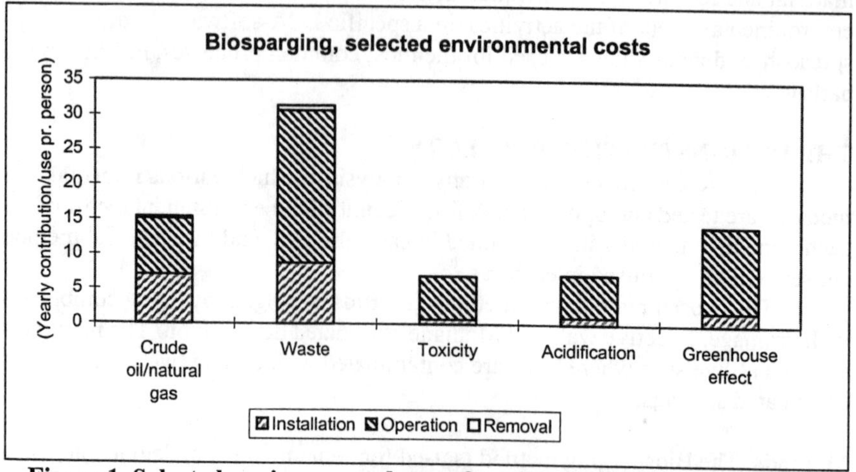

Figure 1. Selected environmental costs for a Biosparging system. Figure showing environmental costs distributed on installation, operation and removal.

Preliminary results of the in situ projects. The function of the Reactive wall is satisfactory. A distinct reduction is seen where the wall passes(app 500 μg/l - 10 μg/l).

The Methane Stimulated Biological Degradation Wall shows no clear reduction in concentrations. This could be due to low oxygen content in the water and the air flow has consequently been increased.

The Biosparging and Bioventing projects do not clearly indicate that degradation takes place. Generally very low oxygen contents are found, and free phase products have been recorded in several borings. The transmission mode during the first period has been pulsating air flow at the bioventing site and one of the biosparging sites while the second biosparging site has been run with continuous air flow.

ACKNOWLEDGMENTS

The project is supported by the LIFE program under the European Economic Community by the European Economic Community, the Danish National Railway Agency, and the Environmental Protection Agency of Denmark. The budget amounts to approximately 2.43 million USD. The purpose of the LIFE program is to contribute to the development and implementation of the European Economic Community policy for environmental issues. The program is administered by the LIFE committee at the European Commission DGXI.

RELATING BIODEGRADATION POTENTIALS TO IN-SITU MICROBIAL COMMUNITY COMPOSITION

D. Ringelberg , Dyntel Corp., Atlanta, GA, USA
E. Perkins, AscI Corp., McLean, VA, USA
L. Hansen, USACE Waterways Experiment Station, Vicksburg, MS, USA
J. Talley, USACE Waterways Experiment Station, Vicksburg, MS, USA
H. Fredrickson, USACE Waterways Experiment Station, Vicksburg, MS, USA

ABSTRACT: A 50 year old diesel fuel contamination plume was used to test the hypothesis that community attributes of the extant microbiota could be directly related to chemical evidence indicating the occurrence of contaminant biodegradation. The co-recovery of petroleum hydrocarbons and bacterial membrane phospholipids, achieved in a single organic solvent extraction, provided the data needed to make the correlation between contaminant chemistry and microbiology. Radio-respirometry verified that the indigenous microbiota was capable of metabolizing the contamination. Two areas within a 5-meter continuous core showed a three-fold decrease in the ratio of hexadecane to pristane, which was taken as a chemical indication of past and/or present biodegradation activity. Phospholipid fatty acid patterns were similar between the two zones and significantly different from those where the hexadecane/pristane ratio was high. The PLFA profiles associated with the lower contaminant ratios reflected the make-up of the organisms (Gram-positive bacteria, actinomycetes and micro-eucaryotes) engaged in the *in-situ* biodegradation of the diesel fuel oils.

INTRODUCTION

Gasoline Alley encompasses an area approximately 1 square mile within the Fort Drum military installation in up state New York. The alley represents subsurface contamination resulting from close to fifty years of fueling activities and leaking under-ground storage tanks. In 1995, the United States Corps of Engineers contracted EA Engineering, Science and Technology to provide a comprehensive assessment report for the site (EA Engineering, Science, and Technology, 1994). EA determined that separate phase product was present in a number of areas throughout gasoline alley with gasoline and diesel fuel being identified as the primary source products for contamination. The area selected for this investigation was used as a fuel dispensing facility from late 1940 through early 1990. Fuel dispensing was halted in 1994 at which time two 25,000 and one 12,000-gallon diesel fuel tanks were removed from the area.

As part of the remedial actions, this study focused on examining the subsurface microbial ecology of area 1595 and the relationships that exist between the extant microbiota and the contamination. Our intent was to define those community attributes that were associated with the *in situ* biodegradation of the petroleum hydrocarbons. By doing so, information would be gained which, ultimately, would enhance the effectiveness of any bioremediation strategy adopted. In order to obtain a quantitative and qualitative picture of the *in situ* microbiota, a direct chemical measure was needed. Ester-linked phospholipid fatty acids (PLFA) were chosen since insight into viable microbial biomass, community composition and metabolic status could be obtained (White and Ringelberg, 1997). Since the typical pathway for the biodegradation of hydrocarbons is via co-oxidative aerobic metabolism (in the pathway hexadecane>naphthalene>pristane>benzanthracene), the ratio of hexadecane/pristane was used to identify areas where contaminant biodegradation had or was occurring. Radiolabeled phenanthrene was then used to verify, *ex-situ*, a capability of the indigenous microbes to biodegrade a recalcitrant petroleum contaminant, with the assumption that more readily utilizable substrates would also be metabolized. Results from these assays formed the criteria on which we based our identification of an *in-situ* hydrocarbon biodegrading microbial community.

MATERIALS AND METHODS

Subsurface material was collected in three 5 ft x 3 in ID split-barrel samplers fitted with a butyrate liner and driven with a 140 lb. hammer dropped from 30 in. Sample material was mostly homogenous and comprised primarily of medium sands with some fines (2). All sampling materials were steam sterilized prior to use. Recovered cores were labeled, capped, sealed in wax and transported to Waterways Experiment Station (Vicksburg, MS) at 5° C in a refrigerated truck.

Radio-respirometry. ^{14}C-labeled acetate (1,2-^{14}C-acetic acid, sodium salt, ICN Pharmaceutical, Inc., Costa Mesa, CA) and phenanthrene (9-^{14}C-phenanthrene, Sigma Chemical Corp., St. Louis, MO) were used in bioslurry radio-respirometry evaluations of the indigenous microbiota. Respirometry flasks were loaded with 10 g of sample and taken up to a final volume of 40 ml with sterile H_2O (including radiolabel). Flasks were run in triplicate, at room temperature, with the evolution of $^{14}CO_2$ being trapped in an alkaline (KOH) solution. The traps were sampled periodically and the $^{14}CO_2$ levels measured by liquid scintillation counting.

PLFAs, sterols and PHs. Soil petroleum hydrocarbons (PHs) and ester-linked phospholipid fatty acids (PLFAs) were recovered by extracting 1 g of soil in 3.5 ml of an organic solvent solution consisting of methylene chloride, methanol and aqueous phosphate buffer (1:2:0.5, CH_2Cl_2:MeOH:PO_4, v:v:v). The soil solvent mixture was then recovered and fractionated as described in White and Ringelberg, 1998, with the following modification. The organic (CH_2Cl_2) phase was recovered and directly applied to a pre-packed silica gel column (0.5 g, Burdick and Jackson, Muskegon, MI). One µl of the CH_2Cl_2 eluent was then injected on a gas chromatograph. Compounds were identified by GC/MS using electron impact ionization (70 eV). Sterols were recovered and quantified as described in White and Ringelberg, 1998.

RESULTS AND DISCUSSION

Contaminant Profiles. Diesel fuel contamination was distributed throughout the continuous core with the greatest concentration occurring at a depth of 3.1 m (figure 1a). Since it has been shown that normal alkanes (hexadecane, octadecane) are removed from the carbon pool sooner than the more recalcitrant methyl branched alkanes (pristane, phytane) the ratio of hexadecane/pristane can be used to estimate *in-situ* biodegradation. Two areas along the continuous core showed depressed ratios of hexadecane/pristane, 1.7 to 2.2 m and 2.7 to 3.5 m (figure 1b). Both zones resided within the plume of PHs bordered on either sides by 500-1000 μg g^{-1} or 3000-5000 μg g^{-1} PHs.

Microbial Community Analysis.

Activity – Material recovered from both the 3.20 and 3.51 m depths contained microbial populations capable of mineralizing ^{14}C labeled acetate, however the extent of acetate mineralization was greater in the 3.2 m core (table 1). This 5-fold difference in general mineralization activity was also reflected in the viable biomass estimates. The larger seed population in material from the 3.20 m depth may explain why significant phenanthrene mineralization was only detected, after 7 days, with material from this location. The microbiota in the core from the 3.2 m depth utilized 8.43% of the added label over the 7-day incubation period (table 1). This result supported the assertion that the area showing a lower hexadecane/pristane ratio represented a zone of *in situ* biodegradation.

Table 1. Chemical and biological characteristics of two Fort Drum subsurface cores.

	3.20 m	3.51 m
Petroleum hydrocarbons (μg g^{-1})	507	13369
Viable biomass (pmol PLFA g^{-1})	10259	4457
Acetate mineralization extent (cumulative %)	29.25	4.45
Phenanthrene mineralization extent (cumulative %)	8.43	0.25

Biomass and Community Composition - Microbial biomass (pmole PLFA g^{-1} soil) ran parallel to contaminant concentrations and a significant positive correlation (Spearman r = 0.56, p = .0005) could be measured between the two variables (figure 1a). A conversion of the PLFA values to cell numbers (assuming 1 pmole PLFA is equivalent to 2.5 x 10^4 cells, Balkwill et al., 1988), indicated the surface soil to contain approximately 4 x 10^8 cells g^{-1}. Although this level of microbial biomass dropped by an order of magnitude at the 0.8 m depth, biomass at 3.7 m (the middle of the contamination plume, ~5,000 μg/g TPH) was slightly greater than that observed at the surface. The significant relationship identified between biomass and TPH concentration suggests that the extant contamination influenced microbial distribution.

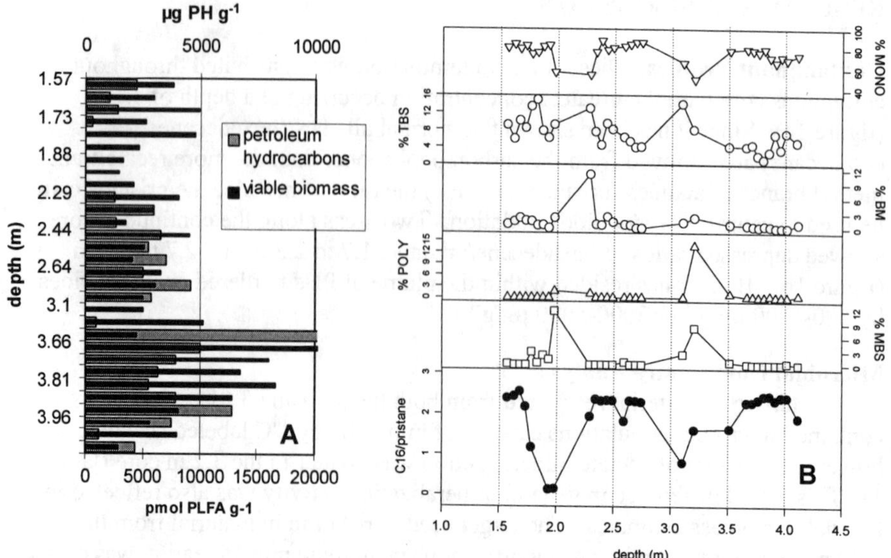

Figure 1. Comparison of microbial community biomarkers to diesel fuel contaminant chemistry throughout a subsurface core from Fort Drum, New York.

The decrease in the hexadecane/pristane ratio coincided with a shift in microbial community composition (figure 1b). Community composition was assessed through the analysis of the individual PLFA patterns obtained from each sample. PLFA were first divided into subsets each reflecting a specific group of microorganism; i.e. Gram-positive bacteria (terminally branched saturated PLFA, TBS), Gram-negative bacteria (monounsaturated PLFA, MONO), actinomycetes (mid-chain methyl branched saturated PLFA, MBS), reducing bacteria (methyl branched monounsaturated PLFA, BM) and micro-eukaryotes (polyunsaturated PLFA, POLY). Increased percentages for all of the PLFA classifications (bacterial groups), except one, occurred in the two zones where the hexadecane/pristane ratio was lowest. The one exception was with the monounsaturated PLFA (Gram-negative bacteria). Monounsaturated PLFA were predominant throughout the continuous core comprising, on average, 74% of the total PLFA detected. This percentage dropped to as low as 50% in the two areas showing decreased hexadecane/pristane ratios. This finding is noteworthy since Gram-negative bacteria, in particular the *Pseudomonas* sp., are consistently recovered from hydrocarbon contaminated sites (Atlas, 1981).

Highly significant correlations ($p<0.01$) were measured between microbial biomarkers indicative of hydrocarbon degrading bacteria and the hexadecane/pristane ratio. Mid-chain branched saturated PLFA, in particular 10me16:0 and 10me18:0, showed the greatest correlation ($r -0.67$) while the relative percentage of terminally methyl branched saturated PLFA was also found to be significantly correlated to the ratio (r –0.53). Within the Gram-positive bacterial classification, *Arthrobacter* and *Bacillus* species have routinely been isolated from diesel fuel hydrocarbon contaminated sites and shown to be capable of utilizing n-alkanes for metabolism (Atlas, 1981). Although not significantly correlated to the biodegradation ratio, increased percentages of the dienoic, 18:2w6 and the 24-methyl sterols (ß-sitosterol, stigmasterol and fucosterol), indicated a micro-eukaryotic presence. These sterols are common to certain fungi (Myxomycota, Eumycota and Oomycetes), slime molds and protozoa. Fungi often exhibit greater hydrocarbon biodegradation activity than bacteria and area consistently recovered from hydrocarbon contaminated sites (Atlas, 1981).

An exploratory analysis of relationships among the recovered membrane lipid profiles from core 1595, revealed the presence of three distinct or unique microbial communities. Application of a K-means hierarchical cluster analysis resulted in the definition of exactly three community profiles of the greatest possible distinction. Significant differences in a number of community attributes were measured between the three communities and are presented in table 2. Group #3 (or community #3) was identified as containing the greatest biodegradative potential, i.e. the lowest hexadecane/pristane ratio. This community was characterized by greater mole percentages of micro-eucaryotic (POLY), actinomycete (MIDBRSAT) and Gram-positive (TERBRSAT) membrane lipids. Community #3 also showed the lowest evidence of long-term metabolic stress (*trans* fatty acid synthesis in relative proportion to *cis* fatty acid synthesis). The fact that a number of different PLFA increased in percentage in the proposed zones of biodegradation activity suggests that these were areas characterized by an increased microbial diversity.

Table 2. Chemical and biological attributes of three Fort drum subsurface microbial communities as defined by a k-means cluster analysis of PLFA profiles.

	Group #1	Group #2	Group #3
Contaminant chemistry			
Hexadecane/pristane	2.2 (0.3)	2.3 (0.2)	0.6 (0.7)
microbial community attributes			
TERBRSAT	5.5 (3.4)	7.4 (2.5)	13.2 (4.9)
MONO	72.7 (7.0)	81.8 (5.1)	63.2 (16.2)
BRMONO	0.3 (0.4)	1.3 (1.3)	1.4 (1.2)
MIDBRSAT	0.2 (0.3)		5.5 (4.6)
POLY	0.3 (0.6)		2.7 (3.9)
Microbial physiological attributes			
Trans/cis	0.20 (0.05)	0.27 (0.06)	

Groups #1 and #2 (or communities #1 and #2) from table 3, showed the least amount of evidence of a biodegradation potential and were distinguished by the greatest percentage of Gram-negative bacterial PLFA (MONO). Both of these communities were also characterized by elevated long-term stress ratios. In addition, the stress ratio (*trans/cis*) correlated positively (r 0.72) with PH concentration. Increased *trans/cis* ratios have been associated with hydrocarbon biodegradation (Ringelberg and White, 1992). However, the study cited did not measure *in-situ* contaminant chemistry, but only total hydrocarbon presence. The results of this study concur that increased ratios of *trans/cis* are associated with hydrocarbon presence and that Gram-negative bacteria are abundant in a hydrocarbon contaminated subsurface. However, the actual occurrence of hydrocarbon biodegradation, at this site, appears not to be related to Gram-negative bacterial physiology, but instead to Actinomycete, Gram-positive bacteria and/or fungi.

Since *in-situ* bioremediation of contaminated soils is a community level phenomenon, where interactions occur not only on a microbe-contaminant level but also on microbe-microbe and microbe-geochemistry levels, an ability to find a community descriptor that is directly related to degradation activity furthers the knowledge needed to successfully remediate the site.

Acknowledgements:

We would like to acknowledge the USACE Baltimore District and the Corp. of Engineers Environmental Quality and Technology Research Program for funding this research in collaboration with the Waterways Experiment Station. We would also like to thank Ms. Margaret Richmond, Mr. Roy Wade, Mr. Glenn Myrick and Mr. Johnny Byrnes for their technical assistance.

References:

Atlas, R.M. 1981. "Microbial degradation of petroleum hydrocarbons: An environmental perspective." *Microbiological Reviews. 45*(1): 180-209.

EA Engineering, Science, and Technology. 1994. *Site-Specific Safety, Health, and Emergency Response Plan (SHERP) for T-1595 Site Characterization Investigation.* EA, Baltimore, MD. November.

Ringelberg, D. and D. White. 1992. *Fatty acid profiles, In Bioremediation of petroleum-contaminated soil on Kwajalein island: Microbial characterization and biotreatability studies* (H.I. Adler, R.L. Jolley, and T.L. Donaldson, eds.) Oak Ridge National Laboratory, Oak Ridge, TN, ORNL/TM-11925:31-36.

White, D.C. and D.B. Ringelberg. 1997. "Utility of the signature lipid biomarker analysis in determining the *in situ* viable biomass, community structure and nutritiona/physiological status of deep subsurface microbiota." In: The Microbiology of the Terrestrial Deep Subsurface (P.S. Amy and D.L. Haldeman, eds.), CRC Press, Boca Raton, FL, p119-136.

White, D.C. and D.B. Ringelberg. 1998. "Signature lipid biomarker analysis." In: Techniques in microbial ecology (R.S. Burlage, R. Atlas, D. Stahl, G. Geesey, and G. Sayler, eds.), Oxford University Press, Inc. New York. P. 255-272.

AUTHOR INDEX

This index contains names, affiliations, and volume/page citations for all authors who contributed to the eight-volume proceedings of the Fifth International In Situ and On-Site Bioremediation Symposium (San Diego, California, April 19–22, 1999). Ordering information is provided on the back cover of this book.

The citations reference the eight volumes as follows:

5(1): Alleman, B.C., and A. Leeson (Eds.), *Natural Attenuation of Chlorinated Solvents, Petroleum Hydrocarbons, and Other Organic Compounds*. Battelle Press, Columbus, OH, 1999. 402 pp.
5(2): Leeson, A., and B.C. Alleman (Eds.), *Engineered Approaches for In Situ Bioremediation of Chlorinated Solvent Contamination*. Battelle Press, Columbus, OH, 1999. 336 pp.
5(3): Alleman, B.C., and A. Leeson (Eds.), *In Situ Bioremediation of Petroleum Hydrocarbon and Other Organic Compounds*. Battelle Press, Columbus, OH, 1999. 588 pp.
5(4): Leeson, A., and B.C. Alleman (Eds.), *Bioremediation of Metals and Inorganic Compounds*. Battelle Press, Columbus, OH, 1999. 190 pp.
5(5): Alleman, B.C., and A. Leeson (Eds.), *Bioreactor and Ex Situ Biological Treatment Technologies*. Battelle Press, Columbus, OH, 1999. 256 pp.
5(6): Leeson, A., and B.C. Alleman (Eds.), *Phytoremediation and Innovative Strategies for Specialized Remedial Applications*. Battelle Press, Columbus, OH, 1999. 340 pp.
5(7): Alleman, B.C., and A. Leeson (Eds.), *Bioremediation of Nitroaromatic and Haloaromatic Compounds*. Battelle Press, Columbus, OH, 1999. 302 pp.
5(8): Leeson, A., and B.C. Alleman (Eds.), *Bioremediation Technologies for Polycyclic Aromatic Hydrocarbon Compounds*. Battelle Press, Columbus, OH, 1999. 358 pp.

KEYWORD INDEX

This index contains keyword terms assigned to the articles in the eight-volume proceedings of the Fifth International In Situ and On-Site Bioremediation Symposium (San Diego, California, April 19-22, 1999). Ordering information is provided on the back cover of this book.

In assigning the terms that appear in this index, no attempt was made to reference all subjects addressed. Instead, terms were assigned to each article to reflect the primary topics covered by that article. Authors' suggestions were taken into consideration and expanded or revised as necessary. The citations reference the eight volumes as follows:

5(1): Alleman, B.C., and A. Leeson (Eds.), *Natural Attenuation of Chlorinated Solvents, Petroleum Hydrocarbons, and Other Organic Compounds.* Battelle Press, Columbus, OH, 1999. 402 pp.

5(2): Leeson, A., and B.C. Alleman (Eds.), *Engineered Approaches for In Situ Bioremediation of Chlorinated Solvent Contamination.* Battelle Press, Columbus, OH, 1999. 336 pp.

5(3): Alleman, B.C., and A. Leeson (Eds.), *In Situ Bioremediation of Petroleum Hydrocarbon and Other Organic Compounds.* Battelle Press, Columbus, OH, 1999. 588 pp.

5(4): Leeson, A., and B.C. Alleman (Eds.), *Bioremediation of Metals and Inorganic Compounds.* Battelle Press, Columbus, OH, 1999. 190 pp.

5(5): Alleman, B.C., and A. Leeson (Eds.), *Bioreactor and Ex Situ Biological Treatment Technologies.* Battelle Press, Columbus, OH, 1999. 256 pp.

5(6): Leeson, A., and B.C. Alleman (Eds.), *Phytoremediation and Innovative Strategies for Specialized Remedial Applications.* Battelle Press, Columbus, OH, 1999. 340 pp.

5(7): Alleman, B.C., and A. Leeson (Eds.), *Bioremediation of Nitroaromatic and Haloaromatic Compounds.* Battelle Press, Columbus, OH, 1999. 302 pp.

5(8): Leeson, A., and B.C. Alleman (Eds.), *Bioremediation Technologies for Polycyclic Aromatic Hydrocarbon Compounds.* Battelle Press, Columbus, OH, 1999. 358 pp.

A

B

C

D

E

F

G

H

I

N

O

P

Q

R

S

T

U

V

W

Z